SYMPOSIUM ON METHODS OF MEASUREMENTS AND RESULTS OF LOWER IONOSPHERE STRUCTURE

Symposium Organized by:

Committee on Space Research — COSPAR and
the "Deutsche Forschungsgemeinschaft"
of the Federal Republic of Germany

Sponsored by:

International Union of Radio Science — URSI
Committee on Space Research — COSPAR
International Association of Geomagnetism and Aeronomy
of the International Union of Geodesy and
Geophysics — IAGA/IUGG

COSPAR

METHODS OF MEASUREMENTS AND RESULTS OF LOWER IONOSPHERE STRUCTURE

Proceedings of the Symposium held in
Constance, F.R.G.

23—26 May 1973

Edited by:
K. RAWER

AKADEMIE-VERLAG · BERLIN

1974

Executive Editor:
A. C. Stickland

Library of Congress Catalog Card Number 73–94253

202 · 100/475/74
Gesamtherstellung: VEB Druckhaus „Maxim Gorki", Altenburg
Bestellnummer: 761 928 1 (6202) · LSV 1435
Printed in GDR

PREFACE

The Symposium was held from 23–25 May 1973 in conjunction with the COSPAR General Assembly at Konstanz (FRG). On the morning of 26 May a discussion meeting was held for all participants; this was also declared to be an open meeting of URSI Panel III.6.1 on the International Reference Ionosphere (IRI), because the main intention of the discussion was the clarification of some critical points concerning this international cooperation.

The same questions had been in mind at the very beginning of plans for this present Symposium. In fact, the idea of such a symposium came up when the IRI group met at earlier COSPAR meetings. The IRI group is an open group headed by a four-person steering committee formed by URSI and COSPAR together, its members being S. A. Bowhill (USA), K. S. W. Champion (USA), E. A. Lauter (GDR) and the undersigned (as chairman). The group has a large number of correspondents and meets on suitable occasions, in particular at COSPAR and URSI assemblies. The need for a deeper and specialized discussion of the lower ionosphere appeared at the assembly of COSPAR at Seattle in 1971. When the idea was given more precision at the Madrid assembly in 1972 it appeared that IAGA was also interested in the subject, and intended to have a symposium of its own, oriented more towards aeronomic interpretation. It was finally agreed that the symposium held at Konstanz should have a programme committee from all three organizations with the followings composition: S. A. Bowhill, E. A. Lauter, J. Ortner (COSPAR), A. D. Danilov, G. Kockarts (IAGA), K. Rawer (Chairman), L. Thomas (URSI). It was agreed that the Konstanz Symposium should have a more "practical" orientation towards the resolution of problems related to the deduction and compilation of significant data from different techniques, while at the symposium to be organized later by IAGA (at Kyoto) the theoretical problems of aeronomy and chemistry of the same height range should have priority. The URSI and COSPAR members of the above programme committee agreed to serve as consultants when the Kyoto meeting was prepared.

Most of the papers presented at the COSPAR Symposium are found in this volume. The discussion of 26 May 1973 may be summarized as follows: Different methods for determining the electron density produce different types of profiles in the lower E and particularly in the D region. Agreement has been reached that the uncertainties of ground-based radio-wave propagation experiments are too great to deduce reliable profiles from this data source alone. *In situ* probes have also been criticized, except that the measurements are cros-checked with onboard radio propagation measurements (e. g. Faraday or Doppler observations). This latter type of combined experiments is therefore particularly recommended for future measurements. Past data obtained by such techniques should have priority as electron density input into IRI. As for the ion density and composition, it is quite certain that negative ions are important in the D region as ion densities are considerably greater than electron density at the lower levels. Unfortunately, for ion composition no agreement could be reached; more experimental determinations are urgently needed.

Thus most of the intended goals have been reached, so that the Konstanz symposium has been very helpful for the task of the IRI group.

We thank the three sponsoring organizations for their help in this context.

K. Rawer
Chairman of Program Committee

Contents List

Discussion Summary

Electron Density Profiles

Lower Ionosphere Structure — Akademie-Verlag, Berlin 1974

IONOSPHERIC PROFILES UP TO 160 KM
A REVIEW OF TECHNIQUES AND PROFILES

E. V. THRANE

Norwegian Defence Research Establishment, Kjeller, Norway

The distribution of electron density with height in the ionospheric D and E regions has been studied by a number of ground-based and rocket techniques. In this review the most important of these methods will be discussed with emphasis on an intercomparison of the measurements and their reported accuracy and resolution. The comparisons will be made at different levels within the height range 50—150 km. The methods considered are wave interaction, partial reflection and A1 absorption measurements from the ground, Faraday rotation and differential absorption as well as probe measurements in rockets. For the E region a review of ionosonde and incoherent backscatter measurements is also included. The validity of the assumptions on which the different methods are based is discussed and the possible sources of error in the analyses reviewed.

1. Introduction

In the height region 50—160 km, which comprises the ionospheric D and E regions, a small number of widely different techniques have been used to study the electron and ion densities. The properties of the atmosphere change very rapidly with height within this range. The atmospheric pressure decreases from 7.6×10^{-1} mb at 50 km to 3.7×10^{-6} at 160 km. The mean free path of a molecule changes from 3×10^{-5} m to 65 m in the same region, whereas that of an electron increases from 10^{-3} m to 200 m. The temperature decreases from 250 °K at the stratopause to 150°K at the mesopause, the coldest region in the earth's environment, and again increases to about 1000°K near 160 km. Within the height range one passes from a region where turbulent mixing of the atmospheric gases is efficient to a region where diffusive equilibrium and laminar flow prevail. The ion production mechanisms as well as the ion loss processes change drastically with height, and the parameter of interest here, the density of free electrons, increases by 5 orders of magnitude from a few tens of electrons per cm^3 at 50 km to 10^5 cm^{-3} near 110 km. The region is very important for the propagation of radio waves from ELF to VHF, and includes that part of the atmosphere in which collisions between the electrons and neutral molecules have a significant effect on radio propagation.

Considering the rapid change of atmospheric properties with height mentioned above, it is not surprising that no single technique for measuring electron densities is adequate throughout the whole region. In the following, ten different methods will be discussed, six ground-based and four rocket techniques. Although these methods are all based on different principles, it is notable that eight out of the ten depend upon the properties of radio waves propagating through the medium through the dependence of the complex refractive index upon the electron density. Only two techniques use the properties of the individual electrons more directly, the incoherent scatter technique and the various types of probes.

Table 1

Experimental Techniques for Determination of Lower Ionosphere Electron Densities
Numbers refer to undisturbed daytime conditions.

Technique	Platform	Useful height range	Estimated min. height resolution	Expected min. error	Time for each measurement
Ionosonde	Ground	>100 km	5—25 km	5%	1 min.
Incoherent scatter	Ground	>90 km	5 km	10%**	5 min.
Partial reflections	Ground	55—90 km	~5 km	25%	>10 min.
Wave interaction	Ground	55—90 km	7—10 km	>10%	>15 min.
Absorption measurements	Ground	60—110 km	10—20 km	*	~15 min.
VLF and LF measurements	Ground	60—90 km	10—20 km	*	*
Faraday rotation and differential absorption	Rocket	65—120 km	2—5 km	10%	5 min.
Probes	Rocket	>70 km	100 m	10%**	5 min.
VLF and LF scalar field measurements	Rocket	65—90 km	2—5 km	15%	5 min.
VLF Doppler	Rocket	>80 km	~5 km	50%	5 min.

* Critically dependent on technique.
** Applies when experiment calibrated with other measurements.

Table 1 lists the methods to be discussed. Emphasis will be on the methods that the author considers most useful, a somewhat subjective choice. There is however one exception, the very important VLF and LF continuous wave methods of probing the lower ionosphere will not be treated in detail, as they are considered elsewhere in this symposium [1].

2. Ionosonde Measurements of Electron Density Distributions

Ionograms contain much information on the electron densities in the E region, and in spite of certain fundamental practical and theoretical limitations of the method, $h'(f)$ recording is an important technique for studying electron density distributions in the upper part of the height range with which we are concerned, 100—160 km. The parameter that is most easily available is, of course, the maximum electron density in the E layer, which is determined from

$$\frac{N_{\max}}{\mathrm{m}^{-3}} = 1.24 \times 10^{10} \left(\frac{f_0\mathrm{E}}{\mathrm{MHz}}\right)^2. \tag{1}$$

If, in a good ionogram, foE is determined with an accuracy of 0.05 MHz, this means that the electron density N_{max} can typically be determined to within an accuracy of a few per cent. Much poorer accuracy must, however, generally be expected when inverting an ionogram to an electron density profile. A successful inversion depends upon a number of factors, and requires a considerable amount of experienced effort. First of all the ionogram must contain the necessary information, that is an $h'(f)$ trace must be continuous over the relevant frequency range and show the normal layers, without distortions due to blanketing Es or off-vertical reflections. The ionogram must be of reasonable quality. The *URSI Handbook of Ionogram Interpretation and Reduction* [2] indicates a minimum accuracy of ± 2 km in virtual height and that frequencies can be read to within $\pm 1\%$. Even for a high quality ionogram, however, systematic errors of several kilometres in height may be present. Care must be taken in the scaling. The choice

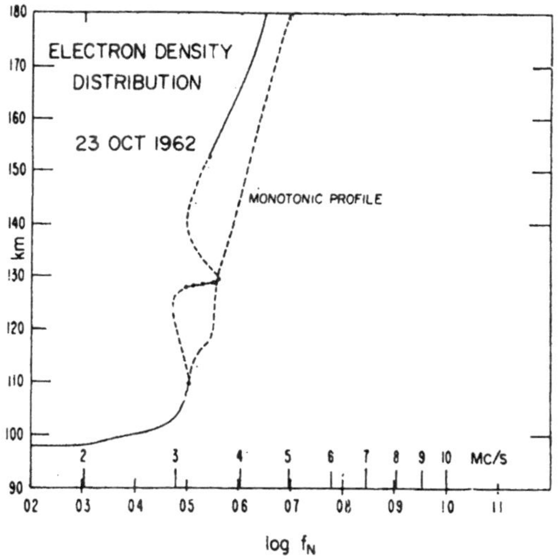

Fig. 1. Profiles of log f_N calculated from ionogram by a monotonic method and using ordinary and extraordinary traces [3].

of sampling intervals to extract the relevant information as well as the choice of computational procedure is important. In principle an electron density profile can be derived uniquely from an $h'(f)$ record provided the electron density increases monotonically with height, and provided an $h'(f)$ trace is available at all frequencies below the greatest height of interest. Since neither condition is fulfilled in practice, the numerical techniques for inversion from $h'(f)$ to $N(h)$ must contain assumptions about (i) the ionization below the lowest height at which an echo is observed, and (ii) the way the electron density varies in the valley between two layers. Both assumptions will limit the accuracy of the derived profile. Although methods exist for minimizing the effects of these uncertainties, unique solutions cannot be found in most cases. Fig. 1 shows an example of $N(h)$ profiles derived from an ionogram by two different techniques, one assuming a monotonic profile and one using both ordinary and extraordinary $h'(f)$ traces to deal with the difficult valley region [3]. It will be seen that the result is critically dependent upon the method above the peak of the E layer. For the lower part of the E region the $N(h)$ profile depends upon the assumptions made about the D region, but useful results can probably be obtained down to about 100 km, when high quality ionograms are available.

3. Incoherent Backscatter

The incoherent backscatter technique has been extensively used to study the F region [4], but in recent years successful efforts have also been made to use the method in the E region [5—7]. For measurements of electron density profiles the large radar antenna sends a pulsed or cw signal (typically at 400 MHz) vertically upwards and records the characteristics of the weak returns scattered from thermal fluctuations in the refractive index of the ionospheric plasma. The received signal-to-noise power ratio is directly proportional to the electron concentration N, to the scattering cross section of an electron σ_e and to the pulse duration τ. The received signal spectrum will be broadened by the thermal motions of the ions and $\sigma = \sigma_e/(1 + T_e/T_i)$ where σ_e is the incoherent electron scattering cross section and T_e and T_i the electron and ion temperatures respectively. For E region work, at least for heights below 140 km, $T_e = T_i$ and $\sigma = \sigma_e/2$, and the measurements can readily be converted into electron concentrations. At lower heights a limit is imposed by the requirement that the radar wavelength $\lambda_0 \gg 4\pi\lambda_D$ where λ_D is the Debye length of the plasma. Both Taylor [7] and Evans and Oliver [5] have shown that useful E region $N(h)$ measurements can be obtained in the altitude range 90—140 km. In order to resolve detailed vertical structure, the pulse length must be considerably smaller than the plasma scale height.

An example of the capabilities of the incoherent backscatter technique in the E region are some results presented by Taylor [7] from the installation at Malvern, England. He uses 400.5 MHz with a pulse width of 33 µs and a peak power of 7 MW. Fig. 2 shows a series of daytime profiles. The great advantage of the technique as compared with ionosonde measurements is that it can see through the peak of the E layer and measure multiple structure and the depth of valleys in the profile. The absolute accuracy of the method is however less than for the ionosonde, and the measurements have to be calibrated at the maximum E region using the operation of a nearby ionosonde. The time required to obtain one profile is 5—10 minutes and the height resolution about 5 km. The errors in N are probably about 10% below 140 km, increasing above this level. The lower limit of detection is quoted as $N \approx 6 \times 10^8\ m^{-3}$ by Taylor [7]. However, equipment sensitivity and unwanted echoes (clutter) can limit the accuracy at the lowest heights. Results obtained below about 90 km should therefore be regarded with caution. Certainly some of the N values given by Evans and Oliver [5] seem unreasonably large in this height range.

It is interesting to note that the multiple structure often recorded in the E region by means of the incoherent backscatter technique can probably explain the multiple cusps often observed on ionograms. Jones and Hart [8] have studied simultaneous records obtained by the two techniques, and Fig. 3 shows (a) an $N(h)$ profile from incoherent backscatter measurements and (b) a comparison of the $h'(f)$ curve computed from the profile with that observed on a nearby ionosonde. The detailed cusp structure observed is quite well reproduced in the computed curve.

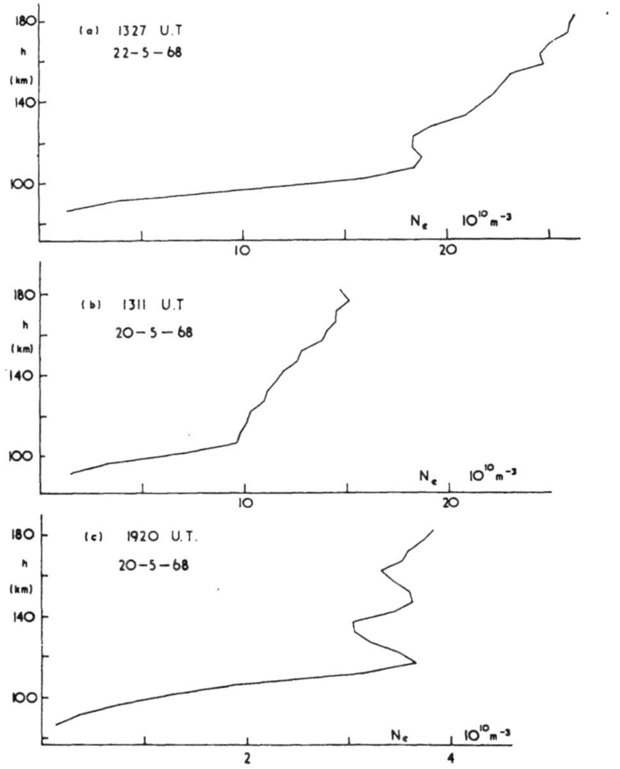

Fig. 2. Daytime profiles from measurements of incoherent scatter [7].

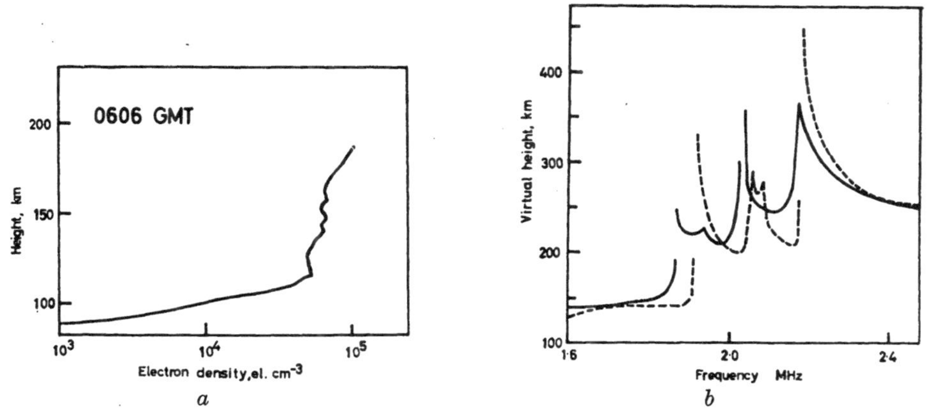

Fig. 3. (a) Electron density distribution from incoherent backscatter; (b) $h'(f)$ record calculated from (a) compared with measured $h'(f)$ values (solid line) [8].

On the whole the incoherent backscatter technique seems to be the best available ground-based method for E region electron density studies, its most serious fault being the complexity and cost of the equipment.

4. The Partial Reflection Technique

A high-frequency radio wave transmitted vertically will be partially reflected from irregularities in the D region. Gardner and Pawsey [9] first proposed that the electron density distribution could be derived by measuring the differential absorption of the two weakly scattered magneto-ionic components as a function of height. Since their first measurements the method has been successfully used by a number of workers to measure both electron densities and collision frequencies in the lowest part of the ionosphere [10—12]. The method has a interesting history. Gardner and Pawsey's interpretation was extremely simple and direct, but was implicitly based on a number of assumptions about the medium and the scattering mechanism. They assumed Fresnel reflection to occur from sharp discontinuities of infinite horizontal extent in the electron density distribution. The received amplitude ratio of the extraordinary and ordinary magneto-ionic components reflected from a height h can then be expressed as

$$\frac{A_x}{A_o}(h) = \frac{R_x}{R_o} \exp\left(-2\int_0^h (\varkappa_x - \varkappa_o)\, dh\right) \tag{2}$$

where $R_{x,o}$ is the Fresnel reflection coefficient and $\varkappa_{x,o}$ the absorption index of the waves. The absorption indices are proportional to electron density, and R_x/R_o depends only on collision frequency. Thus Eq. (1) allows direct derivation of $N(h)$ from $(A_x/A_o)(h)$ provided the collision frequency $\nu(h)$ is known. At low heights where N is small $A_x/A_o \approx R_x/R_o$ and the collision frequency can be determined directly. A typical example of measured A_x/A_o values is shown in Fig. 4 [13].

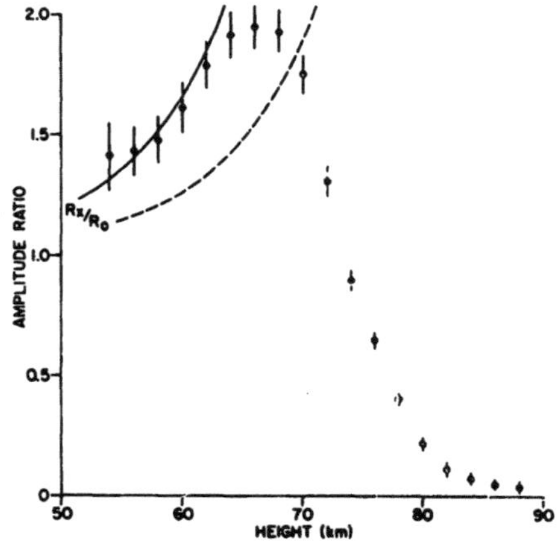

Fig. 4. Measurements of A_x/A_o, Ottawa, 29 October 1971. Solid line shows R_x/R_o curve fitted to measurements [13].

The solid curve is the R_x/R_0 curve fitted to the A_x/A_0 curve at heights where differential absorption is negligible.

Over the years the basic assumptions have been questioned one by one, and the theory and experimental techniques have been extended and improved. However, on the whole Gardner and Pawsey's simple approach has stood up remarkably well. Discussion has centred on the determination of the correct model for the physical source of D region partial reflections. Booker [14] showed that scattering from a volume containing continuous fluctuations of the refractive index does not change the form and magnitude of the ratio R_x/R_0. Flood [15] and Holt [16] considered the effect of differential absorption within the height extent of the scattering volume, but the resulting modifications of the theory were no larger than the present accuracy of the measurements. Recently Cohen and Ferraro [17] have also taken into account the change of reflection characteristics within the scattering volume. They claim that by neglecting this effect one significantly reduces the height resolution of the experiment unless very short pulses (<25 μs) are used. Further work on the height resolution of the experiment is in progress (Coyne and Belrose, private communication).

Another objection to the simple interpretation of the experiment has been raised independently by several authors [18—20], namely that not only irregularities in electron density but also irregularities in collision frequency can contribute to the scattering. Estimates show that serious errors can result if the

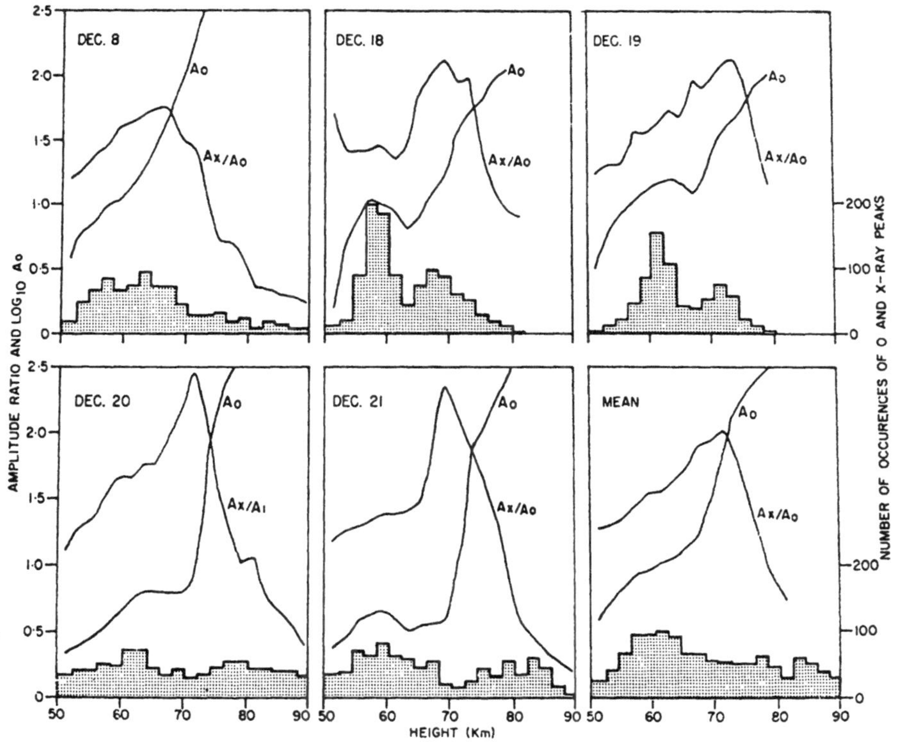

Fig. 5. Partial reflection data at 2.66 MHz at Ottawa, 1961 [21].

right scattering mechanism is not assumed. Recently Belrose et al. [13] using measurements of differential phase as well as differential absorption of the partial reflection conclude that partial reflections are predominantly from irregularities in electron density. However, their conclusion pertains to "normal" A_x/A_o curves of the type shown in Fig. 4. At times the A_x/A_o curves show a curious behaviour that is not consistent with simple theory; Fig. 5 shows a few examples [21]. I feel that the possibility still cannot be excluded that irregularities in collision frequency are important scatterers under some conditions.

The useful height range of the partial reflection experiment depends upon the working frequency and upon ionospheric conditions. Fig. 6 shows results from three different frequencies on a quiet day at Ottawa [21]. The upper useful limit is normally imposed by the increasing importance of off-vertical echoes. At the lowest heights an $N(h)$ profile is obtained [10] by assuming that the electron density is proportional to the ordinary echo amplitude. No completely satisfactory theoretical justification for this procedure has been proposed. The height resolution of the experiment is quite good in the sense that the height from which a particular distinctive echo is returned can normally be determined to within ± 1 km. However the ability to resolve structure in the profile is of necessity limited by the pulse width. Most workers have used 50 μs pulses, corresponding to a height extent of 7.5 km for the scattering volume.

An assessment of the accuracy of the derived electron density profiles is not straightforward, and few workers have published error bars with their profiles. The individual A_x/A_o measurements show great scatter, but with integration times of 10 minutes or more reliable mean values can be established. In a careful study of the diurnal variation of electron density Coyne and Belrose [46] have derived average profiles for different zenith angles for a period of about a week

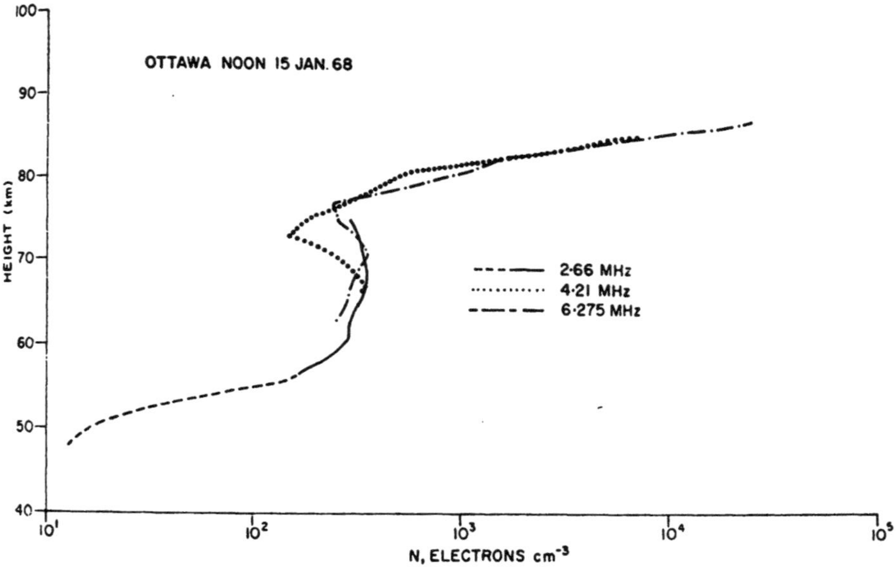

Fig. 6. Electron density profiles at Ottawa obtained from partial reflection measurements at different frequencies [21].

They estimate the accuracy of the mean electron density to be 50% at any one height, though a perturbation of the most probable electron density of more than 20% over a height region of 10 km would be unacceptable. For individual profiles the uncertainties are probably about a factor of 2 at any one height, decreasing rapidly with decreasing height at heights below the maximum of A_x/A_o.

5. The Wave Interaction Technique

The pulsed radio-wave interaction experiment was first used by Fejer [22] to derive electron densities in the D region, and since then groups in Australia, America, Norway and South Africa have used the Fejer technique for this purpose. A recent review of this work is given by Fejer [23]. The principle of the method can most easily be understood from the height—time diagram in Fig. 7 [24].

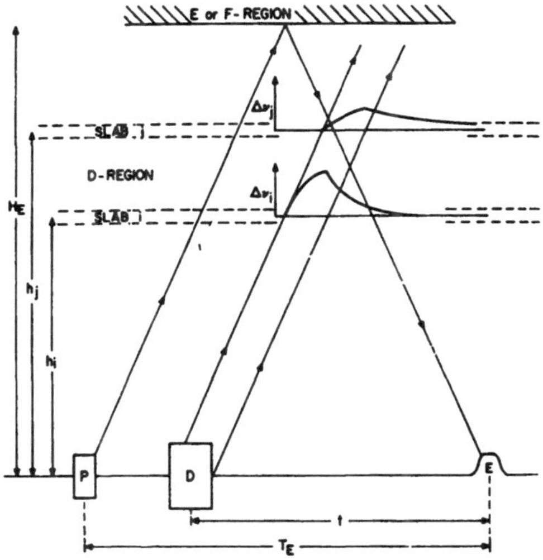

Fig. 7. Time—height diagram illustrating wave-interaction experiment [24]; P, probing pulse; D, disturbing pulse. The time variation of collision frequency is indicated at two heights.

A probing pulse with frequency f_1 is transmitted vertically and is reflected from the E or F region. A powerful disturbing pulse at a different frequency f_2 is transmitted at a time t before the probing pulse arrives at the receiver. On its way up through the D region the disturbing pulse will increase the electron temperature in the medium, thereby increasing the collision frequency. The time variation of the collision frequency at two levels is indicated in the figure. The amplitude and phase of the probing pulse will be perturbed when it passes through the heated region. By choosing a pulse repetition frequency P_1 of the probing wave that is twice that of the disturbing wave P_2, a pulse train of the probing wave will be received in which every second pulse is perturbed. The pulse repetition frequencies must be sufficiently small for the medium to recover completely in the time between the pulses. A detector system can then extract the modulation of phase and/or amplitude with frequency $P_2 = {}^1/_2 P_1$ imposed on the probing wave by the disturbing signal. In the actual experiment the time delay t is altered systematically. When $t < t_p$ (t_p = the travel time of the probing wave) the probing wave meets the disturbing pulse on its way down through the D region; when $t > t_p$ the disturbing pulse precedes the probing pulse up to the reflection level. (It is always assumed that the disturbing wave is strongly attenuated, so that heating by this wave after reflection can be neglected.) Typical "Fejer curves" are shown in Fig. 8 in which impressed amplitude modulation is plotted versus the time delay t [25]. Most experimenters have measured only the first part of this curve.

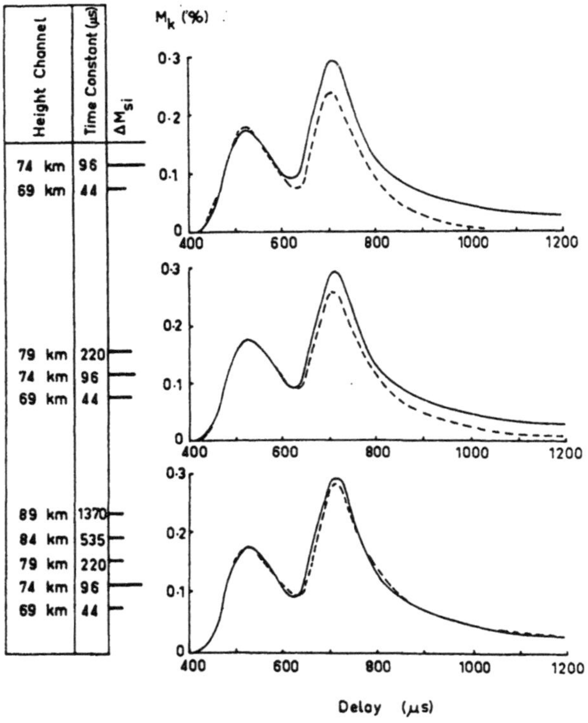

Fig. 8. "Fejer curves" observed at Armidale [25]. Measured impressed modulation versus time delay t is shown. Solid curve, observations; dashed lines, synthesized curves using sums of modulations impressed at different heights (left part of figure).

The theory for the wave interaction experiment is well developed, and is better understood than the theory for the partial reflection experiment. In order to derive electron density profiles from a curve such as the one shown in Fig. 8, the radiated power of the disturbing transmitter, the cooling factor G for the electrons and the collision frequency $\nu(h)$ must be known [25]. The two latter quantities are quite well known, and the disturbing power can be eliminated, for example by using ordinary and extraordinary disturbing waves, or other independent measurements [23]. The derivation of electron densities can be done in two ways: by a trial and error method in which a Fejer curve computed from a chosen model $N(h)$ is fitted to the experimental curve, or by a stepwise method in which the electron density is derived step by step from the measurements, starting at the bottom of the layer. Both methods have their disadvantages: in the first the result may not be unique, in the second great precision in the initial data is required to give a meaningful solution. As has been pointed out by Smith [25] the wave interaction technique inherently suffers from poor height resolution. This is because the heated electrons cool fairly slowly so that the modulation from even a single narrow slab in the ionosphere extends over an appreciable range of delay *t*, or "meeting height".

The measured Fejer curve is a sum of the modulations impressed in the different heights. Coyne [24] has recently discussed the accuracy and height resolution of the experiment and has suggested a stepwise method of analysis that makes optimum use of the data and also provides an index of the height resolution for a particular data set. No completely satisfactory assessment of the height resolu-

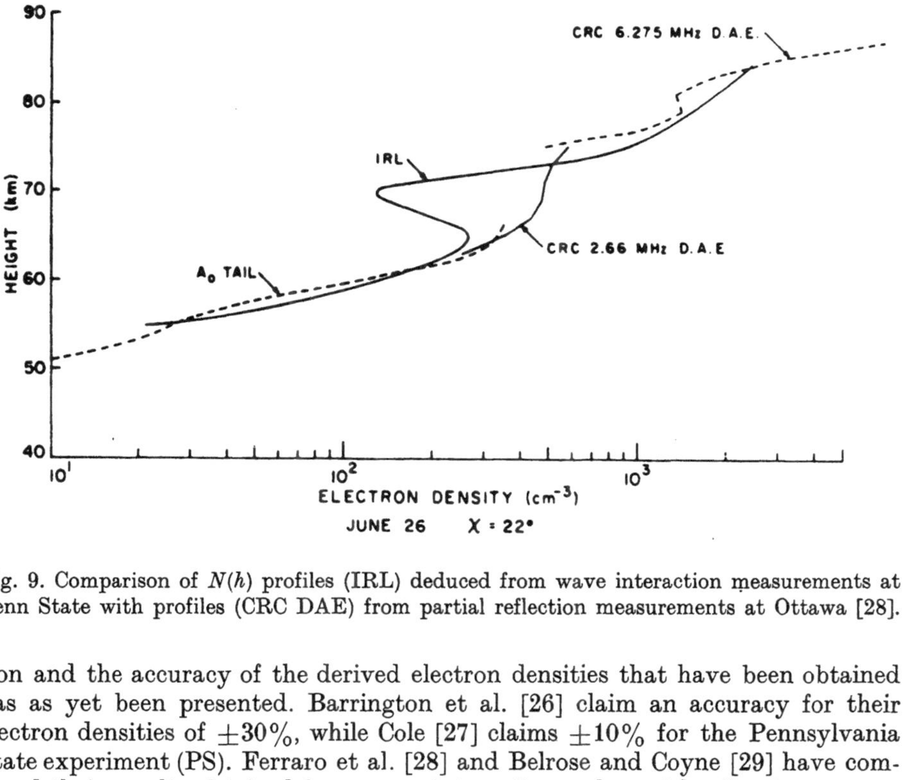

Fig. 9. Comparison of $N(h)$ profiles (IRL) deduced from wave interaction measurements at Penn State with profiles (CRC DAE) from partial reflection measurements at Ottawa [28].

tion and the accuracy of the derived electron densities that have been obtained has as yet been presented. Barrington et al. [26] claim an accuracy for their electron densities of $\pm 30\%$, while Cole [27] claims $\pm 10\%$ for the Pennsylvania State experiment (PS). Ferraro et al. [28] and Belrose and Coyne [29] have compared their results obtained from wave interaction and partial reflections respectively, but have reached no agreement on which results are most reliable. The PS group uses a trial and error method to derive $N(h)$. Fig. 9 shows a comparison of the $N(h)$ profiles obtained by two methods on 26 June 1970 at the Pennsylvania State University and at Ottawa. The wave interaction data yield a characteristic minimum that is not seen by the partial reflection technique. The cause of the discrepancy has not been established. The PS group have very recently revised their data reduction technique (Ferraro, private communication) but the revision has caused no major changes in the derived profiles.

6. Absorption Measurements

Fig. 10 shows the specific absorption, that is the absorption per km per electron, for radio waves in the frequency range 2—30 MHz. The curves depend upon the collision frequency ν, the angular wave frequency ω and the longitudinal component of the electron gyro frequency ω_L. They have maxima at the heights where $\nu \approx \omega + \omega_L$ and the height of the maximum therefore varies with frequency. In

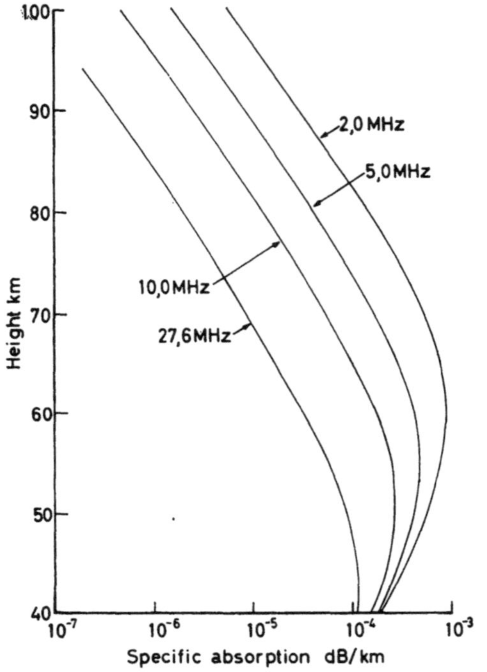

Fig. 10. Specific absorption versus height for four frequencies.

principle D region electron density profiles can be derived from multi-frequency absorption measurements. In practice, however, the method meets with major difficulties. Firstly, if high-frequency waves are reflected from the E or F regions, an appreciable part of the absorption can occur near the reflection level and will depend upon the detailed layer structure near this height. The deviative loss is, of course, particularly pronounced near the critical frequencies. Secondly the height resolution of the experiment is poor, as is evident from Fig. 10. The problem with deviative absorption can sometimes be avoided by using waves that penetrate the ionosphere, either transmitted from satellites or cosmic radio noise from the galaxy. At higher frequencies however, the sensitivity is smaller. In spite of obvious limitations, the A1 and A2 techniques have given useful information about the electron density profiles. Mitra [30] has reviewed this work. Piggott and Thrane [31] have derived normal daytime profiles from A1 measurements, while Little et al. [32] have used the method A2 to study PCA conditions. Basically trial and error methods were used in the analyses and the uniqueness of the solutions can therefore always be questioned.

7. Radio Wave Propagation Techniques in Rockets

Techniques using radio signals transmitted from the ground to a rocket travelling through the lower ionosphere have proved very valuable in studies of the electron density distributions. There is no doubt that in the D region the most accurate information available to date has been obtained in this manner. There are several variations of this experiment [33—38], but the principle of the simplest and most widely used technique is as follows: A linearly polarized HF or MF radio wave is transmitted from the ground and received by a linearly polarized antenna in the rocket. In the ionosphere the two circularly polarized magneto-ionic components of the transmitted wave travel with different velocities and the extraordinary wave is absorbed more rapidly than the ordinary wave. As a result

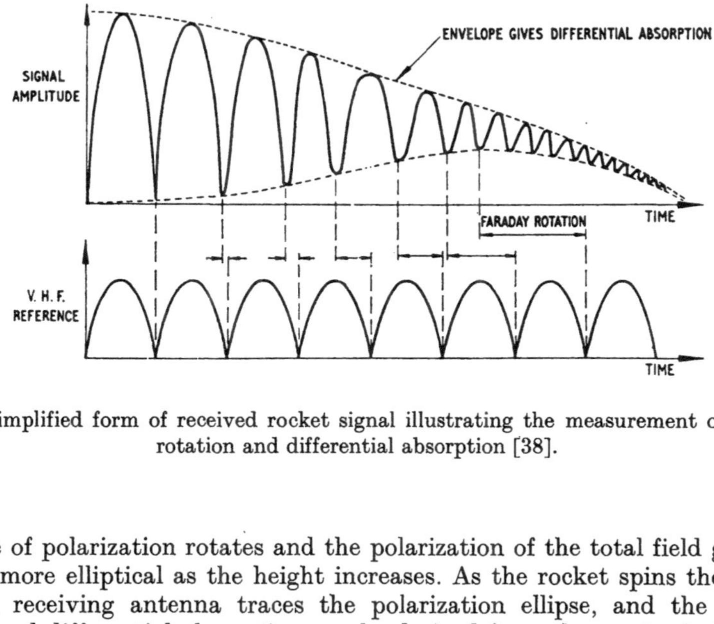

Fig. 11. Simplified form of received rocket signal illustrating the measurement of Faraday rotation and differential absorption [38].

the plane of polarization rotates and the polarization of the total field gradually becomes more elliptical as the height increases. As the rocket spins the linearly polarized receiving antenna traces the polarization ellipse, and the Faraday rotation and differential absorption can be derived from the received signal [38] (see Fig. 11). The Faraday rotation can be measured relative to a VHF reference signal which is not significantly influenced by the medium. The Faraday rotation angle F is given by

$$\frac{dF}{ds} = \frac{\omega}{2c} (\mu_o - \mu_x) \tag{3}$$

where ω is the wave frequency, s the slant range from the transmitter to the rocket and $\mu_{o,x}$ the real parts of the refractive indices for the wave. The differential absorption A in dB is given by

$$\frac{dA}{ds} = 20 \frac{\omega}{c} (\chi_o - \chi_x) \log_{10} e \tag{4}$$

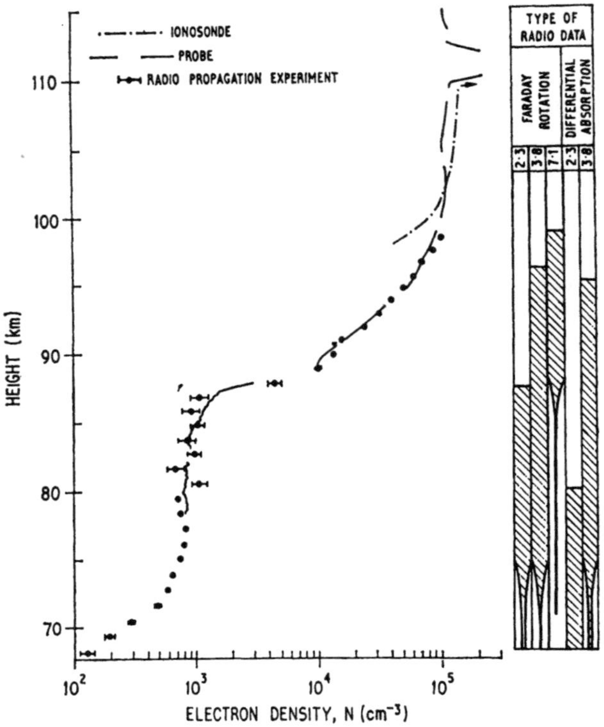

Fig. 12 Electron density profile from rocket measurements [38].

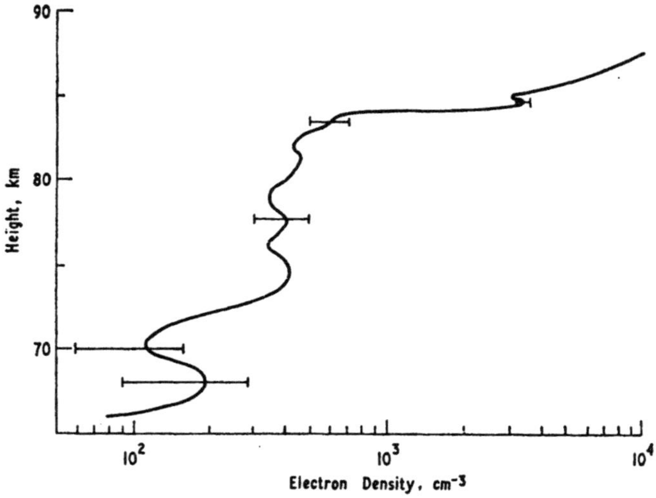

Fig. 13. Daytime electron density profile from amplitude measurements of scalar LF fields [42].

where $\chi_{o,x}$ are the imaginary parts of the refractive indices. If both Faraday rotation and differential absorption measurements are available, both electron density and collision frequency can be derived from the above expressions. χ is a stronger function of the collision frequency than μ. Normally several frequencies are used and, through the choice of frequencies, the experiment can be adapted to different height ranges and ionospheric conditions.

The analysis is based on the assumptions that: (i) the ionosphere is horizontally stratified and does not change throughout the rocket flight, (ii) there is little refraction of the waves travelling to the rocket, and (iii) the slight change in the line of sight to the rocket during the flight has negligible effects. The most critical of these assumptions is probably the first one, which may not hold during disturbed conditions. The Faraday rotation measurements are less sensitive to noise than the differential absorption measurements. Furthermore they are less critically dependent upon the polarization of the transmitted wave which can be influenced by local terrain and conductors near the transmitter antennas. This has the consequence that the electron densities can be determined with greater accuracy than the collision frequencies.

Fig. 12 shows an $N(h)$ profile derived by Bennett et al. [38]. The profile is a composite of measurements at different frequencies and the error bars indicate weighted mean errors. The accuracy is very good, better than 10% at all heights

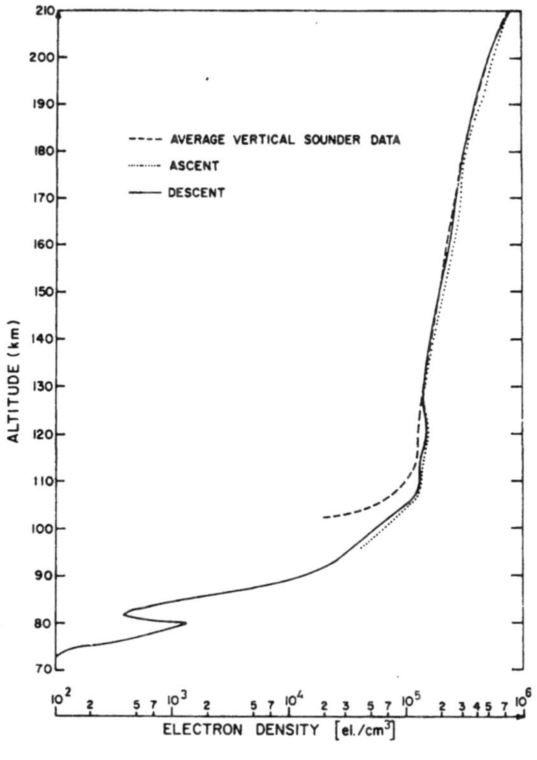

Fig. 14. Electron density profiles (solid and dotted lines) derived from VLF-Doppler measurements at Wallops Island 8 November 1966 [43].

above 70 km. The height resolution seems to be of the order of a few kilometres. This is reasonable in view of the work by Holt et al. [39] who have seen structure of the order of 2 km in Faraday data. Mechtly (private communication) quotes probable errors in electron density of about $\pm 10\%$ at 90 km increasing to a factor of about 3 below 70 km.

In a different type of radio propagation experiment Hall and Fooks [40, 41] and Hall [42] have derived electron densities from rocket measurements of the scalar field of an LF wave transmitted from the ground. In the latter work the phase was also measured (see Fig. 13). A trial and error method has been used to fit a model to the data, and the error bars indicate the range of possible changes in the model. A height resolution of a few kilometres is indicated.

Egeland et al. [43] have measured the Doppler shift of a VLF signal at a moving rocket and derived electron densities in the D and E regions. The VLF signal is transmitted from the ground several hundred kilometres away and the phase of the whistler mode is measured in the rocket. Fig. 14 shows the result of one such measurement. This experiment seems well suited for measurements during disturbed conditions when absorption of HF waves is large. Like the other experiments described above it implicitly depends on the assumption that the ionosphere does not change throughout the flight. It also depends on having a stable VLF transmitter within a suitable range.

8. Probe Measurements

Most of the workers referred to in the preceding section have, in addition to the wave propagation experiments, flown probes on their rockets to measure positive and/or negative current. The probes are, during the flight, moving with supersonic velocity through the ionospheric plasma. In the D region the plasma is collision-dominated with a large Debye length, and no adequate theory exists to derive electron densities from the measured probe currents. Folkestad [44] has discussed the relevant problems for various type of probes. However, one can expect proportionality between probe current and electron density over small height ranges, and by calibrating the probe experiment by the wave propagation data, the rapid response of the probe can be used to measure structure in electron density down to scales of about 100 m. Fig. 12 also shows the height variations in electron density derived from a Langmuir probe [38].

9. Discussion

The most important techniques for measurements of electron densities in the lower ionosphere have been discussed in the preceding sections. One important omission is the use of cw propagation data, mainly VLF and LF radio waves; this theme is discussed by Belrose and Segal [1]. Emphasis has been put on an evaluation of the accuracy and height resolution of the different experiments, based on a discussion of the validity of the assumptions on which the methods are based as well as on the experimental uncertainties. An attempt to summarize the results is given in the form of a diagram in Fig. 15. Here the author's estimate of an error factor for each experiment is indicated along the horizontal axis. The height resolution is indicated by the shaded areas. Each box gives the size of the smallest height structure that can be resolved. The estimates pertain to quiet daytime

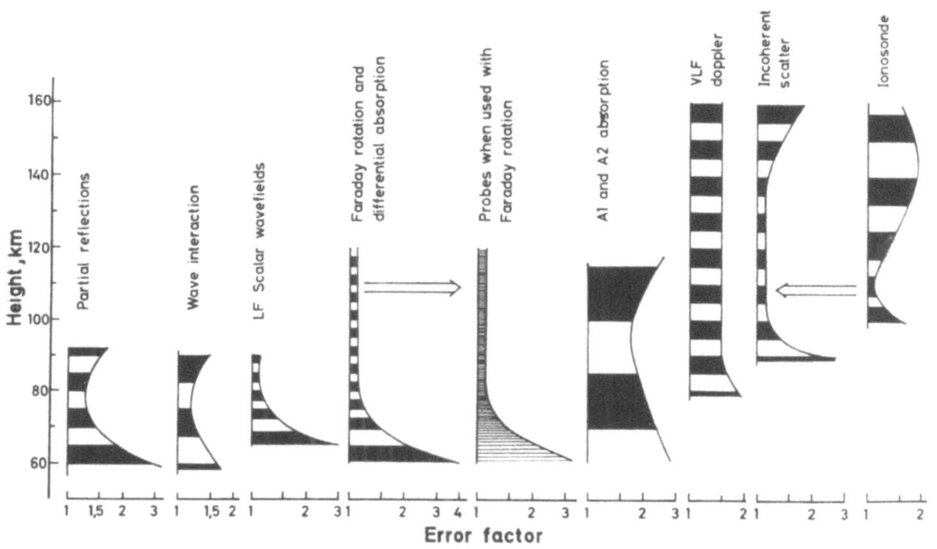

Fig. 15. Error factor and height resolution for various experiments used to determine electron density profiles. The erreor factor indicates probable errors versus height, the shaded slabs indicate minimum height structure that can be resolved. The two horizontal arrows indicate that two of the experiments must be calibrated using a different technique to give the absolute accuracy shown in the figure (see text).

conditions, which represent near optimum conditions for most experiments. As shown in the preceding sections, there are no really reliable estimates of the uncertainties available for most of the experiments. Table 1 lists the author's estimates of the maximum obtainable accuracy and height resolution for each experiment and also the time required for individual measurements. A meaningful comparative study of profiles derived from the different methods is difficult, since very few measurements have been taken at the same time and place. The study made by Booker and Smith [45] is probably the best available at present. In spite of the difficulties Fig. 15 and Table 1 give some useful, semi-quantitative information. We see that at any height above 65 km, using the right technique, it should be possible to measure the electron density to within an accuracy of $\pm 30\%$. It is also evident that for most heights the rocket techniques have superior accuracy and height resolution, and they also cover the widest height range. The ground-based methods, of course, have their real advantage in synoptic studies. However, the height range is limited and there is little overlap between the methods effective in the D and E regions.

References

[1] J. S. BELROSE and B. SEGAL, this volume, p. 77.

[2] W. R. PIGGOTT and K. RAWER, URSI Handbook of Ionogram Interpretation and Reduction, World Data Center A Rep. UAG-23 (1972).

[3] A. K. PAUL and J. W. WRIGHT, in: Electron Density Distribution in Ionosphere and Exosphere, North Holland Publ. Co., Amsterdam 1964 (p. 105).

[4] D. T. FARLEY, J. Atmos. Terr. Phys. **32**, 693 (1970).

[5] J. V. EVANS and M. L. OLIVER, in: COSPAR Symp. on D and E Region Ion Chemistry, Aeronomy Rep. No. 48, University of Illinois 1972 (p. 172).

[6] G. W. ARMISTEAD, J. V. EVANS and W. A. REID, in: COSPAR Symp. on D and E Region Ion Chemistry, Aeronomy Rep. No. 48 University of (Illinois 1972 (p. 331).

[7] G. N. TAYLOR, in: Scatter Propagation of Radio Waves, AGARD Conf. Proc. No. 37, 1968 (paper No. 32).

[8] T. B. JONES and J. W. S. HART, J. Atmos. Terr. Phys. **34**, 387 (1972).

[9] F. F. GARDNER and J. L. PAWSEY, J. Atmos. Terr. Phys. **3**, 321 (1953).

[10] J. S. BELROSE and M. J. BURKE, J. Geophys. Res. **69**, 2799 (1964).

[11] J. B. GREGORY and A. M. MANSON, J. Geophys. Res. **72**, 1073 (1967).

[12] E. V. THRANE, A. HAUG, B. BJELLAND, M. ANASTASSIADES and E. TSAGAKIS, J. Atmos. Terr. Phys. **30**, 135 (1968).

[13] J. S. BELROSE, M. J. BURKE and T. N. R. COYNE, J. Geophys. Res. **77**, 4829 (1972).

[14] H. G. BOOKER, J. Geophys. Res. **64**, 2164 (1959).

[15] W. A. FLOOD, J. Geophys. Res. **73**, 5585 (1968).

[16] O. HOLT, J. Geophys. Res. **74**, 5179 (1969).

[17] D. J. COHEN and A. J. FERRARO, Radio Sci. **8**, 459 (1973).

[18] J. S. BELROSE, I. A. BOURNE and L. W. HEWITT, in: Proc. Conf. on Ground-based Studies of the Lower Ionosphere, Defence Research Board, Ottawa 1967 (p. 125).

[19] W. R. PIGGOTT and E. V. THRANE, J. Atmos. Terr. Phys. **28**, 311 (1966).

[20] A. H. MANSON, J. Geophys. Res. **71**, 3783 (1966).

[21] J. S. BELROSE, J. Atmos. Terr. Phys. **32**, 567 (1970).

[22] J. A. FEJER, J. Atmos. Terr. Phys. **7**, 322 (1955).

[23] J. A. FEJER, J. Atmos. Terr. Phys. **32**, 597 (1970).

[24] T. N. R. COYNE, J. Geophys. Res. **78**, 206 (1973).

[25] R. A. SMITH, in: Proc. Conf. on Ground-based Radio Wave Propagation Studies of the Lower Ionosphere, Defence Research Board, Ottawa 1967 (p. 235).

[26] R. E. BARRINGTON, E. V. THRANE and B. BJELLAND, Canad. J. Phys. **41**, 271 (1963).

[27] A. R. COLE, Scient. Rep. No. 328, NSF Grant GA-1456.

[28] A. J. FERRARO, H. S. LEE and D. J. COHEN, in: COSPAR Symp. on D and E Region Ion Chemistry, Aeronomy Rep. No. 48, University of Illinois 1972 (p. 320).

[29] J. S. BELROSE and T. N. R. COYNE, in: COSPAR Symp. on D and E Region Ion Chemistry, Aeronomy Rep. No. 48, University of Illinois 1972 (p. 326).

[30] A. P. MITRA, J. Atmos. Terr. Phys. **32**, 623 (1970).

[31] W. R. PIGGOTT and E. V. THRANE, J. Atmos. Terr. Phys. **28**, 467 (1966).

[32] G. C. LITTLE, G. M. LERFALD and R. PARTASARATHY, J. Geophys. Res. **68**, 3581 (1963).

[33] J. C. SEDDON, J. Geophys. Res. **63**, 209 (1958).

[34] J. A. KANE, J. Atmos. Terr. Phys. **23**, 338 (1961).

[35] M. JESPERSEN, O. PETERSEN, J. RYBNER, B. BJELLAND, O. HOLT and B. LANDMARK, in: Electron Density Profiles in Ionosphere and Exosphere, North-Holland Publ. Co., Amsterdam 1964 (p. 22).

[36] E. A. Mechtly, S. A. Bowhill, L. G. Smith and H. W. Knoebel, J. Geophys. Res. **72**, 5239 (1967).

[37] M. Jespersen and B. M. Pedersen, J. Atmos. Terr. Phys. **32**, 1859 (1970).

[38] F. D. G. Bennett, J. E. Hall and P. H. G. Dickinson, J. Atmos. Terr. Phys. **34**, 1321 (1972).

[39] O. Holt, J. Tröim and M. Jespersen, J. Atmos. Terr. Phys. **30**, 1231 (1968).

[40] J. E. Hall and J. Fooks, Planet. Space Sci. **13**, 1013 (1965).

[41] J. E. Hall and J. Fooks, Planet. Space Sci. **15**, 1717 (1967).

[42] J. E. Hall, Planet. Space Sci. **21**, 119 (1973).

[43] A. Egeland, G. Bjöntegaard and T. L. Aggson, J. Atmos. Terr. Phys. **32**, 1191 (1970).

[44] K. Folkestad, NDRE Rep. No. 59, Norwegian Defence Research Establishment, Kjeller, Norway 1970.

[45] H. G. Booker and E. K. Smith, J. Atmos. Terr. Phys. **32**, 467 (1970).

[46] T. N. R. Coyne and J. S. Belrose, in: COSPAR Symp. on D and E Region Ion Chemistry, Aeronomy Rep. No. 48, University of Illinois 1972 (p. 342).

Lower Ionosphere Structure — Akademie-Verlag, Berlin 1974

DEFINITION AND USE OF COLLISION FREQUENCIES*

K. SUCHY

Institute for Theoretical Physics, University of Düsseldorf, Düsseldorf, FRG

* *Dedicated to Professor K. Rawer on the occasion of his 60th birthday*

Using Legendre polynominals as weight functions for averaging over deflection angles, singly indicated velocity dependent collision frequencies are defined. With generalized Laguerre polynomials as weight functions for averaging over velocities, doubly indicated collision frequencies as well as "transport integrals" are defined. The transport coefficients for electrons (electrical and heat conductivity as well as the thermal diffusion coefficient) are expressed most economically by these quantities. If the velocity dependence of the (singly indicated) collision frequencies can be approximated by a power law, results for the calculation of the velocity averaged quantities are given.

1. Introduction

To describe the absorption of electromagnetic waves propagating through a magnetized (cold) plasma the concept of a collision frequency was first introduced by Lassen ([1], Eq. 18) for the dispersion formula for such waves. This was the friction constant ν in the equation of motion

$$m(\dot{\boldsymbol{v}} + \nu \boldsymbol{v}) = q(\boldsymbol{E} + \boldsymbol{v} \times \boldsymbol{B}/c\ (\varepsilon_0\mu_0)^{1/2}) := q\boldsymbol{E} + m\boldsymbol{\omega}_{\mathrm{B}} \times \boldsymbol{v} \tag{1.1}$$

for the mean electron velocity $\boldsymbol{v}$ under the influence of the electric wave field $\boldsymbol{E}$ and the ambient magnetic field $\boldsymbol{B}$, i.e. the earth's magnetic field for the ionospheric plasma. A "mean free path" theory allows us to interpret this ν as a mean collision frequency [2].

The solution of the equation of motion (1.1) for time harmonic waves $\boldsymbol{E} \sim \exp(-i\omega t)$ yields an expression for the conductivity tensor $\overleftrightarrow{\sigma}(\omega; \omega_{\mathrm{N}}^2, \boldsymbol{\omega}_{\mathrm{B}}, \nu)$ in Ohm's law

$$\boldsymbol{j} = \overleftrightarrow{\sigma} \cdot \boldsymbol{E} \tag{1.2}$$

which in turn via Maxwell's equations enters the dispersion equation for space harmonic

$$\boldsymbol{E} \sim \exp\left(i \frac{\omega}{c_0} \boldsymbol{n} \cdot \boldsymbol{r}\right),$$

viz.

$$\det\left[n^2 \overleftrightarrow{I} - \boldsymbol{n}\boldsymbol{n} - \overleftrightarrow{I} - \frac{i\overleftrightarrow{\sigma}}{\omega\varepsilon_0}\right] = 0. \tag{1.3}$$

The solution of this equation for $n^2 = \boldsymbol{n} \cdot \boldsymbol{n}$, i.e. for the square of the refractive index n, is the well-known Appleton-Lassen dispersion formula.

The absorption index $\chi := |\operatorname{Im} \boldsymbol{n}|$, taken from this formula, describes the wave absorption the better the smaller ν is compared with ω. This was sufficient as long as the wave frequencies ω exceeded ν appreciably and the accuracy of the absorption measurements was not very high. In the late fifties rocket experiments [3] with low wave frequencies ω were made, the accuracy of which was high enough to render the Appleton-Lassen formula insufficient for the interpretation of these measurements.

Kinetic gas theory must be used to improve the expressions for the conductivity tensor $\overleftrightarrow{\sigma}(\omega; \omega_N{}^2, \boldsymbol{\omega}_B)$. This improves the dispersion equation (1.3) and its solution $n^2 = n^2(\omega; \omega_N{}^2, \boldsymbol{\omega}_B)$. Wyller was the first to calculate improved dispersion formulae for the lower ionosphere. He used two different methods [4,5] which will be described in a generalized manner in this paper.

2. Transport Collision Frequencies

Kinetic gas theory needs collision frequencies more accurately defined as the friction constant ν from the "mean free path" theory used in Eq. (1.1). This results in the fact that a variety of different collision frequencies are needed to describe the results of the kinetic gas theory.

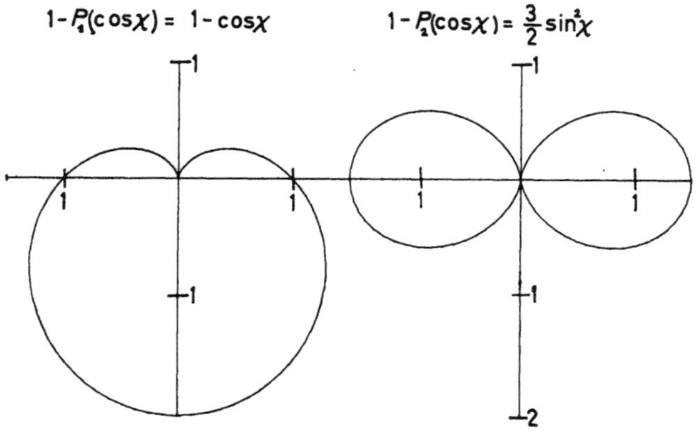

Fig. 1. Polar diagrams of $1 - \mathrm{P}_1(\cos\chi) = 1 - \cos\chi$ and $1 - \mathrm{P}_2(\cos\chi) = 3/2 \sin^2\chi$.

Usually the frequency of collisions of a particle of species j with particles of species k is considered as the product of the relative speed g_{jk} times the collisional cross section $Q(g_{jk})$ times the number density N_k of the colliding particles. But stripping collisions with low deflection angle χ are less influential than collisions with $\chi \simeq \pi/2$. To account for this the kinetic gas theory [6] introduces weight functions $1 - \mathrm{P}_l(\cos\chi)$ to define "transport collision frequencies"

$$\nu_{jk}^{(l)}(g) := N_k g_{jk} \int dQ_{jk} [1 - \mathrm{P}_l(\cos\chi)]. \tag{2.1}$$

The $P_l(\cos \chi)$ are Legendre polynomials with $P_l(1) = 1$. Fortunately only those $\nu^{(l)}(g)$ with low l-values ($\geqq 1$) are needed for the conductivity. The first of these, $\nu^{(1)}(g)$, is often called "collision frequency for momentum transfer". Polar plots of $1 - P_1(\cos \chi)$ and $1 - P_2(\cos \chi)$ are given in Fig. 1.

3. Averaged Transport Collision Frequencies

In a gas all relative speeds g_{jk} will occur between colliding particles. But the speeds are distributed in velocity space with a certain probability. If all particle species had a (local partial) Maxwellian distribution the relative speeds would be distributed like $\exp(-g^2/\hat{g}^2)$ with

$$\hat{g}_{jk} := \left(\frac{2kT_j}{m_j} + \frac{2kT_k}{m_k}\right)^{1/2} \tag{3.1}$$

as the "most probable relative speed". But in this case no current would exist. However, already a weak deviation from this "local partial thermal equilibrium" (which is described by these Maxwellian distributions) allows for currents. Therefore kinetic gas theory [7] introduces weight functions

$$\exp(-\gamma^2)\gamma^{2l+1}L_m^{(l+1/2)}(\gamma^2) \tag{3.2}$$

to define "averaged transport collision frequencies"

$$\nu^{(lm)}(\hat{g}) := \frac{(-1)^m m!}{(m+l+1/2)!} \int_0^\infty d\gamma^2 \exp(-\gamma^2)\,\gamma^{2l+1}\,L_m^{(l+1/2)}(\gamma^2)\nu^{(l)}(\hat{g}\gamma) \tag{3.3}$$

with

$$\gamma := g/\hat{g} \text{ and } \left(m + l + \frac{1}{2}\right)! = \Gamma\left(m + l + \frac{3}{2}\right). \tag{3.4}$$

The $L_m^{(l+1/2)}(\gamma^2)$ are generalized Laguerre polynomials, sometimes called Sonine polynomials when used in kinetic gas theory [8].

The use of the weight functions (3.2) ensure that the $|\nu^{(lm)}|$ decrease with increasing m, because they are the coefficients of the expansion of $\nu^{(l)}(\hat{g}\gamma)$ in orthogonal polynomials $L_m^{(l+1/2)}(\gamma^2)$. Upon comparing various $\nu^{(lm)}$ the superscript m serves as a measure for the magnitude of the corresponding $\nu^{(lm)}$.

Fortunately, as with the $\nu^{(l)}(g)$ which are only needed for low l-values, the $\nu^{(lm)}(\hat{g})$ are only needed for low m-values, too. It is $\nu^{(1,0)}(\hat{g})$ which has to replace the friction coefficient ν in the Appleton-Lassen formula when applied for weak collisional absorption, i.e. for $\nu \ll \omega$. (This is calculated for several gases by Itikawa [9] who named it "effective collision frequency" ν_{eff}.) To improve the Appleton-Lassen formula by the next better approximation the $\nu^{(lm)}$ with $l = 1, 2$ and $m = 0, 1, 2$ are needed (see Eqs. (5.1), (5.2) below).

4. Most Probable Transport Collision Frequencies

To calculate the $\nu^{(lm)}(\hat{g})$ from the $\nu^{(l)}(g)$ with the definition (3.3) to any described accuracy the $\nu^{(l)}(g)$ must be known for all relative speeds $0 \leqq g \leqq \infty$. But due to the weight functions $\exp(-\gamma^2)\gamma^{2l+1}L_m^{(l+1/2)}(\gamma^2)$ only a certain speed range around the maxima of the weight functions is really relevant. Thus for most purposes one can approximate the (measured or calculated) curves $\nu^{(l)}$ vs g around the maximum of the weight functions. This is certainly allowed as long as the $\nu^{(l)}(g)$ are known only with limited accuracy [10]. In most cases the "transport cross sections" $Q_{jk}^{(l)}(g) := \nu_{jk}^{(l)}(g)/gN_k$ are given instead of the $\nu^{(l)}(g)$. Then the functions $\hat{g}\gamma^{2l+2} \exp(-\gamma^2) \mathrm{L}_m^{(l+1/2)}(\gamma^2)$ are the appropriate weight functions to calculate the $\nu^{(lm)}$ from the $Q^{(l)}(g)$.

To approximate $Q^{(1)}(g)$ for the calculation of $\nu^{(1,0)}(\hat{g})$ the curve $Q^{(1)}$ vs g is relevant only in the range of the maximum of the weight function $\hat{g}\gamma^4 \exp(-\gamma^2)$. Figs. 2 to 4 show transport cross sections $Q^{(1)}(g)$ for electron-air collisions and electron-CO_2 collisions. In Fig. 5 the weight function $\hat{g}^{-3}g^4 \exp(-g^2/\hat{g}^2)$ is plotted with $\hat{g}^2 \simeq 2kT_e/m_e$ as parameter. Comparison of this figure with Figs. 2, 3 for

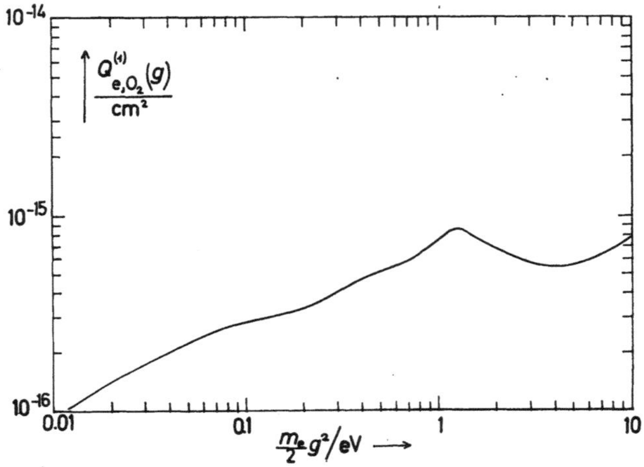

Fig. 2. Transport cross section $Q^{(1)}(g)$ for elastic electron-oxygen collisions [26—28].

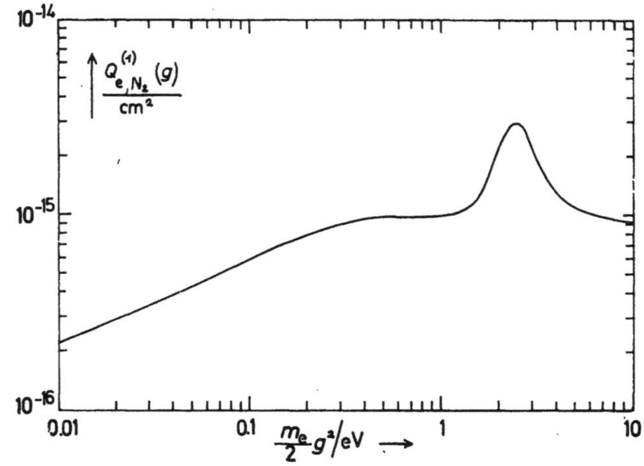

Fig. 3. Transport cross section $Q^{(1)}(g)$ for elastic electron-nitrogen collisions [29].

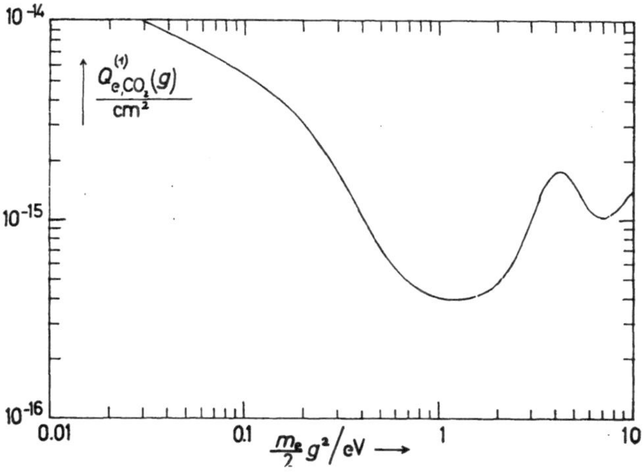

Fig. 4. Transport cross section $Q^{(1)}(g)$ for elastic electron-CO_2 collisions [26].

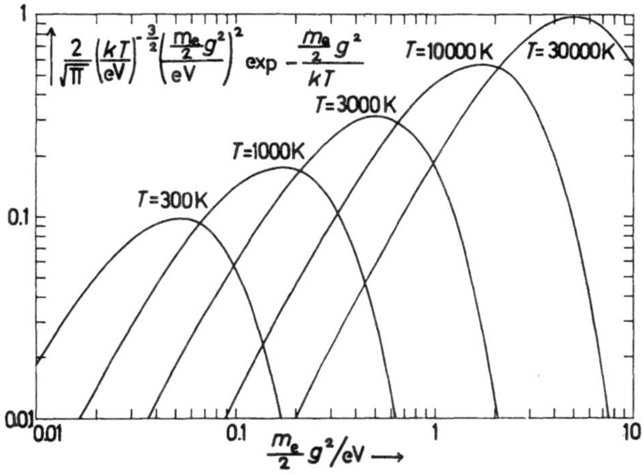

Fig. 5. Weight functions for the velocity averaging of the transport cross section $Q^{(1)}(g)$ to obtain the averaged transport collision frequency $\nu^{(10)}(T)$.

electron-air collisions shows that for $100\,K < T_e < 1200\,K$ the $\nu^{(1)}(g)$ can be approximated by a power law

$$\nu^{(1)}(g) = \nu^{(1)}(\hat{g})\gamma^{\delta} \tag{4.1}$$

with $\delta = 2$ [11]. Fig. 4 for electron-CO_2 collisions reveals a good approximation with $\delta = 0$ for $100\,K < T_e < 800$ K [12]. The factor $\nu^{(1)}(\hat{g})$ may be interpreted as a "most probable transport collision frequency" because $\hat{g}$ (3.1) is the "most probable relative speed". For the joint matching of a sum of collision frequencies (with different species) having different δ values for the same temperature range, procedures are given by Schunk and Walker [13] and by Suchy and Rawer ([14], Sect. 4.4).

The power law approximation

$$\nu^{(l)}(g) = \nu^{(l)}(\hat{g})\gamma^{\delta} \tag{4.2}$$

has the great advantage that the averaged transport collision frequencies $\nu^{(lm)}(\hat{g})$ (3.3) can now be calculated, yielding ([14], Eqs. 4.2, 4.3)

$$\frac{\nu^{(l0)}(\hat{g})}{\nu^{(l)}(\hat{g})} = \frac{\left(l + \frac{1}{2} + \frac{\delta}{2}\right)!}{\left(l + \frac{1}{2}\right)!} \tag{4.3}$$

$$\frac{\nu^{(lm)}}{\nu^{(l0)}} = \frac{\delta(\delta - 2)(\delta - 4) \ldots (\delta - 2m + 2)}{(2l + 1 + 2m)(2l + 1 + 2m - 2)(2l + 1 + 2m - 4) \ldots (2l + 1 + 2)}. \tag{4.4}$$

A plot of $\nu^{(1m)}/\nu^{(10)}$ for $0 < m \leqq 3$ with $-3 \leqq \delta \leqq +5$ as parameter ([14], Fig. 1) shows the decrease of $|\nu^{(lm)}|$ with increasing m. For $\delta = 0$ one has $\nu^{(l0)} = \nu^{(l)}(\hat{g})$ and all $\nu^{(lm)}$ vanish for $m \geqq 1$.

5. The Moment Method of Kinetic Gas Theory

In the kinetic gas theory there exist several methods for the calculation of the electrical conductivity. Two of them are relevant for a cold plasma like that of the lower ionosphere. The "moment method" was introduced by Maxwell and extended by Grad [15]. It yields an infinite system of coupled equations; two of them are essentially the equation of motion and the balance for the heat flux. For a plasma without bulk motion these equations are for electrons [16]

$$\frac{\partial \boldsymbol{J}_e}{\partial t} - \boldsymbol{\omega}_B \times \boldsymbol{J}_e + \frac{\partial p_e}{\partial \boldsymbol{r}} - q_e N_e \boldsymbol{E}$$

$$= -\boldsymbol{J}_e \sum_k \nu_{eh}^{(10)} - \boldsymbol{Q}_e \frac{m_e}{kT_e} \sum_h \nu_{eh}^{(11)} + O\left(\frac{m_e}{m_h}, \nu^{(l2)}\right) \tag{5.1}$$

$$\frac{\partial \boldsymbol{Q}_e}{\partial t} - \boldsymbol{\omega}_B \times \boldsymbol{Q}_e + \frac{5}{2} \frac{p_e k}{m_e} \frac{\partial T_e}{\partial \boldsymbol{r}}$$

$$= -\boldsymbol{J}_e \frac{5}{2} \frac{kT_e}{m_e} \sum_h \nu_{eh}^{(11)} - \boldsymbol{Q}_e \left[\frac{1}{3} \nu_{ee}^{(20)} + \sum_h \left(\nu_{eh}^{(10)} + 2\nu_{eh}^{(11)} + \frac{7}{2} \nu_{eh}^{(12)}\right)\right]$$

$$+ O\left(\frac{m_e}{m_h}, \nu^{(l2)}\right). \tag{5.2}$$

Here $\boldsymbol{J}_e = \frac{m_e}{q_e} \boldsymbol{j}_e$ and $\boldsymbol{Q}_e$ are the mass flux and the heat flux of the electrons, while p_e is the (scalar) electron pressure. The subscript h refers to heavy particles, i.e. neutrals and ions.

Neglecting the coupling terms on the right-hand sides, which are of the order $\nu^{(11)}/\nu^{(10)}$, one obtains

$$\overleftrightarrow{\sigma}^{(0)} = \omega_N{}^2\varepsilon_0 \left\{\left[-\,i\omega + \sum_h \nu_{eh}^{(10)}\right]\overleftrightarrow{I} - \omega_B \times \overleftrightarrow{I}\right\}^{-1} \tag{5.3}$$

for the electrical conductivity tensor and

$$\overleftrightarrow{\varkappa}^{(0)} = \frac{5}{2}\,\frac{p_e k}{m_e}\left\{\left[-i\omega + \frac{\nu_{ee}^{(20)}}{3} + \sum_h\left(\nu_{eh}^{(10)} + 2\nu_{eh}^{(11)} + \frac{7}{2}\,\nu_{eh}^{(12)}\right)\right]\overleftrightarrow{I} - \omega_B \times \overleftrightarrow{I}\right\}^{-1} \tag{5.4}$$

for the heat conduction tensor in this "weak absorption approximation". Comparison of (5.1) with (1.1) shows that the friction coefficient ν in the Appleton-Lassen formula has to be replaced by $\Sigma_h\, \nu_{eh}^{(10)}$ in this approximation.

The next approximation $\sigma^{(1)}$ retains the coupling terms of order $\nu^{(11)}/\nu^{(10)}$ and leads to a more accurate dispersion equation ([14] Eq. (7.5)). For the special case of a fully ionized hydrogen plasma this dispersion equation was derived by Wyller [5]. Further approximations $\sigma^{(n)}$ can be obtained by calculating the omitted terms of order $\nu^{(12)}/\nu^{(10)}$ and higher in the equations (5.1), (5.2). Then additional balance equations for higher order moments have to be derived too. For the case of a Lorentz plasma, where electron-electron collisions can be completely neglected, the exact expression for $\overleftrightarrow{\sigma}$ can be calculated and compared with the approximations $\overleftrightarrow{\sigma}^{(n)}$ for $\nu^{(l)}(g)$ of the form $\nu^{(l)}(\hat{g})\gamma^{\delta}$ (4.2). This has been done by Schirmer and Friedrich [17]. It turn out that for $\delta > 0$ the convergence (with increasing n) decreases with increasing δ, whereas for $-3 < \delta < 0$ the approximation $\sigma^{(1)}$ gives already very good results (see also [14], Fig. 2).

Since $\nu^{(1)}(g)$ for electron-air collisions in the lower ionosphere can be approximated by a power law with $\delta = 2$ the convergence of the moment method is too slow in this case. Another method of the kinetic gas theory is better suited. It should, however, be noted that in the higher ionosphere where Coulomb collisions with $\delta = -3$ are dominant, the approximation $\overleftrightarrow{\sigma}^{(1)}$ gives satisfactory accuracy.

6. The Lorentz Method

For the treatment of collisions between partners of very different masses the heavy particles can be assumed as being "at rest". More strictly speaking their velocity distribution functions can be assumed as Maxwellian with very small widths, i.e. with $2kT_h/m_h \to 0$. This shrinking of velocity distribution functions to Dirac δ-functions allows considerable "short cuts" in the methods of kinetic gas theory. It must be stressed, however, that this "Lorentz method" (which was considerably refined by Allis [18]), is only applicable if the electron-electron collisions can be neglected in comparison with the collisions of electrons with heavy particles. From Eq. (5.2) the condition is

$$\nu_{ee}^{(20)} \ll 3\nu_{eh}^{(10)} \tag{6.1}$$

for the applicability of the Lorentz method. It is certainly satisfied in the lower ionosphere.

In contrast to the moment method as described in the preceding section the Lorentz method does not yield coupled balance equations for fluxes like Eqs. (5.1), (5.2) but gives directly the transport coefficients like electrical and heat conductivity, thermal diffusion coefficient and viscosity. These (tensorial) transport coefficients are given as velocity averages of the tensors ([19] Eqs. (104.6) to (104.8))

$$\overleftrightarrow{\beta}^{(l)}(C) := \left\{\left[-i\omega + \sum_{\mathrm{h}} \nu_{\mathrm{eh}}^{(l)}(C)\right] \overleftrightarrow{I} - \boldsymbol{\omega}_{\mathrm{B}} \times \overleftrightarrow{I}\right\}^{-1}$$

$$= \beta_0^{(l)}(C)\hat{\boldsymbol{B}}\hat{\boldsymbol{B}} + \frac{\beta_{+1}^{(l)}(C) + \beta_{-1}^{(l)}(C)}{2}\left[\overleftrightarrow{I} - \hat{\boldsymbol{B}}\hat{\boldsymbol{B}}\right] + i\frac{\beta_{+1}^{(l)}(C) - \beta_{-1}^{(l)}(C)}{2}\,\hat{\boldsymbol{B}} \times \overleftrightarrow{I} \tag{6.1}$$

with the three complex eigenvalues

$$\beta_s^{(l)}(C) = \frac{1}{-i(\omega - s\omega_{\mathrm{B}}) + \sum_{\mathrm{h}} \nu_{\mathrm{eh}}^{(l)}(C)}, \qquad s = 0, \pm 1. \tag{6.2}$$

Here C is the electron speed.

7. Transport Integrals and Transport Coefficients

For the velocity averaging the weight functions (3.2) with $\gamma := C/\hat{C} = C/(2kT_{\mathrm{e}}/m_{\mathrm{e}})^{1/2}$ are again useful. The moduli of the complex "transport integrals"

$$\beta_s^{(lm)}(\hat{C}) := \frac{(-1)^m m!}{(m + l + 1/2)!} \int_0^\infty d\gamma^2 \exp(-\gamma^2)\, \gamma^{2l+1}\, \mathrm{L}_m^{(l+1/2)}(\gamma^2) \beta_s^{(l)}(\hat{C}\gamma) \tag{7.1}$$

decrease with increasing m as do the $|\nu^{(lm)}|$ in § 3 above.

For weak absorption $\Sigma_{\mathrm{h}}\, \nu_{\mathrm{eh}}^{(l)}(C) \ll \omega - s\omega_{\mathrm{B}}$ holds. Binomial expansion of $\beta_s^{(l)}(C)$ and the use of the definition (3.3) for the $\nu^{(lm)}$ yields

$$\beta_s^{(lm)} \simeq \frac{1}{-i(\omega - s\omega_{\mathrm{B}}) + \Sigma_{\mathrm{h}}\, \nu_{\mathrm{eh}}^{(lm)}} \tag{7.2}$$

for weak absorption.

The expressions $\beta_s^{(l)}(C)$ of (6.2) were eigenvalues of the tensor $\overleftrightarrow{\beta}^{(l)}(C)$ (6.1). Likewise the transport integrals $\beta_s^{(lm)}$ (7.1) are eigenvalues of the "transport tensors"

$$\overleftrightarrow{\beta}^{(lm)}(\hat{C}) := \beta_0^{(lm)}(\hat{C})\, \hat{\boldsymbol{B}}\hat{\boldsymbol{B}} + \frac{\beta_{+1}^{(lm)}(\hat{C}) + \beta_{-1}^{(lm)}(\hat{C})}{2}\left[\overleftrightarrow{I} + \hat{\boldsymbol{B}}\hat{\boldsymbol{B}}\right]$$
$$+ i\,\frac{\beta_{+1}^{(lm)}(\hat{C}) - \beta_{-1}^{(lm)}(\hat{C})}{2}\,\hat{\boldsymbol{B}} \times \overleftrightarrow{I}. \tag{7.3}$$

With these tensors the Lorenth method yields the following coupled set of equations for the current density $\boldsymbol{j}_e$ and the heat flux $\boldsymbol{Q}_e$ of the electrons ([19], Sect. 105 with $N_e^{-1}\,\partial N_e/\partial\boldsymbol{r}$ replaced by $p_e^{-1}\,\partial p_e/\partial\boldsymbol{r} - T_e^{-1}\,\partial T_e/\partial\boldsymbol{r}$):

$$\boldsymbol{j}_e = \overleftrightarrow{\sigma}\cdot\left(\boldsymbol{E} - \frac{1}{q_e N_e}\frac{\partial p_e}{\partial\boldsymbol{r}}\right) + q_e N_e \overleftrightarrow{D}_T\cdot\frac{-\partial T_e}{T_e\,\partial\boldsymbol{r}} \tag{7.4}$$

$$\boldsymbol{Q}_e = q_e N_e \overleftrightarrow{D}_T\cdot\left(\boldsymbol{E} - \frac{1}{q_e N_e}\frac{\partial p_e}{\partial\boldsymbol{r}}\right) - \overleftrightarrow{\varkappa}\cdot\frac{\partial T_e}{\partial\boldsymbol{r}}. \tag{7.5}$$

The transport coefficients $\overleftrightarrow{\sigma}$, $\overleftrightarrow{\varkappa}$ and $\overleftrightarrow{D}_T$ for electrical and heat conductivity and the thermal diffusion coefficient are expressed as follows by the transport tensors (7.3):

$$\overleftrightarrow{\sigma} = \varepsilon_0\omega_N{}^2\overleftrightarrow{\beta}^{(10)} \qquad \overleftrightarrow{D}_T = \frac{5}{2}\,\frac{kT_e}{m_e}\,\overleftrightarrow{\beta}^{(11)} \tag{7.6}$$

$$\overleftrightarrow{\varkappa} = \frac{5}{2}\,\frac{p_e k}{m_e}\left[\overleftrightarrow{\beta}^{(10)} + 2\overleftrightarrow{\beta}^{(11)} + \frac{7}{2}\overleftrightarrow{\beta}^{(12)}\right] \tag{7.7}$$

The occurrence of the thermal diffusion coefficient $\overleftrightarrow{D}_T$ in both coupled equations (7.4), (7.5) verifies the Onsager reciprocity relations. It couples electric (or mass) and heat fluxes. The strength of the coupling is of the order $\beta_s{}^{(11)}/\beta_s{}^{(10)}$ analogous to the order $\nu^{(11)}/\nu^{(10)}$ of the coupling terms in Eqs. (5.1), (5.2).

For weak absorption the tensors $\overleftrightarrow{\sigma}$ (7.6) and $\overleftrightarrow{\varkappa}$ (7.7) take the values of the tensors $\overleftrightarrow{\sigma}^{(0)}$ (5.3) and $\overleftrightarrow{\varkappa}^{(0)}$ (5.4), respectively. The latter have been obtained by the moment method in lowest order approximation. Of course, the electron-electron collision frequency $\nu_{ee}^{(10)}$ in $\overleftrightarrow{\varkappa}^{(0)}$ (5.4) does not occur in $\overleftrightarrow{\varkappa}$ (7.7) because of the Lorentz approximation (6.1).

8. Transport Integrals for Power Law Collision Frequencies

If the transport collision frequencies $\nu_{eh}^{(l)}(C)$ (2.1) in $\beta_s{}^{(l)}(C)$ (6.2) are approximated by a power law $\nu^{(l)}(C) = \nu^{(l)}(\hat{C})\gamma^\delta$ (4.2), then the transport integrals $\beta_s{}^{(lm)}(\hat{C})$ (7.1) can be represented by tabulated Dingle integrals for $-3 \leqq \delta \leqq +3$. To show this one expresses the $\beta_s{}^{(lm)}$ by quantities ([19], Eq. 106.3)

$$T_s{}^{(l|\mu)}(\hat{C}) := \frac{1}{(\mu + 1/2)!}\int_0^\infty d\gamma^2 \exp(-\gamma^2)\,\gamma^{2\mu+1}\beta_s{}^{(l)}(\hat{C}\gamma), \tag{8.1}$$

viz. [20]

$$\beta_s{}^{(lm)} = \sum_{\mu=0}^{m}(-1)^{m+\mu}\binom{m}{u}T_s{}^{(l|l+\mu)}. \tag{8.2}$$

Approximating $\sum_\nu \nu_{eh}^{(l)}(C)$ by $\nu^{(l)}(\hat{C})\gamma^\delta$ (§ 4) one has

$$T_s{}^{(l|\mu)}(\hat{C}) = \frac{1}{\nu^{(l)}(\hat{C})}\,B_{\mu,\delta}\left(\frac{\omega - s\omega_B}{\nu^{(l)}(\hat{C})}\right) \tag{8.3}$$

with the dimensionless quantities

$$B_{\mu,\delta}(y) := \frac{1}{(\mu + 1/2)!} \int\limits_0^\infty d\gamma^2 \exp(-\gamma^2)\, \gamma^{2\mu+1} (-iy + \gamma^\delta)^{-1}. \tag{8.4}$$

These $B_{\mu,\delta}(y)$ can easily be related to the Dingle integrals [21, 22] for $-3 \leqq \delta \leqq +3$, see [19], sect. 108.

The electron conductivity $\overleftrightarrow{\sigma} = \varepsilon_0 \omega_N{}^2 \overleftrightarrow{\beta}^{(10)}$ (7.6) has been calculated for the special case $\delta = 2$ (electron-air collisions, § 4) by Phelps [23]. With these values for $\overleftrightarrow{\sigma}$ Sen and Wyller [4] calculated refractive indices from the general dispersion equation (1.3). Curves $\nu^{(10)}\beta_s{}^{(10)}$ versus $(\omega - s\omega_B)/\nu^{(10)}$ for the more general cases $\delta = 0, \pm 1, \pm 2$ are given in [24]. The normalization with $\nu^{(10)}$ instead of $\nu^{(1)}(\hat{C})$ to express the results has been used to demonstrate the weak absorption approximation (7.2).

9. Summary

To express the results of the kinetic gas theory for the calculation of transport coefficients (e.g. electrical and heat conductivity) several types of collision frequencies must be introduced. The use of orthogonal polynomials for the weight functions in the averaging over deflection angles and velocities allows us to express the transport coefficients with as few of these collision frequencies as possible, see Eqs. (5.3), (5.4), (7.6), (7.7). The coupling between electrical (or mass) and heat fluxes can be expressed by one single quantity ($\nu^{(11)}$ in Eqs. (5.1), (5.2) and $\overleftrightarrow{\beta}^{(11)}$ in Eqs. (7.4), (7.5), which allows a quick estimate of the strength of this coupling.

To illustrate the influence of the introduced transport collision frequencies (3.3) and transport integrals (7.1) on the refractive index, the dispersion equation (1.3) must be solved. For the electrical conductivity tensor $\overleftrightarrow{\sigma}$ the expression $\varepsilon_0 \omega_N{}^2 \overleftrightarrow{\beta}^{(10)}$ (7.6) can be used in the lower ionosphere. After the expansion of the determinant only the three eigenvalues $\beta_s{}^{(10)}$ (7.1) with $s = 0, \pm 1$ appear in the dispersion equation together with $\cos\theta := \hat{\boldsymbol{B}} \cdot \hat{\boldsymbol{n}}$ [25], which is complex for oblique propagation. The weak absorption approximation leads back to the Appleton-Lassen formula for normal incidence and to the Booker quartic for oblique incidence. Full account of electron-air collisions yields the Sen-Wyller formula for normal incidence. To the author's knowledge an analogous numerical discussion of a correspondingly generalized Booker quartic is still missing.

References

[1] H. Lassen, Elektr. Nachr.-Techn. **4**, 324 (1927).
[2] G. Burkhardt, Annln Phys. **5**, 373 (1950).
[3] J. A. Kane, J. Geophys. Res. **64**, 133 (1959).
[4] H. K. Sen and A. A. Wyller, J. Geophys. Res. **65**, 3931 (1960).
[5] A. A. Wyller, in: Proc. 5th Int. Conf. on Ionization Phenomena in Gases, Munich 1961, North-Holland Publ. Co., Amsterdam 1962 (Vol. I, p. 940).
[6] J. C. Maxwell, in: Scientific Papers, University Press, Cambridge 1890 (Vol. 2, p. 681, in particular p. 688).

[7] K. SUCHY, in: Proc. 9th Int. Conf. on Phenomena in Ionized Gases, Bucharest 1969, Academiei Republ. Soc. Romania, Bucharest 1969 (p. 397).
[8] S. CHAPMAN and T. G. COWLING, The Mathematical Theory of Non-uniform Gases (2nd edn), University Press, Cambridge 1952, sect. 7.5.
[9] Y. ITIKAWA, Planet. Space Sci. **19**, 993 (1970).
[10] I. P. SHKAROFSKY, T. W. JOHNSTON and M. P. BACHYNSKI, The Particle Kinetics of Plasmas, Addison Wesley, Reading (Mass.) 1966, Figs. 5-6 to 5-13.
[11] A. V. PHELPS and J. L. PACK, Phys. Rev. Lett. **3**, 340 (1959).
[12] J. L. PACK et al., Phys. Rev. **127**, 2084 (1962).
[13] R. W. SCHUNK and J. C. G. WALKER, Planet. Space Sci. **18**, 1535 (1970).
[14] K. SUCHY and K. RAWER, J. Atmos. Terr. Phys. **33**, 1853 (1971).
[15] H. GRAD, Communs. Pure Appl. Math. **2**, 331 (1949).
[16] K. SUCHY, in: Proc. 11th Int. Conf. on Phenomena in Ionized Gases, Prague 1973, Czechoslovak Akad. Sci., Inst. of Physics, Praque 1973 (p. 272).
[17] H. SCHIRMER and J. FRIEDRICH, Z. Physik **151**, 375 (1958), Fig. 1.
[18] W. P. ALLIS, in: Encyclopedia of Physics, Springer, Heidelberg 1956 (Vol. 21, p. 323, Chap. III).
[19] K. SUCHY, Ergebn. exakt. Naturwiss. **35**, 103 (1964).
[20] K. RAWER, in: Encyclopedia of Physics, Springer, Heidelberg 1972 (Vol. 49/4, p. 1, last equation).
[21] R. B. DINGLE et al., Appl. Sci. Res. **B 6**, 144, 155 (1957).
[22] R. B. DINGLE, Proc. Roy. Soc. A **244**, 456 (1958).
[23] A. V. PHELPS, J. Appl. Phys. **31**, 1723 (1960).
[24] K. SUCHY, Radio Sci. **7**, 871 (1972), Fig. 2.
[25] K. RAWER and K. SUCHY, in: Encyclopedia of Physics, Springer, Heidelberg 1967 (Vol. 49/2, p. 1, Sect. 6ε).
[26] R. D. HAKE and A. V. PHELPS, Phys. Rev. **158**, 70 (1967).
[27] M. N. MENTZONI, J. Res. Nat. Bur. Stand. **69 D**, 213 (1965).
[28] G. E. VEACH et al., Bull. Amer. Phys. Soc. **11**, 496 (1966).
[29] A. G. ENGELHARDT et al., Phys. Rev. **135**, A 1566 (1964).

Lower Ionosphere Structure — Akademie-Verlag, Berlin 1974

TENTATIVE E REGION EMPIRICAL ELECTRON DENSITY PROFILES*

by S. Ramakrishnan and K. Rawer

Arbeitsgruppe für physikalische Weltraumforschung, Freiburg, FRG

In order to get representative $N(h)$ profiles for IRI, profiles compiled by Maeda have tentatively been chosen for the region between 80 km and the height hmE. The profiles are now described by means of a more suitable function. This description is compared with special ionosonde measurements of high quality. Above hmE valley profiles for mid-latitude have been established from incoherent scatter data obtained at Great Malvern (UK). It is proposed that low latitude incoherent scatter stations take similar measurements so that E region and valley profiles can be established. Classical sounding though giving helpful qualitative indications seems to be unable to provide unique results in the valley zone.

1. Introduction

The "International Reference Ionosphere" (IRI) is a project sponsored by URSI and COSPAR, for establishing a set of tables, from which representative values of plasma parameters in the earth's ionosphere corresponding to typical geophysical conditions can be found as functions of height. Since reference data are now often needed in connection with computing facilities, descriptions of profiles of the main plasma parameters by means of empirical equations seemed desirable. The idea of such description derived from statistically significant collections of experimental data has been accepted in principle.

We use here as basic information mean electron densities below the height hmE of maximum E region electron density (NmE) as found in the descriptive literature, and electron density profiles determined by means of incoherent scatter measurements above hmE. At present the superiority of these ground-based measurements over the other methods for describing the E region valley can be taken for certain. The profiles have been described by empirical mathematical functions, different below and above hmE. The reliability of important profile parameters taken from the mean E region profiles is critically reviewed and the urgent need for more data is pointed out.

---

* The co-author wishes to dedicate his effort to Professor K. Rawer on the occasion of his 60th birthday.

2. Choice of E Region Data

For the region below hmE, Maeda [1, 2] has collected most of the available rocket and ground-based measurements, sorted them into different groups and established mean electron density profiles corresponding to different solar activities, solar zenith angles and seasons. We have normalized some of his results and tried to derive empirical mathematical functions in terms of hmE and NmE.

For the valley region between hmE and the bottom of the F region, the only two methods that do provide reasonably good data are incoherent scatter and rocket measurements, but not classical sounding. After compiling the incoherent scatter measurements of Malvern (UK), provided by G. N. Taylor, we studied mid-day profiles for the years 1968–1971 and established three mean profiles corresponding to summer, winter and equinoxes.

3. Empirical Functions

The mean E region profiles of Maeda [1, 2] are first normalized against NmE and hmE. A simple parabolic function was found to be not good enough. We therefore first fitted a parabola to the "nose" of the profile, i.e. to the peak region, and determined the relevant semi-thickness parameter (YmE). Using this parameter the electron density distribution is described by an equation of the type

$$\frac{N}{N\mathrm{mE}} = 1 - \left[\frac{h\mathrm{mE} - h}{Y\mathrm{mE}\left[1 + \dfrac{a(h\mathrm{mE} - h)}{h\mathrm{mE} - h_{\mathrm{min}}}\right]}\right]^n$$

where a and n are constants determined to provide a good fit. h_{min} is the minimum height of the E region. A simple parabolic function means $a = 0$ and $n = 2$. One of the summer profiles could be described in this manner. All the other

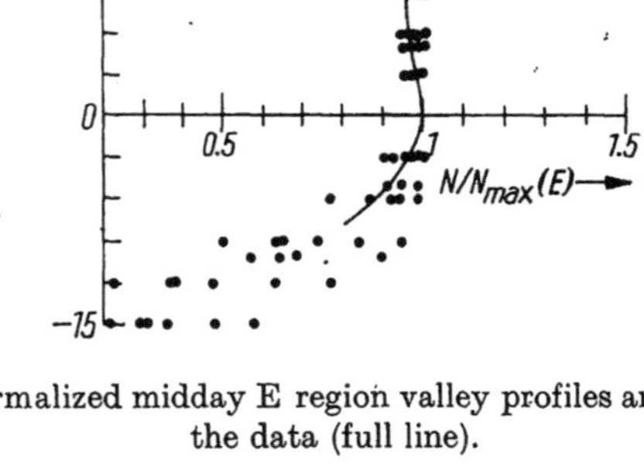

Fig. 1. Scatter plot of normalized midday E region valley profiles and our polynomial fitted to the data (full line).

profiles had to be described with other parameter values, i.e. by slightly "deforming" the zero order parabolic approximation. The effect of the parameter a can be described as varying the effective thickness with height while the power n is used to provide good overall agreement with data.

Even around noon most Malvern incoherent scatter data exhibit a shallow valley. We have arranged the mid-day profiles in three seasonal groups and produced mean profiles from the three sets of scatter plots. Fig. 1 shows such a scatter plot for mid-day hours together with our polynomial fit (5 coefficients). These are to be derived also for other hours of the day.

4. Results and Discussion

Fig. 2 shows one sample of the Maeda profiles, from which we have tried to determine values of hmE. Since it is difficult to determine a cusp in these data hmE has been defined as the height corresponding to NmE as also given by Maeda. It should be emphasized that we do not intend to use his NmE values in the future

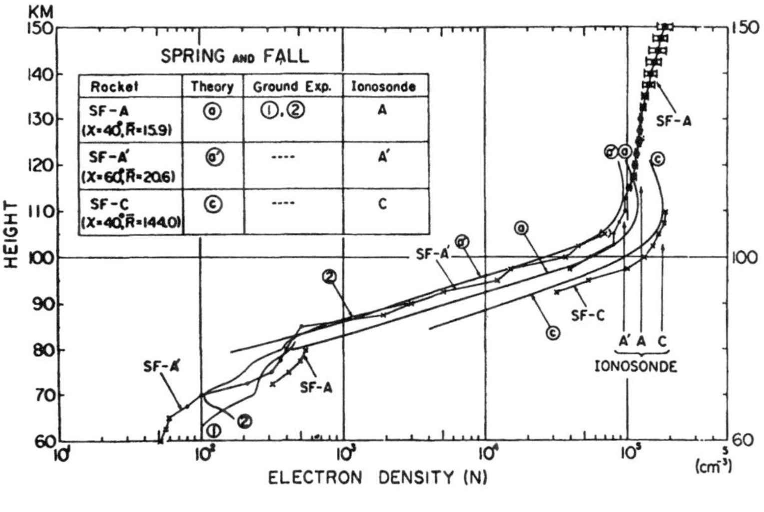

Fig. 2. An example of E region profiles compiled by Maeda [2].

but something like the ITS computer program "f0E". With this criterion, we find rather high hmE values, of the order 130 km, for spring and autumn profiles at low solar activity, and similarly for the summer profile of medium solar activity. This result appears to be doubtful. The "nose semi-thickness" YmE calculated for the various Maeda profiles shows a minimum value of about 10 km for winter at low solar activity. With the same activity, the summer season has a semi-thickness of about 20 km. The profiles for spring and autumn at low solar activity showing large values of hmE have an average semi-thickness of about 24 km. Contrary to theoretical conclusions of Taubenheim [3], Whitehead [4] insisted that his observed data show an increase in E region thickness from winter to summer. It seems that this point merits further discussion with better and more measured data.

In the height range above *h*mE mid-day profiles show a deeper valley in winter than in summer. There is not appreciable difference of valley thickness during these two seasons but a wider trough occurs at the equinoxes with a shallow shape, the depth being similar to that found in summer. We feel that a statistically significant collection of lower E region and valley profiles from all incoherent scatter stations would be of enormous value for IRI. The scientific community is invited to provide such evidence, and any other reliable data which could be used for establishing reliable profiles.

Acknowledgments

The authors thank K. I. Maeda and, in particular, G. N. Taylor, from whom the only systematic collection of E and F region profiles measured by incoherent backscatter technique could be obtained until now. ALGOL programs provided by D. Bilitza are acknowledged with thanks. Part of this work has been supported by a grant of Bundesministerium für wissenschaftliche Forschung, FRG.

References

[1] K. MAEDA, J. Geomag. Geoelect. **22**, 4 (1970).
[2] K. MAEDA, Space Research XII, 1229 (1973).
[3] J. TAUBENHEIM, J. Atmos. Terr. Phys. **33**, 1481 (1971).
[4] J. D. WHITEHEAD, J. Atmos. Terr. Phys. **35**, 183 (1973).

Lower Ionosphere Structure — Akademie-Verlag, Berlin 1974

ELECTRON DENSITY PROFILES FROM IONOGRAMS: COMPARISONS WITH ROCKET PROFILES

J. W. Wright and A. K. Paul

National Oceanic and Atmospheric Administration, Environmental Research Laboratories, Boulder, Col., USA

From a series of ten rocket flights at Wallops Island conducted by the University of Illinois, detailed electron density profiles are available in the altitude range 60—120 km. Concurrent ionograms from a nearby conventional ionosonde of good sensitivity and precision have been reduced to $N(h)$ profiles by the NOAA methods. Special attention is given to the needs for accurate ionosonde calibration, proper ionogram interpretation, and corrections for underlying ionization, "valley" ionization, and effects of lateral gradients. Proper corrections in the present cases yield profiles by ionogram inversion which agree to within a few per cent in density, and a few hundred meters or less in height, with the rocket results.

1. Introduction

From a sustained program of rocket flights at Wallops Island, the University of Illinois has published a series of ten electron density profiles of the lower ionosphere representing daytime conditions over a wide range of solar activity [1]. Concurrently, a standard analog ionosonde of good sensitivity was in operation nearby, providing an opportunity to conduct a detailed comparison between the two methods. A discussion of the ionogram inversion problems, methods of minimizing errors, and ways to future improvement will be published more fully elsewhere; here we shall illustrate the essential conclusions with several selected examples.

The necessary mathematical properties of the inversion process were formulated by Paul [2] leading to the development and publication of several practical systems for the purpose [3—5]; an independent approach, also satisfying the necessary mathematical conditions, has been developed by Titheridge [6]. An adaptation of the Paul-Howe-McKinnis inversion method, for automatic inversion of digital ionograms, has recently been given by Wright et al. [7]. From the practical standpoint there are, however, a number of additional necessities which have received less attention. Obviously, precision of the observations affects the outcome of the inversion method, although not in a simple way [8]. More seriously, "unobserved" regions of the ionosphere which nevertheless contribute retardation to the echoes of the ionograms, as often by the D region in daytime, the E valley, and the night-time E region, can cause large and systematic errors if proper corrections are not made. Methods for this purpose have been described [3, 9, 10] which usually depend upon joint use of the ordinary and extraordinary magneto-ionic components to obtain the required additional information for an "unobserved ionization" correction. Paul and Smith [11] have discussed the generalized "information content" of the o/x components for this purpose and have concluded that they may control at most one parameter describing the unobserved ionization; under some circumstances (e.g. at magnetic dip angles near 52°) even this control is absent. However, practical correction methods (even for one parameter) depend also on the formulation of specific models, and the model constraints may control the solution at least as much—and sometimes more than—the echo information available.

There are two additional problems to consider in assessing profile accuracy, both closely related to the highly variable nature of the ionosphere itself. Important lateral variations often exist and are known to be in motion and to vary with time. The echoes registered by an ordinary ionosonde then fail to represent exclusively the altitude variation; uncontrolled errors are introduced in the virtual "heights" through variable obliquity. Secondly, *definition* of the *intended* profile is complicated by the spatial and temporal variability itself. It is in the nature of radio sounding that some spatial averaging occurs in the reflection process; this is essentially uncontrollable, but it might be measurable and hence subject to specification. The experiment permits virtually unlimited control of its time-averaging properties, however, and it is a simple matter to superimpose successive ionograms in a series to see if the details form a progressive sequence, or (instead) fluctuate about some central tendency. This is often done to obtain monthly "median" profiles [12], and can also be used to assess the representativeness of an individual profile.

2. Consideration of Ionogram Inversion Problems by Example

The detailed rocket profiles of [1] have been provided to us by Mechtly (private communication). They are divided into two groups of 5 during solar minimum (1964—1965; smoothed relative sunspot numbers 10—18) and 5 during solar maximum (1968—1969; sunspot numbers 51—139). In our figures these are referred to as Q-1 ..., A-1 ... respectively (essentially as in [1]) for easy reference. Not all the examples are discussed as several of the cases clearly demonstrate more than one feature of the total problem.

2.1. Accuracy Criteria; Calibrations, Obliquity Effects

For aeronomical work we shall define the acceptable accuracy of an electron density profile as 5% of a scale height, where the accuracy of an individual profile is in question, and as 5% of the estimated standard deviation of electron density at a fixed height, where the accuracy of an average profile is of concern. Errors of more than about twice these values tend to be misinformative, and therefore counterproductive.

Conventional ionosondes pose measurement-accuracy problems most typical of which are those arising through echo pulse-delay effects in the receiver, the echo-amplitude dependence of these delays and obliquity effects. Pulse delay effects can be calibrated, but unless the delay is a weak function of echo amplitude, the absence of amplitude information on the ionogram renders the calibration of little use. In the Wallops Island ionosonde employed here, the receiver delay varies with amplitude ± 0.8 km about an offset (nominal calibration) which, unless carefully monitored, wanders slowly with system temperature and other conditions more significantly than with the amplitude dependence itself. Under these circumstances, the best approach is a statistical assessment of the receiver delays using the echo information from ionograms near the time of interest; this can be done by comparison of the "heights" of successive multiple echoes. In the present data sample these errors have been estimated in this manner; they range from -13 to $+5$ km. The individual values are marked on the figures, and obviously corresponding errors of these magnitudes would apply throughout the profiles if appropriate calibrations and corrections were not made.

Multiple echo comparisons from successive ionograms (or at widely separated frequencies from a given ionogram) are sometimes found in disagreement by more than ± 1.0 km; this must be attributed to obliquity effects. Successive reflections do not arrive, on the average, from the same direction, nor does the principal (first) echo usually arrive from directly overhead. This represents a direct source of error in the use of h' for profile inversion, and complicates the inference of the ground-pulse offset as well. These effects can be measured with a digital ionosonde [13], but without corrections for such effects h' errors of one or two km must be anticipated.

2.2. Consequences of No Underlying Ionization Corrections

Since the virtual height even at the lowest frequency of the ionogram always exceeds the real height, failure to perform some kind of underlying ionization correction always results in a profile estimate which is too high. In the figures, an arrow marks the "unextrapolated start" (i.e. the lowest observed virtual height, corrected, however, for receiver delay) at which one would initiate the profile calculation if no extrapolation or underlying ionization correction were performed. In this data sample the resulting height errors (at the variable starting density) range between $+3$ and $+14$ km, with the majority near 6 or 7 km, i.e. about one scale height in this region. These height errors are accompanied by errors of the electron density gradient; it might be expected that the starting retardation errors would diminish quickly as the inversion proceeds to higher densities, but in fact the height and gradient errors propagate each other and diminish only slowly through the profile. In general, the entire E region may be one-half to one scale height too high throughout, if no starting corrections are performed; such errors are clearly of the counterproductive type.

2.3. Unaided Underlying Ionization Corrections

When the ionogram shows a substantially linear increase of virtual height over a considerable frequency range, it is a reasonable assumption that an approximately linear gradient also exists in the profile over the corresponding range of electron densities; furthermore, it is reasonable to expect that this gradient prevails to lower, unobserved densities as well. These assumptions provide the basis to accomplish a correction for underlying ionization with two unknown quantities, the constant gradient and the starting height.

Our ionogram inversion system includes a starting model based on these principles [5]. Any number of virtual heights within a frequency range where the variation of h' with $\log f$ is found to be reasonably linear are combined in a least-squares system of equations to determine z' the unknown constant slope, and h_n the real height at the nth (highest) plasma frequency included in the model. The individual deviations of the observed virtual heights from those of the model (with the determined values of z' with h_n) are easily obtained and their rms value (divided by $n - 2$) yields the standard deviation of the least-squares estimate. When this quantity is comparable with the virtual height error ($\sim$1 km) it may be presumed that an adequate model choice has been made, at least according to the information content of the measurements. This does not, however, assure that the correction for underlying ionization is satisfactory. It must also be determined that the values of the two unknowns do not vary significantly as the low-density limit of the model is varied. When this is so, it may be concluded that the model is valid over a sufficiently great range of densities to account fully for the group retardation experienced near the frequency f_n (and above) in traversing this region.

In Fig. 1 (Q-1, 16 April 1964) only a very minimal amount of h' information is available in the five initial points to start the calculation. This interval of 2:1

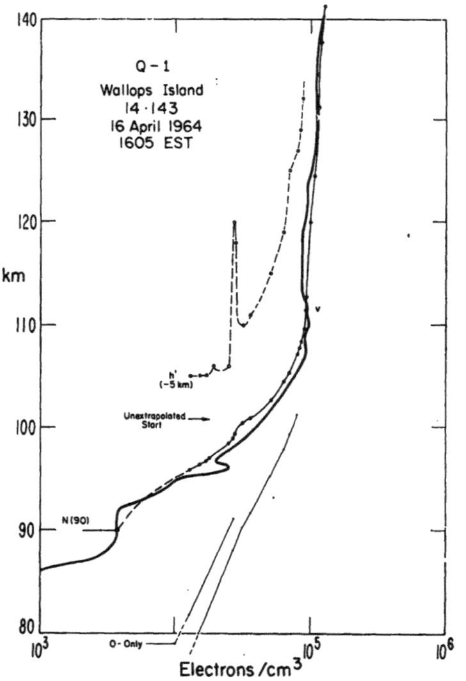

Fig. 1 (Q-1). Heavy curve: rocket profile. Dashed curve connecting O-observations: ionogram virtual heights (O-only). Light curve: ionogram inversion profile, proceeding from N(90) start. Lower straight lines: O-only constant z' starts, for two arbitrary extrapolation ranges. "Unextrapolated start" shows where profile inversion would commence after correcting for receiver delay (—5 km) but without underlying ionization correction.

in density is limited at the lower end by ionosonde sensitivity (an older C-3 ionosonde was in use at that time), and at the upper end by significant variations in profile shape. If an attempt is made to apply the constant-gradient starting procedure, the result is found to depend strongly on the (lower) density limit of extrapolation, as shown by the "O-only" examples. Both are serious over-corrections; the true profile lies roughly midway between these attempts and the result ("unextrapolated start") of ignoring the underlying ionization completely. Although the standard deviation of the constant-slope start is satisfactory (± 1.2 km), the slope itself and the height of the model each depend strongly on the arbitrary extrapolation range. A second example of this same behavior is shown in Fig. 2

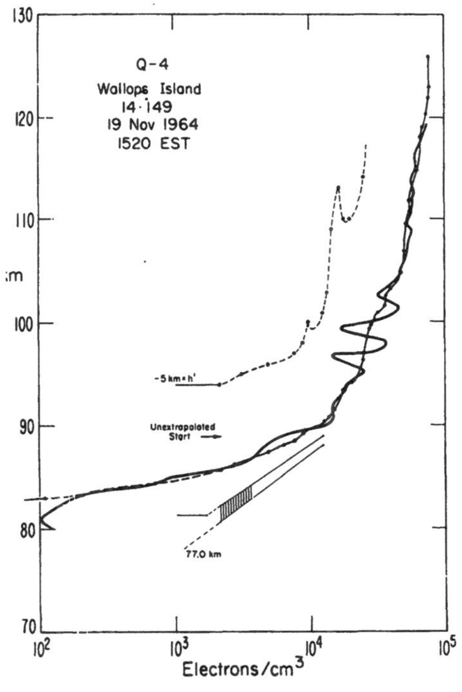

Fig. 2 (Q-4). Coding similar to Fig. 1. No X-echoes available.

(Q-4, 19 Nov. 1964); here the errors are less serious since (a) the virtual height variation is more nearly linear in the starting range (1st 4 points), (b) this range is larger than in the previous example (nearly 4:1 in density), and (c) the starting slope itself is larger and therefore (within the available measurement precision) relatively better determined. The starting slope and height are now less variable with the arbitrary range of extrapolation, and at least it can be concluded that the extrapolation yields an improvement over the "unextrapolated start" which remains some 3 km too high over the first 10:1 increase of density (i.e. to about 1.5×10^4 cm^{-3} on the abscissa). Nevertheless, the performance of the constant-slope model in this example cannot be considered satisfactory. The model height decreases about 1 km per 2:1 increase of the range of low-density extrapolation, and without the rocket results as a guide, no decision could be made to select the best extrapolation range. To be acceptable, any model must yield results which are independent of arbitrary decisions.

This is achieved with the present model when the (density) range over which the model can be fitted, is sufficiently great. In Fig. 3a (A-5a, 15 Sept. 1965), despite moderate (± 1 km) fluctuations of h', a significant linear increase is evi-

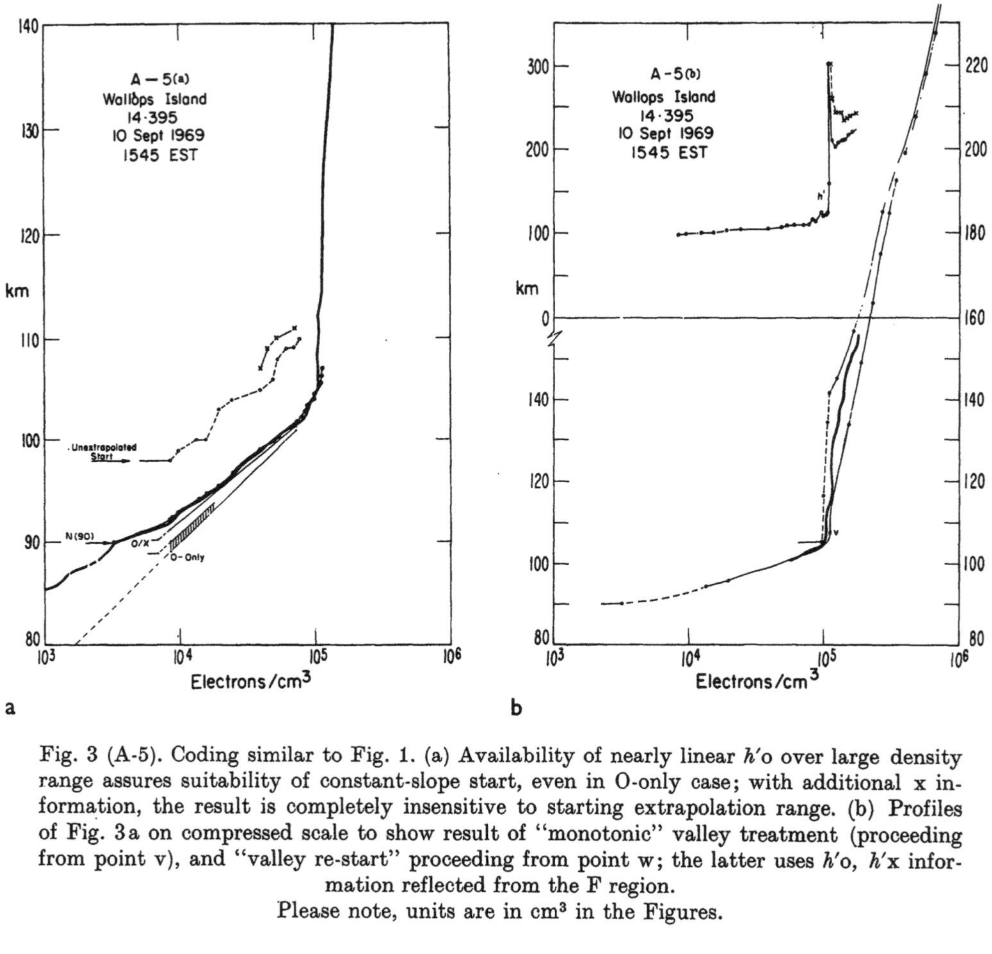

Fig. 3 (A-5). Coding similar to Fig. 1. (a) Availability of nearly linear h'o over large density range assures suitability of constant-slope start, even in O-only case; with additional x information, the result is completely insensitive to starting extrapolation range. (b) Profiles of Fig. 3a on compressed scale to show result of "monotonic" valley treatment (proceeding from point v), and "valley re-start" proceeding from point w; the latter uses h'o, h'x information reflected from the F region.

Please note, units are in cm³ in the Figures.

dent over a range of 27:1 in density. When 13 values of h' in this interval are combined in the least-squares solution, the result is nearly independent of the extrapolation range; at 4×10^{10} m^{-3} the starting model height varies by only 0.2 km over a factor of 8:1 in the extrapolation range, and the starting slope is similarly insensitive. Moreover, the starting slope and height are very nearly correct, as judged by comparison with the rocket profile in the range (2.5—4.0) $\times 10^{10}$ m^{-3}. Below 88 km (not plotted in Fig. 3) the rocket profile shows that the density becomes nearly constant (at 3×10^2 cm^{-3} on the abscissa) over a considerable height range (74—86 km); thus the constant slope model is very inappropriate below about 2×10^9 m^{-3} (90 km) in this example, but this has only negligible effect on its performance over the considerable range up to 4×10^{10} m^{-3} (100 km) where it provides a good description of the profile shape. It may be noted that the rocket profile in this example has a distinct positive curvature in the range $(0.2-3) \times 10^{10}$ m^{-3} which the constant-slope model of course fails to represent; we shall return to this point subsequently.

2.4. Joint o/x Constant-Slope Corrections

The least-squares constant-slope starting procedure is applicable whether the h' values are taken from the ordinary or the extraordinary echo traces of the ionogram or any mixture of o and x, provided that the required magnetoionic coefficients for each interval of the profile are calculated using the refractive index expressions appropriate to the o or x mode of each h' entering the system. D-region absorption of the extraordinary mode is greater than for ordinary-ray echoes, particularly at the frequencies commonly reflected from the lower E region. When such echoes are observed, however, a satisfactory correction for underlying ionization may sometimes be obtained with h'o,x data available within a more limited density range than is possible with h'o (or h'x) alone. In case A-5a, the O-only start, while certainly satisfactory, shows some dependence on the range of extrapolation to lower densities; in the O/X start, the result is not only improved,

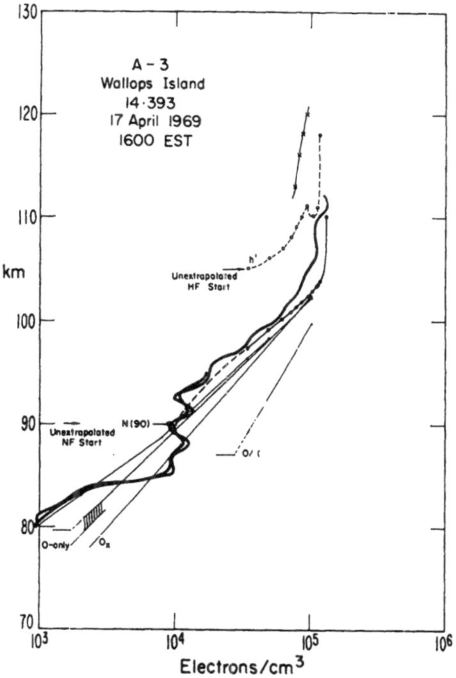

Fig. 4 (A-3). Coding similar to Fig. 1. Various combinations of frequency range and O-only, O/X data employed in constant-slope start. Note significant "ledge" on rocket profile near 10^{10} m^{-3}.

but is entirely independent of the extrapolation range. Fig. 4 (A-3, 17 April 1969) provides examples of both satisfactory and unsatisfactory application of this procedure. The constant slope solution labeled O/X combines the full range of ordinary-ray observations (including two echoes obtainable only by the low-frequency (0.25 MHz) capability of the Wallops Island ionosonde) together with

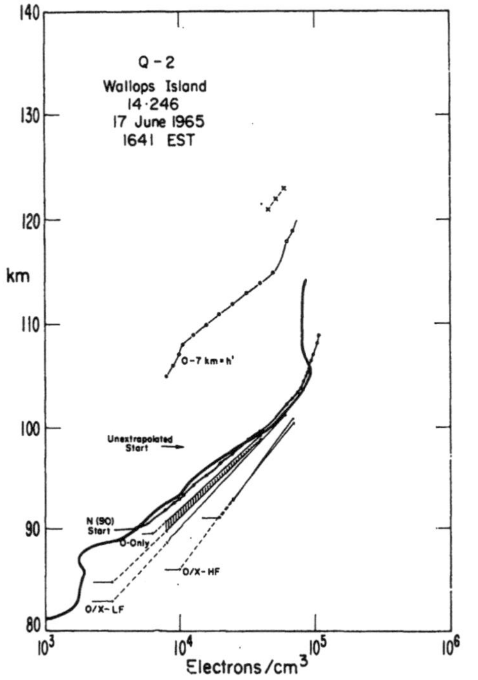

Fig. 5 (Q-2). Coding similar to Fig. 1. Virtual height observations by extraordinary mode (X) tare also available; their use with O-mode echoes at high frequencies only (O/X-HF), or with low-frequency O-mode echoes (O/X-LF) in constant-slope starting models is shown.

the higher-frequency extraordinary-ray observations. The rexult is about the same as the o-only starting models, but is more stable: essentially the same z' and h_n are obtained whatever the extrapolation range. If the low-frequency ordinary-ray data are excluded from the solution (o/x) then the result is rather too low to be satisfactory, but it retains stability against arbitrary choices of extrapolation range.

In Fig. 5 (Q-2) the stability of the O/X starts is more clearly evident.

2.5. Aided Underlying Ionization Corrections

The foregoing examples have shown that when the profile is of simple form over a sufficient density range—and when the ionogram provides accurate data from this range—then a satisfactory fit can be obtained to a simple model including an extrapolation range, which accurately starts the inversion analysis. Aside from the obvious problem that this "fit" cannot *always* be performed because of inadequate information, the procedure has the additional defect that structural detail of interest in the fitting range of the profile must be averaged away to determine the starting model. It is worthwhile to examine what might be achieved if the starting problem could be avoided in this form, i.e. if all the virtual heights could be used in the same manner to obtain direct information on the local form of the profile.

If one point $(N_0(h))$ were known on the profile below the lowest-reflection-density virtual height observed, then a single interval (of variable slope, for reasons to be discussed below), requiring no additional information, would achieve the necessary underlying ionization correction. The principle is illustrated by Q-1, Q-2, Q-4, A-3 (Figs. 1, 2, 4, 5): with the density at 90 km from the rocket profile introduced as a known starting height, the inversion proceeds without further difficulty to produce the plotted curves. We have plotted the results of this starting procedure for the complete inversion of each ionogram in the figures of this paper (rather than plotting the continuations of other starting methods) since in this way other details of the comparisons can be shown without the variable complications of the starting procedures.

It will be noted in Q-1 (Fig. 1) that the ionogram profile begins with a discontinuity or "corner" at 90 km. With one point known (or reliably assumed) from which to start, it is *only* the slope or the variation of the slope which is unknown up to the first observed point, and a more flexible connection by an interval of constant second derivative (z'') is warranted. This model involves two unknowns (the slopes at the beginning and end of this interval) since it is not desirable to assume continuity of slope at the starting height where (to the left) it is naturally zero. We therefore consider z'' constant over two observed intervals and permit a discontinuity of slope at the starting height. In virtually all of the cases illustrated here this starting procedure initiates an extremely close agreement between the ionogram and rocket profiles. In cases Q-1 and A-3 (Figs. 1, 4) the procedure "selects" a negative z'' in the starting interval while positive curvatures result in cases Q-2, Q-4. The rocket profiles show that this is the correct adaptation except in case Q-2. The disagreement in case Q-2 may be dismissed as too small to matter; the example does show, however, that with a starting interval of less than 2:1 in density (below the initial h' at the abscissa value 7.8×10^3 cm^{-3}), a full correction is achieved for the retardation accumulated at and above the corresponding sounding frequency. Since the ionogram profile remains some 0.7 km too low over a considerable (6.5:1) range above the starting height, we may accept this as an indication of the maximum error attributable to the starting process in this case. The starting error may well be smaller, with temporal changes, obliquity, or other measurement errors being more significant throughout the profile.

It is, of course, obvious that the $N_0(h)$ starts illustrated here are heavily favored to succeed because the density at the starting height is known precisely for each case from the rocket profiles. To be generally successful as a starting method, this supplementary information must be available with sufficient accuracy from a source which is not extravagantly more elaborate than the ionosonde itself. The two rather independent questions are then: (a) what can be considered "sufficient" accuracy for this method?; (b) what sources of additional information are available?

We do not yet have an answer to the first question, but only some initial indications based upon the present data sample. It is necessary, of course, that N_0 at the chosen starting height be *less* than the reflection density of the lowest observed echo frequency. It should automatically follow that the observed virtual height lies significantly above the starting height. Our tests suggest that N_0 should lie in the range 0.1—0.5 times the density for reflection of the lowest frequency at which an echo is observed; the virtual height should be at least several times the estimated measurement error *greater than* the starting height. It is not the function of this paper to propose a "best" approach to the development of an $N_0(h)$ model; indeed, it is doubtful that sufficient statistical information is yet available to accomplish this satisfactorily as a fully general-purpose tabular model. We may note, however, that two median-value profiles (quiet and active sun, respectively) derived from the ten rocket profiles by Mechtly et al. [1] lie within better than a factor of two of the density fluctuations in the constituent profiles. Thus, if the solar zenith angle dependence of the density at, say, 70, 80, 90 km, etc. were established, a standard model for the $N_0(h)$ starting process might be practical. Some efforts toward the development of a model of this kind are now under way (N. Smith, private communication). Another approach, involving a less variable dependence on the vagaries of a fixed model, is to include at least a rudimentary D region measurement capability within the ionosonde.

2.6. E Valley Problems

At the E layer peak the small density gradient imposes very large retardation on radio frequencies reflected nearby; often above the peak the slope remains small, or the profile becomes actually re-entrant. This "valley" creates another region of "unobserved ionization" requiring special treatment in ionogram inversion methods. Even when the profile is not actually re-entrant, one may not know *a priori* that this is so if absorption precludes observation of very large virtual heights; in such cases attempts at a valley correction is required if only to prove that the valley errors are small.

It is fairly widely appreciated that some independent information on the valley distribution is available from the F region ordinary and extraordinary echo traces of the ionogram, and methods for their joint use in this problem have been discussed in [3, 5, 14]. A study of the general information content of the o and x retardation functions by Paul and Smith [11] has shown that independently of any model assumptions describing the valley, there is information to determine at most one parameter of the valley distribution. If adequate o *and* x information is not available, of course, then not even one parameter of a possible valley can be obtained. Even then, a special procedure may be used to obtain the highest possible monotonic profile above a lower-layer peak, when a valley may in fact exist. This is achieved by permitting a slope discontinuity at the peak density, and solving for a constant z'' over two subsequent intervals. Case Q-1 (Fig. 1) provides a first example. At the point v, a constant z'' is assumed over two intervals, thereby permitting z' to assume a large value to the right of this point. Despite the shallow valley, the result agrees closely with the rocket profile at 125 km and above. Another example is shown in A-5b (Fig. 3b). (The D region portion of the rocket profile has been suppressed; it lies too close to the ionogram result in this range to be distinguished on the scales used.) Over the 105—115 km range, the rocket profile displays a nearly constant density, and the monotonic ionogram profile (proceeding above the permitted discontinuity at v) is predictably too low.

With accurate o, x information available in the echoes from above the valley, a more complicated model may be determined by the observations. Our procedure for this purpose was discussed briefly in [5] and more recent developments of it will be given elsewhere. The solution is again by the method of least squares, from any adequate sample of the o, x virtual height observations over which a reflection-height model of constant z'' may be appropriate. The analysis determines this parameter together with six other unknowns describing an "equivalent profile distribution" [15]. The least-squares solution is slightly weighted toward physically plausible values of these parameters, in order to compromise the excessive flexibility of this model with the limited information provided by the ionogram.

A-5b (Fig. 3b) demonstrates the result of this "re-start" procedure by the curve labeled w. The dashed curve below w represents the "equivalent distribution" determined by the six model parameters below w, while the height at w and the constant z'' over a short interval above w represent the corrected starting profile. Although profile detail is averaged away in this short starting range (compare, for example, the minor cusps in h'_o, h'_x near 1.1×10^{11} m^{-3} with corresponding details of the rocket profile), the general effect of the valley correction is quite satisfactory.

3. General Assessment and Conclusions

The examples shown here clearly indicate the need for careful treatment of virtual height observations in performing inversions to obtain electron density profiles; they also show that high accuracy can be achieved if accurate calibrations are carried out, and if the ionosonde provides adequate sensitivity and good resolution in time, virtual height, and radio frequency. Variable success must be expected of limited starting models, but when correct starts are achieved, the detailed variations of the rocket and ionogram results are in very close agreement. This is self-evident from most of the cases represented in this data sample, but case Q-4 (Fig. 2) deserves further comment. Near 100 km, the rocket profile shows an intense, oscillatory structure which is almost absent in the ionogram profile. It must first be noted that in this particularly early case (1964) a much older ionosonde was in use which lacks the height resolution and amplitude sensitivity needed for accurate work. Of more significance to the present case, however, is evidence from the time-series of ionograms (each 5 minutes) of rapid variations in the structure near 100 km; abundant "oblique echoes" on the ionograms show that much lateral structure is also present, but of course these analog observations provide no quantitative information on the lateral structure itself. Thus, the single ionogram profile in Fig. 2 may be more representative of the average vertical structure present, while the rocket profile provides a much more interesting, but possibly atypical, sample.

It is necessary to comment that the present results do not employ the most modern, accurate, and informative developments of the radio vertical sounding methods. In the Dynasonde [16], many improvements are available which can increase the accuracy, time resolution and information content of $N(h)$ profiles while providing these and other measurable quantities (e.g. wind-profiles in the E region) in real time. The Dynasonde combines computer control of the sounder functions with high-resolution digitization of all meaningful parameters of each echo. Echo recognition, noise rejection, angle-of-arrival computation and profile inversion (as examples) are performed in real time. Thus essentially all the accuracy and reliability limitations of the ground-based ionospheric measurements discussed in the present paper are open to substantial further improvement.

Acknowledgment

This work was supported under NASA Contract No. W-13298 and NOAA research project No. 21J30720.

References

[1] E. A. MECHTLY et al., J. Atmos. Terr. Phys. **34**, 1899 (1972).
[2] A. K. PAUL, Arch. Elek. Übertragung **14**, 468 (1960).
[3] A. P. PAUL and J. W. WRIGHT, J. Geophys. Res. **68**, 5413 (1963).
[4] J. R. DOUPNIK and E. R. SCHMERLING, J. Atmos. Terr. Phys. **27**, 917 (1965).
[5] H. H. HOWE and D. McKINNIS, Radio Sci. **2**, 1135 (1967).
[6] J. E. TITHERIDGE, Radio Sci. **2**, 1169 (1967).
[7] J. W. WRIGHT et al., Radio Sci. **7**, 1033 (1972).
[8] J. W. WRIGHT and G. SMITH, Radio Sci. **2**, 1119 (1967).
[9] J. E. TITHERIDGE, J. Atmos. Terr. Phys. **17**, 110 (1959).
[10] K. DAVIES and A. K. SAHA, Electron Density Profiles in the Ionosphere and Exosphere, Pergamon Press, New York, N.Y. 1962.
[11] A. K. PAUL and G. SMITH, Radio Sci. **3**, 163 (1968).
[12] A. R. LAIRD et al., submitted to Radio Science.
[13] A. K. PAUL et al., submitted to J. Atmos. Terr. Phys.
[14] T. L. GULELEVA, Geomagn. i Aeron. **12**, 488 (1972).
[15] L. A. MANNING, Proc. IRE **37**, 599 (1949).
[16] J. W. WRIGHT, Proc. IEEE **57**, 481 (1969).

Lower Ionosphere Structure — Akademie-Verlag, Berlin 1974

DOWNWARD TRANSPORT OF NIGHT-TIME Es LAYERS INTO THE LOWER E REGION AT ARECIBO

J. F. Rowe, Jr.

Arecibo Observatory, Arecibo, Puerto Rico, USA

Incoherent scatter observations at the Arecibo Observatory have shown that night-time electron density enhancements are formed in the upper E region and move down to merge with a stable Es layer lying between 105 and 110 km. The phenomenon is strongest in winter months, and is accompanied by the presence of an underlying Es layer near 90 km. These observations provide support for a night-time "corkscrew" mechanism which sweeps metallic ions down from higher altitudes and deposits them between 105 and 110 km. A statistical study of the wind shears necessary to maintain night-time Es layers suggest that strong layers lying below 110 km are composed primarily of long-lifetime metallic ions.

Full paper is appearing in the *Journal of Atmospheric and Terrestrial Physics* under the above title.

Lower Ionosphere Structure — Akademie-Verlag, Berlin 1974

CONSTRUCTION OF D REGION ELECTRON DENSITY PROFILES BY COMBINED USE OF GROUND-BASED REFLECTION AND SATELLITE-BASED TRANSMISSION MEASUREMENTS

L. CAIRO[a], C. ALTMAN[b] and E. FIJALKOW[c]

[a]Groupe de Recherches Ionosphériques, Saint-Maur, France,
[b]Department of Physics, Technion-Israel Institute of Technology, Haifa, Israel
[c]Department of Physics and Astronomy, University of Tel-Aviv, Israel

VLF field measurements on satellites in optical sight from ground-based transmitters yield ionospheric transmission coefficients as a function of angle of incidence and of azimuth. These, when taken in conjunction with ground-based measurements of VLF and LF reflection (or conversion) coefficients, permit an overall ionospheric electron-density distribution to be determined.

Reflection measurements at different frequencies and incident angles give an overall picture by looking at essentially different levels of the ionosphere. However, overall transmission coefficients are often more sensitive to angular changes than reflection coefficients, so that with transmission measurements electron-density profiles may be checked. Using a generalized multiple-reflection full-wave method, the contribution of each ionospheric level to the overall transmission and reflection (or conversion) coefficients can be identified and modifications introduced where required.

With the FR 1 satellite field-strength measurements in the near zone of the transmitter the D region night-time profiles of Deeks for different seasons have been re-evaluated. The original profiles must be modified, especially the low-lying ionization below 90 km, in order to obtain overall agreement.

1. Introduction

Most of the available ground-based measurements of reflection and conversion coefficients ($_{||}R_{||}$ and $_{||}R_{\perp}$) in the VLF and lower LF ranges, involve mid-latitudes (51°N—55°N) during a period (1947—1949) of high sunspot activity, with mean sunspot numbers $R = 151$, 136 and 135 respectively. The FR 1 satellite gave magnetic field strength data of the 16.8 kHz transmission from FUB (Sainte Assise) in France at 48.5°N geographic latitude during a moderate and a high sunspot year (1966 and 1968). Thus the ground-based and satellite data here analysed represent mid-latitude results at somewhat different phases of the sunspot cycle.

To determine the overall electron-density distribution at night our strategy will be as follows. We assume, together with Deeks [1] and Mechtly and Smith [2], a very sharp gradient near 89 km. First we determine the amount of low-lying ionization below this gradient, using VLF oblique incidence data. Then we determine the details of the steep gradient, using VLF reflection data at steep incidence, and also the east-west, west-east transmission differences at oblique incidence. Finally we adjust the profile above 90 km using the reflection data at 30—70 kHz.

2. The Satellite Field-Strength Data

The FR 1 measurements have been described by Storey [3], and the VLF field-intensity data reduction by Cairo [4]. The theoretical computations have taken into account the following factors: The point of entry of the ray at the lower E layer is determined for each satellite position by means of a ray-tracing calculation, which also yields the spreading loss (due to divergence of the ray wave-packet) of roughly 1.7 dB to the south and 1.3 dB to the north of the transmitter. An absorption loss of 1 dB in the F region is assumed [5]. We also correct for the horizontal southward gradient in electron density at the satellite height, 750 km [6].

Given the point of entry of the ray in the lower ionosphere, and assuming a radiated power of 20 kW we determine the field strength at the base of the ionosphere. The remaining unknown quantity, the D region transmission coefficient T_v for a vertically polarized incident wave, is then calculated by means of a generalized multiple-reflection full-wave method [7]. In Fig. 1 T_v is plotted as

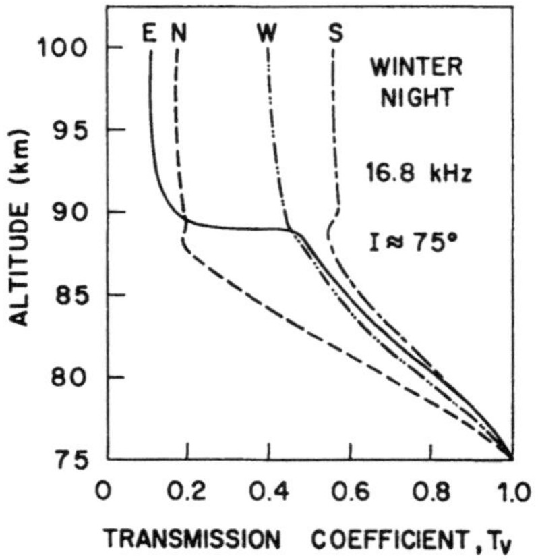

Fig. 1. Transmission coefficient T_v versus altitude for vertically polarized incident waves. Angle of incidence I is close to 75°. E, N, W and S refer to the direction of propagation.

a function of altitude for a winter-night profile due to Deeks [1], given in Fig. 3. The altitude dependence of T_v is to be understood as follows. The ionospheric profile is built up from below as elementary layers are added at the top, each with its predetermined electron density, and T_v is recalculated at each step or height.

In Fig. 2 the magnetic field strength (in units of dB above 5×10^{-8} A m^{-1}) is plotted for various positions of the satellite. The FR 1 measurements are sampled within rectangles of 0.5° latitude and 1° longitude, and then averaged over areas of $2° \times 2°$. The information density is rather low, especially in summer, and two of the $2° \times 2°$ squares in summer are seen to be void of information. "Night" was taken to commence one hour after ground sunset, and to terminate one hour before ground sunrise.

The computed field strengths at the satellite in Fig. 2 assume either the sunspot maximum profiles of Deeks (Fig. 3), or two profiles marked C.A.F. (1) and C.A.F. which are proposed hereinafter.

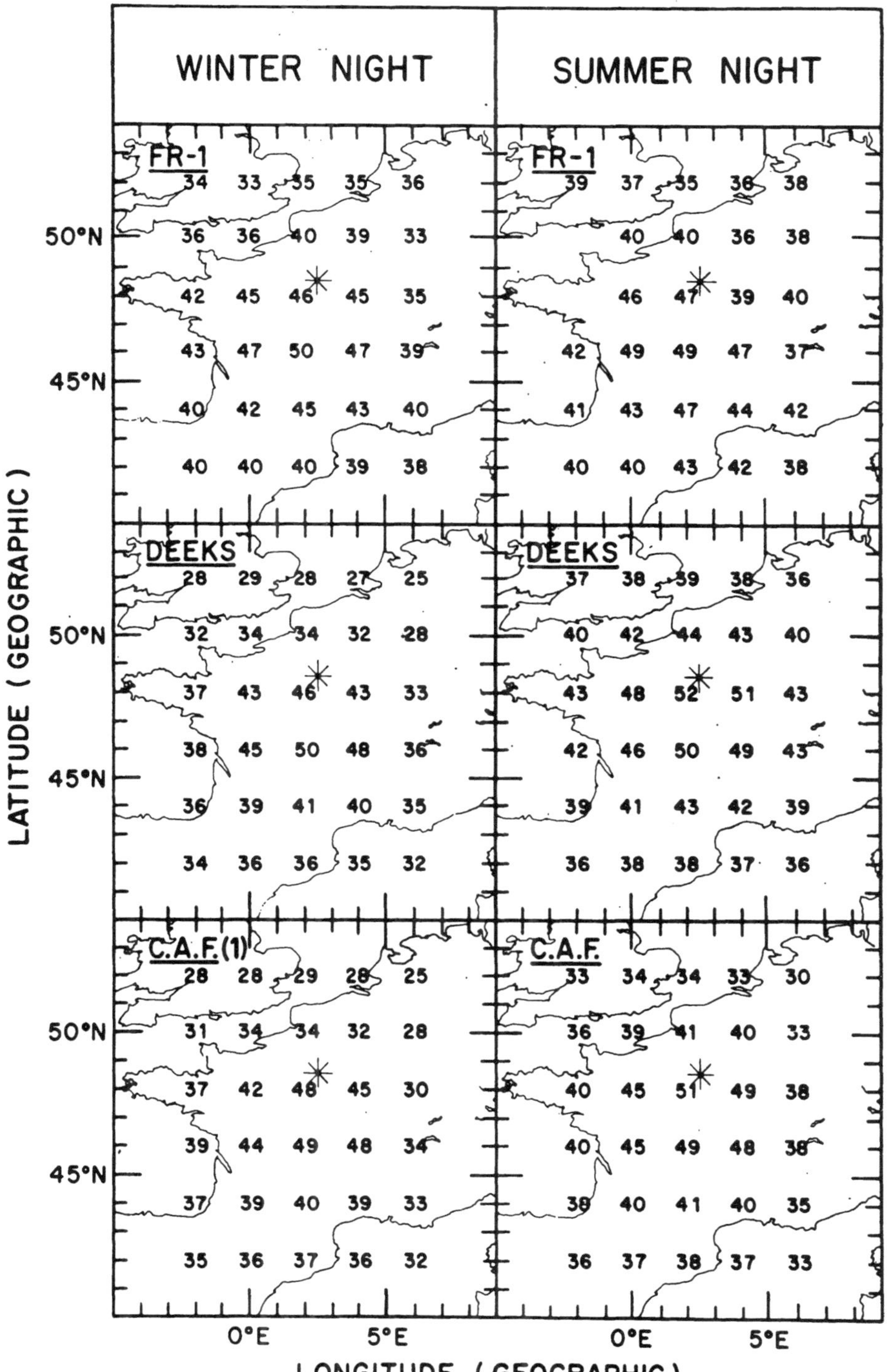

Fig. 2. Measured (FR 1) and computed values of magnetic field strength (in dB above 5×10^{-8} A m^{-1}) at 750 km altitude for various satellite locations. The transmitter (F.U.B.) marked*.

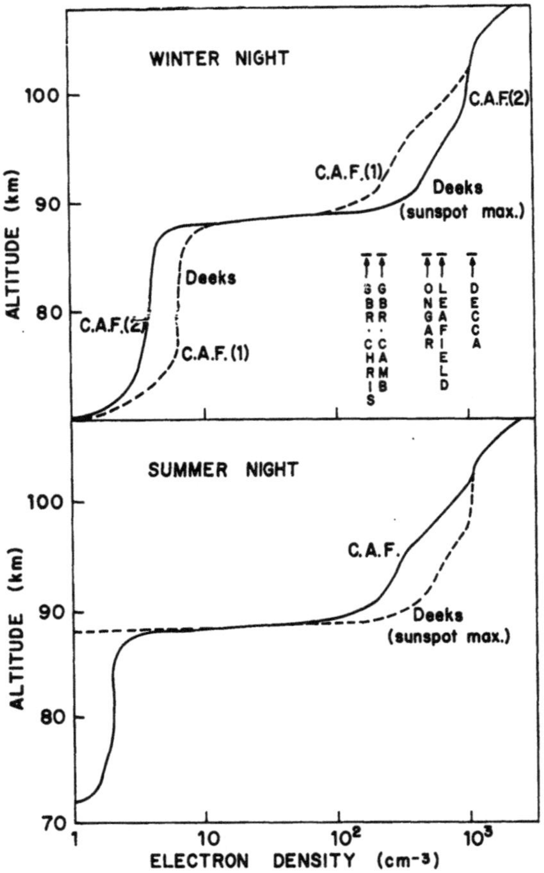

Fig. 3. Night-time electron-density distributions. The distributions C.A.F. (1) (winter) and C.A.F. (summer) agree best with observations.

3. The Ground-based Reflection Measurements

The stations providing data of reflection and conversion coefficients during the period 1947—1949 are summarized in Table 1. "Azimuth" refers to the bearing of the receiver from the transmitter, measured east of magnetic north.

Table 1
List of Ground Stations

Transmitter	Receiver	f (kHz)	Distance (km)	Azimuth (°)	Reference
GBR (Rugby)	Cambridge	16	90	111	[8]
GBR (Rugby)	Christchurch	16	186	199	[9]
GBR (Rugby)	Aberdeen	16	535	2	[10]
GLC (Ongar)	Cambridge	30	54	2	[11]
GIX (Leafield)	Cambridge	43	122	79	[11]
Decca	Cambridge	70	103	98	[12]

4. Determination of the Low-lying Ionization

4.1. The Summer Night Ionization

The existence of low-lying ionization is manifested primarily by the absorption of obliquely incident waves, especially when the wave normal makes a large angle with respect to the geomagnetic field. Thus oblique northward propagating waves will be much more strongly absorbed than southward (see for instance the variation of T_v below 88 km in Fig. 1 in the N and S cases).

Comparing the FR 1 measurements at high and low latitudes (Fig. 2) we may infer immediately the presence of low-lying ionization during both summer and winter. The Deeks summer profile, which has no underlying ionization, shows no north-south effect at all.

In summer we have also the Rugby-Aberdeen measurement of $_{\parallel}R_{\parallel}$ ($\simeq 0.55$). Almost any profile lacking low-lying ionization yields values of $_{\parallel}R_{\parallel}$ in excess of 0.7. The C.A.F. summer profile, Fig. 3, brings $_{\parallel}R_{\parallel}$ down to 0.54. The shape of this portion of the profile is based on the information [10, 13] that for oblique southward propagation of 16 kHz from GBR there is a rapid change of polarization for incident angles I between 65° and 70°, the wave being left circularly polarized for lesser angles. Writing the wave polarization $\varrho = {_{\parallel}R_{\parallel}}/{_{\parallel}R_{\perp}} = m \exp(i\phi)$, we find from the C. A. F. (summer) profile that ϕ takes the values 64°, 57°, −131°, −141° for southward propagation at angles of incidence I of 60°, 65°, 70° and 75° respectively. We conclude that the low-lying summer sunspot maximum ionization is represented satisfactorily by the C.A.F. curve of Fig. 3.

4.2. The Winter Night Ionization

The Rugby-Aberdeen ground-based measurements were unfortunately not pursued in winter. We have however the FR 1 north-south differences (Fig. 2) which are smaller than those given by the Deeks winter profile. We conclude that the amount of low-lying ionization in winter has been overestimated by Deeks. The relatively small summer-winter differences observed by FR 1 at all latitudes support this conclusion. On the basis of the transmission and reflection computations, described presently, we shall propose the profile C. A. F. (2) as yielding the best overall agreement with observations.

4.3. The East-West Effect

In the absence of low-lying ionization there is almost no east-west effect (see, for instance, the computations for the Deeks summer profile in Fig. 2). Roughly speaking, the function of the low-lying ionization is to rotate the plane of rotation of the vertically polarized incident waves by different amounts for east-west and west-east propagation, and the reflecting layer near 89 km will then have different reflection coefficients for the differently polarized waves. The effect is clearly demonstrated by the eastward (E) and westward (W) propagating curves of T_v in Fig. 1. The Deeks winter profile gives a much smaller east-west effect than observed by FR 1. The C. A. F. (1) modification remedies this drawback but is inconsistent with the north-south differences, and especially with the reflection measurements discussed next.

5. Determination of the Upper D Region Profile

In Fig. 3 we have marked the electron densities near which the various signals undergo their main reflection processes. The reflection-conversion coefficients will be particularly sensitive to the electron-density gradients at and near these regions. The full-wave method gives us the contribution of each section of the profile, and permits the use of a trial and error process based on intelligent guessing. The available data on night-time reflection and conversion coefficients is summarized in Table 2, together with computed values using the C.A.F. (summer) and C.A.F. (2) (winter) profiles. Except for GBR–Aberdeen the quantity most accurately determined is ${}_{||}R_{\perp}$, whereas ${}_{||}R_{||}$ is determined through a rough estimate of the polarization $\varrho = {}_{||}R_{||}/{}_{||}R_{\perp} = m \exp(i\phi)$.

Table 2

${}_{||}R_{\perp}$ and ${}_{||}R_{||}$ for Summer and Winter Nights
Computed values are based on the C.A.F. (summer) and C.A.F. (2) (winter) profiles.

(a) Summer Night

Transmission	${}_{\vert\vert}R_{\perp}$		${}_{\vert\vert}R_{\vert\vert}$ or m	
	Measured	Computed	Measured	Computed
GBR—Cambridge	0.52	0.51	$m \simeq 1$	0.48
GBR—Christchurch	0.47—0.50	0.54	—	—
GBR—Aberdeen	—	—	0.55	0.54
Ongar—Cambridge	~0.2	0.25	$m \simeq 1$	0.24
Leafield—Cambridge	~0.3	0.38	$m \simeq 0.8$	0.25
Decca—Cambdrige	0.3—0.4	0.36	—	—

(b) Winter Night

Transmission	${}_{\vert\vert}R_{\perp}$		${}_{\vert\vert}R_{\vert\vert}$ or m	
	Measured	Computed	Measured	Computed
GBR—Cambridge	0.41	0.44	$m \simeq 1$	0.47
GBR—Christchurch	0.47—0.50	0.45	—	—
Ongar—Cambridge	~0.3	0.33	$m \simeq 1$	0.27
Leafield—Cambdrige	0.34	0.31	$m \simeq 0.8$	0.24
Decca—Cambridge	0.3—0.4	0.35	—	—

We note in conclusion that the agreement is generally good, and extends also to $\phi = \arg \varrho$, not shown in Table 2. Small changes in the profiles lead to large deviations in one or more of the coefficients. Nevertheless the profiles are not unique, and additional information containing the height parameter, even implicitly as in the Hollingworth ground patterns, should be used to complete the analysis.

Acknowledgments

Part of the work reported here was performed while one of the authors (C.A.) was one leave of absence at the Groupe de Recherches Ionosphériques, in Saint Maur, France, in the FR 1 satellite group. We are indebted to Mr. J. Hieblot, director of the G.R.I. and to Dr. L. R. O. Storey, head of the FR 1 satellite group, for their hospitality and cooperation.

References

[1] D. G. Deeks, Proc. Roy. Soc. A **291**, 413 (1966).
[2] E. A. Mechtly and L. G. Smith, J. Atmos. Terr. Phys. **30**, 363 (1968).
[3] L. R. O. Storey, Space Research VII, 588 (1967).
[4] L. Cairo, Note Technique GRI/NTP/101 (1972).
[5] M. P. Aubry, J. Atmos. Terr. Phys. **30**, 1161 (1968).
[6] K. L. Chan and L. Colin, Proc. IEEE **57**, 990 (1969).
[7] C. Altman and H. Cory, Radio Sci. **4**, 459 (1969).
[8] T. W. Straker, Proc. IEE **102 C**, 122 (1955).
[9] R. N. Bracewell, Proc. IEE **99**, **IV**, 217 (1952).
[10] W. C. Bain, R. N. Bracewell, T. W. Straker and C. H. Westcott, Proc. IEE **99**, **IV**, 250 (1952).
[11] R. N. Bracewell, J. Harwood and T. W. Straker, Proc. IEE **101**, **IV**, 154 (1953).
[12] K. Weekes and R. D. Stuart, Proc. IEE **99**, **IV**, 38 (1951).
[13] R. N. Bracewell, K. G. Budden, J. A. Ratcliffe, T. W. Straker and K. Weekes, Proc. IEE **98**, **III**, 221 (1951).

Lower Ionosphere Structure — Akademie-Verlag, Berlin 1974

MEAN D REGION ELECTRON DENSITY PROFILES DERIVED BY COMBINATION OF ROCKET AND RADIO WAVE PROPAGATION DATA

J. Bremer and W. Singer

Academy of Sciences of the German Democratic Republic, Observatory of Ionospheric Research of the Central Institute for Solar-Terrestrial Physics, Kühlungsborn, GDR

Mean D region electron density profiles are derived for medium latitudes, low solar activity ($R \leqq 30$), and summer conditions at solar zenith angles 40°, 60°, 70° and 80°. Generalized averages of about 80 published profiles were adapted to experimental data of radio-wave propagation in the LF, MF and HF range and to ionosonde observations of critical frequencies f_0E. The profiles are in good agreement with those of Maeda, and of Mechtly et al. They do not show a distinct CR-D layer at 60—65 km and allow a fairly good interpretation of the LF, MF and HF radio propagation data and their dependence on solar zenith angle.

An attempt has been made to derive mean electron density profiles of the lower ionosphere which are compatible with results of direct measurements by rocket-borne methods or by partial reflection or cross modulation experiments as well as with other ground-based radio wave propagation data. Such profiles, valid for summer conditions, low solar activity ($R \leqq 30$) and medium latitudes, are presented in this paper.

In order to derive these profiles, a preliminary set of mean profiles, for different solar zenith angles, was first determined by averaging about 80 individual profiles published in the literature [1]. Then these preliminary mean profiles were checked as to how far they were compatible with the results of ionospheric absorption and phase height measurements which are available for a broad spectrum of different frequencies. The calculations necessary for this purpose were made according to the method of full-wave equations in the LF and MF range [2, 3], whereas in the HF range the phase-integral method [4] was used. The collision frequency profile was assumed on the basis of Groves' tables of atmospheric pressure [5] and agrees very well with the models by Lodato and Mechtly [6] and that used by Bain and Harrison [7]. It was found that the original mean electron density profiles had to be changed only relatively slightly and within the margins of error, in order to achieve the desired adaptation to the absorption and phase height results. Therefore, it is felt that these changes may be considered as an improvement of the original mean profiles, making them suitable to be proposed as standard profiles for more general use.

The adapted mean profiles obtained in this way for the solar zenith angles 40°, 60° ,70° and 80° are presented in Fig. 1 in comparison with results of other authors. The hatched region contains the majority of rocket-measured electron density profiles collected by Maeda [8] for solar zenith angles greater than 60°.

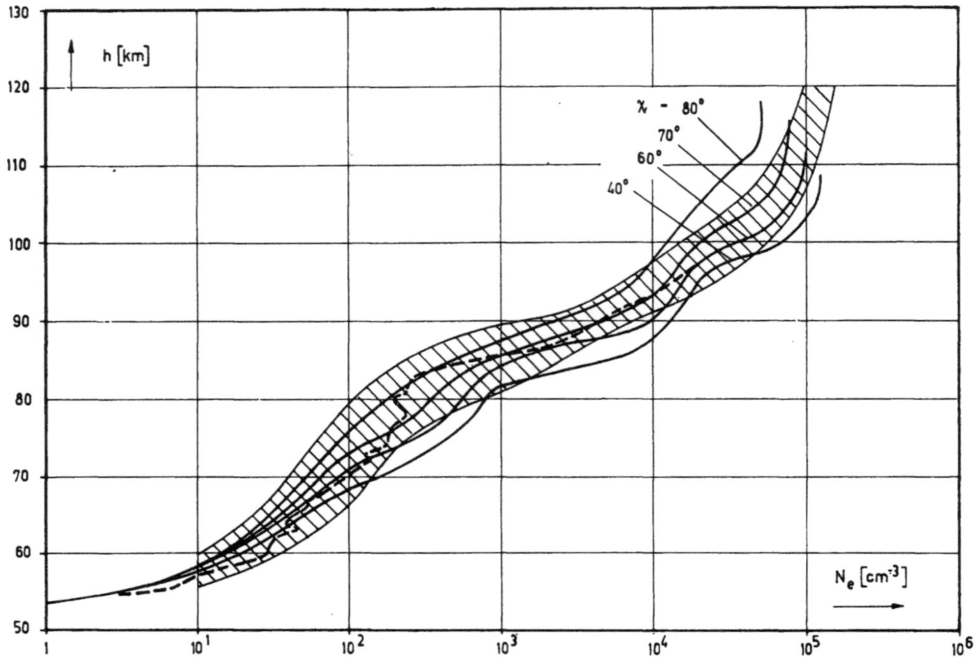

Fig. 1. Mean electron density profiles of the lower ionosphere for middle latitudes and low solar activity ($R \leqq 30$). Full lines, profiles for different solar zenith angles in summer, as derived in this paper; hatched region, envelope of profiles collected by Maeda [8] at $\chi \geqq 60°$; dashed line: profile from Mechtly et al. [9] at $\chi = 60°$.

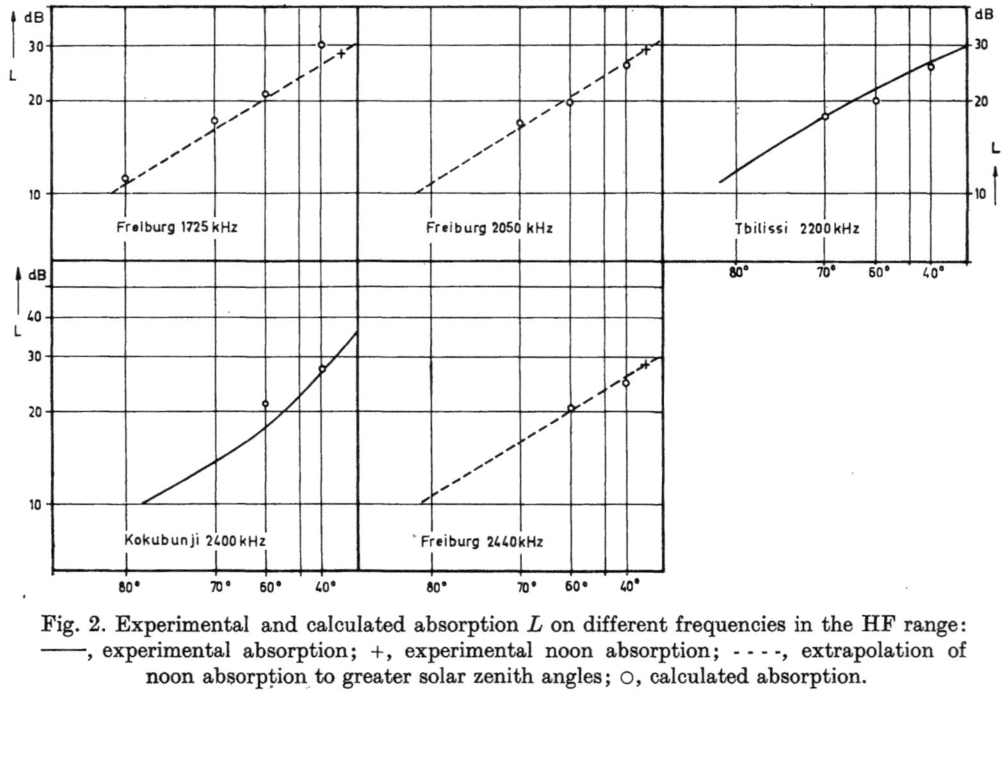

Fig. 2. Experimental and calculated absorption L on different frequencies in the HF range: ——, experimental absorption; +, experimental noon absorption; - - - -, extrapolation of noon absorption to greater solar zenith angles; ○, calculated absorption.

The dashed curve represents the average of 5 rocket measurements at $\chi = 60°$ by Mechtly et al. [9]. As can be seen, the agreement is quite satisfactory.

In the following, it will be demonstrated how the derived electron density profiles in Fig. 1 agree with the different kinds of radio wave propagation data. Absorption measurements according to the A 1 method at about 2 MHz [10—12] were used to secure the upper part of the profiles above 95 km (mainly from 100 km to 110 km). The maximum electron density of the E region was determined from f_0E measurements of the ionosonde station Juliusruh [13] and the height of this maximum from an empirical formula of Robinson [14]. In Fig. 2 the experimental absorption data are represented together with the corresponding theoretical

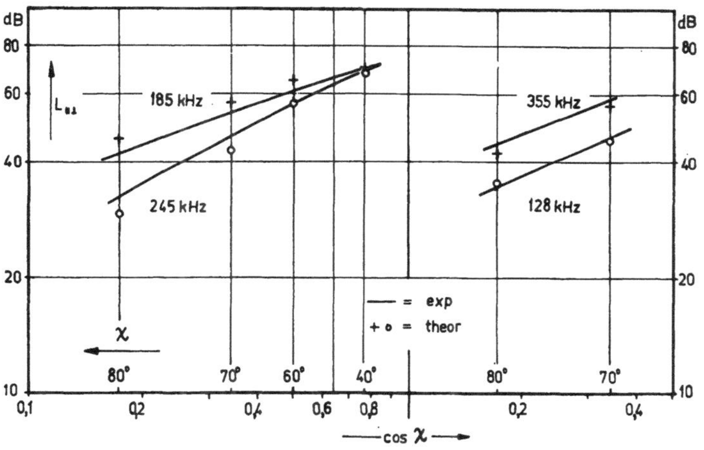

Fig. 3. Experimental and calculated absorption $L_{\|\perp}$ on different frequencies in the LF range.

values which were calculated by means of the phase-integral method with the mean electron density profiles of Fig. 1. As may be seen, the agreement is always within ± 4 dB. In the case of Freiburg, however, this can be said with certainty only for the values near $\chi = 40°$, because only noon values (crosses) were available from this station. An extrapolation to greater solar zenith angles has been attempted by means of a $\cos^n \chi$ law with $n = 0.6$ (dashed lines).

Below 95 km down to about 70 km the electron density profiles of Fig. 1 can be tested by means of the results of absorption measurements in the LF and MF range (A 3 method [15, 16]). In Fig. 3 the experimental absorption on four different measuring paths with frequencies between 100 kHz and 400 kHz [17, 18] are represented together with the corresponding theoretical data which were obtained from full-wave calculations with the electron density profiles of Fig. 1. The agreement is quite good, the differences are less than 5 dB.

The results of phase height measurements in the LF region on two very long measuring paths [18, 19] were mainly used for investigation of the lower part of the electron density profiles between about 85 km and 60 km. The most important conclusion to be drawn from these investigations is that the experimental results can be interpreted only with profiles without a distinct CR-D layer as postulated by Deeks [20] and others near about 65 km. In Fig. 4 the experimental results and the theoretical phase heights obtained from full-wave calculations with the proposed electron density profiles are represented. The calculated values differ from the experimental data by less than 2 km. On the other hand, calculations with a CR-D layer would give phase heights by at least 5 km less than the corresponding experimental data.

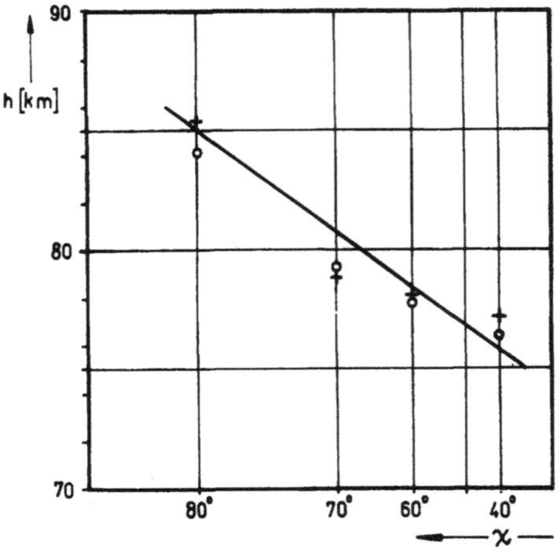

Fig. 4. Experimental (——) and calculated (+ ○) phase heights on two different measuring paths in the LF range.

Thus, the derived mean electron density profiles of Fig. 1 are in accordance not only with other mean profiles from direct measurements but also with radio-wave propagation data of different kinds and their dependence on solar zenith angle, even at very different frequencies.

The adaptation of the electron density profiles to the results of radio wave propagation measurements, of course, depends on the assumed profile of collision frequency. Therefore, we investigated the influence of changes of collision frequency within plausible limits and found that changes of electron density by not more than 20% were sufficient to obtain adaptation again. Furthermore, we investigated the influence of the seasonal variation of collision frequency. According to the pressure data of Groves [5] the collision frequency is greater in summer than in winter at heights below 84 km, whereas above this height the relation is reversed. First calculations with the winter profile of collision frequency show an increase of HF absorption and a decrease of LF phase height, i.e. tendencies in accordance with experimental observations. This result points to a considerable contribution of the collision frequency to the problem of winter anomaly of the lower ionosphere. On the other hand, strong changes of the electron density profile have also to be expected in winter in order to explain the phenomenon of winter anomaly. But in winter the electron density is obviously extremely variable from day to day, and therefore it seems to be problematic to propose mean standard electron density profiles for this season. Nevertheless, detailed investigations in this respect will be continued.

References

[1] W. Singer and J. Bremer, Geomag. i Aeron. **12**, 54 (1972).
[2] M. L. V. Pitteway, Phil. Trans. Roy. Soc. London A**257**, 219 (1965).
[3] T. A. Seliga, J. Inst. Telecom. Engrs **12**, 198 (1966).
[4] C. Altman, Radio Sci. **69D**, 511 (1965).
[5] G. V. Groves, US Air Force Cambridge Res. Lab. Rep. AFCRL-70-0261 (1970).
[6] R. F. Lodato and E. A. Mechtly, Aeronomy Rep. No. 45, University of Illinois, Urbana (1971).
[7] W. C. Bain and M. D. Harrison, Proc. IEE **119**, 790 (1972).
[8] K. Maeda, J. Geomag. Geoelect. **21**, 557 (1969).
[9] E. A. Mechtly, S. A. Bowhill and L. G. Smith, J. Atmos. Terr. Phys. **34**, 1899 (1972).
[10] Ionosphärendaten, Absorptionsmessungen, Ionosphären-Institut Breisach, H. 11—14 (1964/65).
[11] IQSY data for ionospheric absorption at Kokubunji, National Committee for the IQSY, Tokyo (1966).
[12] D. K. Kvavadze, M. I. Tevdorashvili and K. I. Tukashvili, Ionospheric Researches No. 17, 154, Moscow 1969.
[13] Beobachtungsergebnisse, Heinrich-Hertz-Institut Berlin-Adlershof (1964/65).
[14] B. J. Robinson, Rep. Prog. Phys. **22**, 241 (1959).
[15] E. A. Lauter, Thesis, Universität Rostock (1950).
[16] R. N. Bracewell et al., Proc. IEE **98**, Pt. III, 221 (1951).
[17] E. A. Lauter and P. Nitzsche, J. Atmos. Terr. Phys. **29**, 533 (1967).
[18] Unpublished data of Observatory for Ionospheric Research, Kühlungsborn (1964/65).
[19] E. A. Lauter and G. Entzian, Schriftenreihe des NKGG der DDR, Reihe II, H. 1, 67 (1966).
[20] D. G. Deeks, Proc. Roy. Soc. A **291**, 413 (1966).

Lower Ionosphere Structure — Akademie-Verlag, Berlin 1974

AVERAGE NIGHT-TIME ELECTRON-DENSITY PROFILES BETWEEN 87 AND 115 KM DERIVED FROM ABSORPTION MEASUREMENTS AT 10 FIXED FREQUENCIES FROM 250 TO 1000 kHz AT TSUMEB (19°S)

W. ELLING

Max-Planck-Institut für Aeronomie, Institut für Ionosphären-Physik, Lindau/Harz, FRG

Results of absorption measurements carried out in 1964 at Tsumeb, S.W.A., at ten fixed frequencies between 250 and 1000 kHz, are presented and interpreted by theory. The calculations yielded the night-time E region (87—115 km) electron-density profiles of the background ionization (averages over 14 nights), in which various Es layers were embedded. The electron density N decreased continuously with time t after sunset in the entire height range. The corresponding recombination coefficients α, defined by the equation $dN/dt = -\alpha N^2$, decreased from 9×10^{-8} cm^3 s^{-1} at 87.5 km to 6×10^{-9} cm^3 s^{-1} at 115 km. At heights ranging from 87.5 to 97 km α was proportional to atmospheric pressure. Different recombination coefficients were obtained for the two periods from 1800 to 2400 and 0000 to 0600. All these results suggest that different physical processes occurring in the night-time E region depend not only on height but also on time.

ELECTRON-DENSITY PROFILES OBTAINED FROM MF SOUNDING NEAR THE SOUTHERN TROPIC

E. H. Gruschwitz

Max-Planck-Institut für Aeronomie, Institut für Ionosphärenphysik, Lindau, FRG

In 1965 (IQSY) MF-pulse sounding was carried out at Tsumeb (19.2° S 17.7° E) at ten frequencies in the range from 255 to 950 kHz. The diurnal variation of reflection heights and of the attenuation of the reflected pulses at 520 kHz will be presented as an example.

Taking into account all the measurements electron-density profiles were calculated by use of a thin-film-optics full-wave analysis. As a result, the diurnal variation of electron density was obtained for the height range from 50 to 100 km. The six profiles to be presented are averages over a period of eight days in April 1965. In the daytime, at heights of about 75 and 87 km, steep gradients were found as a typical feature in the electron-density profiles.

1. Diurnal Variation of Reflection Heights and of the Attenuation of the Reflected Pulses

In 1965 MF-pulse sounding was carried out by Elling [1] at Tsumeb (19.2°S 17.7°E) at ten frequencies in the range from 255 to 950 kHz. Method A 1 for measurement of absorption of radio waves in the ionosphere was applied. The reflections were observed on an oscilloscope screen and photographically recorded on films. From these films reflection heights and amplitudes of the reflected pulses were analysed for a period of eight days in April 1965 [2]. The diurnal variation of reflection heights is shown in Fig. 1. The heights plotted are mean heights representing the individual echoes distributed in height intervals of 10 km centred at the mean heights shown.

Reflections from the E region are observed nearly independently of the time of day from a distinct region with mean heights of 96 km. During the day, at the lower boundary of the E region, increased reflection levels develop that suppress and blanket reflections from the upper region. Starting with sunrise the heights of lower E reflections descend from the upper region, attaining their day-time values about four hours later. Thereafter mean reflection heights remain nearly constant until half an hour before sunset when the heights quickly ascend into the upper region. The mean day-time value (between 1000 to 1730 MET, MET = 15°EST) is 87 km for the ordinary component of the wave. Reflections from the D region are observed beginning at about one hour past sunrise until sunset. Their heights are independent of the time of day with the exception of the initial and final phases. The mean height for the ordinary component is 75 km. For the extraordinary component the virtual reflection heights in the lower E region are

1.5 km higher and in the D region 1.5 km lower than for the corresponding ordinary components.

From the night-time Es layer recordings reflections of radio waves which had not suffered attenuation on propagation through the ionosphere were taken as a reference. The attenuation of an individual pulse is now given in respect to

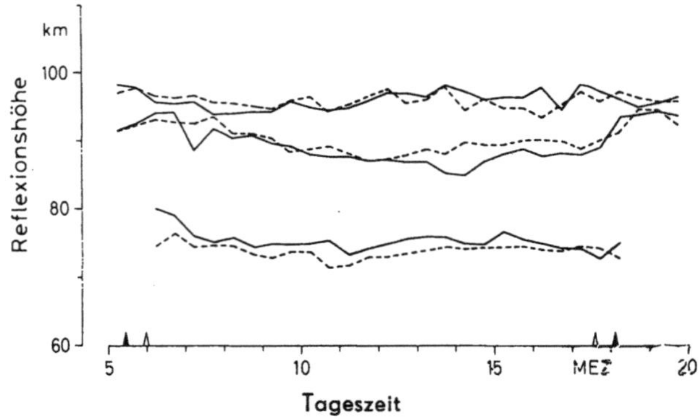

Fig. 1. Diurnal variation of mean reflection heights for D, lower E, and E region echoes: —— ordinary component, - - - - extraordinary component. The arrows on the time axis indicate the mean times of sunrise and sunset at ground (white), and at a height of 100 km (black), respectively, during the period of measurement.
Reflexionshöhe = reflection height, Tageszeit = time of day (MEZ = MET = 15° EST)

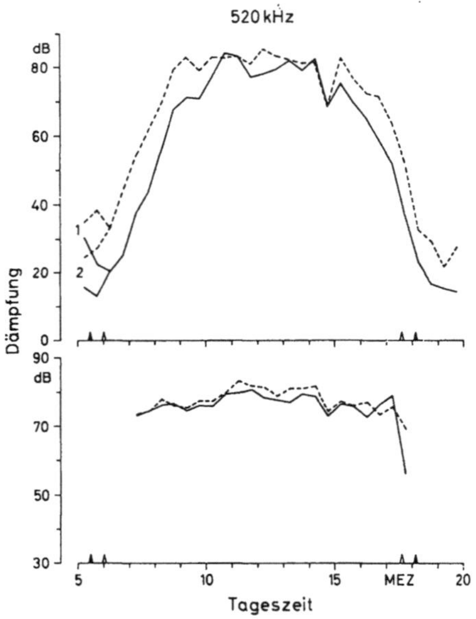

Fig. 2. Diurnal variation of attenuation of reflections from D region (bottom) and E region (top) at 520 kHz, as an example.
Lines and arrows as in Fig. 1. 1, reflections from lower E region; 2, from upper E region. Lines without a number are mean values from both reflection levels.
Dämpfung = attenuation, Tageszeit = time of day (MEZ = MET).

these reference reflections. In Fig. 2 the attenuation of the reflected pulses at 520 kHz is shown as an example.

Except for occasionally observed reflections during the initial or final state of D region formation, i.e. before 0730 or after 1700 MET, reflections from the D region are of nearly equal strength independent of the time of day (Fig. 2, bottom). According to the frequency used, attenuation is 60 to 80 dB. Reflections from the E region show a clear diurnal variation (Fig. 2, top): before sunrise they are very intense and become weaker as the sun ascends. After some time of very high attenuation during noon hours, in the range of 60 to 90 dB according to frequency and magneto-ionic component of the radio wave, reflections become stronger again. After sunset they attain magnitudes comparable with that in the early morning. A clear variation of attenuation with the cosine of the solar zenith angle is observed for E region reflections. The ordinary and the extraordinary reflections behave very similarly apart from weaker extraordinary than ordinary E region reflections in the upper frequency range.

2. Electron Density Profiles from Model Calculations applied to the MF-Sounding Experiment

For interpretation of the MF-sounding results from Tsumeb a method of model calculations was developed [3]. This thin-film-optics full-wave analysis divides the ionosphere into thin homogeneous layers of 100 m thickness. Multiplication of the transfer-coefficient matrices along the propagation path of the wave yields the electric intensities of the ordinary and the extraordinary component at the receiver site.

Various model profiles of electron density were tested by these full-wave calculations. From comparison of the reflected pulses obtained from the experiment and those from model calculations the diurnal variation of electron density for the height range from 50 to 100 km was determined. The electron-density profiles of Fig. 3 correspond well with the diurnal variations of reflection heights (Fig. 1) and of the attenuation of the reflected pulses (Fig. 2).

The diurnal variation of electron density starts from the night-time profile shown in Fig. 3. Reflections here are produced by Es layers above about 90 km or by total reflection in the E region. The Es layers are not plotted in Fig. 3. After sunrise electron density chiefly increases below about 80 km. This means an increase of attenuation as is seen from Fig. 2 (top). On the other hand the increase of E region electron density causes a decrease of E region total-reflection heights as is seen from Fig. 1. Reflections from Es layers above the lower boundary of the E region become weaker and also occur less frequently because the Es layers disappear in the increased E region electron density.

At about oq h MET the ionosphere below about 80 km is fully established. A steep gradient, with increase of electron density by almost a factor ten within a few 100 m, is present in the electron-density profile at a mean height of 75 km throughout the day causing reflections of nearly constant intensity, as is seen from Figs. 1 and 2 (bottom). The ionosphere below this gradient remains constant until sunset. The gradient itself changes its magnitude very quickly and also varies its height within an interval of about 10 km centred at the mean height of 75 km. This means the lower ionosphere is leaky and consists of electron clouds varying their heights and densities within a few minutes or even seconds.

The ionosphere above about 80 km shows a clear diurnal variation with an increase of electron density until noon and a decrease thereafter. At a mean height of 87 km a second steep gradient exists at the lower boundary of the E region. It is here that the lower E region reflections seen in Fig. 1 are caused. Below this gradient there is a trough in the electron-density profile that is filled up until noon and later deepens again. The variation of attenuation with the solar zenith angle seen in Fig. 2 (top) is caused by the diurnal variation of electron density in the height interval between the two gradients of Fig. 3. This is concluded from

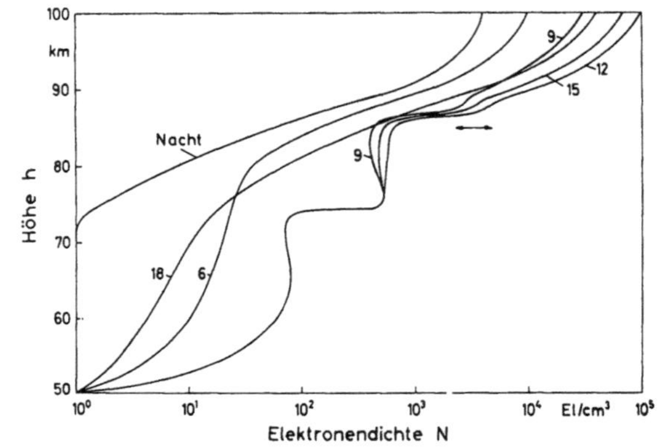

Fig. 3. Diurnal variation of electron density N at heights h between 50 and 100 km determined from comparison of measurements and model calculations. Nacht = night-time profile; numbers give time in MET.

the fact that the attenuations of the pulses reflected from different heights in the E region are nearly the same. Another conclusion from the comparison of experiment and model calculations is that the lower boundary of the E region is not dense but leaky. This means that the electron clouds with a steep gradient at their lower boundary are only limited in horizontal extent. Reflections of MF radio waves therefore are caused mostly by a flat gradient of electron density as seen in the transition region above the steep gradient indicated by the arrow in Fig. 3. In the E region, also, very rapid variations of the gradient occur. Reflection heights are distributed over a 10 km interval centred at 87 km.

After noon the electron density decreases with descending sun and at about 18 h MET gradually turns over to the night-time profile, Analogous to the electron production in the morning, electrons disappear very quickly after sunset in the height region below about 80 km. Above about 80 km the diurnal variation is not symmetrical relative to the solar zenith angle. This is seen from Fig. 2 (top) where attenuation at the same solar zenith angle is higher in the afternoon than in the forenoon. A delay time of about 35 minutes was determined for the variation of attenuation with the solar zenith angle. Therefore, in Fig. 3, a higher electron density is seen in the region above about 80 km at afternoon hours than at comparable forenoon hours.

Discussion of the electron-density profiles of Fig. 3 is concluded with the emphatic warning that the profiles shown are averages over a period of eight days.

The fine structure of the electron density on the other hand varies rapidly in time and space. But the steep gradients at heights of about 75 and 87 km are a typical feature of the electron-density profiles in April 1965 at Tsumeb. From an intense study of MF-pulse sounding in comparison with model calculations it is concluded that the fine structure of electron-density profiles obtained by rockets is also valid only for a few seconds and for a few 100 m of horizontal extent.

References

[1] W. ELLING, in: Ground-based Radio Wave Propagation Studies of the Lower Ionosphere, Ottawa 1967, Defence Research Telecommunications Establishment, Ottawa 1967 (p. 614).
[2] E. H. GRUSCHWITZ, Z. Geophys. **39**, 779 (1973).
[3] E. H. GRUSCHWITZ, Radio Sci., in press.

Conclusions from Propagation Experiments

Lower Ionosphere Structure — Akademie-Verlag, Berlin 1974

ON THE INTERPRETATION OF CW PROPAGATION DATA FOR LONG RADIO WAVES

J. S. Belrose and B. Segal

Communications Research Centre, Department of Communications, Ottawa, Canada

This paper deals with techniques of ionospheric probing using ground-based facilities and frequencies in the range 3 to 300 kHz. The purpose of the paper is to review briefly the theoretical grounds on which this probing technique depends, to review the methods employed for determining D region electron density distributions from CW propagation data, and to assess the limitations of the results obtained, using model and actual propagation data. It is concluded that long-wave propagation offers an unique and sensitive probe for studying the lowest region of the ionosphere.

1. Introduction and Early History

Low and very low frequency ($f < 300$ kHz) radio waves have been used, since the birth of radio communication, to study the ionospheric D region, but only recently has progress been made in exploiting this probing technique to measure electron density height profiles. There are abundant data that are useful for synoptic studies, and that show the sensitivity of the propagation parameters to small ionization changes at low heights in the D region, but there are few published observational results that extend over a sufficiently wide frequency range that they can be used to derive electron density height profiles. There are also mathematical difficulties in interpreting these data. As we shall show, these have largely been overcome, but only recently.

Most of the profiles that have been derived from LF propagation, showing development of ionization during dawn, diurnally, or over the epoch of the solar cycle, have utilized averaged data recorded at Cambridge, England, during the period 1947—1956. Others have made use of the observations of Dr. Lauter and his colleagues at Kuhlungsborn from 1951 to the present. While one can find in the literature a few early attempts at interpreting long-wave propagation in terms of plausible $N(h)$ profiles, e.g., the work of Nertney [1] who tried to interpret extensive series of 150 kHz vertical incidence pulse transmissions, it was not before the work of Deeks [2] that much attention was given to the idea of VLF/LF propagation being a useful electromagnetic probe for the study of D region ionization.

Deeks' profiles almost immediately attracted wide interest but it is difficult to assess the reliability and uniqueness of his profiles. In the view of one of the present authors (J. S. B.) the data used by Deeks may not have been sufficiently reliable to allow the detailed analysis which was made. (It was this author's data, collated with those of others, that was used by Deeks.) The $N(h)$ profiles derived by Deeks seemed to be too much like idealistic models to represent reality. The diurnal variation of electron densities derived by Deeks, and shown in Fig. 1,

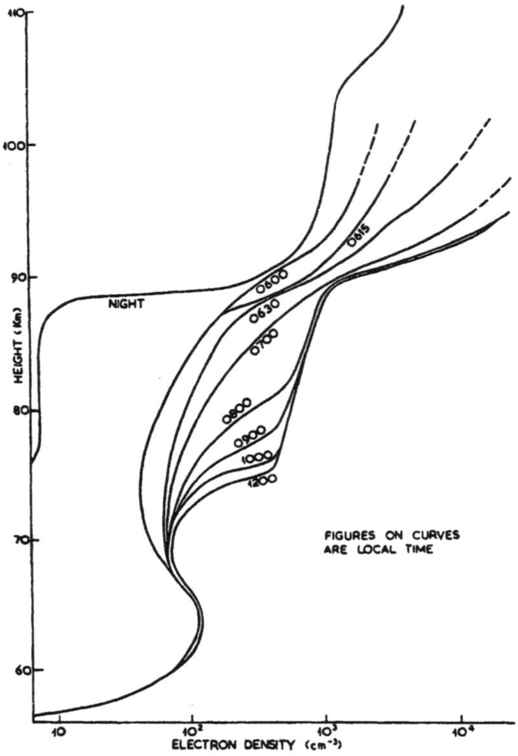

Fig. 1. Diurnal development of ionization. From Deeks [2].

displayed more clearly and distinctly the three strate, the C, D and E layers than is revealed by other experimental observations. The sunspot cycle variation (Fig. 2) showed exactly the diurnal change expected from theory, i.e. the D layer was more heavily ionized during solar maximum years as would be expected due to the increase in solar X-rays, and the C layer, consistent with the known solar cycle change of galactic X-rays, was more heavily ionized in solar minimum years. Such a variation is also deduced from rocket-borne experiments [3].

The night-time profiles of Deeks were particularly interesting since they showed that he needed a lightly ionized ledge ($N < 10$ electrons cm^{-3}) below about 87 km, to explain the propagation data. Further analysis of the same data by Thomas and Harrison [4] gave further weight to the reality of such a ledge (Fig. 3), but the ledge deduced by these authors was more restricted in height and more densely ionized by an order of magnitude. Rocket experiments [5—7] as well as other ground-based experiments [8] have shown evidence for such a ledge. Belrose [9] from a study of the mean nocturnal variation of VLF phase height suggested that

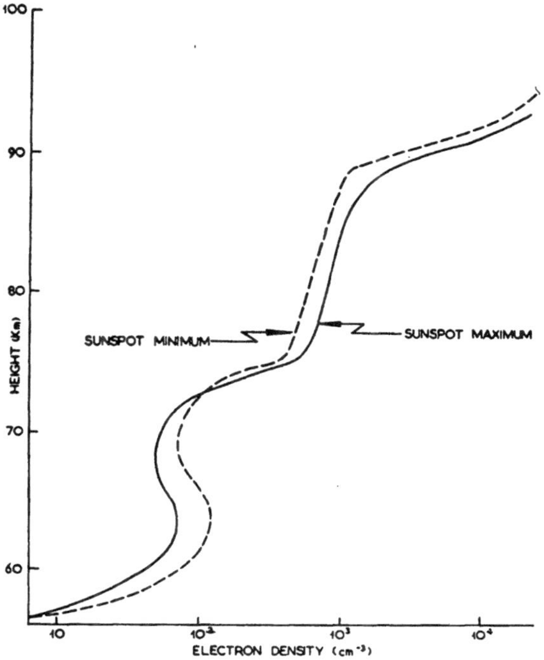

Fig. 2. Sunspot cycle variation of electron densities. From Deeks [2].

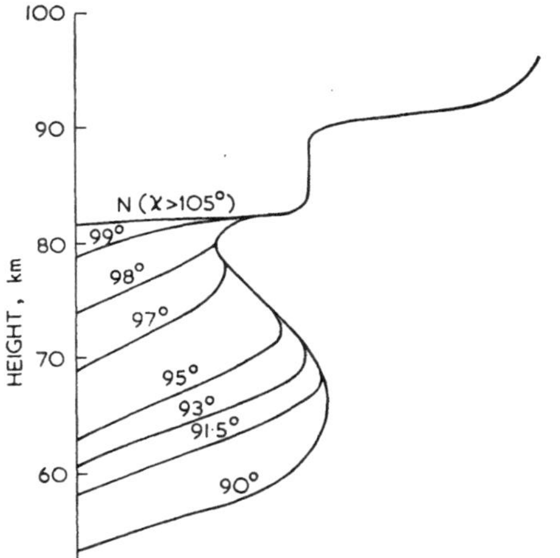

Fig. 3. Development of low-level ionization during sunrise. From Thomas and Harrison [4].

the ledge was under solar control, even for negative solar elevation angles. This fact seems to be revealed (Fig. 4) by a series of rocket-borne experiments conducted on a single night [10]. Thomas and Harrison [4] (Fig. 3) also showed a development of the D region during dawn, which was similar in gross detail to that deduced from wave interaction measurements [8].

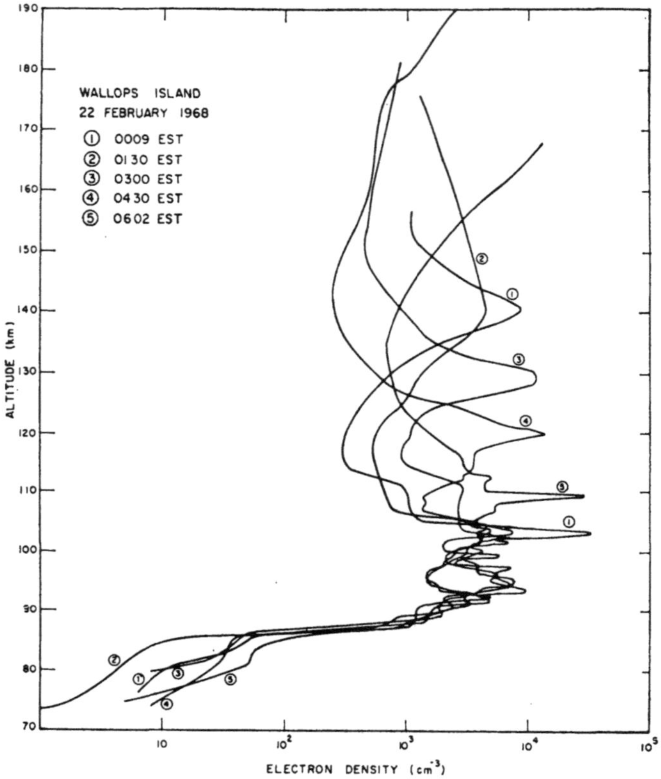

Fig. 4. Electron density variations during night-time as measured by a rocket-borne Langmuir probe. From Smith [7].

Uncertainties over the reliability of the Deeks and Thomas profiles can be found in subsequent work. For example, Bain and May [11] showed differences between profiles derived by them, based mainly on early Hollingworth interference pattern measurements at 16 kHz in solar cycle maximum years, and what they thought Deeks would have derived (Deeks did not actually show a summer profile for solar maximum). Bain and Harrison [12], on analysing recent 16 kHz Hollingworth interference pattern data [13] found they had to reject their earlier profile. Differences are found [14] between profiles derived by British workers and Russian workers, even though both these groups were analysing the same VLF/LF propagation data. This difference is shown in Fig. 5 which is from the work of Krasnushkin and Knyazeva [15]. The dashed curve labelled C is Bain's earlier profile for summer solar maximum conditions, which should be compared with profile B derived by Krasnushkin for similar solar conditions. Curve A represents summer noon in solar minimum years, curve D winter noon averaged over the 11-year solar cycle and curve E the average night profile.

Crain [16] has recently reviewed the technique of ionospheric probing with long wavelength radio waves. The work was done for an URSI Working Group on EM probing of the Upper Atmosphere. While the paper was a good review

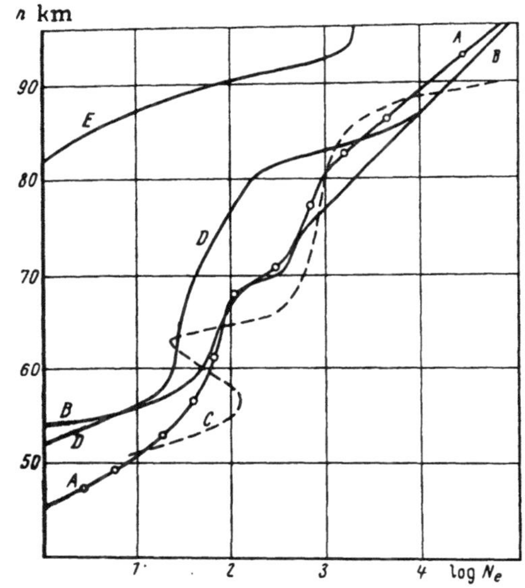

Fig. 5. Lower ionosphere profiles of Krasnushkin and Knyazeva [15]. See text for identification of curves.

of the state-of-the-art at that time, no profiles derived from propagation data were given, and consequently the work did not try to justify any of the results which had been obtained by the researches reviewed. At the same time, Gordon [17] referred to long waves as being a useful *monitor* technique, rather than a technique which *measured* parameters such as electron density and collision frequency. There has been considerable progress made in this field since Crain's review, and a number of papers have been presented at various scientific meetings: on steep incidence VLF/LF sounding by Hildebrand and his colleagues, by Moler and his colleagues in Arizona, by Helms using data obtained in the Antarctic, and by Lewis and his colleagues for propagation paths in the vicinity of Rome, N.Y. and at Natal, Brazil (at the geomagnetic equator). Most of these researches were known to the group participating in the Crain review, but they were, as they are now, essentially unpublished.

This paper will report on the present state-of-the-art of derivation of $N(h)$ profiles from VLF/LF propagation data (we will focus our attention on propagation distances < 1000 km), it will try to establish the reliability of the deduced profiles, and finally an attempt will be made to evaluate the potential of this probing technique and recommendations concerning the future of such work will be put forward. Before going into this, a brief summary of the methods used in calculating the reflection of long waves from the ionosphere will be given and some of the difficulties and uncertainties in the work will be discussed. These include the choice of a collision frequency model and the effective height of reflection which is needed to interpret oblique incidence propagation data.

2. Calculation of Long Wave Reflection Coefficients

The interaction of the lower ionosphere with waves having lengths of 3 to 30 km makes the use of full-wave theory a virtual necessity. Several mathematical formulations exist for the numerical evaluation of ionospheric reflection. These differ greatly in detail depending upon the ionospheric model or analogy from which the method is derived, and the physical problems which their authors sought to investigate. In all of the methods, however, the continuous ionosphere is replaced by a fictitious one consisting of thin horizontal slabs. The method of Johler and Berry [18—21] considers a plane wave incident on a horizontally stratified ionosphere from below. The ionosphere is assumed to be homogeneous within each of the horizontal layers which are chosen to be thin in comparison with the wavelength and with the structure of the electron density profile. The wave is mathematically decomposed into the two linearly independent magneto-ionic modes (Ordinary and Extraordinary). The two wave components are assumed to propagate independently within the individual layers. At each layer boundary, however, the reflection, transmission and mode inter-conversions for both upgoing and downcoming waves are calculated. The different wave contributions are suitably recombined to give the conventional reflection coefficients below the ionosphere.

The second method in common use today is that due originally to Budden [22] although many of the computer programs in actual use owe their parentage to Pitteway [23]. In this method, the usual magnetoionic modes are not considered. Instead, one considers those polarizations which, after traversing the ionosphere, would possess the greatest and the smallest relative intensities. These are called the penetrating and non-penetrating modes, respectively. Since the modes are defined in terms of their behaviour at great heights, the analysis must start there and proceed downward. The solutions are integrated numerically and may be combined linearly at the bottom of the ionosphere to give the usual vertical and horizontal polarizations. The choice of penetrating and non-penetrating modes is defended on the grounds that these are more closely related to the physics of wave propagation within the ionosphere than are the ordinary and extraordinary polarizations.

In practice, we have found the Johler-Berry program to compare unfavourably with that of Budden and Pitteway. Partly as a result of being written in a less sophisticated programming language, the Berry program requires far greater computer core storage and is several times slower than Pitteway's. More seriously, however, we have found, using several versions of Berry's program, that they occasionally predict physically impossible reflection coefficients.

Like Pitteway, Inoue and Horowitz [24, 25] were motivated by considerations of propagation processes in the ionosphere. These workers selected as the independent modes of propagation within the ionosphere those waves having fast and slow phase velocities in the upward and downward directions. Using this approach they studied the variation in wave polarization with altitude as well as the nature of the mode coupling. By rejecting the usual assumption of homogeneous slabs and in its place allowing the propagation matrices to vary linearly from one slab boundary to the next, they claim to have greatly increased the allowable slab thickness with a corresponding reduction in computation time.

In a formulation derived from optics, Altman and Cory [26, 27] sum the infinite series of internal reflection and transmission coefficients at the boundaries of each elemental slab. By initially restricting the analysis to vertical propagation and neglecting mode coupling at the slab boundaries, the program offered a considerable speed advantage over other techniques in cases where the model limitations were acceptable. The program was later generalized to include mode coupling and allow for oblique incidence [28].

Bjontegaard [29, 30], in a novel approach, has attempted to make allowance for contributions to the received signal from the entire reflecting layer and not just from the path mid-point. By means of an approach inspired by Huygen's principle he breaks the ionosphere into elemental regions and computes the reflected contribution from each element using conventional full-wave methods. He then adds the reflected contributions from all elements visible to both the transmitter and the receiver. Fig. 6 shows the results of such a calculation for "one-hop" and "two-hop" reflected waves for the electron density profile shown in the upper left-hand corner of the figure, compared with results obtained for reflection at the path mid-point. The calculations were made for a frequency of 20 kHz. It is seen that in this case the simpler geometrical model gives satisfactory results for distances up to 500—700 km. The approach, however, permits allowance to be made for variation in the ionospheric parameters in three dimensions and also permits account to be taken of non-isotropic characteristics of both transmitting and receiving antennas [30].

3. The Collision Frequencies employed in Full-wave Analyses

Since the laboratory measurements of Crompton, Huxley and Sutton [31] and Phelps and Pack [32] showed that in nitrogen the collision frequency is proportional to electron energy, various workers have investigated the effect this would have on the Appleton-Hartree magnetoionic theory, where collision frequency is treated as a real constant. Molmud [33], Shkarofsky [24] and Sen and Wyller [35] have dealt with the problem (the formulation of Sen and Wyller is the one most frequently used), and have given formulas for computing ionospheric refractive indices in which a parameter ν_m, the collision frequency for monoenergetic electrons (calculated from gas pressures), is employed. Such formulas can be directly used at HF, where ray theory can be employed. Deeks [36] tried to justify the use of an "effective" collision frequency ν_{eff} so chosen that it minimized the differences between wave fields and wave polarization calculated on a so-called generalized theory and the classical theory. He found that choosing $\nu_{eff} \approx {}^5/_2\nu_m$ where $\nu_m \ll |\omega - \omega_H|$ and $\nu_{eff} \approx {}^3/_2\nu_m$ where $\nu_m \gg |\omega - \omega_H|$ gave acceptable agreement although the generalized theory predicted slightly less absorption at all heights.

Johler and Harper [37] have considered the effects of energy dependent collisions on the reflection and transmission of radio waves from and travelling within the ionosphere, and have concluded that the reflection coefficients are not drastically changed by the introduction of energy dependent collisions. Horo-

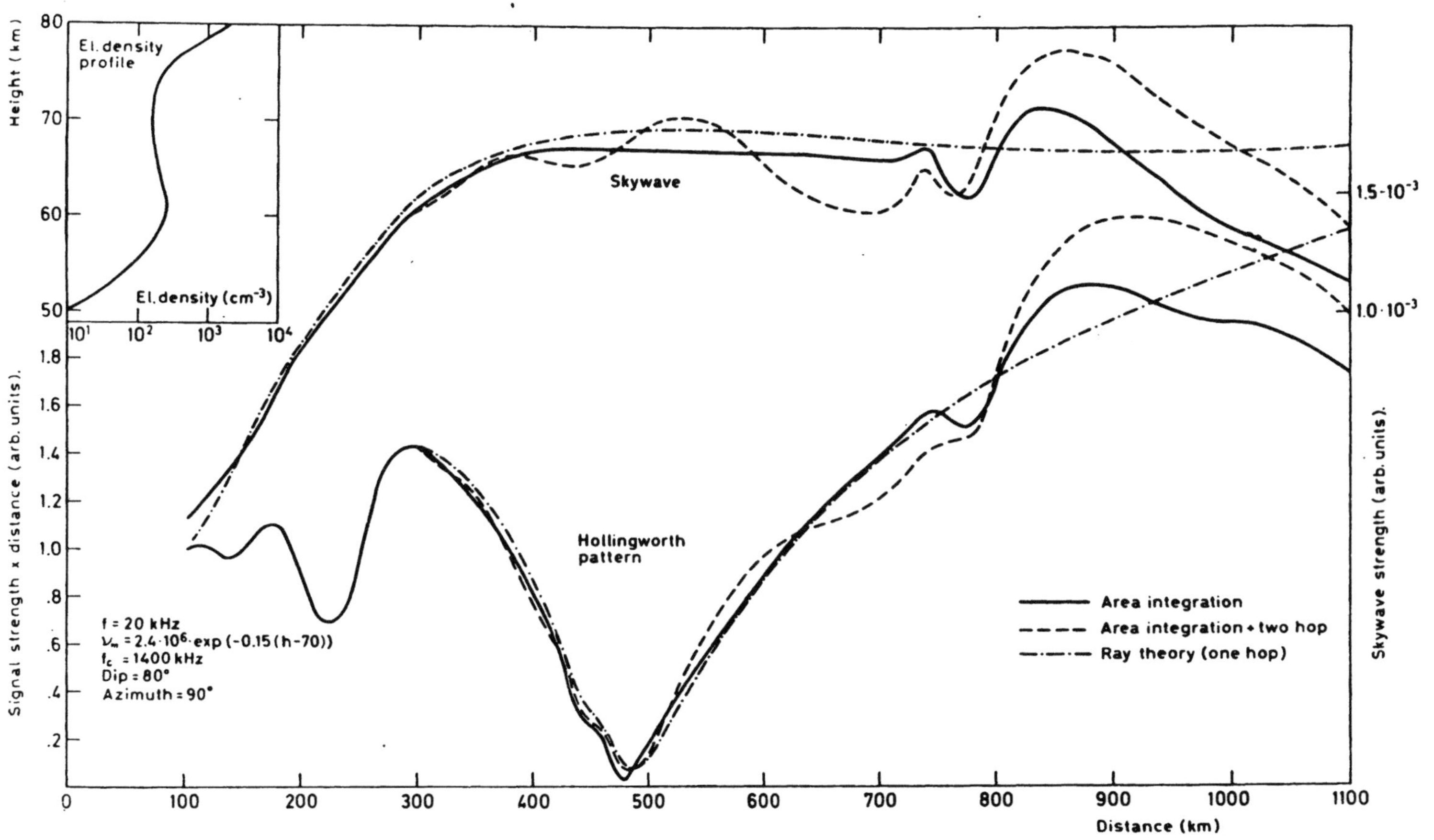

Fig. 6. 20 kHz field patterns computed using Huygen's principle. From Bjontegaard [31].

witz [38], on the other hand, claims that the reflection coefficients obtained using a generalized theory were as much as 25% greater than those obtained using an "effective" collision frequency. Budden [39] found that the use of $\nu = {}^5/_2\nu_m$ in the Appleton-Hartree formulas produced only small errors in the refractive index for frequencies down to 16 kHz. While he showed that for low heights there is a significant error in the wave polarization, he could not state that this would lead to errors in the calculation of reflection coefficients using full-wave

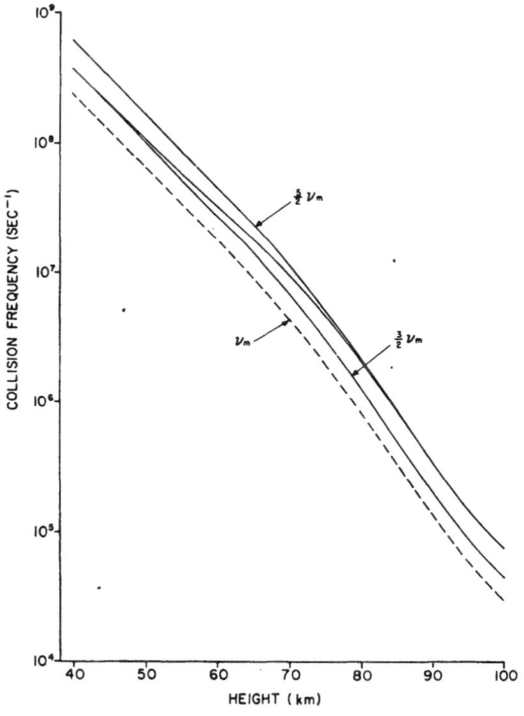

Fig. 7. Monoenergetic and "effective" collision frequencies.

methods. In fact, Budden [40] suggests that further refinement is probably not justified in practical calculations in view of the uncertainties in the collision frequency profile. He did, however, caution against the unsubstantiated application of ungeneralized theory to VLF propagation near the magnetic equator or at grazing incidence for east-west propagation at any latitude.

In a comparison between the reflection coefficients for two profiles calculated using generalized and non-generalized full-wave theory, Wand [41] showed discrepancies which increased with frequency although the use of a $\nu_{eff} = {}^3/_2\nu_m$ to ${}^5/_2\nu_m$ profile provided generally acceptable results.

Consistent with other workers, an effective collision frequency will be employed in the work to be reported, in which ν_{eff} is a curve that changes smoothly from ${}^3/_2\nu_m$ at low heights to ${}^5/_2\nu_m$ at high heights as shown in Fig. 7.

4. The "Effective" Height of Reflection

In analysing the propagation of very long waves over great distances, one generally considers the ionosphere and the earth's surface as forming the boundaries of a waveguide. For distances less than about 1500 km or for frequencies greater than about 100 kHz this approach requires consideration of large numbers of modes propagating in the earth-ionosphere waveguide. On the other hand, the problem is conveniently analysed in terms of a wavehop concept [20, 21] where the ionospheric reflection coefficients, the ground wave and the propagation path geometry are combined to yield field strength as a function of distance. Among the parameters required in this analysis is the "height" at which the waves are reflected. Allowing that full-wave theory must be employed in analyzing the reflection of long waves it is most unfortunate that in applying the results of such analysis to real problems one must associate the reflection process with some reference height in the ionosphere. There are various ways of defining an "effective" reflection height, each one giving a different result. One definition which was commonly used until recently was the "phase height" [42]. For plane polarized incident and reflected waves, this is the reference height for which the phase of the reflection coefficient is 180°. There are at least two objections to this definition. First, except at grazing incidence, there is no basis for associating the reflection process with a 180° phase reversal. Second, because of the cycle ambiguity in the phase of a wave, there is a corresponding ambiguity in the phase height. Berry and Chrisman [20] have suggested that the height at which the E-vector amplitude falls rapidly be adopted as the reflection height. In our experience, this works well only occasionally. Generally, the amplitudes decay too gradually to provide a satisfactory result. On other occasions the amplitudes oscillate or decay erratically at heights which, by simple inspection, could not be accepted as reflection levels.

The approach we have used is similar to Berry's in that it relates to the magnitudes of the field components in the ionosphere; however, it has never provided us with a value which has been unreasonable or physically unacceptable. What is done is to combine the E and H components to give the Poynting vector for the upgoing non-penetrating mode. The magnitude of this vector appears always to be well behaved and to decay fairly rapidly at reasonable heights.

Field and Engel [43], in studying nuclear perturbations of the ionosphere, used a criterion which gives the region where the rate of reflection of the upgoing wave is maximized. Crain and Booker [44] and Booker and Crain [45] use a phase integral approach to obtain reflecting heights related to but somewhat higher than those of Field and Engel. Wait [46] derived even higher reflection levels by increasing the height of a Gaussian perturbation until it had little effect on the reflected wave. The work of Crain [16] and Wait [46] demonstrates clearly what is intuitively evident: that reflection does not occur at a single height but over a region of some considerable extent.

A definition which appears to be gaining in popularity today is the triangulation height based on the principle of stationary phase [4, 11, 12]. This is a geometrical optic approach which is valid only where there is a single strong reflecting layer. Its application assumes that the phase at a fixed receiving antenna of the ionospherically reflected sky wave oscillates rapidly with varying angle of incidence except over a small range of angles for which it remains stationary. The use of computer techniques to search for an angle of incidence for which the rate of change of phase vanishes unfortunately obscures the fact that the necessary conditions stated above rarely hold true. The phase of the sky wave reaching the receiver (as computed by full-wave techniques) typically varies only slowly with angle of incidence, sometimes changing by less than one radian while the angle of incidence varies over as much as 10 degrees. For propagation out to distances of about 1000 km this variation may correspond to a height range covering the entire C, D and E layers. Thomas and Harrison [4] acknowledge that the stationary phase method breaks down when there is more than one angle for which the rate of change of phase vanishes; in fact, the technique is invalid *unless* the received phase changes rapidly over many cycles except for a specific angle of incidence. A further disturbing feature is that calculations are usually made for only one component of the reflected wave, the normal or the abnormal component (the components polarized in the plane and perpendicular to the plane of propagation). Except in rare cases these two components predict different "reflection heights". Some of these difficulties with the stationary phase technique will be further discussed in § 5.

5. Some Results of Numerical Calculation of the Reflection of Long Waves from Model D Region Electron Density Profiles

In order to gain a better insight into the reflection properties of the ionosphere, particularly during sunrise, we have done calculations [47] for both steep and oblique incidence employing a set of sunrise electron density profiles, shown in Fig. 8, which are similar to those measured by Smith et al. [8] at Armidale, Australia (30°36′ south geographic, $\Lambda = -39°$ magnetic latitude). The observa-

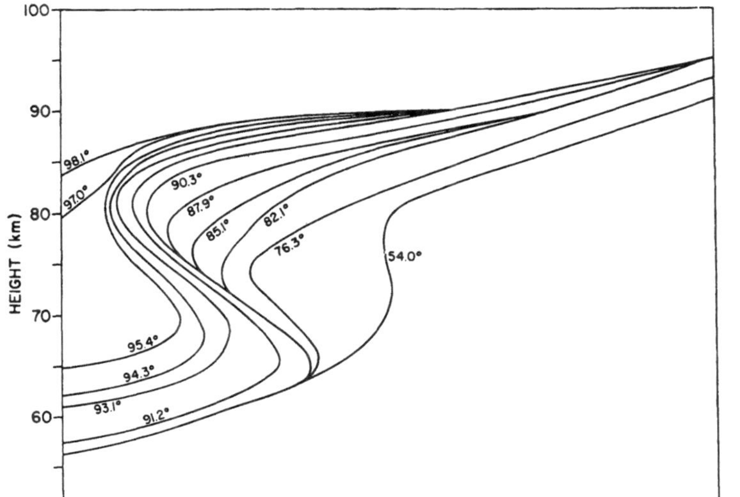

Fig. 8. Diurnal variation of electron densities after Smith et al. [8].

tions were made between June 1964 and January 1965. We have no long-wave propagation data at the appropriate latitudes for comparison. Our purposes are to see what can be learned about the reflection properties of the ionosphere at VLF and LF by numerically calculating the reflection coefficients and their phases, and to see how well these calculations agree with the overall features of LW propagation observations made at other latitudes.

5.1. Reflection of VLF Radio Waves at Steep Incidence

Using a full-wave integration technique, calculations were made for steep incidence reflection of 16 kHz waves. The calculations were made for an angle of incidence $\phi_i = 30°$ corresponding to the path Rugby-Cambridge. The magnitudes

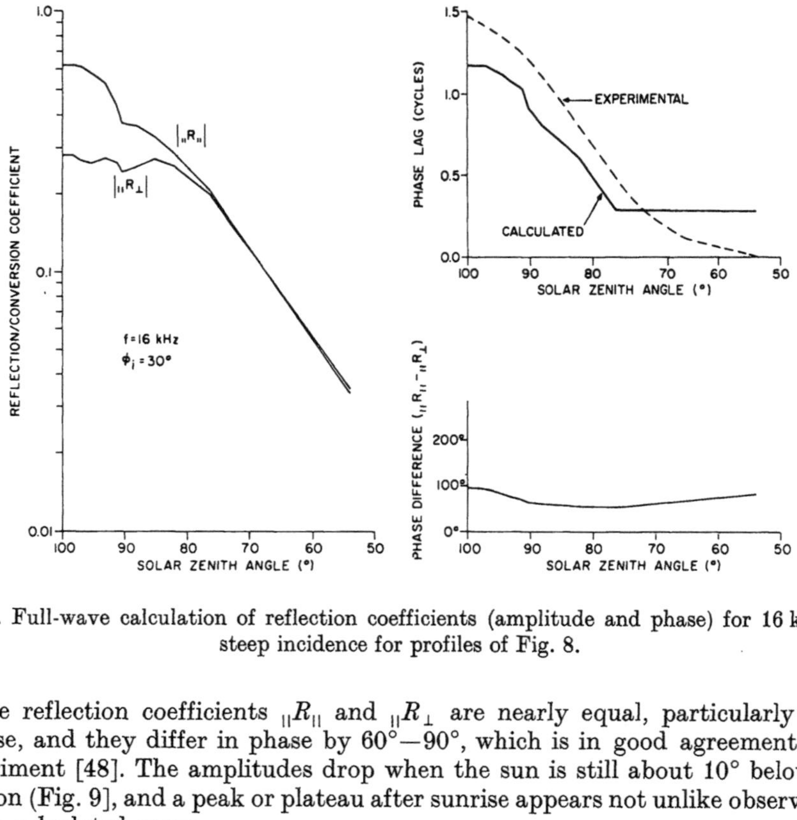

Fig. 9. Full-wave calculation of reflection coefficients (amplitude and phase) for 16 kHz at steep incidence for profiles of Fig. 8.

of the reflection coefficients $_{\parallel}R_{\parallel}$ and $_{\parallel}R_{\perp}$ are nearly equal, particularly after sunrise, and they differ in phase by 60°—90°, which is in good agreement with experiment [48]. The amplitudes drop when the sun is still about 10° below the horizon (Fig. 9], and a peak or plateau after sunrise appears not unlike observation in the calculated curve.

5.2. Reflection of VLF/LF Radio Waves at Oblique Incidence

Using a full-wave integration method and employing a stationary phase technique, calculations for $_{\parallel}R_{\parallel}$ were made for the oblique reflection of 16 kHz from the sunrise profiles. Figs. 10 to 13 give results of these calculations. Fig. 10 shows the angle at which waves should be incident on the lower ionosphere according to a stationary phase calculation, to reach a receiver at the distances shown. It is clear that there is no systematic relation between the various curves, as would be intuitively expected. One would expect that as the reflection height decreased during dawn the angle of incidence for a fixed distance would increase; in fact the angle of incidence for the distance range 500—1000 km first decreases and

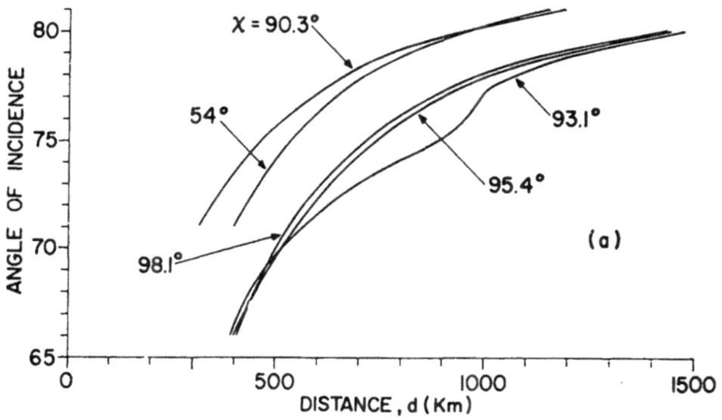

Fig. 10. Angle of incidence for 16 kHz propagation over given range for profiles of Fig. 8.

then increases. The profile for $\chi = 93.1°$ is distinctly different from the other curves. Fig. 11 shows the triangulation heights corresponding to these angles of incidence. According to a simple wave-hop model one would expect that the "reflection" height would be relatively independent of angle of incidence, or that this height would decrease as angle of incidence (distance) is increased. This is the situation for the night-time ($\chi = 98.1°$) and the daytime ($\chi = 54°$) profiles,

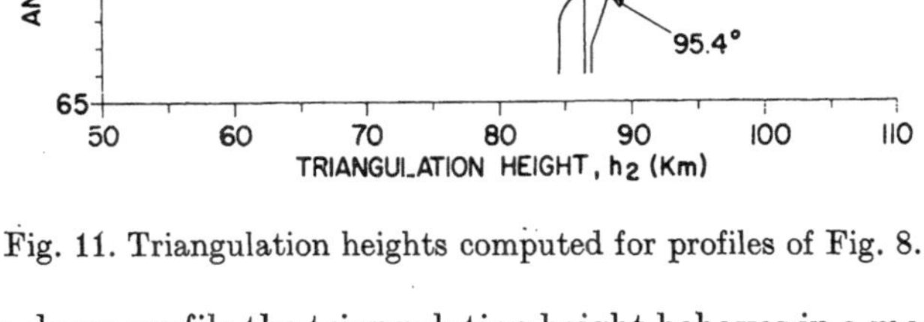

Fig. 11. Triangulation heights computed for profiles of Fig. 8.

but for the pre-dawn profile the triangulation height behaves in a most unexpected way. The curve corresponding to the $\chi = 93.1°$ profile gives triangulation heights which are quite unreasonable. The curve for $\chi = 90.3°$ suggests reflection heights less than those during daytime ($\chi = 54°$).

Fig. 12 shows that there is a minimum in the magnitude of the reflection coefficient for the $\chi = 93.1°$ profile for a certain range of angles of incidence. No such minimum is found for the conversion coefficient $_{\|}R_{\perp}$ which decreases, as expected, for increase in angle of incidence. For the triangulation heights shown, and the phase of the downcoming wave calculated by the full-wave programme, the effective phase change on reflection can be deduced, and these, for the various angles of incidence and profiles are shown in Fig. 13. The two profiles ($\chi = 93.1°$ and $90.3°$) that show peculiar triangulation heights also show rapid variation with

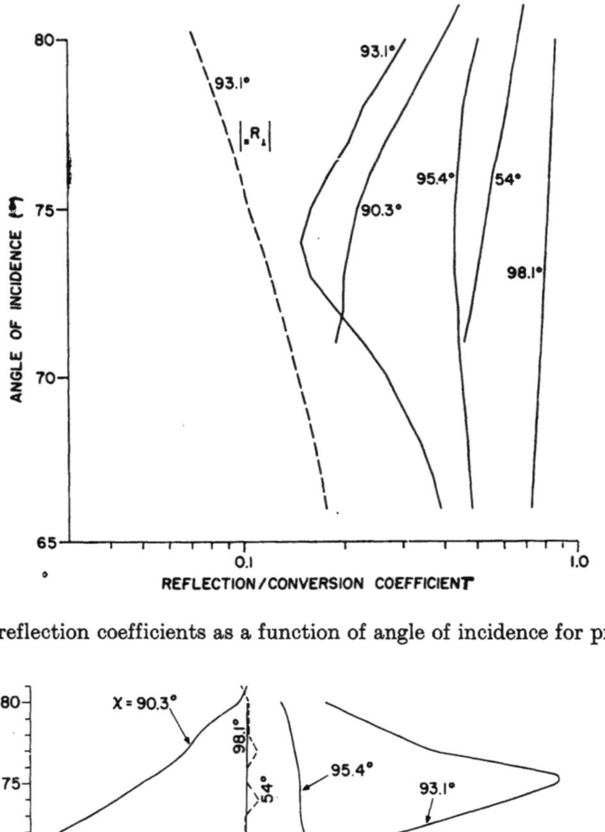

Fig. 12. 16 kHz reflection coefficients as a function of angle of incidence for profiles of Fig. 8.

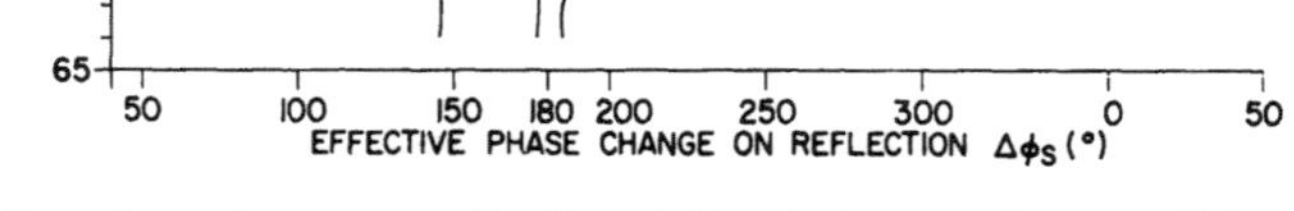

Fig. 13. Effective phase change on reflection at the stationary phase condition for profiles of Fig. 8.

changing angle of incidence of effective phase change on reflection. The night-time ($\chi = 98.1°$) and daytime ($\chi = 54°$) profiles are characterized by $\sim 180°$ phase change on reflection, which is in agreement with experiment [47].

Fig. 14 shows calculations for 70 kHz waves at near grazing incidence on the ionosphere ($\phi_i = 80°$) calculated for reflection from the $N(h)$ profiles of Fig. 8. The results show a complicated change of polarization and a deep minimum in the amplitude of the reflection coefficient during dawn; and a large polarization change between night-time and daytime. These facts are in reasonable agreement with experiment [47]; however, the polarization of LF waves obliquely reflected from the ionosphere is very difficult to measure.

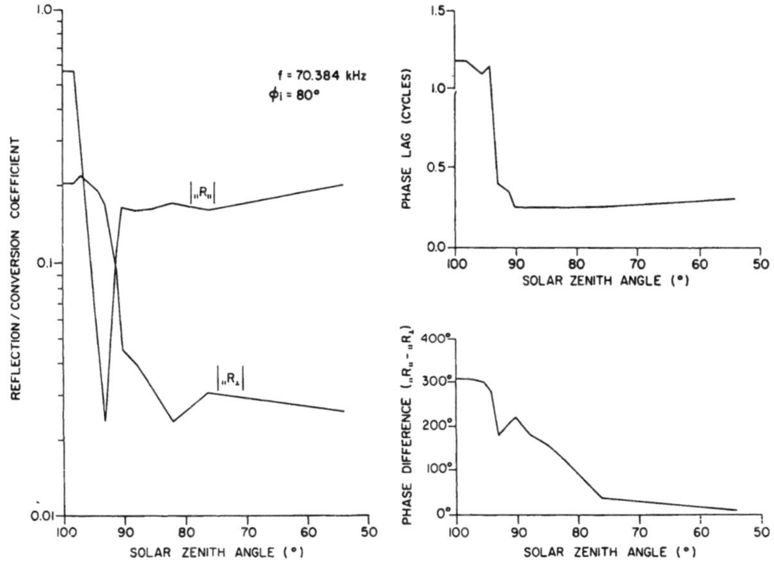

Fig. 14. 70 kHz reflection coefficients for profiles of Fig. 8 at oblique incidence.

5.3. Polarization Changes associated with a Solar Flare

At local noon on 1 March 1962 a solar flare occurred that resulted in a sudden ionospheric disturbance at the time observations were being made employing the partial reflection systems at Ottawa [49]. The changes in the reflection coefficients, their phases, and the wave polarization were calculated from the derived electron density profiles for 16 kHz at steep incidence ($\phi_i = 30°$) and 70.4 kHz at oblique incidence ($\phi_i = 80°$). At 16 kHz it was found that the conversion coefficient was reduced from 0.035 to 0.024, and the change in phase height of 10.2 km corresponds with the actual change in height of the base of the ionospheric profiles. The polarization changed from circular to linear. At 70.4 kHz the calculated reflection coefficient *increased* from 0.2 to 0.28, the calculated phase change amounted to 13.5 km and the polarization ellipticity increased from 7.7 to 25. All these facts are in reasonable agreement, at least in gross detail, with observation [47].

6. Recent Results on Electron Density Distribution from Long Wave Propagation Data

Although efficient techniques exist for accurately calculating the reflection coefficients or the received fields for given geophysical parameters, the inverse is not true. It is not possible, given a set of reflection coefficients, to compute directly the corresponding electron density profile. This must be done by trial and error, which, combined with the scarcity of reliable propagation data, explains the questioned reliability of $N(h)$ profiles deduced from LF/VLF measurements. Where a given experimental technique is to be applied regularly in a given way, some degree of automation is possible. Thus, Morfitt [50, 51] used tables of reflection coefficients for extensive sets of hypothetical profiles as an aid in analysing multi-frequency VLF propagation over the Hawaii to California circuit. Helms [52] proposes to establish similar sets of tables to help in the routine analysis of multi-frequency VLF soundings of the D region carried out near Byrd Station, Antarctica, during the period 1966—1970.

In general, however, one makes successive adjustments on the basis of personal judgement until a reasonable fit to experimental data is obtained. Thus Deeks [2] adjusted different portions of his profiles in order to fit measured values of reflection coefficients for various "effective" frequencies. Bain and May [11] made further adjustments to Deeks' profiles in order to bring them into agreement with the 16 kHz ground interference patterns of Budden et al. [53] and Weekes [54]. Bain and Harrison [12] proceeded in the same way to fit the lower portion of the profile to the ground interference pattern of Smeathers [13], while the upper portions were adjusted to fit other data.

What is perhaps not as well known is that Krasnushkin and his co-workers proceed in a completely analogous manner to derive their profiles [55]. The details of their analysis appear to be unnecessarily complicated. Below 65 km they derive the profile on the basis of ground interference patterns. They do this by deriving from the interference patterns the attenuation and phase velocities for the two lowest modes propagating in the earth-ionosphere waveguide. They then find the *linear* electron density profile which gives the best fit to these modal parameters [56]. It is not at all surprising therefore that their results do not show the low-altitude structure deduced by British workers [14]. Above 65 km they adjust an exponential model to optimize the fit to published steep-incidence LF reflection coefficients.

Shellman [57—59] uses a computer to analyse the sensitivity of steep incidence reflection data to perturbations at different levels in the electron density model. Starting with an exponential profile and proceeding through successive approximations, the profile is adjusted subject to a constraint which establishes a compromise between profile structure and goodness of fit.

We can conclude from earlier dicussions that profiles based on "averaged propagation data" are difficult to defend, firstly because the various data were not obtained with the same precision, were not observed at the same place or time, and secondly because the profiles cannot be compared with ones measured by other techniques. The following discussion attempts to better justify the use of VLF/LF propagation as a probing technique by using model and actual propagation data.

6.1. Analysis of Hollingworth (oblique incidence) Patterns

The observation of uncontaminated sky wave relatively close to a transmitter poses serious experimental problems. Nearly half a century ago, Hollingworth [60] outlined the method, which still bears his name, for studying the ionosphere by means of the CW propagation of long waves. The ground measured field pattern as a function of distance from the transmitter provides information regarding not only the amplitudes of the ground and sky waves but also their relative phase differences.

On 26 June 1965 the Royal Aircraft Establishment flew a jet airplane in a direct line from Ottawa, Ontario, to Churchill, Manitoba, a distance of some 1900 km. The aircraft acrried a low-frequency receiver set up to record the field strength of an 80 kHz CW transmission from a transmitter located at Ottawa. Partial reflection equipments were operated at both Ottawa and Churchill during the period of the flight which was near local noon and an attempt was made to compare the observed long-wave field pattern with the electron density profiles deduced from the partial reflections.

The first step in the analysis of the data consists of determining the effective ground constants for the path. Fig. 15 shows the observed interference pattern

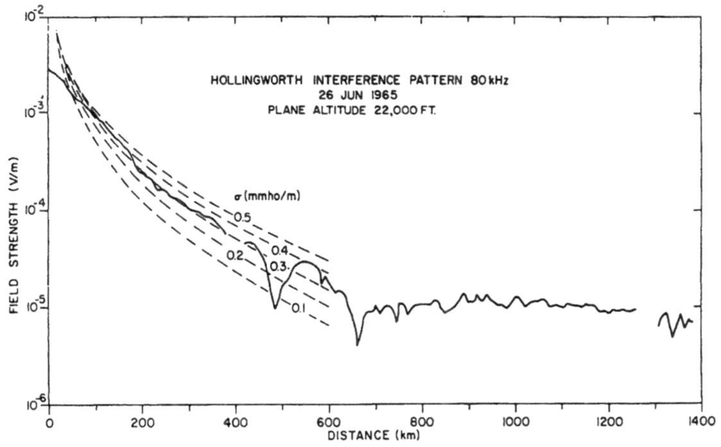

Fig. 15. 80 kHz Hollingworth pattern and ground-wave field strength curves.

out to 1400 km and theoretical ground-wave field strengths for various values of conductivity. The scale for the measured field strengths has been adjusted to correspond to the estimated 400 watts of radiated power. All computations were corrected for the 6.7 km (22000 ft) altitude of the aircraft. We see that a conductivity $\sigma = 3.0 \times 10^{-4}$ mho m^{-1} fits the observed data very closely between 90 and 450 km and this value was used in the subsequent computations. Considerable differences existed between the measured profiles at the two extremities and neither one could explain the observed Hollingworth pattern. Fig. 16 shows the computed fields for the Ottawa profile compared with the experimental one. Note that the standing wave patterns are almost completely out of phase with one another.

Starting from scratch an attempt was made to deduce a profile which would reasonably explain the observed pattern. By trial and error and a little judgment it took eleven attempts to arrive at the agreement with experiment seen in Fig. 17. The quality of fit is to some extent a subjective matter and others might undoubtedly have chosen slightly different criteria. We placed far greater emphasis on maintaining the locations of the maxima and minima than on precisely fitting the amplitudes which had greater uncertainty attached to them.

The deduced profile compared with the two obtained by partial reflection is shown in Fig. 18. The deduced profile is not shown above 70 km since that region influenced the sky wave only for steep incidence where the sky wave is swamped by the much stronger ground wave. Below about 67 km the field pattern becomes

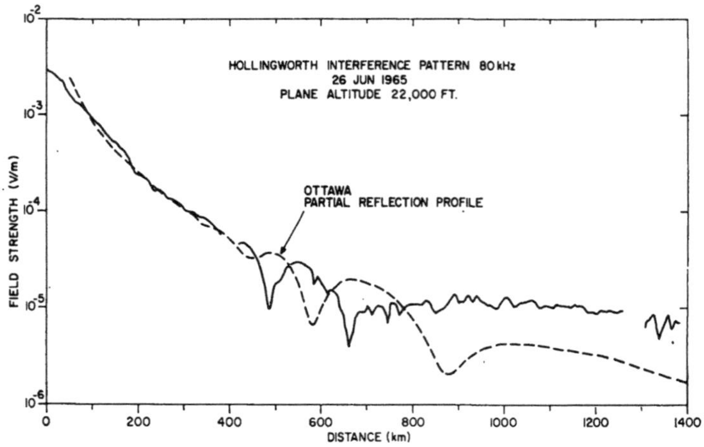

Fig. 16. Interference pattern calculated from partial reflection profile measured at time of aircraft flight.

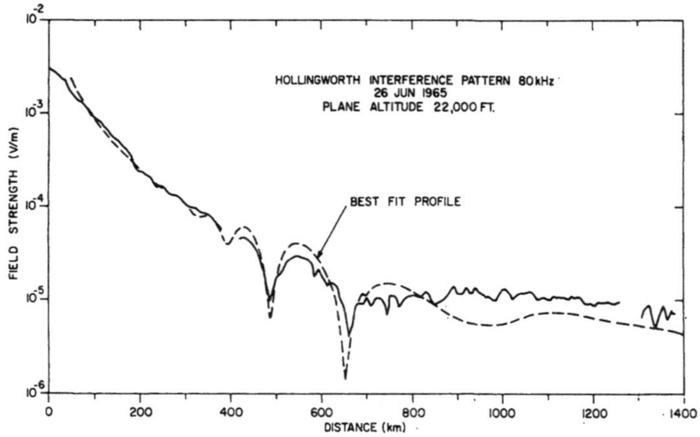

Fig. 17. Interference pattern calculated from profile giving best fit to observed data.

very sensitive to minor perturbations in the profile shape. For example, if the slope of the gradient near 64 km is altered by raising or lowering one end by 0.4 km, not only are the wave field amplitudes beyond 500 km altered but the locations of the minima in the interference pattern are shifted by up to 25 km. The ledge from 59 to 63 km influences the fields at distances greater than 650 km. A 10% change in ionization in this region will produce a 40% change in field strength at a range of 900 km. The sensitivity to perturbations below 59 km is somewhat reduced, but we are now looking at electron densities of less than

30 cm^{-3} and one cannot alter the structure significantly without producing noticeable changes in field strength. Nothing much can be said about the profile below 51 km. The sensitivity of the partial reflection experiment decreases as height and electron density decrease, and below a certain height (60 km in the case of the Ottawa profile) the curve (shown by a dashed line) is extrapolated in pro-

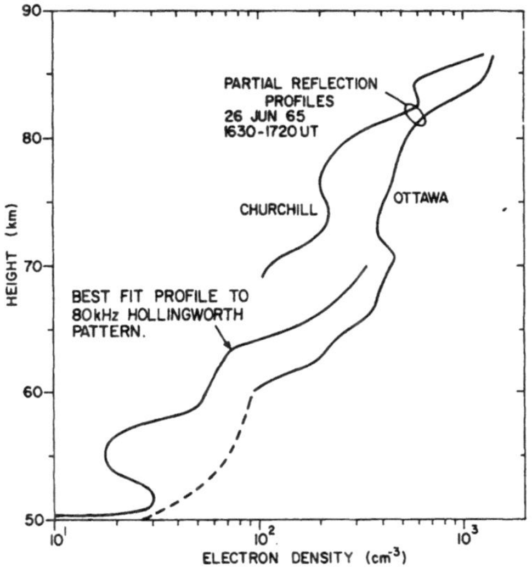

Fig. 18. Comparison of "best-fit" profile with partial reflection profiles measured at extremities of flight.

portion to echo amplitude rather than being computed for the two magnetoionic modes. It is concluded that long-wave propagation offers a more sensitive probe for studying the lowest region of the ionosphere (this will be later readdressed).

Bain and Harrison [12] have recently published the results of an attempt to analyse 16 kHz ground interference patterns measured by Smeathers [13] during the summer of 1969 and to compare the results with those obtained from a rocket flight [61] which took place on one of the days for which the Hollingworth pattern was measured. The rocket flight took place at 1500 UT while the Hollingworth patterns were measured at approximately 1100 UT so that a correction had to be applied to the rocket deduced profile in order to allow for the difference in solar zenith angles (44° and 30°). This had the effect of increasing the overall ionization somewhat but left unresolved the ionization below 68 km. The deduced profile shown by the continuous line in Fig. 19 is compared with the rocket profile shown by the broken line.

In order to fit the measured interference pattern, Bain and Harrison had to further increase the ionization between 67 and 70 km. They also added a pronounced C layer at 59 km which provided a sharp reflecting gradient above 64 km. Above 70 km the profile was modified so as to fit selected measurements made at Cambridge more than 20 years previously. There was thus at best only a 2 km height interval over which the deduced and measured profiles could be compared. No indication was given of the relative sensitivities of different height regions and one would really like to know exactly how much of the profile can be

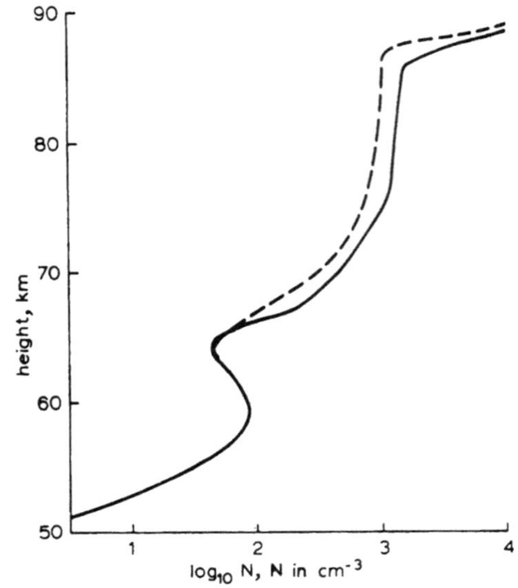

Fig. 19. Ionization profiles as measured by rocket (dashed curve) and as deduced from propagation data (solid curve). From Bain and Harrison [12].

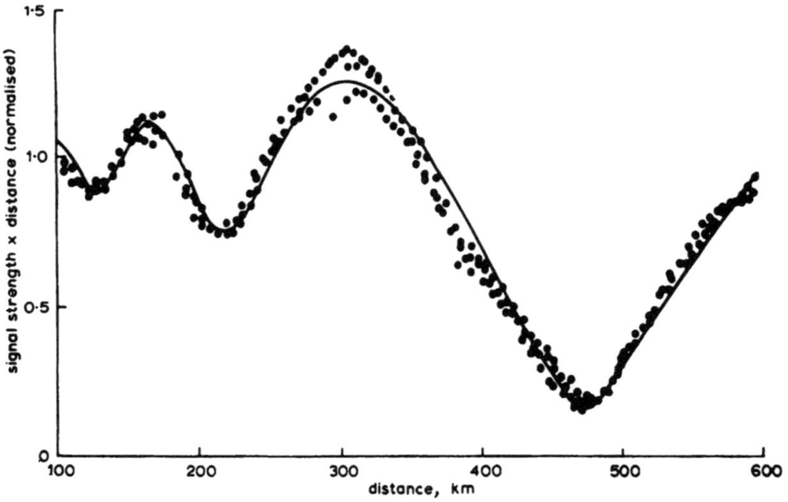

Fig. 20. Hollingworth pattern computed for profile of Fig. 19 compared with observations of Smeathers [13]. From Bain and Harrison [12].

supported on the basis of the Hollingworth pattern. One also wonders why the computed ground interference pattern (seen in Fig. 20) was compared with the measurements obtained over six days rather than with the pattern for the day in question. The calculated fields in fact agree somewhat better with those measured six days earlier. One also wonders why the data between 600 and 800 km were not taken into account since this would have permitted a more sensitive synthesis of the lowest portion of the $N(h)$ profile.

6.2. Medium Range, Multi-frequency

Another useful technique is to measure, as nearly simultaneously as possible, the reflection coefficients at a number of distinct frequencies in the LF/VLF bands [51, 62, 63]. On 16 September 1971 the LF transmitter at Ottawa was tuned sequentially to eight logarithmically spaced frequencies. Absolute field strengths were measured at distances of 20 km and 665 km. The former measurements were used to compute the radiated power which varied from less than 2.5 watts at 23.7 kHz to nearly 1.4 kW at 184.0 kHz. The signal at 23.7 kHz was

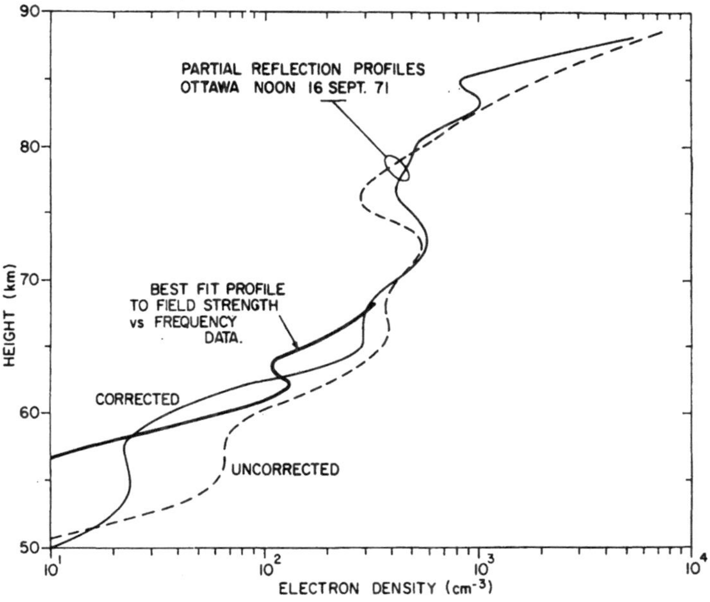

Fig. 21. Synthesized and measured profiles for 16 September 1971.

too weak at the receiving site to measure reliably on a standard field strength meter (Stoddard NM12AT), hence data from only seven frequencies were used in the analysis.

Starting with eight exponential profiles, theoretical field strengths were calculated for each of the seven frequencies. From this, we were able, in four trials, to fit the observed field strengths at the distant receiver with a mean error of less than 10%. This profile is seen in Fig. 21 along with two deduced from the partial reflection experiment at Ottawa — a preliminary one, and one deduced more recently with added corrections for the finite pulse duration. While there is a general agreement between the deduced best fit profile to the observed field strengths and the corrected partial reflection profile in the height range 57—67 km, the profile that best fits the propagation data contains structure in the 60—65 km height range not observed by the partial reflection sounder, and the partial reflection profile exhibits a ledge below 67 km that was not required to explain the propagation data. The presence of such a ledge at low heights in the D region

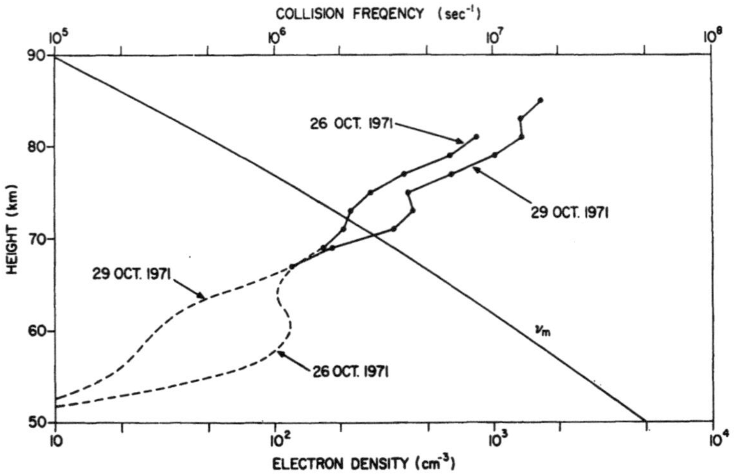

Fig. 22. Partial reflection profiles for selected days in October 1971. From Belrose et al. [64].

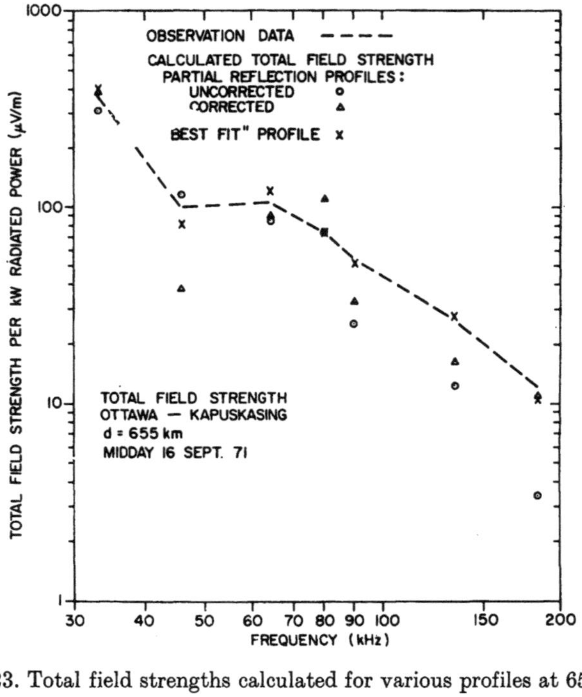

Fig. 23. Total field strengths calculated for various profiles at 655 km.

is an intriguing feature of the partial reflection data. Fig. 22 shows data for two days in October 1971, one showing a ledge and one not. The profile at these heights is shown by a broken line since it has been determined from the ordinary wave echo amplitude and not from the differential amplitudes which provide a measure of the electron density at greater heights. One would like to know how sensitive VLF/LF propagation data are to a lightly ionized ledge below 67 km, since while the magnitude of the electron densities within the ledge may not be correct, the observational data clearly show that there is ionization in this height region [64]. A further ten trials indicated that the agreement with observed fields could be improved very slightly if we added a gradient at the top of the profile shown.

Table 1

Normalized Field Strengths at 655 km calculated for Profile of Fig. 21

Frequency (kHz)	Ground wave field (μv m^{-1})	Sky wave field (μv m^{-1})
184.0	1.9	12.3
130.8	6.8	20.3
89.7	28.0	37.0
80.0	41.2	47.9
65.1	74.1	73.9
46.0	146.0	138.0
32.8	215.0	218.0

The field strengths measured and computed for the profiles just seen are shown in Fig. 23. The fields have been normalized to a constant radiated power of 1 kW at all frequencies. It is clear that the total field strengths calculated for the "best fit" profile better match the observed field strengths than do those calculated from the "corrected" partial reflection profile. Table 1 gives the calculated values for sky-wave and ground-wave field strengths for the "best fit" profile. The ground-wave field strengths were calculated for a ground conductivity of 3.0×10^{-4} mho m^{-1} as determined from the aircraft flight previously reported. The total field for $f > 100$ kHz at a distance of 655 km in middle Europe in September would be predominantly ground wave.

6.3. Analysis of Steep Incidence Data

The analyses used by various researchers to deduce electron density height profiles that are consistent with observational VLF/LF data have been based on a "change-and-try" procedure; that is, the parameters to be compared with experiment are calculated for an assumed $N(h)$ profile, these are compared with observation, the profile is altered, and the parameters recalculated. This procedure is continued until acceptable agreement is reached between calculation and experiment. While this seems at first to be a tedious process, the various workers have become adept in this method of analysis and, as we have seen, the deduced profiles show quite detailed structure, which each researcher claims is needed to explain the propagation data. The reliability and uniqueness of the deduced profile is, however, rather difficult to establish. A method of profile inversion for deducing D region electron density distributions from propagation data has recently been published [57, 58] and results from this inversion technique will be described.

It may be of some interest, however, at the outset, to show how Deeks arrived at his profiles. The initial reasoning was based on rather simple arguments which followed a suggestion of Piggott that partial reflections were significant from two levels in the ionosphere: where $X \approx Y$ and $Y \approx Z$ [65]. Typical values of $Y = Z$, and lines for $X = Y$ and $X = Z$ at 16 and 70 kHz are shown in Fig. 24. If we look at the spread of heights from which various workers think 70 kHz waves are reflected, and also the $X = Z$ line and $X = Y$ line for 70 kHz, we can draw a box in which the electron densities responsible for the main reflections at 70 kHz

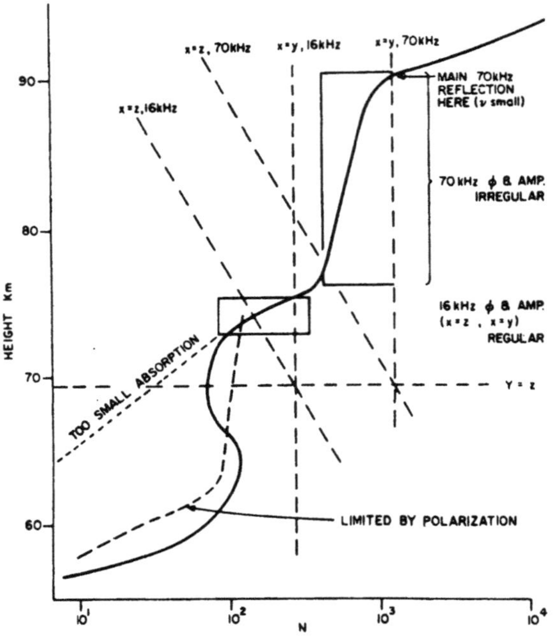

Fig. 24. Deduction of ionization profile from long-wave reflection data. From May [65].

must lie. We now transfer our attention to 16 kHz. The reflection height for the conditions Deeks considered is thought to be about 73 or 74 km. If we draw a box at this height, it contains both the $X = Y$ and $X = Z$ for 16 kHz. We now go downward in height. We cannot have the profile too far to the left because the absorption would be too low at both frequencies, however, we cannot have the profile too far to the right because 16 kHz waves would not penetrate up to 73 or 74 km, and we must not have the electron density gradient too great or the downcoming wave would be linearly polarized. The observational results, although not very accurate, are consistent with the downcoming wave at 16 kHz being circularly polarized. Obviously one cannot proceed too far with such arguments, but the simple model does provide some insight into the problem.

Krasnushkin and his co-workers at the USSR Academy of Sciences have claimed to be able to deduce uniquely the electron density profile from a full LF and VLF spectrum of steep-incidence reflection coefficients or its equivalent [15]. In an effort to explore these claims more fully, we initiated a co-operative programme with Professor Krasnushkin in mid 1970. Towards this end, we selected, quite arbitrarily, a smooth winter daytime profile which had been obtained from our routine partial reflection study of the ionosphere at Ottawa. We then computed the full-wave reflection coefficients which one would expect to result from such a profile for relatively steep (30°) incidence. Fig. 25 shows the spectrum so obtained. We were somewhat puzzled and concerned by the large amount of structure present on the curve; however, we sent the entire plot to Professor Krasnushkin, who assured us that the "wiggles" were of no importance and he supposed, as we do, that these "wiggles" were due to "nuances of an instan-

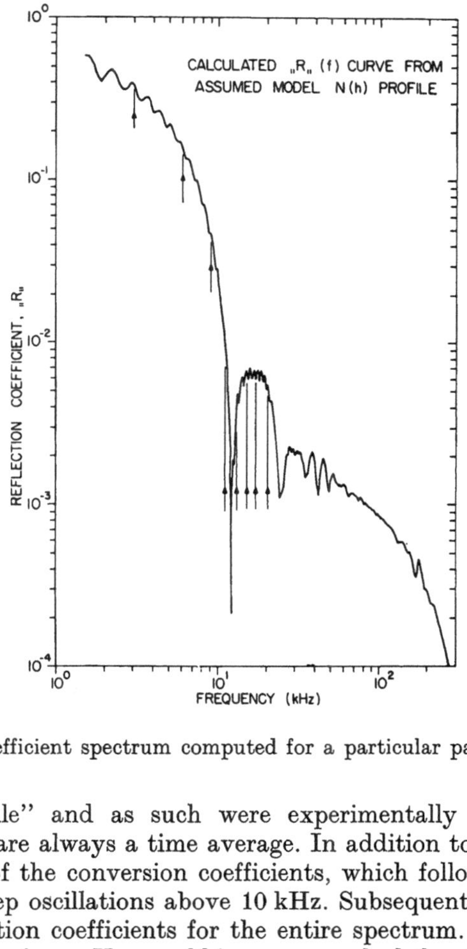

Fig. 25. Reflection coefficient spectrum computed for a particular partial reflection profile.

taneous $N(h)$ profile" and as such were experimentally unrealistic because observational data are always a time average. In addition to these data we sent him the spectrum of the conversion coefficients, which followed a similar curve but without the deep oscillations above 10 kHz. Subsequently, we sent him the phases of the reflection coefficients for the entire spectrum. Unfortunately, the co-operation with Professor Krasnushkin never reached the stage where we could state whether he could or could not deduce a reasonable fit to the original undisclosed profile.

In November 1972 we sent the same data (i.e. amplitude and phase vs frequency) plus the collision frequency model used to Dr. Shellman of the Naval Electronics Laboratory Center in San Diego. On the basis of data taken from these curves at eight frequencies in the range 3—20 kHz (the frequencies chosen are marked by vertical arrows on the curve), Shellman [58, 66] was able to produce the result seen in Fig. 26 which also gives some insight into the method of profile inversion used by Dr. Shellman. The initial or "starting" profile is the best-fit exponential labelled "0". Using a computer controlled linear perturbation theory, structure was gradually added to the profile as demanded by the propagation data until the final profile was obtained after 31 iterations. The deduced profile and the original model profile are shown in Fig. 27. Shellman estimated the fit to be $\pm 10\%$ between 58 and 75 km. Note the near perfect fit over this range and beyond. The height of the minimum was located to within ± 1 km. We believe that part of the discrepancy between the curves is due to the fact

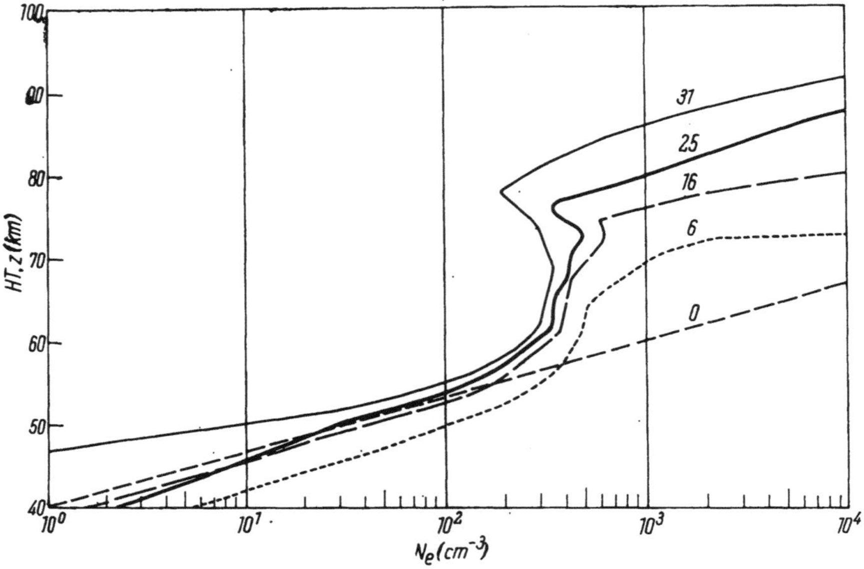

Fig. 26. Selected iteration steps in profile search using computed data (those for reflection coefficient are given in Fig. 25). From Shellman [66].

that Shellman did not start his profile search from a sufficently great height. Although most of the wave energy is reflected below 70 km, we often find it necessary—as we have here—to start the numerical integration at 100 km and 10^5 or 10^6 electrons cm^{-3}.

Being very much encouraged by these results, we attempted to extend the investigation one step farther. Having already synthesized the $N(h)$ profile needed to explain the 80 kHz Hollingworth interference pattern, we wished to see how well the multi-frequency VLF analysis could resolve the low-level profile structure. The profile deduced from the 80 kHz propagation data and a similar profile in which the variations below 65 km have been "smoothed out" are shown in Fig. 28. The complex reflection coefficients for these two profiles were then computed at ten spot frequencies ranging from 8 to 30 kHz using an exponential collision frequency profile (as used in the waveguide models of Wait and Spies [67]). Bear in mind that with both amplitudes and phases this represents a total of 40 data values for each profile. The next two figures show the results of Shellman's efforts [66]. For the smoothed profile (Fig. 29) we note the excellent fit over the range 53 to 87.5 km. Between 60 and 75 km the two curves agree to the accuracy with which one can read points on a graph. For the structured profile (Fig. 30) the fit is equally good above 53 km. It should be remembered that below

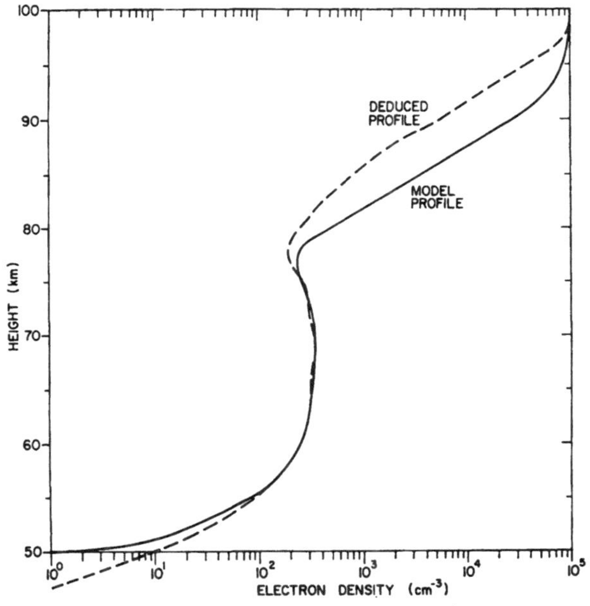

Fig. 27. Comparison of actual and deduced profile found from computed data.

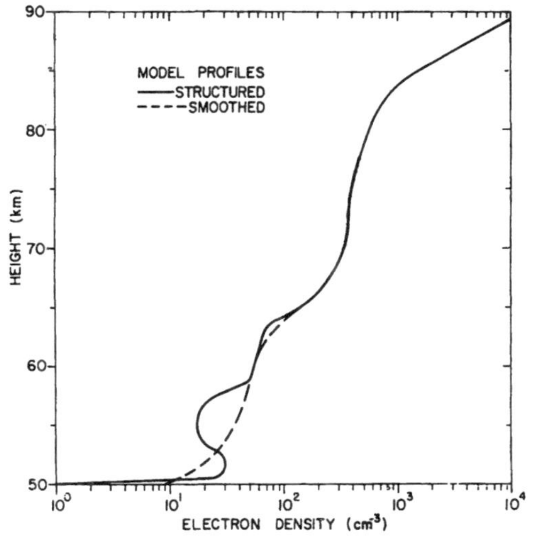

Fig. 28. Model profiles with and without low-level structure.

53 km there are less than 30 electron cm^{-3}. In addition to the phenomenally good fit to the electron density profiles, Shellman correctly deduced the collision frequency profile used. It is worth noting that in this analysis the greatest errors were those predicted by the error analysis. Fig. 31 shows the percentage probable error in electron densities realistically predicted by Shellman. The predicted error

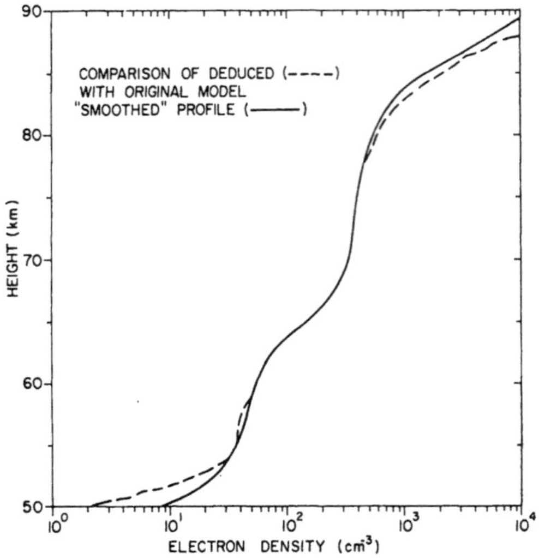

Fig. 29. Comparison of deduced profile with smooth model of Fig. 28.

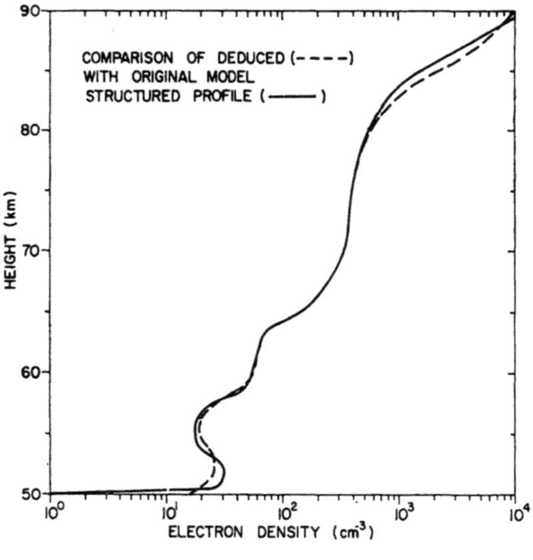

Fig. 30. Comparison of deduced profile with structured model of Fig. 28.

exceeds 10% except for a 13 km height range. On the basis of this prediction, the major structural differences between the two profiles of Fig. 28 would be unresolvable; indeed, Shellman felt at that time that the differences he had deduced were spurious (private communication, 1973). The actual profile errors, shown roughly in Fig. 31, are seen to be less than 10% over a height range exceeding 25 km. Similarly, large probable errors were predicted for collision fre-

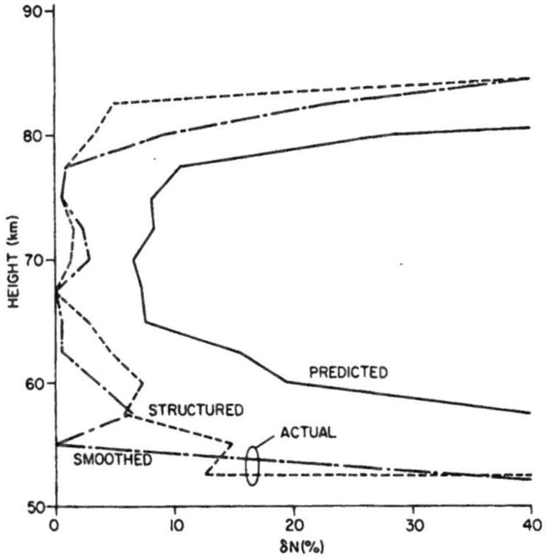

Fig. 31. Estimated probable error and actual errors for deduced profiles of Figs. 29 and 30.

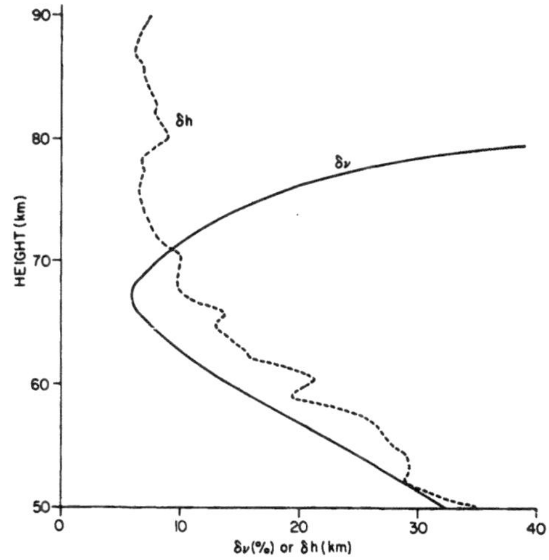

Fig. 32. Estimated electron density height resolution and probable error in collision frequency for profiles of Figs. 29 and 30.

quency and for height resolution as shown in Fig. 32. In fact, the height resolution was everywhere better than 1 km for both profiles and the collision frequency profile was as correct as one could determine from a graph in one case and in the other case tapered from a 3—4% error at 90 km to the correct value at 50 km.

Unfortunately, Shellman [58] was unable—at least in his early attempts—to successfully analyse some LF/VLF steep incidence measurements taken by Moler in Arizona. Here, even after allowing a displacement of some 7 km due to a phase error in the original data, the deduced "best fit" exponential collision frequency profile (Fig. 33) is about an order of magnitude too large. Note that in this case, Shellman was working from only 19 as compared with 40 data points. However, in view of the accuracy and precision with which Shellman was able to deduce our profiles from synthesized reflection parameters, we suspect that the origin of the errors in the case of the Arizona experiment more likely lies with the

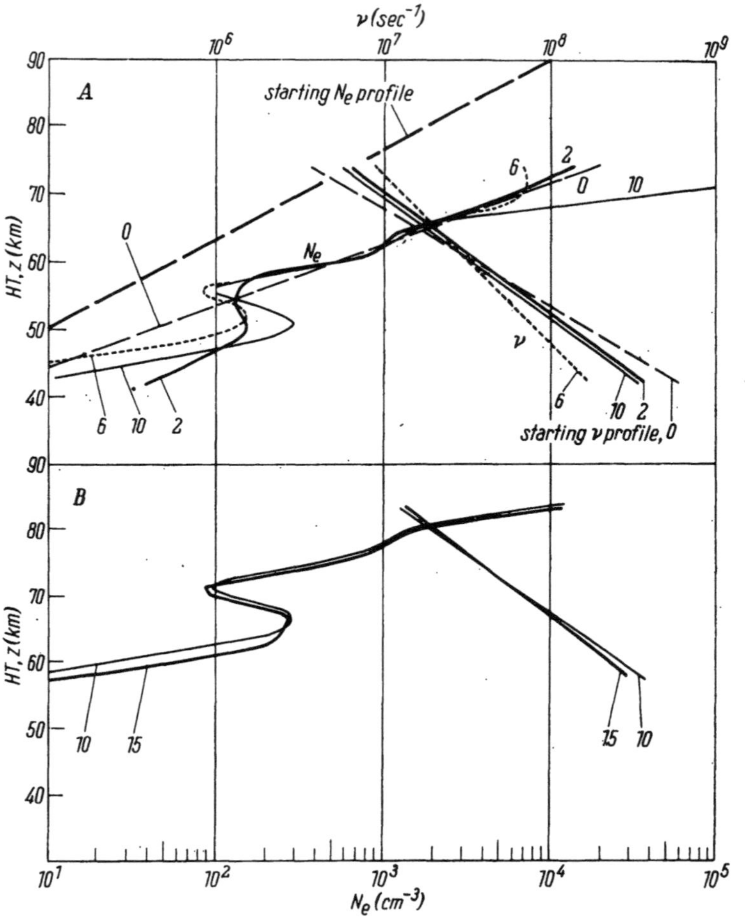

Fig. 33. Profiles deduced from actual steep-incidence observations. From Shellman [58].

original measurements than with the analysis. It is also worth noting that this was not a CW measurement. The observation of ionospheric reflections at vertical incidence uncontaminated by ground wave was made possible by the use of short pulses containing only a few RF cycles. It is also noteworthy that this profile, in accord with the profiles deduced by the British workers, exhibits a sharp minimum near 70 km, and while further investigations are needed to resolve the discrepancy, the qualitative feature of a minimum in the profile above a C layer is believed to be real [66].

7. Evaluation of use of Long Wave Propagation Data as a Probe to study the Ionospheric D Region

When Krasnushkin and Kolesnikov [68] in the USSR and later Deeks [36] in Britain published their $N(h)$ profiles based on accumulated LF/VLF propagation data they did not, in our opinion, produce any useful information concerning the ionosphere. It matters little that both groups claimed to have produced *the* electron density profile—both the Russian and British workers later altered their profiles quite significantly! The true significance of this work lay in the fact that after several decades of experimental long-wave observation, the time had arrived when one could hope to deduce structural details about the electron density profile in the lowest ionosphere from propagation data.

Some of the work presented here for the first time demonstrates that given suitable, precise propagation data it is possible to deduce profiles with great precision in the region of the ionosphere below that in which other ground-based techniques are most sensitive. Gordon [17] had expressed the hope that long-wave propagation measurements may give D region profiles. We believe the fulfilment of this wish to be at hand. There are several questions, however, which should be answered.

Firstly, there is the question of collision frequency model. Assuming agreement is reached on how the mean collision frequency profile behaves, there is no clear indication of when the use of generalized magnetoionic theory is necessary, nor of the magnitude of the errors which result from its neglect. There is about as much fragmented evidence to suggest that the energy dependence of collision frequency should be considered as there is evidence suggesting that errors resulting from the use of a suitably chosen monoenergetic collision frequency are negligible. Theoretical differences in the refractive indices or in characteristic wave polarizations must first be related to discrepancies in deduced $N(h)$ profiles before one can judge their importance.

We have seen that given sufficient suitable data, precise electron density information can be deduced. If the inversion of long-wave propagation data is to be done to any great extent, however, some indication is needed of how many or how few data are actually required and how the reliability of the deduced profiles is altered. Closely related, clear evidence is needed regarding the sensitivity and probable errors in $N(h)$ profiles deduced from propagation data. In the long run it is counter-productive to publish profiles which extend to heights where the sensitivity of the inversion process or the accuracy of the observational data will not support such extrapolations. On the other hand, it appears to us on the basis of our cooperative endeavours that Shellman may, quite possibly, be overestimating his probable error and underestimating his profiles' resolution—producing a picture more pessimistic than needs be.

Finally, we hope that the effort involved in converting propagation data to electron densities will not be wasted in utilizing different experimental observations which are unrelated either in space or in time. It is all too easy to take large amounts of data and have some deduced parameter fall within their combined scatter plot. If the scatter is large and representative of the individual data, one should be wary. If, on the other hand, the scatter is a result of temporal or spatial variation then one should analyze the data individually in order to reveal the underlying ionospheric processes.

One intriguing feature of profiles deduced from long-wave propagation data is the evidence for a ledge of ionization at heights below about 65 km. There is some evidence in other data that a ledge of ionization is to be found at these heights. It is interesting to note that one rocket-borne LF propagation experiment [69] showed evidence for such structure. Since we have concluded that long-wave propagation offers an unique and sensitive probe for studying the lowest regions of the ionosphere, we should look forward to future work giving evidence for ionospheric structure at these low heights.

The number of workers in this field is, however, small and diminishing—our own research program in this area is now terminating. This leaves a greater burden on those remaining to ensure a fruitful conclusion to their efforts. We look forward to their success.

References

[1] R. J. Nertney, J. Atmos. Terr. Phys. **3**, 92 (1953).
[2] D. G. Deeks, Proc. Roy. Soc. A **291**, 413 (1966).
[3] E. A. Mechtly, S. A. Bowhill and L. G. Smith, J. Atmos. Terr. Phys. **34**, 1899 (1972).
[4] L. Thomas and M. D. Harrison, J. Atmos. Terr. Phys. **32**, 1 (1970).
[5] J. S. Belrose and I. A. Bourne, in: Proc. of Conf. on Ground-Based Radio Wave Propagation Studies of the Lower Ionosphere, Def. Res. Board (Canada), 1967 (p. 79).
[6] J. E. Hall and J. Fooks, Planet. Space Sci. **13**, 1013 (1965).
[7] L. G. Smith, J. Atmos. Terr. Phys. **32**, 1247 (1970).
[8] R. A. Smith, T. N. R. Coyne, R. G. Loch and I. A. Bourne, in: Proc. Conf. on Ground-Based Radio Wave Propagation Studies of the Lower Ionosphere, Defence Research Board (Canada), 1967 (p. 335).
[9] J. S. Belrose, Nature **214**, 660 (1967).
[10] J. S. Belrose, in: Univ. of Illinois Aeronomy Rep. No. 48, 1967 (p. 132).
[11] W. C. Bain and B. R. May, Proc. IEE **114**, 1593 (1967).
[12] W. C. Bain and M. D. Harrison, Proc. IEE **119**, 790 (1972).
[13] J. R. Smeathers, Proc. IEE **118**, 515 (1971).
[14] W. C. Bain, J. Atmos. Terr. Phys. **31**, 755 (1969).
[15] P. Y. Krasnushkin and T. A. Knyazeva, Geomagn. and Aeron. **10**, 789 (1970).
[16] C. M. Crain, J. Atmos. Terr. Phys. **32**, 551 (1970).
[17] W. E. Gordon, J. Atmos. Terr. Phys. **32**, 459 (1970).
[18] J. R. Johler and J. D. Harper, J. Res. Natl. Bur. Stds. **66 D**, 81 (1962).
[19] J. R. Johler, Proc. Inst. Radio Engrs. **50**, 404 (1962).
[20] L. A. Berry and M. E. Chrisman, National Bureau of Standards Rep. No. 8889 (1965).
[21] L. A. Berry and J. E. Herman, Institute for Telecommunication Sciences Res. Rep. No. 11 (1971).
[22] K. G. Budden, Proc. Roy. Soc. A **227**, 516 (1955).
[23] M. L. V. Pitteway, Phil. Trans. Roy. Soc. London A **251**, 219 (1965).
[24] Y. Inoue and S. Horowitz, Radio Sci. **1**, 427 (1966).
[25] Y. Inoue and S. Horowitz, Radio Sci. **1**, 957 (1966).
[26] C. Altman and H. Cory, in: M. F., L. F. and V. L. F. Radio Propagation, IEE, London 1967 (p. 98).
[27] C. Altman and H. Cory, Radio Sci. **4**, 449 (1969).
[28] C. Altman and H. Cory, Radio Sci. **4**, 459 (1969).
[29] G. Bjontegaard, in: Proc. VLF Symp., Sandefjord, Norway, Norwegian Institute of Cosmic Physics (1972).
[30] G. Bjontegaard, to be published.
[31] R. W. Crompton, L. G. H. Huxley and D. J. Sutton, Proc. Roy. Soc. A **218**, 507 (1953).
[32] A. V. Phelps and J. L. Pack, Phys. Rev. Lett. **3**, 340 (1959).

[33] P. MOLMUD, Phys. Rev. **114**, 29 (1959).
[34] I. P. SHKAROFSKY, RCA Victor Research Lab. Rep. 7-801, Montreal (1960).
[35] H. K. SEN and A. A. WYLLER, J. Geophys. Res. **65**, 3931 (1960).
[36] D. G. DEEKS, J. Atmos. Terr. Phys. **28**, 839 (1966).
[37] J. R. JOHLER and J. D. HARPER, National Bureau of Standards Tech. Note 1964 (1963).
[38] S. HOROWITZ, Air Force Cambridge Research Lab. Rep. 71-0237 (1971).
[39] K. G. BUDDEN, J. Res. Natl. Bur. Stds. **69 D**, 191 (1965).
[40] K. G. BUDDEN, Alta Frequenza **38**, 1967 (1969).
[41] I. C. WAND, J. Atmos. Terr. Phys. **31**, 1233 (1969).
[42] W. R. PIGGOTT, M. L. V. PITTEWAY and E. V. THRANE, Phil. Trans. Roy. Soc. London A **257**, 243 (1965).
[43] E. C. FIELD and R. D. ENGEL, Proc. IEEE **53**, 2009 (1965).
[44] C. M. CRAIN and H. G. BOOKER, J. Geophys. Res. **69**, 4713 (1964).
[45] H. G. BOOKER and C. M. CRAIN, Rand Corporation Memorandum RM-5249-PR (1967).
[46] J. R. WAIT, Radio Sci. **69 D**, 865 (1965).
[47] J. S. BELROSE, in: AGARD Lecture Series XXIX (1968).
[48] R. N. BRACEWELL, K. G. BUDDEN, J. A. RATCLIFFE, T. W. STRAKER and K. WEEKES, Proc. IEE **98** (Pt III), 221 (1951).
[49] J. S. BELROSE and E. CETINER, Nature **195**, 688 (1962).
[50] D. G. MORFITT, Naval Electronics Laboratory Center Rep. TR 1798 (1971).
[51] D. G. MORFITT, Naval Electronics Laboratory Center Rep. TR 1854 (1973).
[52] W. J. HELMS, unpublished correspondence (1972).
[53] K. G. BUDDEN, J. A. RATCLIFFE and M. V. WILKES, Proc. Roy. Soc. A **171**, 188 (1939).
[54] K. WEEKES, Proc. IEE **97** (Pt III), 100 (1950).
[55] P. Y. KRASNUSHKIN, unpublished correspondence (1970).
[56] R. B. BAYBULATOV and P. Y. KRASNUSHKIN, Geomagn. and Aeron. **6**, 819 (1966).
[57] C. H. SHELLMAN, Radio Sci. **5**, 1127 (1970).
[58] C. H. SHELLMAN, Naval Electronics Laboratory Center Rep. TR 1856 (1973).
[59] C. H. SHEDDY, unpublished paper (1968).
[60] J. HOLLINGWORTH, J. IEE **64**, 579 (1926).
[61] F. D. G. BENNET, J. E. HALL and P. H. G. DICKINSON, J. Atmos. Terr. Phys. **34**, 1321 (1972).
[62] D. A. WULFING and V. E. HILDEBRAND, Naval Ordnance Laboratory Rep. No. 722 (1967).
[63] E. A. LEWIS, J. E. RASMUSSEN and P. A. KOSSEY, unpublished correspondence (1972).
[64] J. S. BELROSE, M. J. BURKE, T. N. R. COYNE and J. E. REED, J. Geophys. Res. **77**, 4829 (1972).
[65] B. R. MAY, in: Proc. Conf. on Ground-Based Radio Wave Propagation Studies of the Lower Ionosphere, Defence Research Board (Canada), 1967 (p. 656).
[66] C. H. SHELLMAN, private correspondence (1973).
[67] J. R. WAIT and K. P. SPIES, National Bureau of Standards Tech. Note 300 (1964).
[68] P. Y. KRASNUSHKIN and N. L. KOLESNIKOV, Aerospace Information Rep. 62-207, Library of Congress, Washington (1962).
[69] J. E. HALL, Planet. Space Sci. **21**, 119 (1973).

Lower Ionosphere Structure — Akademie-Verlag, Berlin 1974

DETERMINATION OF D REGION ELECTRON DENSITY DISTRIBUTIONS FROM RADIO PROPAGATION DATA

C. H. SHELLMAN

Naval Electronics Laboratory Center, San Diego, Calif., USA

A method of profile inversion is described which enables deduction of D region electron density and (exponential) collision frequency profiles given sufficiently accurate propagation data. Phase must be known as well as amplitude. Long path data, in the form of field strengths vs distance, may also be used. The profile is found by an iterative technique in which the total curvature in the profile is allowed to increase in steps to improve the fit to data. Each step requires a perturbation-type solution in which the full-wave solution of Budden is analytically differentiated with respect to perturbations in the profile, thus making a profile inversion scheme possible.

An error analysis is based on the mathematical equivalence of the smoothing technique to the use of smoothing filters, the widths of which correspond to the resolution in the profile. The iteration may be terminated when the proper trade-off between error width and resolution is obtained. The inversion technique has been tested by the use of "data" sets computed for profiles not revealed to the author. In one case the "data" was computed by a worker at another research facility. In each case the fit to the original profile was within limits predicted by the error analysis.

Lower Ionosphere Structure — Akademie-Verlag, Berlin 1974

RESULTS OBTAINED FROM ROCKET BORNE-FIELD STRENGTH MEASUREMENTS*

Y. V. Ramanamurty and K. Rawer

Arbeitsgruppe für physikalische Weltraumforschung, Freiburg, FRG

Using field strength records of terrestrial broadcast transmitters received by passive radio-frequency spectrometers in ascending rockets, and assuming a model profile for the collision frequency profile, the electron density profile is deduced up to about 87 km height. Ray theory of radio-wave propagation in an anisotropic and collision dominated plasma is applied. The measurements refer to daytime and the frequencies were 610 and 701 kHz. After correcting for the radiation pattern of the transmitting antenna, computations are made using both the generalized Sen-Wyller and the classical Appleton-Lassen dispersion formulae. The results of the computations exhibit very low electron density between 60 and 75 km as compared with model profiles reported in the literature.

1. Introduction

A method of determining electron densities in the D region is the swept-frequency radio spectrometer technique suggested by Rawer and his colleagues [1, 2]. The experiment is just to monitor the field strength of radio waves transmitted by ground-based broadcast LF and MF broadcasting transmitters by a receiver in an ascending rocket. The results presented in this paper refer to the earliest measurements of this type obtained in the first scientific payload launched in 1954 with a Véronique rocket from Hammaguir (31°N 2°W) in the Sahara desert. While the method of measurement [1] is straightforward, the reduction of the measurements to deduce an electron density profile of the lower ionosphere is difficult and complicated. We present in this paper some of our attempts in this direction and the results obtained so far.

Rawer and Argence [2] have outlined methods of attacking the present problem. The method we have adopted in the present investigation follows from their approach. We limit our present investigation to measurements on medium-frequency radio waves.

2. Interpretation of Measurements

Since we have to compare field strength at different altitudes we must first determine the vertical radiation pattern of the ground-based transmitting antenna at the frequencies used. For this purpose, classical methods are used [3, 4] taking

* The co-author wishes to dedicate his effort to Professor K. Rawer on the occasion of his 60th birthday.

account of the influence of ground losses with standard parameters. As expected, the results of calculation showed maximum transmitted radiation at a few degrees of elevation above the radio horizon. In the following, only data at altitudes above the height of maximum radiation at the launching place are considered for interpretation. Knowing the vertical radiation pattern, the measurements are corrected for the decrease in field intensity with increasing elevation angle. When the rocket enters the ionosphere, radiation reaching the rocket undergoes attenuation and refraction, unfortunately in a complicated way because the medium is anisotropic. The complexity of the problem arises from the fact that the two phenomena are interdependent. A solution is to be found by trial and error, i.e. by iterative methods.

The method of approach in an outline form is as follows. The ionosphere is divided into a number of slabs. First, the electron density in the first slab is given a value; the theoretically expected absorption index is then calculated using the generalized Sen-Wyller dispersion relation. Because of the anisotropic nature of the medium, the wave normal direction and the ray direction are in general not coincident in the ionosphere. To start with, the wave normal direction is identified with the ray direction in the first slab. The electron density in the first slab is changed until agreement is obtained between the calculated and measured absorption indices within a prescribed limit (in the present case about 20%). Having obtained this, the refraction effect of the ray in the anisostropic plasma is calculated using Haselgrove's ray tracing technique [5] applied to the present problem. Knowing the initial values of electron density, wave-normal directions and ray coordinates at the lower boundary, the differential equations relating the position coordinates of the ray and the wave-normal directions are numerically integrated to estimate their values in the upper slab. The process of matching the measured absorption index with the theoretically expected one in the lower slab and calculating ray coordinates and wave-normal directions in the upper slab is repeated up to about 87 km, the height up to which field strength data are available.

For the computation of absorption indices it is necessary to assume a collision frequency profile. We have used that given by Thrane and Piggott [6] for the appropriate season. In the height region of interest the numerical values are not much different from those used by Kane [7] in the analysis of rocket data obtained at low latitudes. Theoretical calculations are made using both the generalized Sen-Wyller and the classical Appleton-Lassen formula. Application of these two dispersion relations leads to different results. Budden [8] has pointed out that the difference can be minimized by adapting the collision frequency values used in both relations.

3. Results and Discussion

3.1. Preliminary model calculations not only indicated very small numerical values of electron density at certain heights but also showed pronounced maxima and minima. Model calculation of absorption indices for the two magneto-ionic modes showed that both components must be considered. Therefore polarization fading cannot be excluded and may have provoked minima and maxima of composite field strength at certain altitudes. Since we did not consider this aspect but interpreted field strength as if the ordinary component was alone present, the deduced electron densities would be in considerable error especially at heights where the two magneto-electronic modes are of comparable magnitude. For this reason we neglect points where the comparison would result in an electron density below 10^7 m^{-3}.

3.2. Fig. 1 shows the profiles of electron density obtained independently with measured values on 610 and 701 kHz (transmitters at Rabat and at Fez, both in Morocco). At local noon the solar zenith angle was about 41°. The results show

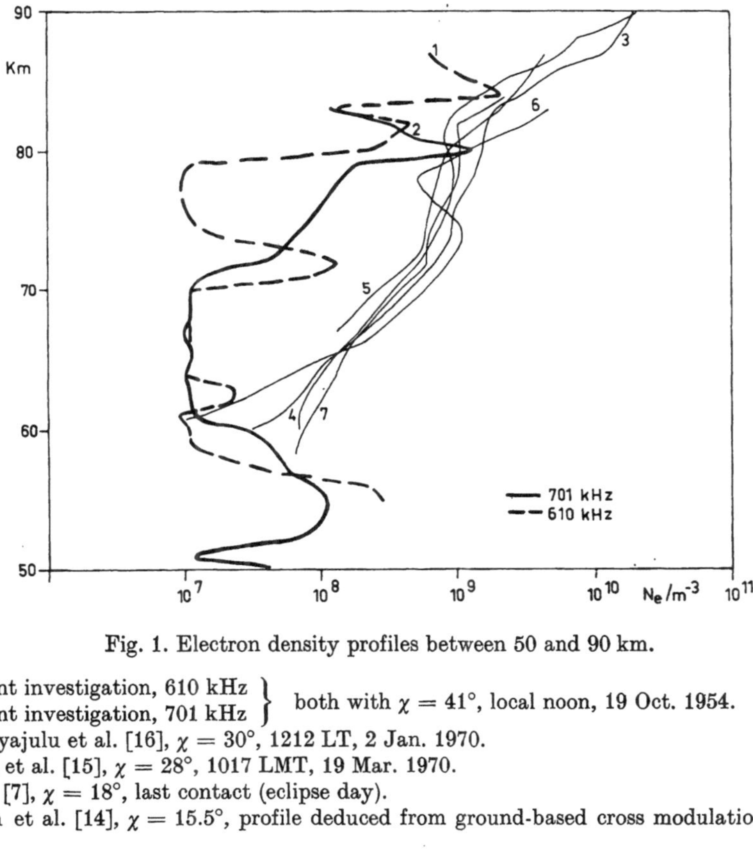

Fig. 1. Electron density profiles between 50 and 90 km.

1. Present investigation, 610 kHz } both with $\chi = 41°$, local noon, 19 Oct. 1954.
2. Present investigation, 701 kHz } both with $\chi = 41°$, local noon, 19 Oct. 1954.
3. Somayajulu et al. [16], $\chi = 30°$, 1212 LT, 2 Jan. 1970.
4. Aikin et al. [15], $\chi = 28°$, 1017 LMT, 19 Mar. 1970.
5. Kane [7], $\chi = 18°$, last contact (eclipse day).
6. Smith et al. [14], $\chi = 15.5°$, profile deduced from ground-based cross modulation experiment.
7. Kane [13], $\chi = 14°$, near local noon, 8 Mar. 1968.

good agreement among themselves up to a height of about 70 km. It must be emphasized that the experimental data input is incremental attenuation, so that the absolute transmitted power is not important. Since we get the input data after correcting for the (computed) radiation diagram this latter input is important at low elevation angles. Taking the results as they are we find rather large electron density near 55 km, followed by a minimum around 65 km. It is known that electron density in the lowest regions is produced by cosmic rays and in the year 1954, which was a quiet sun year, the production rate due to this source of ionization would be maximum. In view of this, the maximum near 55 km is interesting. There is some disagreement between 70 and 80 km where both curves come together again.

3.3. In Fig. 1 some profiles obtained by other authors who used other techniques are also shown (referring to low latitudes). These show substantially higher electron densities below 80 km. This might perhaps be due to the fact that the data refer to a sunspot minimum period and to a greater solar zenith angle. However, independent of the theories used, the very low densities found between 60 and 70 km are due to the fact that no serious decrease of field strength with height occurred there on either frequency. Since this was so an explanation by interference patterns is hard to accept*.

4. Limitations of the Adopted Method of Interpretation

The main limitations are the following:

(1) Haselgrove's ray tracing technique, which utilizes the real part only of the complex refractive index, is not suitable for the present problem although the collision term is retained in the present investigation. If ray tracing cannot be avoided it should be made in the complex plane [9—11].

(2) Our problem is one of oblique incidence. Precisely speaking the known dispersion relations are valid at vertical incidence only. One needs the vertical component of the propagation vector (***k***-vector) in addition to Bouguer's law which applies for the horizontal component only. For this purpose, a general complex form of the dispersion equation similar to Booker's quartic (which is valid for the unabsorbed Appleton-Lassen case only) has to be taken as the starting point. This approach should be preferable to our present one in which the dispersion relation in the form given correctly by Sen-Wyller [12] for vertical incidence propagation is used, and then the obliquity of the ray path is taken into account by noting that the direction of extinction (or attenuation) will be vertical when the stratification is horizontal.

(3) In the present investigation, only the predominant magneto-ionic mode (the "ionic" one) is considered, but the other mode should also be taken into account.

In a final interpretation of the results, all the above limitations have to be removed. Work is in progress in this direction.

5. Potentialities of the Experimental Technique

Since our technique makes use of the existing commercial broadcasting stations, no ground installation is needed for transmission. By using transmitters located at different distances and locations from the rocket launching station, cross checks could be obtained. Other cross checks could be obtained from co-ordinated ground-based propagation measurements (A3 absorption measurements, LF/VLF phase and amplitude measurements). The vertical field strength profile might be of practical interest to communication engineers.

The present technique using medium and low frequency radio waves has good sensitivity; it should allow detection of fine structure in the lower D region, since the local absorption coefficient at these frequencies is very high, so that electron

* Note added in proof: E. A. Lauter said in the discussion that this type of profile is typical for the seasonal transition period in late autumn.

densities of the order of 10^7 m^{-3} could be detected. Therefore, the technique appears to be very promising for study of the lower part of the D region which is difficult to explore with other techniques.

Acknowledgments

One of us (Y. V. R.) is grateful to the Alexander von Humboldt Foundation for the award of a Postdoctoral Fellowship during the course of the research work reported here.

References

[1] H. MENDE, K. RAWER and E. VASSY, Annls Géophys. **13**, 231 (1957).
[2] K. RAWER and E. ARGENCE, Space Research II, 1220 (1961).
[3] H. BREMMER, Terrestrial Radio Waves, Elsevier, Amsterdam 1949.
[4] K. RAWER, Taschenbuch der Hochfrequenztechnik, Springer, Berlin 1968 (p. 644).
[5] J. HASELGROVE, The Physics of the Ionosphere, Phys. Soc., London 1954 (p. 355).
[6] E. V. THRANE and W. R. PIGGOTT, J. Atmos. Terr. Phys. **28**, 721 (1966).
[7] J. A. KANE, GSFC Rep. No. X-615-69-499 (1969).
[8] K. G. BUDDEN, Radio Sci. **69D**, 191 (1965).
[9] K. SUCHY, J. Plasma Phys. **8**, 53 (1972).
[10] K. G. BUDDEN and P. D. TERRY, Proc. Roy. Soc. A **321**, 275 (1971).
[11] R. M. JONES, Radio Sci. **5**, 793 (1970).
[12] H. K. SEN and A. A. WYLLER, J. Geophys. Res. **65**, 3931 (1960).
[13] J. A. KANE, Planet. Space Sci. **17**, 609 (1969).
[14] R. A. SMITH, T. N. R. COYNE, R. G. LOCK and I. A. BOURNE, Proc. Conf. on Ground-Based Radio-Wave Propagation Studies of the Lower Ionosphere, Defence Research Board, Ottawa 1967 (p. 355).
[15] A. C. AIKIN, Y. V. SOMAYAJULU and M. B. AVADHANULU, J. Atmos. Terr. Phys. **34**, 1483 (1972).
[16] Y. V. SOMAYAJULU, M. B. AVADHANULU, K. S. ZALPURI and S. C. GARG, Space Research XI, 1131 (1971).

Lower Ionosphere Structure — Akademie-Verlag, Berlin 1974

THE USE OF VLF PROPAGATION MEASUREMENTS FOR STUDIES OF MAGNETOSPHERIC AND METEOROLOGICAL INFLUENCES ON THE LOWER IONOSPHERE

T. A. Potemra

Applied Physics Laboratory, The Johns Hopkins University, Silver Spring, Md, USA

Variations in the phase of stable-frequency midlatitude VLF transmissions are used to study the D region as it reacts to magnetospheric and stratospheric influences. Observations of night-time VLF phase disturbances associated with minor geomagnetic disturbances ($\sim 10\gamma$) are presented. These data support the view that small fluxes of energetic electrons ($E > 40$ keV) are often precipitated on the atmosphere and can provide a significant ionization source in the night-time midlatitude D region. Daytime VLF phase values for a midlatitude transmission path during an 8-month period are also presented showing high variability during the winter months, but relative stability during the rest of the year. Previous correlations of VLF phase values with atmospheric planetary waves suggest that this winter variability in VLF phase is related to the winter variability of atmospheric circulation. These data are offered as further evidence for the coupling of D region ionization to meteorological changes in the stratosphere.

1. Introduction

Long-distance VLF (3—30 kHz) transmissions have been used for many years to study a variety of ionospheric disturbances related to solar X-rays, solar energetic particles, aurora, and geomagnetic activity [1—10]. Recent claims have even been made for the detection of night-time ionization effects due to cosmic X-ray sources with VLF transmissions [11—14], although this is a controversial issue [15, 16].

More recently, satellite, rocket, and balloon measurements, as well as other ground-based radio transmission techniques such as riometers, have been used in connection with the VLF transmissions to make quantitative evaluations of D region ionization models, especially during polar cap absorption (PCA) events [6, 9]. The ionization effects of solar protons, electrons, and alpha-particles and the time varying cutoff region for PCA particles have been investigated in this manner [9, 17]. Auroral effects have also been extensively studied with VLF transmissions and are shown to be a more sensitive indicator of ionization enhancements than, for example, forward scatter HF transmissions [18].

The purpose of this paper is to present the results of recent analyses of two types of D region disturbances using long-distance VLF transmissions at middle latitudes (below $\sim 60°$ geomagnetic latitude). First, night-time disturbances to the phase of VLF transmissions associated with minor magnetic activity ($< 10\gamma$)

are presented which suggest that small fluxes of energetic electrons, >40 keV, are often precipitated on the night-time ionosphere. These data demonstrate the sensitivity of the VLF to very small D region disturbances. Second, variations in the daytime phase of cesium-controlled VLF transmissions are presented for an 8-month period in 1970–1971, which add to the evidence presently available for the influence of atmospheric planetary waves on the D region [19–27]. The data presented here demonstrate the potential use of VLF transmissions for analysis of magnetospheric and meteorological influences on the midlatitude D region.

2. VLF Transmission Paths

Fig. 1 shows some midlatitude VLF transmitter and receiver locations in the United States and Great Britain. Transmitter locations are indicated by the solid circles and labelled with their international designation. The Omega (Ω) trans-

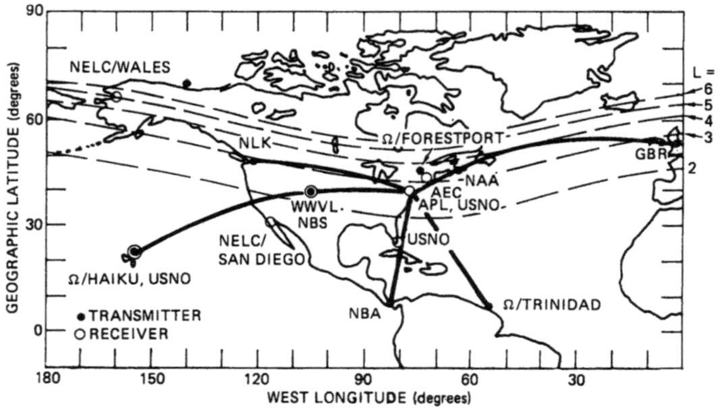

Fig. 1. VLF transmission paths monitored at APL near Washington, D.C. with projections of magnetic L shells at 100 km altitude [43]. The following receiving facilities are also shown: US Naval Electronics Laboratory Center (NELC) in San Diego, California and Wales, Alaska, US Army Electronics Command (AEC) at Fort Monmouth, New Jersey, US Naval Observatory (USNO) in Washington, D.C., Miami, Florida, and Hawaii, and National Bureau of Standards (NBS) in Boulder, Colorado. Transmitter specifications are listed in [39].

mitter at Forestport, New York, was closed on 29 November 1972. The receiver locations are denoted by the open circles and are in some instances located nearby transmitters, such as in Hawaii and Boulder, Colorado. For clarity, only the midlatitude paths monitored by the Applied Physics Laboratory (APL) near Washington, D.C. are indicated on this figure. The other receiving facilities shown in Fig. 1 also monitor most of the transmissions monitored by APL. Many transpolar VLF transmission paths, not shown in this figure, have been used in PCA and auroral studies [6, 28–30], for example, NLK to Thule, Greenland, Omega in Aldra, Norway, to the Army Electronics Command in Fort Monmouth, New Jersey. Fig. 1 illustrates the network of midlatitude VLF transmission paths that can be used for geophysical studies.

VLF transmissions are often analyzed as waves propagating in the waveguide formed by the earth's surface and the D region, at about 70 km altitude for undisturbed daytime conditions and at 85 km during the undisturbed night [18, 31]. The VLF phase change corresponding to a specific change of ionization in the D region has been treated in many past studies and the results will be applied in the following sections, but the details of these calculations will not be duplicated here [29, 30].

3. Night-time VLF Phase Disturbances associated with Minor Magnetic Activity

Fig. 2 illustrates a type of phase disturbance associated with small geomagnetic field changes often observed during the night. The horizontal, H, component of the magnetogram at Fredericksburg in Fig. (2 a) shows a small ($< 10\gamma$) but definite increase at 0605 UT on 5 January 1970. Phase advances, which can be explained by a lowering of the effective earth-ionosphere VLF waveguide height

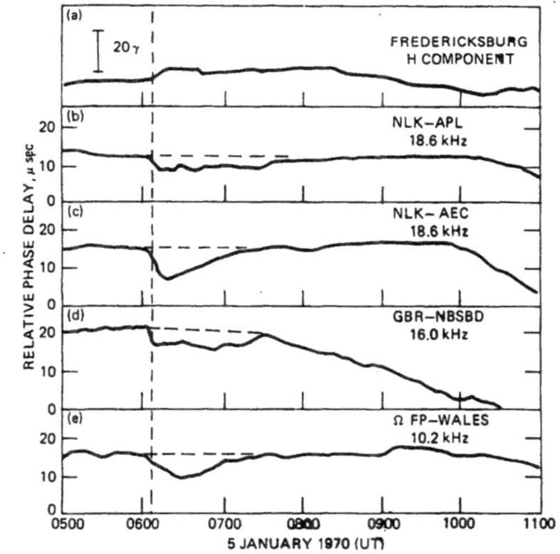

Fig. 2. (a) Horizontal, H, component of the Fredericksburg magnetogram and relative phase delay observed over the following VLF transmission paths, (b) NLK (Jim Creek, Wash.) to APL, (c) NLK to US Army Electronics Command (AEC) at Fort Monmouth, (d) GBR (Rugby, England) to National Bureau of Standards, Boulder, Colorado (NBSBD), and (e) Omega (Forestport, New York) to US Naval Electronics Center (NELC) in Wales, Alaska.

due to an enhancement of ionization, were observed on several different VLF paths beginning at the same time as the magnetic disturbance and are shown in Fig. 2 (b)—(e). Similar VLF phase disturbances, also associated with small magnetic disturbances, were observed at least on 20% of the night-time periods of VLF phase transmissions received at APL in 1970.

Phase disturbances similar to the one shown in Fig. 2 were detected on a number of midlatitude VLF transmissions during a substorm on 2 January 1971 [10]. The latter VLF phase disturbances occurred at the same time as enhancements in broadband VLF emissions and bremsstrahlung X-ray observations at Siple Station, Antarctica ($L = 4$, 60° geomagnetic latitude). These data were presented as evidence for the precipitation of energetic electrons ($E > 40$ keV) at midlatitudes due possibly to a cyclotron resonance between energetic trapped electrons and whistler-mode VLF waves. This event was part of a very complex substorm involving many other magnetospheric and ionospheric phenomena and is still under investigation [32].

The similarity of the VLF disturbances shown in Fig. 2 to those studied during the 2 January 1971 substorm, and the association of these disturbances with magnetic activity, suggest that ionization enhancements due to energetic electron precipitation is the cause of the former VLF phase disturbances. This suggestion is supported by many studies on the importance of precipitating energetic electrons to the midlatitude lower ionosphere [33—36], but, in particular, by a recent series of rocket measurements at South Uist, Scotland ($L = 3.5$, 58° geomagnetic latitude) [37]. Fluxes of energetic electrons > 40 keV were measured during these rocket flights and were observed to be highly variable, ranging from $J(>40 \text{ keV}) = 4$ electrons $\text{cm}^{-2}\,\text{s}^{-1}\,\text{sr}^{-1}$ during magnetically quiet conditions to $J(>40 \text{ keV}) = 300000$ electrons $\text{cm}^{-2}\,\text{s}^{-1}\,\text{sr}^{-1}$ during a magnetic storm. Analysis of these rocket data indicate that the precipitating energetic electrons are a dominant source of the night-time D region on $15 \pm 11\%$ of the nights at solar minimum rising to $35 \pm 20\%$ of the nights at solar maximum.

It is difficult to estimate accurately the electron flux necessary to produce an observed VLF phase disturbance such as shown in Fig. 2, because of a number of uncertainties, including the relative portion of the VLF path affected and the effects of second-order VLF modes. However, if previous calculations [10] for the NLK-ALP path are applied here, and if it is assumed that the entire transmission path is uniformly disturbed and that only the first-order VLF mode is important, the following estimates can be made. The 5 µs maximum phase disturbance to the NLK-APL path in Fig. 2 (b) can be produced by a range of electron fluxes with energy spectra of the form $J(>E) \sim \exp - (E/E_0)$ from $J(>40 \text{ keV}) = 20$ electrons $\text{cm}^{-2}\,\text{s}^{-1}\,\text{sr}^{-1}$ for $E_0 = 100$ keV to $J(>40 \text{ keV}) = 120$ electrons $\text{cm}^{-2}\,\text{s}^{-1}\,\text{sr}^{-1}$ for $E_0 = 30$ keV. These fluxes are within the range expected to precipitate from compilations of rocket and satellite data [35, 36]. The VLF transmissions are available as a continuous measurement of the ionosphere (in comparison, for example, with the rocket observations) and are more sensitive to these small electron fluxes than riometer absorption or bremsstrahlung X-ray measurements [10, 29].

Preliminary work [38] indicates that observations of disturbances to several VLF transmission paths (such as the network in Fig. 1) can be used to investigate the spatial extent of these events during different magnetic conditions.

4. Daytime VLF Phase Variations associated with Atmospheric Planetary Waves

Fig. 3 shows the daytime relative phase for the GBR (Rugby, England) transmission received at the US Naval Observatory (USNO) in Washington, D.C. during the 8-month period from 1 September 1970 to 30 April 1971. The magnitude of a 10 µs phase change (equivalent to a 67° phase shift) is indicated in this figure. A 10 µs phase avance may be caused by a uniform 5.5 km lowering of the daytime VLF waveguide height near 70 km [27]. The NLK transmitter oscillator and USNO receiver oscillator were both under control of cesium atomic standards and the advantage of this is demonstrated in Fig. 3. During the 8-month period shown in this figure, the average daytime phase has drifted approximately 20 μs (equivalent to less than a half-cycle), which represents a frequency stability of better than one part in 10^{12}. Long-term oscillator stabilities of this magnitude

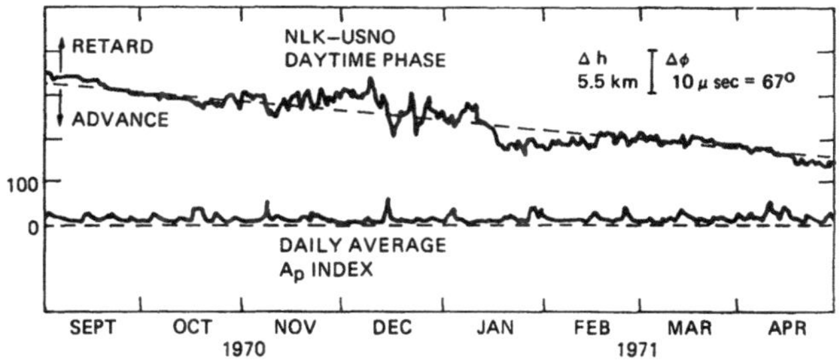

Fig. 3. Daytime (local noon at Washington, D.C.) phase received at the US Naval Observatory (USNO) from NLK (Jim Creek, Washington, 18.6 kHz) and the daily equivalent planetary amplitude, *Ap* [44].

were not available before the use of cesium atomic standards and severely limited the use of VLF transmissions for analysis of seasonal variations [39].

The NLK-USNO daily phase is highly variable during the winter months (varying as much as 10 μs in some cases) as shown in Fig. 3, but shows relatively small variations during the spring and autumn months (and also during the summer months not shown here).

The daily equivalent planetary amplitude *Ap* [40] is adopted here as the daily over-all index of magnetic activity and is also plotted in Fig. 3. Although there are clear examples of phase advances (due presumably to a lowering of the VLF waveguide height) associated with severe enhancements in the *Ap* index, the seasonal VLF phase variation does not appear to be entirely related to the magnetic activity. The winter variability of VLF phase in Fig. 3 is reminiscent of meteorological observations which show that atmospheric circulation is much more variable in winter than in summer [22].

Many observations support the view that ionization variations in the D and E regions are coupled to meteorological variations in the stratosphere [19, 41]. These observations have been almost exclusively determined from ground-based MF and HF (>1 MHz) radio transmissions and pressure and temperature variations in the stratosphere [20—26]. The possible influence of atmospheric planetary waves on VLF phase variations has been recently studied by Cavalieri et al. [27] and a figure from this analysis is presented in Fig. 4. This figure shows the running 5-day average daytime phase observed for the GBR (Rugby, England) to APL transmission, the height variations of the 10 mb isobaric surface at 50° N geographic latitude and 90° W geographic longitude, and the daily equivalent planetary amplitude *Ap*. The GBR-APL phase variations in this figure are correlated with the 10 mb geopotential height variations when the latter is delayed by about 14 days (correlation coefficient = 0.67). Good correlations of this same VLF data were found with 10 mb variations at several other longitudes (with different delays) [27], with height variations of a constant electron density surface in the E region (determined from ionograms [22]), and with f_{min} variations determined from absorption measurements [25, 26]. The effects of the PCA and magnetic activity in Fig. 4 was shown to have a small

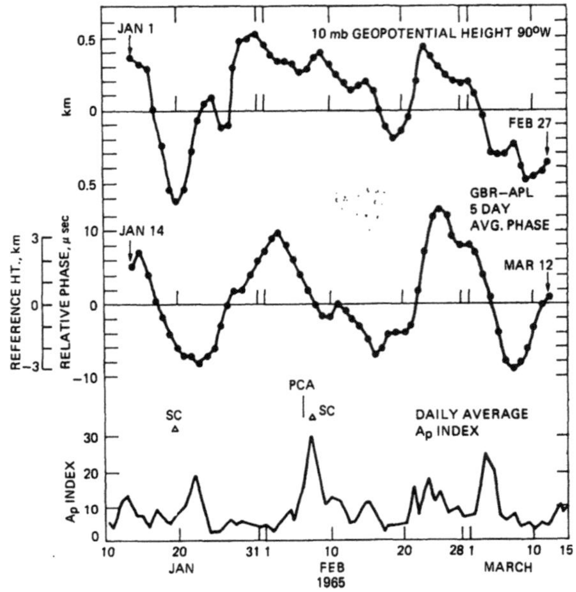

Fig. 4. The 10 mb geopotential height corresponding to the sum of the first three Fourier harmonics at 50° N geographic latitude and 90° W geographic longitude, the running 5-day average of the daytime relative phase delay observed for the GBR-APL transmission path and the daily equivalent planetary amplitude *Ap* [44]. Taken from Cavalieri et al. [27].

effect on long-term variations in the GBR-APL phase in comparison with the planetary waves. This combination of data provided evidence for the existence of a westward-tilted eastward-travelling planetary wave during the first three months of 1965.

The results of the study by Cavalieri et al. [27] suggest that some of the winter variability in the NLK-USNO phase shown in Fig. 3 is due to atmospheric planetary waves. The same network of VLF transmission paths described earlier can be used for further studies of the stratosphere-ionosphere coupling. For example, the VLF phase for paths directly south of APL (NBA to APL and Ω/Trinidad to APL in Fig. 1) show relatively small variations during the entire year and support the suggestion for a geographic latitude "cut-off" for the D region variation due to planetary waves [42].

5. Conclusions

The use of long-distance VLF transmissions as an ionospheric probe is limited by the fact that localized disturbances (small in spatial extent compared with the path length) are difficult to interpret. But, as shown here, the phase variations of cesium-controlled midlatitude ($<60°$ geomagnetic latitude) VLF transmissions can be very effective for the study of short-period ($\sim$ hours) night-time disturbances in the ionosphere that are associated with small magnetic disturbances ($\sim 10\gamma$) and of long-period (seasonal) planetary-scale variations. The threshold electron fluxes that can disturb the VLF are below the night-time sensitivity of most other detectors of energetic electrons including riometer absorption and bremsstrahlung X-ray measurements. The VLF transmissions are available in large networks as a continuous monitor of the D region, in comparison with short-period rocket observations. Also, the VLF is affected by a smaller altitude range of ionization in the D region, in comparison with MF and HF absorption measurements, and is, therefore, a more direct measure of small changes (± 1 km) in the daytime effective height of the ionosphere near a 70 km altitude.

For these reasons, and as demonstrated here, stable-frequency VLF transmissions can serve as a useful tool for the study of low-intensity electron precipitation related to magnetospheric processes and stratospheric-ionospheric coupling due to planetary waves.

Acknowledgments

This work was supported by the Naval Ordnance Systems Command, Department of the Navy, under Contract N00017-72-C-4401.

The author thanks D. J. Cavalieri, R. J. Deland, T. J. Rosenberg and F. A. Saus for permission to present some of their preliminary results. He also thanks R. F. Gavin and B. W. Shaw for the APL phase data and G. M. R. Winckler and M. Raines for the VLF data from the US Naval Observatory in Washington.

References

[1] L. Espenshied, C. N. Anderson and A. Bailey, Proc. Inst. Radio Engrs **14**, 7 (1926).
[2] R. N. Bracewell, K. G. Budden, J. A. Ratcliffe, T. W. Straker and K. Weekes, Proc. IEE, Part 3, **98**, 221 (1951).
[3] D. D. Crombie, Proc IEEE **53**, 2027 (1965).
[4] J. S. Belrose and L. Thomas, J. Atmos. Terr. Phys. **30**, 1397 (1968).
[5] P. Kauffman and M. H. Paes de Barros, Solar Phys. **9**, 478 (1969).
[6] S. Westerlund, F. H. Reder and C. Abom, Planet. Space Sci. **17**, 1329 (1969).
[7] R. H. Doherty, Radio Sci. **6**, 639 (1971).
[8] S. D. Despande and A. P. Mitra, J. Atmos. Terr. Phys. **34**, 299 (1972).
[9] A. J. Zmuda and T. A. Potemra, Revs. Geophys. Space Phys. **10**, 981 (1972).
[10] T. A. Potemra and T. J. Rosenberg, J. Geophys. Res. **78**, 1572 (1973).
[11] S. Ananthakrishnan and K. R. Ramanathan, Nature **223**, 488 (1969).
[12] P. J. Edwards, G. J. Burtt and F. Knox, Nature **222**, 1053 (1969).
[13] C. J. Chilton and J. S. Crary, Radio Sci. **6**, 699 (1971).
[14] J. F. Svennesson, F. Reder and J. Crouchley, J. Atmos. Terr. Phys. **34**, 49 (1972).
[15] B. Burgess and T. B. Jones, Nature **224**, 680 (1969).
[16] G. A. Baird and R. J. Francey, J. Geophys. Res. **77**, 1966 (1972).
[17] T. A. Potemra and A. J. Zmuda, J. Geophys. Res. **77**, 6916 (1972).
[18] S. Westerlund and F. H. Reder, J. Atmos. Terr. Phys. **35**, 1475 (1973).
[19] S. A. Bowhill, Annls IQSY, **5**, 83 (1969) (M.I.T. Press, Cambridge, Mass.).
[20] J. B. Gregory and A. H. Manson, J. Atmos. Terr. Phys. **31**, 703 (1969).
[21] H. Schwentek, Space Research IX, 405 (1969).
[22] G. M. Brown and D. C. Williams, J. Atmos. Terr. Phys. **33**, 1321 (1971).

[23] E. A. Lauter and J. Taubenheim, Space Research XI, 1005 (1971).
[24] L. Thomas, J. Atmos. Terr. Phys. **33**, 157 (1971).
[25] R. J. Deland and R. M. Friedman, J. Atmos. Terr. Phys. **34**, 295 (1972).
[26] R. J. Deland and D. J. Cavalieri, J. Atmos. Terr. Phys. **35**, 125 (1973).
[27] D. J. Cavalieri, R. J. Deland, T. A. Potemra and R. F. Gavin, J. Atmos. Terr. Phys. in press (1974).
[28] T. A. Potemra, A. J. Zmuda, C. R. Haave and B. W. Shaw, J. Geophys. Res. **72**, 6077 (1967).
[29] T. A. Potemra, A. J. Zmuda, C. R. Haave and B. W. Shaw, J. Geophys. Res. **74**, 6444 (1969).
[30] T. A. Potemra, A. J. Zmuda, B. W. Shaw and C. R. Haave, Radio Sci. **5**, 1137 (1970).
[31] R. J. Johler, in: Phase and Frequency Instabilities in Electromagnetic Wave Propagation, Technivision Services, England 1970 (p. 154).
[32] D. L. Carpenter, L. J. Lanzerotti and T. J. Rosenberg, Trans. Amer. Geophys. Un. **54**, 413 (1973).
[33] A. H. Manson and M. W. J. Merry, J. Atmos. Terr. Phys. **32**, 1169 (1970).
[34] G. S. Ivanov-Kholodny, Annls Géophys. **26**, 575 (1970).
[35] T. A. Potemra and A. J. Zmuda, J. Geophys. Res. **75**, 7161 (1970).
[36] T. A. Potemra, in: Physics and Chemistry of Upper Atmospheres, D. Reidel Publ. Co., Dordrecht, Holland 1973 (p. 67).
[37] M. P. Gough and H. L. Collin, J. Atmos. Terr. Phys. **35**, 835 (1973).
[38] F. A. Saus and T. J. Rosenberg, private communication (1973).
[39] F. H. Reder and S. Westerlund, in: Phase and Frequency Instabilities in Electromagnetic Wave Propagation, Technivision Service, England 1970 (p. 103).
[40] G. Rostoker, Rev. Geophys. Space Phys. **10**, 935 (1972).
[41] Nature Physical Science **234**, 102 (1971).
[42] B. Schäning, J. Atmos. Terr. Phys. **35**, 1003 (1973).
[43] R. E. Wiley and F. D. Barish, Tech. Rep. AFWL-TR-69-144, Air Force Weapons Lab., Kirtland AFB. N. Mex. 1970.
[44] U. S. Dept. of Commerce NOAA Solar Geophysical Data.

Lower Ionosphere Structure — Akademie-Verlag, Berlin 1974

CHECKING MODELS OF THE LOWER IONOSPHERE WITH EXPERIMENTAL DATA ON SCHUMANN RESONANCES*

E. Fellmann

Siemens A.G., München, FRG

**Dedicated to Prof. Rawer on the occasion of his 60th birthday*

Atmospheric noise in the 4—40 Hz frequency range has been recorded simultaneously at two widely spaced stations. The parameters Q, frequency and amplitude characterizing the Schumann resonance modes have been obtained by computer treatment of the digitized spectra of integrated noise. The characteristics of Schumann resonances are determined by the lower ionosphere properties *and* by the source-receiver geometry. The variation of the resonance parameters as a function of the source-receiver distance has been calculated with several models of electron and ion density profiles in the lower ionosphere. The observed systematic diurnal variation of the mode parameters can be explained reasonably by the diurnal variation of the worldwide source distribution. The ion contribution to the ionospheric conductivity in the height range 30—60 km must be taken into consideration. Some recent observations of cluster ions in this region may further improve the agreement of theoretical with experimental data.

The natural electromagnetic noise in the frequency range between 4 and 40 Hz has been recorded simultaneously at two widely spaced stations in France and in the southern Indian Ocean [1]. The signals were received by three orthogonally orientated antennas and subsequently subjected to spectral analysis. Fig. 1 shows the typical diurnal evolution of the spectra obtained from the two horizontal magnetic components of the noise signal. These spectra show clearly the first five resonance modes excited in the earth-ionosphere cavity. The maxima of these resonances are situated near 8, 14, 20, 26 and 32 Hz respectively. In order to analyse quantitatively the first two resonance modes, the spectra were digitized and then smoothed by a computer treatment which consisted in matching the analytical function

$$V = a \left\{\left[1 - \left(\frac{f}{f_0}\right)^2\right]^2 + \frac{1}{Q^2}\left(\frac{f}{f_0}\right)^2\right\}^{-1/2} \tag{1}$$

to the experimental points. This function (1) permitted the determination of the parameters V_{max} (amplitude of the maximum), f_{max} (frequency for which $V = V_{max}$) and Q (quality factor) characterizing the resonance modes. A spectrum of the signal being integrated over 45 minutes of recording was analysed for every hour of the day. In order to monitor the systematic diurnal variations of the three resonance parameters, these were averaged, for each hour, over the recordings of one week. Fig. 2 shows the representative parameters for the signals recorded,

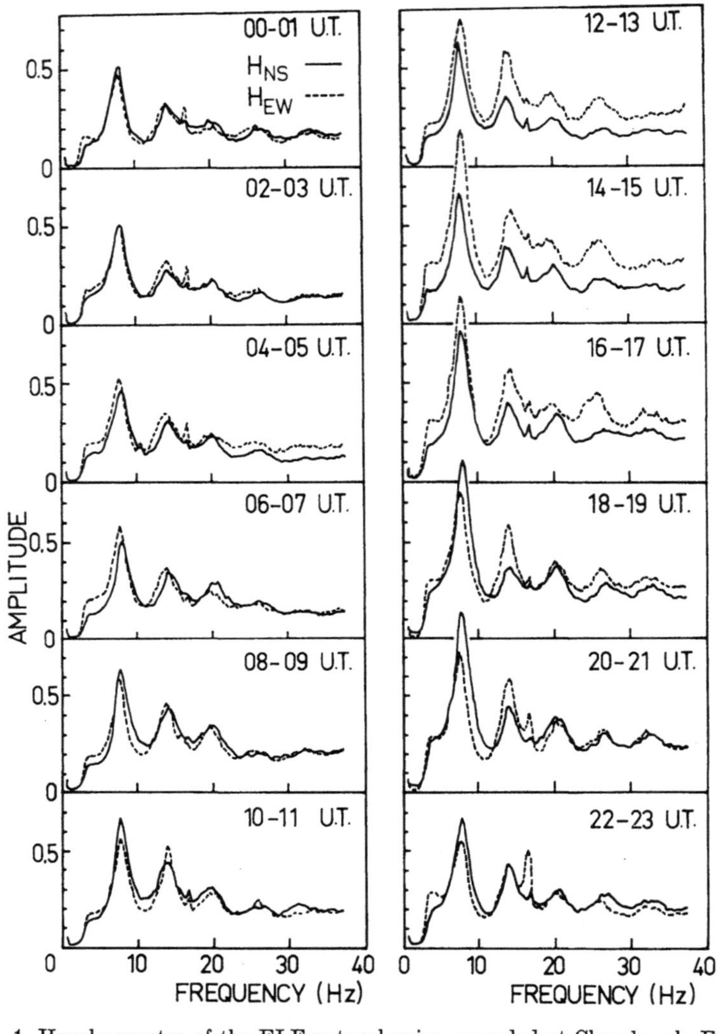

Fig. 1. Hourly spectra of the ELF natural noise, recorded at Chambon-la-Forêt.

in Chambon-la-Forêt (48°01′ N 2°26′ E) in France, during February 1971. The curves differ considerably from one component to the next and from one resonance mode to the next.

The ELF electromagnetic waves emitted from a vertical electric dipole into the earth-ionosphere cavity propagate essentially in the TM_0 mode and may be represented, in accordance with [2], by the equations:

$$E_R = \frac{iIds}{4\pi\omega\varepsilon_0 ha^2}\,\nu(\nu+1)\sum_{n=0}^{\infty}\frac{2n+1}{n(n+1)-\nu(\nu+1)}\,\mathrm{P}_n(\cos\theta) \tag{2}$$

$$H_\Phi = -\frac{Ids}{4\pi ah}\sum_{n=0}^{\infty}\frac{2n+1}{n(n+1)-\nu(\nu+1)}\,\frac{d}{d\theta}\,\mathrm{P}_n(\cos\theta) \tag{3}$$

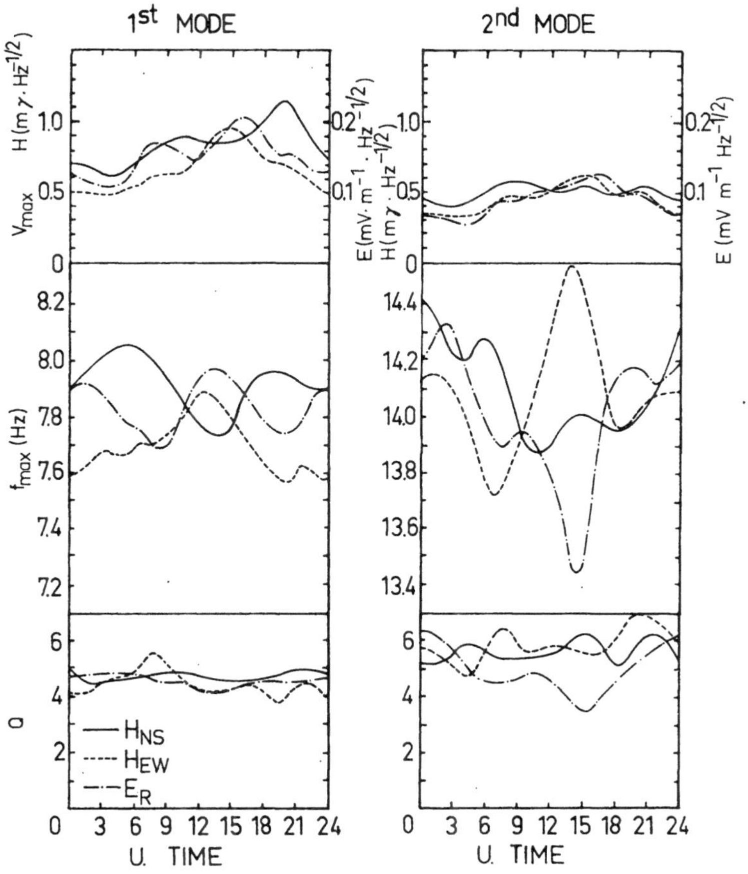

Fig. 2. Curves of the average diurnal variation of the resonance parameters, observed at Chambon-la-Forêt in February 1971.

where the different parameters have the following significance: Ids = current moment of the emitting dipole; h = height of the ionospheric reflecting layer; a = radius of the earth; θ = angular separation between emitter and receiving station, measured at the earth's centre; $P_n(\cos\theta)$ = Legendre polynomial; ν = parameter characterizing the ionospheric propagation properties; $\nu(\nu+1) = (k_0aS)^2$; $\mathrm{Re}\, S = c/v_\Phi$; $\mathrm{Im}\, S = 5.49\alpha\ (\mathrm{dB/Mm})^{-1}(\mathrm{f/Hz})^{-1}$; k_0 = wave number; c = absolute velocity of light; v_Φ = phase velocity; α = attenuation in dB per 1000 km.

It is obvious from Eqs. (2) and (3) that the fields of the ELF signals are determined by the ionospheric properties *and* the source-receiver geometry. The signal forming one spectrum of Fig. 1 is composed of a large number of atmospherics radiated from thunderstorm discharges distributed across the earth. In order to study the ionospheric properties by means of the Schumann resonance data, one must first obtain information on the source distribution and evaluate its influence on the resonance parameters. The exact source distribution not being

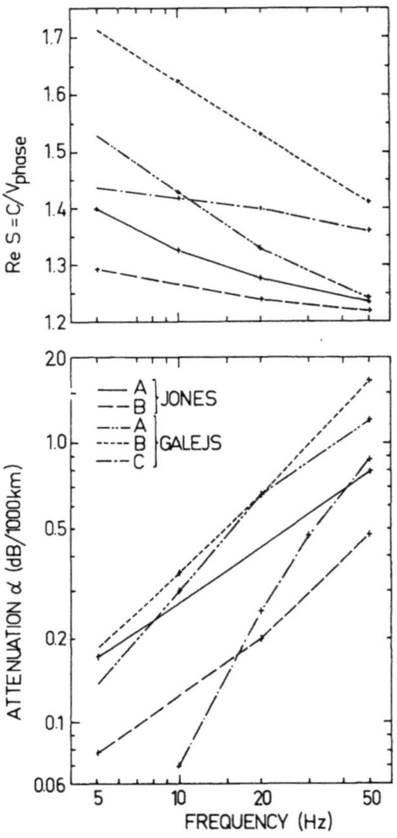

Fig. 3. Ionospheric propagation parameter ν as a function of frequency. Comparison of different ionospheric models.

known, it has been estimated on the basis of thunderstorm activity charts published in the *Handbook of Geophysics* [3]. Then the variation of the resonance parameters as a function of the source-receiver distance was calculated with several models of electron and ion density profiles in the lower ionosphere. Fig. 3 shows the ionospheric parameters ν calculated by Jones [4, 5] and by Galejs [6, 7] for different ionospheric models. The resulting resonance parameters as a function of distance are shown in Fig. 4. These resonance parameters have been computed with Eqs. (2) and (3) where the ionospheric parameters ν of Fig. 3 have been introduced.

The most valuable resonance parameter, showing clear and systematic diurnal variations in Fig. 2, is the frequency f_{max}. It is for this reason that an attempt has been made to deduce the diurnal frequency curves on the basis of a diurnal variation pattern of the average distance between the most powerful thunderstorm centres and the receiving station. These artificially constructed frequency curves were then compared with the experimentally observed curves of Fig. 2, which were most closely matched by the model designated "Jones A". This model

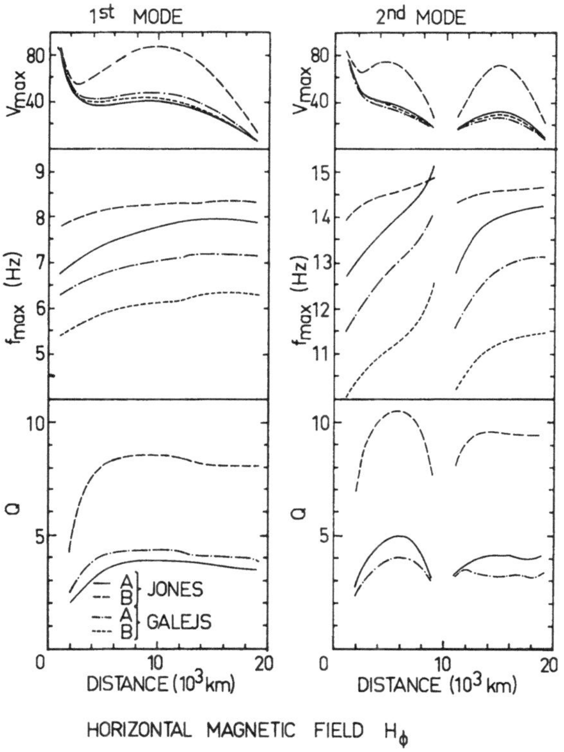

Fig. 4. Resonance parameters as a function of distance for different ionospheric models.

takes into consideration the ion contribution to the ionospheric conductivity at heights below 56 km. It was calculated using electron density $N_e(h)$ and collision frequency $\nu_{en}(h)$ profiles given by Deeks [8] for heights above 56 km, and with ion density $N_i(h)$ and collision frequency $\nu_{in}(h)$ profiles given by Cole and Pierce [9] for heights below 56 km.

However, the resonance frequencies as a function of distance obtained by this model seem to be slightly too low and the Q factors are considerably too low compared with the experimentally observed values. Our knowledge of the number densities and molecular weights of electrically charged particles in the height region between 30 and 60 km is still rather incomplete. Recent mass spectrometric measurements in sounding rockets down to heights of 65 km made during the night and completed by model calculations [10—12] resulted in total positive and negative ion densities $N_i^-/\mathrm{m}^{-3} \approx N_i^+/\mathrm{m}^{-3} \approx 10^9$ and in average molecular weights of $\overline{\mathrm{MW}}(N_i^+) \approx 35$ and $\overline{\mathrm{MW}}(N_i^-) \approx 60$. When extrapolating these values down to heights of 30 km and supplementing them by daytime ion profiles of Rose and Widdel [13], which indicate number densities N_i/m^{-3} between 10^8 and 10^9 for both positive and negative ions in the height range 60—40 km, one must conclude that the influence of ions has been overestimated by Cole and Pierce. They have assumed $N_i/\mathrm{m}^{-3} \approx 5 \times 10^9$ and $\overline{\mathrm{MW}}(N_i^+) \approx \overline{\mathrm{MW}}(N_i^-) = 29$.

Computations of the ionospheric parameter P using these recent data of ion densities and molecular weights have not been made so far, but it is suggested that these data should give still better correspondence between calculated and observed resonance data. Applying the method explained above, Schumann resonance data seem to be a valuable aid for checking models of the D region which have been established by other techniques.

Acknowledgments

This analysis was completed during a stay at the Groupe de Recherches Ionosphériques in Issy-les-Moulineaux, France, where the author held an ESRO Research Fellowship. The author is very grateful to Dr. R. Gendrin for many stimulating discussions.

References

[1] E. Fellmann, Thesis, Universität des Saarlandes, Saarbrücken 1973.
[2] J. Galejs, Terrestrial Propagation of Long Electromagnetic Waves, Pergamon Press, Oxford 1972.
[3] Handbook of Geophysics, Macmillan Co., New York 1960.
[4] D. L. Jones, Radio Sci. **5**, 803 (1970).
[5] D. L. Jones, J. Atmos. Terr. Phys. **29**, 1037 (1967).
[6] J. Galejs, Radio Sci. **5**, 1041 (1970).
[7] J. Galejs, J. Geophys. Res. **75**, 2529 (1970).
[8] D. G. Deeks, Proc. Roy. Soc. A **291**, 413 (1966).
[9] R. K. Cole and E. T. Pierce, J. Geophys. Res. **70**, 2735 (1965).
[10] F. Arnold et al., J. Atmos. Terr. Phys. **33**, 1169 (1971).
[11] F. Arnold and D. Krankowsky, J. Atmos. Terr. Phys. **33**, 1693 (1971).
[12] D. Krankowsky et al., Radio Sci. **7**, 93 (1972).
[13] G. Rose and H. U. Widdel, Radio Sci. **7**, 81 (1972).

Lower Ionosphere Structure — Akademie-Verlag, Berlin 1974

VLF PULSE IONOSOUNDER MEASUREMENTS OF THE REFLECTION PROPERTIES OF THE LOWER IONOSPHERE

P. A. Kossey, J. E. Rasmussen and E. A. Lewis

Air Force Cambridge Research Laboratory, Bedford, Mass., USA

The reflection properties of the lower ionosphere at very low frequencies were studied using an oblique pulse ionosounder with a 1—2 km balloon-supported antenna. Measured values for the normal and the converted reflection coefficients of the ionosphere from 6 to 45 kHz are presented as functions of incidence angle and azimuth, along with phase and group heights of reflection. Exponential models of the D region are given, for which full-wave computations gave results in good agreement with the experiments. Results are described for measurements obtained at a mid-geomagnetic latitude and at the geomagnetic equator.

1. Introduction

The US Air Force Cambridge Research Laboratory has developed a very low frequency (VLF) oblique pulse ionosounder to study ionospheric reflection properties, which depend in a complex manner on such factors as frequency, electron density and collision frequency profiles, incidence angle, and the propagation direction relative to the earth's geomagnetic field. The pulse ionosounder was specifically designed in order to be able to obtain a wide variety of experimental information related to these factors.

The basic feature of the system is the radiation of a very short pulse which has a spectral content in the 6—50 kHz range. The pulse is short enough that, even at distances a few hundred kilometers from the transmitter, the ground wave arrives, and passes the receiver before the arrival of the sky waves. Thus, reflection data are obtained which are free of ambiguities arising from ground and sky wave interferences. At the receiver both the normal ($\|$) and the converted ($\perp$) sky wave components are sensed, individually, using an orthogonal pair of loop antennas.

The pulses are radiated (399 times a second) from a vertical antenna so that the system has a uniform azimuthal radiation pattern. This allows measurements to be made simultaneously in any direction from the transmitter. For the data described in this report, the transmitting antenna consisted of 1—2 km of wire, supported by a spherical balloon.

Data are recorded in both the time and frequency domains. The frequency data are obtained using standard phase lock tracking receivers, tuned to one of the sixty-five or more usable frequencies contained in the radiated pulses. The amplitudes and phases of the ground wave and the two components of the sky wave are individually recorded. The time domain data consists of recordings of the individual ground and sky wave pulses, which are then Fourier analyzed to obtain the amplitude and phase information over a wide frequency range. The interpretation of these data provides information on ionospheric reflection heights, group delays, reflection coefficients, and sky wave polarization.

2. Mid-Geomagnetic Latitude Results

Fig. 1 shows typical ground and ∥ sky wave pulses recorded at a midlatitude location (Onyx, Cal., USA), at a distance of 170 km to the west of the transmitter. Fig. 1 (a) shows typical daytime pulses, and Fig. 1(b) shows a typical night-time recording, with the sky wave further delayed and having a considerably larger amplitude than in the daytime.

The variability of the night-time ionosphere is illustrated in Figs. 1(c)—(f). Fig. 1(c) shows a sky wave of twice the usual length, as if there were two reflections of nearly the same amplitude, but from different heights. Sometimes at night, when successive oscillograms such as these are compared, an 'extra' sky wave is seen to develop, descend, and merge with the usual sky wave in such a way as to temporarily cancel it out. In Fig. 1(d), taken near sunset, the sky wave

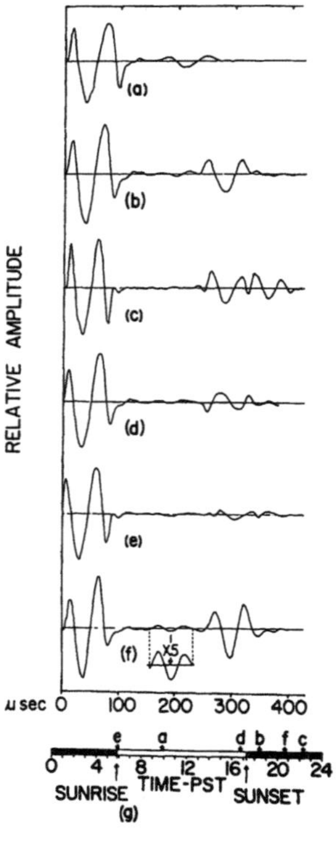

Fig. 1. Midlatitude waveform examples, 12 Oct. 1969: (a) day, (b) night, (c) extended night waveform, (d) inverted night-time waveform, (e) weak night sky wave, (f) night-time D layer reflection.

appears inverted when compared with other examples, while in Fig. 1(e), taken near sunrise, the sky wave is unusually weak. Occasionally at night a very weak sky wave arrives from about the height of the daytime reflections, as shown in Fig. 1(f), while the main sky wave appears to arrive from the usual night-time height of reflection.

The corresponding oscillograms for the $\perp$ sky wave component did not exhibit the variability observed in the night-time $\parallel$ sky wave pulses, and even on those occasions near sunrise and sunset when the $\parallel$ sky waves were very weak, as in Fig. 1(e), the $\perp$ sky waves tended to maintain their usual appearance.

It is convenient to describe the phase data phenomenologically by simply attributing the observed phase differences between the sky and ground waves at the receiver to a difference in propagation distance. This difference can be used to determine geometrically the height of a fictitious plane mirror which would give the same phase difference between the reflected and direct waves as that actually abserved. Fig. 2 shows phase 'mirror' heights of reflection obtained in this manner from $\parallel$ sky wave data obtained by Fourier analyses of the pulses shown in Fig. 1(a), (b). One reflection height, at about 72 km in the day and 92 km at night, conveniently describes the data from 10—45 kHz, and is not greatly in error even at 6 kHz.

Group 'mirror' heights are also shown in Fig. 2. These were determined from the group delays observed between the $\parallel$ sky and ground waves at the receiver. These group delays, found by numerically taking the derivatives of the phase difference between the sky and ground waves, with respect to frequency, were relatively constant over the 10—45 kHz range, and gave group heights of reflection in close agreement with the somewhat more precisely determined phase heights.

Experimental values for the normal (${}_{\parallel}R_{\parallel}$) and conversion (${}_{\parallel}R_{\perp}$) reflection coefficients of the ionosphere were determined from the ground and sky wave amplitude data, after application of a number of simple adjustment factors to

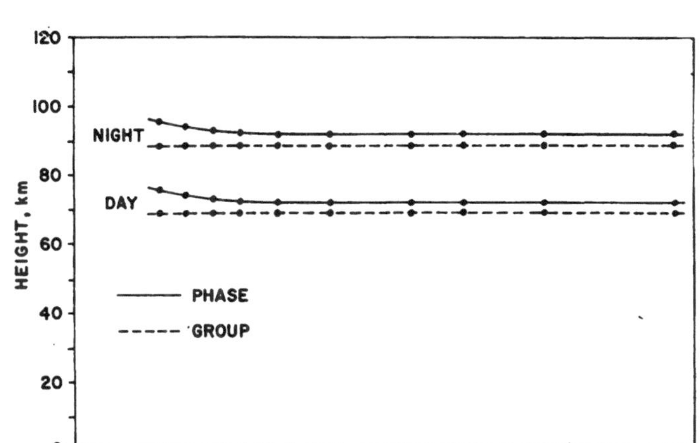

Fig. 2. Midlatitude ionospheric phase and group reflection heights, 12 Oct. 1969. $d = 170$ km.

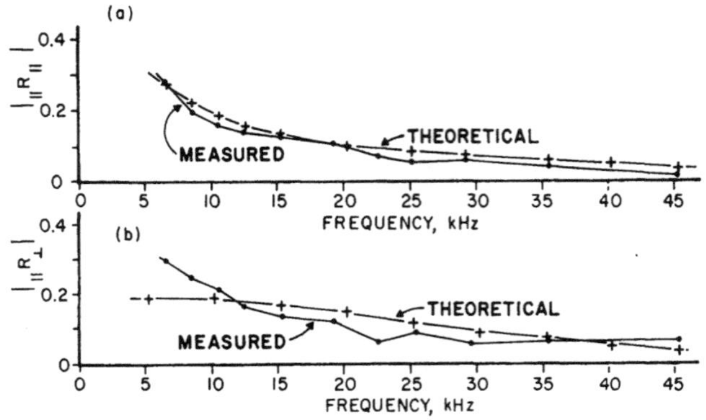

Fig. 3. Midlatitude ionospheric reflection coefficients, 12 Oct. 1969. $d = 170$ km.

the data to account for such factors as the antenna's radiation pattern, the effects of a finite ground conductivity, the ground reflections at the receiver, etc. These coefficients are shown in Fig. 3, together with theoretical values computed with a full-wave computer program [1], using an exponential model of the ionosphere characterized by $\beta = 0.5\ \text{km}^{-1}$ and $h = 70$ km in the notation of Wait and Spies [2]. The agreement between the experimental and computed values was quite good, the absolute decibel differences between them being only 1.4 dB for the normal coefficients and only 2.2 dB for the conversion coefficients.

For the same propagation path, night-time reflections were about 50% stronger than those in the day, and winter reflections were stronger than summer values by 40—80%, depending on frequency. In general the normal reflection coefficients for paths to the east of the transmitter were about 50% stronger than those to the west, over 170 km propagation paths. This directional effect is consistent with theoretical expectations [3].

3. Equatorial Results

At the geomagnetic equator in Brazil the east-west differential was slightly larger than that observed at midlatitudes, although the phase and group heights of reflection at the two locations were comparable. The most dramatic difference between the equatorial and midlatitude measurements was observed in the E—W and W—E directions from the transmitter, where the conversion coefficient $_{\parallel}R_{\perp}$ was identically zero at the equator, at all distances from the transmitter. This contrasts with the midlatitude results where, for example, the conversion and normal coefficients were about equal for a 170 km path to the west of the transmitter (see Fig. 3). In the N—S and S—N directions at the equator the conversion was comparable with that observed at midlatitudes. Fig. 4 illustrates the equatorial observations where, for distances about 150 km from the transmitter, the conversion is appreciable to the north, but none is evident to the east, where there is no component of the geomagnetic field present to move the wave-excited electrons in a direction perpendicular to the plane of propagation.

Fig. 5 shows experimental values of the normal reflection coefficient as a function of incidence angle (or range, basically) at a frequency of 20 kHz, as deduced

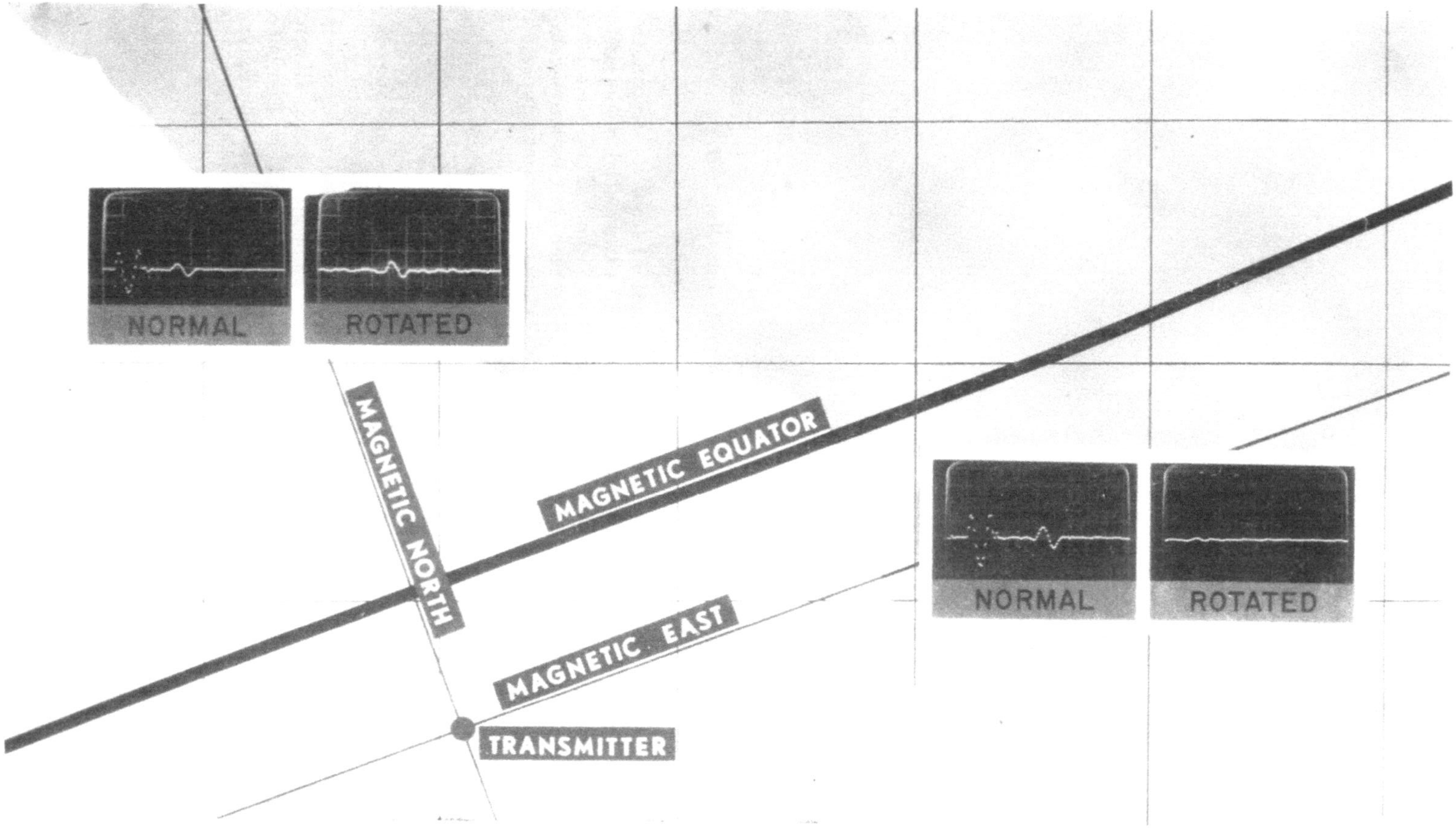

Fig. 4. Equatorial waveform examples, May 1971.

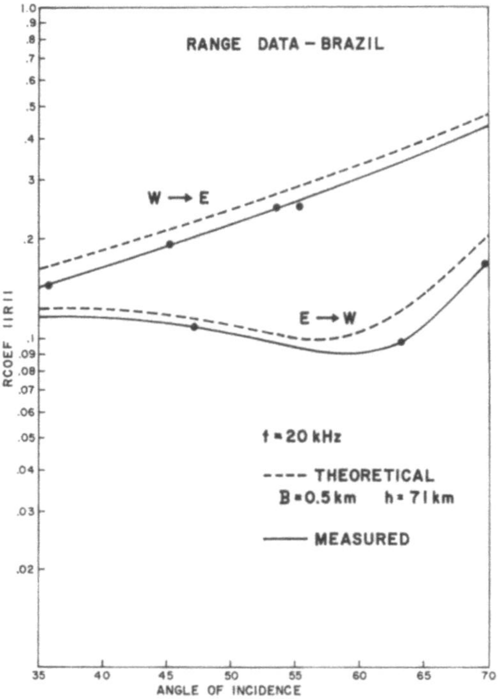

Fig. 5. Equatorial ionospheric reflection coefficients, May 1971.

from data obtained at the equator, in cardinal directions from the transmitter. Also shown are theoretical values calculated using an exponential model of the ionosphere characterized by $\beta = 0.5\ \text{km}^{-1}$ and $h = 71$ km. The general tendency of the coefficients to increase with range can be seen, but for propagation to the west, both the experimental and theoretical values show what appears to be a quasi-Brewster angle effect for incidence angles around 60—65°.

4. Conclusions

Measurements show that the VLF reflectivity of the daytime ionosphere is much more stable than at night. The results also show that phase and group delays can be described by equivalent mirror heights of reflection valid over the 6—45 kHz range, and that the variations of the normal and conversion coefficients with frequency, geomagnetic latitude, season, etc., are appreciable. Computed values of the reflection coefficients, using exponential models of the ionosphere, were in good agreement with the experimental observations.

References

[1] Y. INOUE and S. HOROWITZ, Radio Sci. **3**, 957 (1968).
[2] J. R. WAIT and K. P. SPIES, NBS Technical Note 300 (1964).
[3] D. D. CROMBIE, J. Geophys. Res. **12**, 110 (1958).

Lower Ionosphere Structure — Akademie-Verlag, Berlin 1974

MULTI-FREQUENCY LONG-WAVE VERTICAL SOUNDING OF THE LOWER IONOSPHERE*

B. W. Reinisch[a] and G. S. Sales[b]

[a]Lowell Technological Institute Research Foundation, Lowell, Mass., USA
[b]Air Force Cambridge Research Laboratories, Bedford, Mass., USA

**Dedicated to Professor K. Rawer on the occasion of his 60th birthday*

Vertical pulse sounding at 30, 60, 180 and 525 kHz at Quabbin Reservoir, Mass., USA, reveals diurnal and seasonal variations in structure, reflectivity and height of the lower ionosphere. A solid state wideband amplifier feeds 10 kW peak power into a ground-laid six-element antenna. A special digital sampling and integration scheme selects median values of amplitude and phase received at perpendicular loop antennas.

Day- and night-time D-region model profiles are fitted to the observed data by matching group height and reflection coefficient. Since the group retardation is small the iteration process is simple and fast.

Measurements on a winter anomaly day, 16 January 1973, are compared with normal days. The three lower frequencies show little effect in echo amplitude and height.

1. Introduction

In order to provide continuous monitoring of the D layer over a long time period, a vertical sounding system operating at four low frequencies was designed and built. Operation began during March 1972 at a remote site some 96 km from the Cambridge Research Laboratories. The four frequencies 30, 60, 180 and 525 kHz were chosen so that the expected reflection points would be separated by approximately 5 km through the D region in the daytime. This was tested using various models of the D region and applying the "full wave" solution program [1] to compute reflection coefficients as a function of frequency. For any model of electron density vs height the group height is calculated from the continuous wave program by using closely spaced frequencies and calculating $\Delta\phi/\Delta f$ from the complex reflection coefficients with phase ϕ. In general, the group heights show approximately 1—3 km retardation over the expected $X = 1 + Y$ level reflection point (so-called z-condition for the "electronic" polarization). $X = f_N^2/f^2$, $Y = f_B/f$, where f_N = plasma frequency, f_B = gyrofrequency. The modeling of the D region described in § 4 uses both the group height and the magnitude of the reflection coefficient to fit the model to the data. The combination of both group height and reflection coefficient provides a degree of uniqueness to the model which would not be possible from reflection coefficients alone. Certainly the accuracy of the calculated electron profile varies from good for those areas where the above reflection condition applies for the four frequencies, to less good for the lower lying ionization and to poor for the electrons above the highest reflection point.

2. Instrumentation

For reliable operation at a relatively remote site, the system was designed to be completely solid state including the 10 kW peak power final amplifier. The system transmits pulses with a variable width up to 150 μs and a variable repetition rate of up to 200 Hz. The transmitting antenna [2] consists of six parallel-terminated dipoles spaced 67 m apart and running 1000 m end to end. The dipoles are placed directly upon relatively low conducting ground which reduces the mutual coupling between elements and results in a total input impedance of approximately 100 Ω. The terminations were provided to increase the bandwidth of the antenna.

The receiving antennas are two broadband loops oriented parallel and perpendicular to the transmitting antenna and separated by about 700 m from the transmitting array. The signals from the two loops are processed and stored separately.

The receiving system consists of a pulse receiver which automatically tunes as the operating frequency is changed and 40 pairs of digital quadrature sampling gates are generated to provide height coverage, in the usual mode of operation, of from 67.5 km to 247.5 km in intervals of 4.5 km. The quadrature gates allow the phase and amplitude of the echoes to be measured. The digital system uses a unique integration technique [3] which allows up to 40 dB of signal-to-noise improvement which is necessary considering the small signals expected from D region reflections. After integration the data are stored on magnetic tape in digital form for automatic computer processing.

3. Data Processing

The digital magnetic tapes which store approximately nine days of data are computer processed to provide two basic outputs. The first is a compressed intensity modulated picture of group delay vs time for both polarizations for a given frequency. A sample of these data for the frequency of 525 kHz is shown in Fig. 1. The data sampling time at the sounder is two minutes, i.e. every two minutes the complete cycle of four frequencies is sampled. For the compressed display in Fig. 1, every adjacent six samples or 12 minutes are averaged to produce one vertical line and the whole day can be reduced to one page size. The digital font system allows the amplitude to be directly scaled from these charts. However, to process a large quantity of data, it was necessary to develop a computer program for the automatic processing of this type of data. This code is described in [4] and is useful for this and other "Digisonde" types of data. The program has two basic parts; first it locates the echoes and second determines the echo amplitude and group delay. This is done automatically and a tape of nine days of data can be processed in approximately 15 minutes. A sample of this output is shown in Fig. 2 where a week's data are superimposed. These weekly median plots are used to determine the variation of group height and reflection coefficient for the purpose of generating the best fit D region models.

→

Fig. 1. Low frequency data, Quabbin, Massachusetts, 525 kHz.

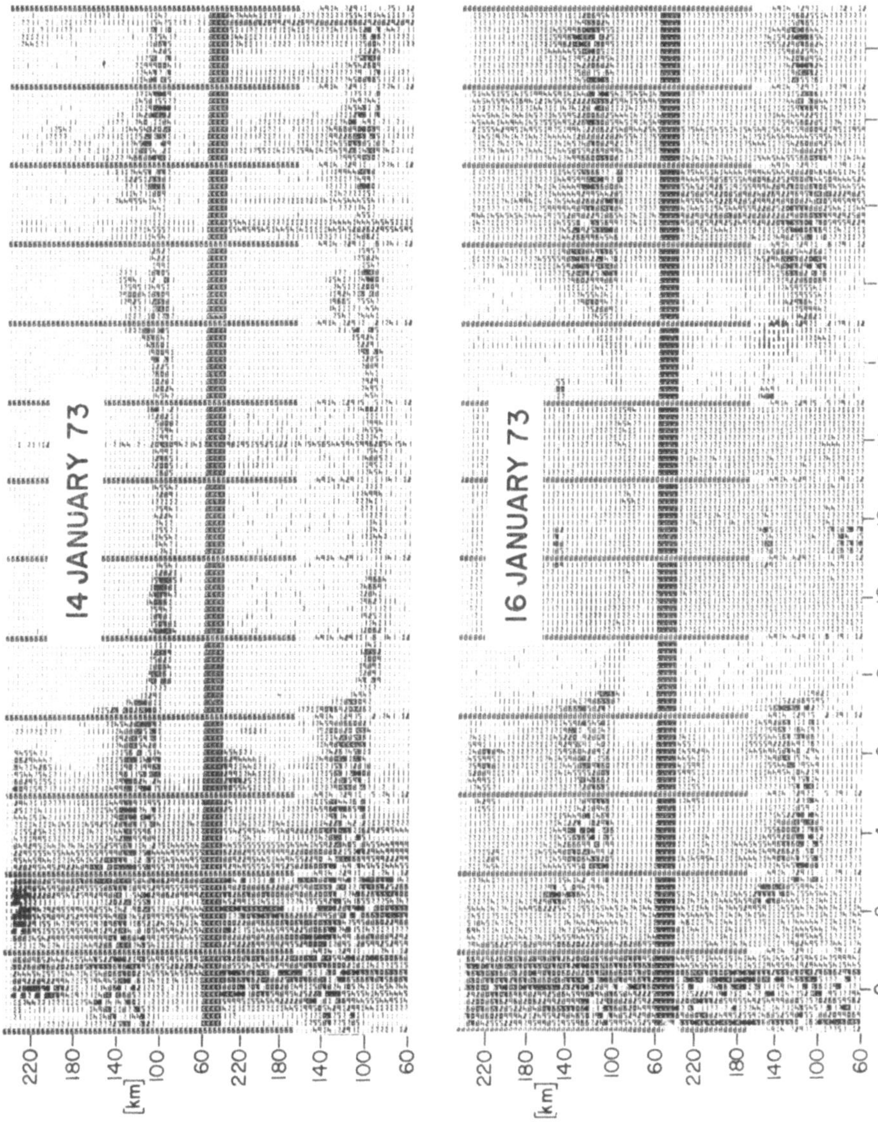
14 JANUARY 73
16 JANUARY 73
[km]
220
180
140
100
60
h[EST]
0
2
4
6
8
10
12
14
16
18
20
22
00

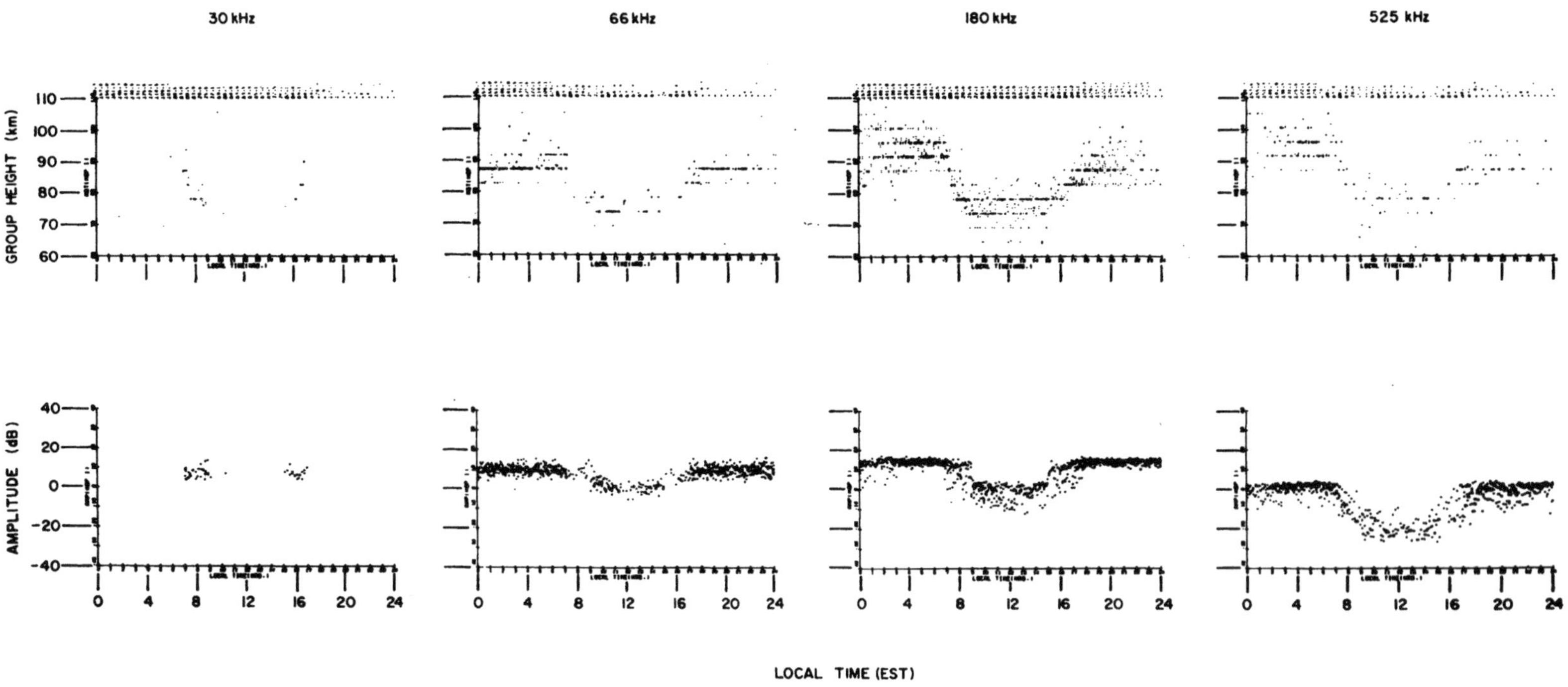

Fig. 2. Low frequency data, summary plot, Quabbin, 11—19 January 1973.

4. Analysis

4.1. General

The determination of the electron density profile from low-frequency pulsed soundings is basically an iterative process using a full-wave solution computer code which has been adapted for a computer graphics center. The median group height data is used to specify the electron density and its gradient in the region where $X = 1 + Y$ for the frequencies 30, 60, 180 and 525 kHz. For a gyro-frequency of 1400 kHz and a dip angle of 72°, the following reflection levels are used (shown in Table 1). With data for all four frequencies, the D region model between 5×10^8 and 1×10^{10} m^{-3} is reasonably well specified, i.e. within a few kilometers. It then remains to vary the lower part of the profile below $N_e = 5 \times 10^8$ m^{-3} to obtain a fit to the measured reflection coefficients. To improve the lower D region modeling, it would be useful to extend the range of sounding down to lower frequencies, for example, 10 kHz produces reflections at $N_e = 1.6 \times 10^8$ m^{-3}.

Table 1

Reflection Levels

f (kHz)	N_e ($X = 1 + Y$) (m^{-3})
30	5×10^8
66	1×10^9
180	3×10^9
525	1×10^{10}

The concept of using a specific electron density to characterize the reflection of long waves is supported by the results of the full-wave solution comparing the calculated group height and the $X = 1 + Y$ level. Horowitz [5] ran tests for frequencies down to 6 kHz and only at these very low frequencies did the group height depart significantly from the model height given by $X = 1 + Y$.
This is obviously not true when a peak of a layer is approached or when the gradient of electron density with height becomes small.

4.2. Data

As an example of particular interest the tape containing the data from the period 11 January 1973 to 19 January 1973 was selected for processing. The day of 16 January 1973 has been categorized as a day of anomalous absorption [6]. Two of the days during this period are shown in Fig. 1 for the 525 kHz frequency: a relatively normal day, 14 January, and the anomalous day, 16 January. For each of the two days, the lower half is the normal or parallel to the transmitting antenna data while the upper half is the abnormal or perpendicular data. The ordinate is group delay converted to kilometers with a digital increment of 4.5 km permitting a height resolution of approximately 1.5 km. During the daytime on 14 January 1973 a single echo return is observed corresponding to $X = 1 + Y$.

At a reflection height of 85 km (all reflection heights are measured to the leading edge) during the day the collision frequency is sufficiently large so that the $X = 1$ condition no longer exists. However, at night when the reflection height rises to around 105 km both reflection conditions exist and it is possible to observe both reflections. Often during the night the electron density falls below 3×10^9 m^{-3} and no reflection can be observed in the 95–105 km height range at 525 kHz. As an example of this, the 16 January 1973 data shows that before 0200 the night-time E-layer reflection has disappeared at 525 kHz and then at 0200 the E layer reforms from above. Many of these events are observed at night at 525 kHz.

Before discussing the anomalous day of 16 January, we can see the normal behavior of the D region and lower E layer from the summary plots in Fig. 2. The amplitude data are relative and it is not possible to compare data from different frequencies on this figure. However, the data are calibrated using multiple reflections before profile modeling is attempted. The effects of system saturation, particularly at 180 kHz at night, is evident, making the analysis more difficult.

A set of profiles was generated to fit the summary plots during the sunrise period. The group height data determine the general shape of the upper part of the profile while the tail or lower part is adjusted to fit the reflection coefficients. The detailed shape of the tail is not determined, the reflection coefficient is sensitive mainly to the electron content of the layer below the reflection height. For this analysis the shape of the tail was dictated by low-frequency oblique sounding observations with reflection heights near 65 km [7]. The numbers attached to each curve in Fig. 3 are solar zenith angles. The symbols attached to the upper curves identify the profiles for $\chi = 98°$ (triangle), $\chi = 93°$ (square) and $\chi = 90°$ (cross) to indicate that there are actually three curves. The heavy dots in Fig. 3 show the group heights calculated for the profiles to which the small arrows are pointing. Group retardation effects are limited to about 5 km. In

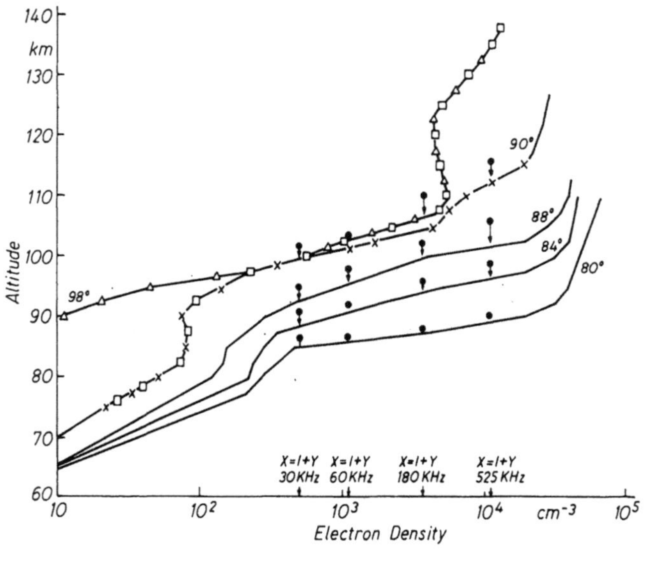

Fig. 3. Electron density profiles, January 1973.

general the reflections for the three frequencies below 525 kHz are relatively simple involving only the $X = 1 + Y$ level. For 525 kHz and solar zenith angles greater than 90° the reflection switches between $X = 1$ (the o-condition for the "ionic" polarization) at $N_e = 3.5 \times 10^9$ m^{-3} and $X = 1 + Y$ at $N_e = 10^{10}$ m^{-3} controlled by the shape of the electron density profile near the peak of the night-time E layer. It often happens that both reflections are present simultaneously.

The variation of signal strength and group height for 16 January 1973, the day of anomalous absorption, can be seen in the bottom half of Fig. 1. The most obvious difference between the normal day of 14 January 1973 and 16 January is the sharp decrease in signal in the morning at almost exactly 90° solar zenith angle. The reflection at 525 kHz remained weak for the rest of the day until sunset. After sunrise the amplitude difference between the normal and disturbed day decreases until at noon there is an excess absorption of approximately 6 dB. The group heights measured on 16 January are depressed by about 7 km at 525 kHz and less at the lower frequencies. The absorption at the lower frequencies is not significantly increased. A slight lowering of the D layer ionization together with some compression at its bottom part can explain the observations.

5. Summary

Low frequency pulse sounding of the lower ionosphere is an effective means of providing continuous monitoring of D region variations. These data can be used as a background against which direct measurements, by rocket-borne probes, can be compared. Future work involves a complete analysis of the seasonal and diurnal variations as well as special events that affect the D region.

References

[1] S. Horowitz and Y. Inoue, Radio Sci. **1**, 957 (1966).
[2] G. S. Sales and A. Ganio, AFCRL Rep. No. 70-0671 (1970).
[3] K. Bibl and R. Vargas-Vila, AFCRL Rep. No. 71-0107 (1971).
[4] K. Dieter and G. S. Sales, AFCRL Rep. (to be published) (1974).
[5] S. Horowitz, AFCRL Rep. No. 71-0237 (1971).
[6] C. F. Sechrist, personal communication.
[7] A. D. Watt, VLF Engineering, Pergamon Press, New York 1967.

Lower Ionosphere Structure — Akademie-Verlag, Berlin 1974

A PARTIAL REFLECTION EXPERIMENT SYSTEM

A. HAUG

The Auroral Observatory, University of Tromsø, Norway

The partial reflection experiment installation at the Auroral Observatory, Tromsø, has been redesigned in order to achieve better height resolution, the inclusion of phase detection, and system control and fast data analysis. Different antenna systems have been considered; the final choice is guided by the demands for height resolution, band width and suitable antenna pattern characteristics. The data processing system, consisting of a small computer with the necessary peripherals, will perform (i) time control of the sounding system, (ii) digitizing and noise filtering, (iii) statistical analysis and final calculation of experimental outputs such as electron density profiles, (iv) graphical display and transfer to magnetic tape of selected intermediate and final data. The data processing system is designed as a general purpose data acquisition system which may be connected to different types of experiments. The redesigned system is now being tested and will be in operation during 1974.

Lower Ionosphere Structure — Akademie-Verlag, Berlin 1974

RECENT WORK ON THE ANALYSIS OF PARTIAL REFLECTION DATA

J. S. Belrose and T. N. R. Coyne

Communications Research Centre, Department of Communications, Ottawa, Canada

Two questions have long needed answering in connection with the partial reflection experiment. First and most important, what are the irregularities responsible for the partial reflections (are they irregularities in electron density or collision frequency); and secondly what is the effect of the limited height resolution of the sounder?

To this end partial reflection observations involving measurement of differential-phase as well as differential-amplitude, and the amplitude of echoes for different pulse widths, have been recorded at Ottawa. The results of the measurements of differential-phase and differential-amplitude (J. S. Belrose, M. J. Burke and T. N. R. Coyne, *J. Geophys. Res.* **77**, 4829 (1972)) and the pulse width work (T. N. R. Coyne and J. S. Belrose, *J. Geophys. Res.* **78**, 8276 (1973)) were reviewed together.

Variations in Lower Ionosphere Electron Densities

Lower Ionosphere Structure — Akademie-Verlag, Berlin 1974

THE TEMPORAL AND GEOGRAPHICAL VARIATIONS OF D REGION ELECTRON CONCENTRATIONS

L. THOMAS

S.R.C., Radio and Space Research Station (*now* Appleton Laboratory), Ditton Park, Slough, UK

Our knowledge of the height distribution of electron concentrations in the D region has been acquired from rocket-borne radio and probe experiments, and ground-based measurements using the partial-reflection and cross-modulation techniques. Although certain characteristic features have been recognized, such as the steep gradient of ionization at heights between 80 and 90 km, the data obtained from a limited number of locations are insufficient to describe spatial and temporal variations of D region electron concentrations. However, supplementary information has been provided by recordings of VLF and LF radio-wave transmissions and measurements of radio-wave absorption at MF and HF.

The purpose of this paper is to review the present information on the temporal and spatial variations of D region electron concentrations. The diurnal, seasonal, latitudinal and solar-cycle variations are described in turn with particular attention to the anomalous behaviour at middle latitudes during winter. In addition, the changes associated with geomagnetic storms are described. The possible causes of the variations are mentioned.

1. Introduction

Radio techniques, ground-based and rocket-borne, provide the most accurate measurements of electron concentrations in the D region. The ground-based methods of measurement make use of the partial reflection of radio waves from D region irregularities [1—3] or the cross modulation between two signals propagating simultaneously [4—6]. The rocket-borne experiments examine the Faraday rotation of the plane of polarization of a linearly polarized transmitted signal and the differential absorption of the ordinary and extraordinary modes of propagation; a dc probe is sometimes included to provide information on the detailed height distribution of electron concentration [7]. Unfortunately these ground-based and rocket-borne methods of measurement have been confined to relatively few sites and the data obtained are not adequate to describe the morphology of the D region in any detail. Ground-based recordings of VLF or LF transmissions [8, 9] and measurements of radio-wave absorption at MF or HF [10, 11] have provided a continuous monitor of D region conditions and have helped to describe the temporal and spatial variations. The interpretation of this type of data is, however, not straightforward and can be misleading. Thus at VLF the radio wavelength is comparable with the scale length of the height variation and a full-wave approach needs to be adopted in the interpretation of propagation data [12, 13]; at MF effects of small changes of electron concentration near the reflection level on the total radio-wave absorption [14] can complicate the use of absorption data. Nevertheless these experiments represent a useful supplement to the more direct methods of measuring electron concentrations.

The purpose of this paper is to review the information available on D region electron concentrations and their temporal and spatial variations. Only changes occurring over times of several hours or longer are considered, the shorter-term changes occurring during solar flares [15], solar eclipses [16], and during certain nights [17] being omitted. In § 2 the main features of the height distributions of electron concentration during day and night are considered and the diurnal changes observed at different sites are described. Radio-wave propagation data have indicated that the diurnal variations are anomalous during and following certain geomagnetic storms, probably as a result of changes in atmospheric composition or of ionization produced by precipitated high-energy particles; these storm-associated effects are described in § 3. The seasonal and latitudinal changes shown by the D region are outlined in § 4. Measurements of electron concentrations and radio-wave propagation data have shown anomalous variations of the D region at middle latitudes in winter; three aspects of the winter-time behaviour are described in § 5. The limited information available on the secular variation of D region ionization is presented in § 6. Finally, in § 7 the uncertainties in our present information are emphasized and some of the major gaps in our knowledge are underlined.

2. Diurnal Changes

2.1. Height Distributions in Daytime and Night-time

A comparison of measurements carried out with a similar technique at different locations in daytime reveals that the height distributions of electron concentration have the same general form, the differences arising chiefly from differences in solar zenith angle, χ. This is illustrated in Fig. 1 which shows the results of rocket-borne measurements carried out by Mechtly and Smith [18] and Bennett et al. [19] using the combined Faraday-rotation and dc probe experiment at Wallops Island, USA, and South Uist, Scotland, and by Aikin et al. [20] using a radio-wave absorption experiment at Thumba, India. It is seen that all three distributions show a rapid increase of electron concentration with height at the lower heights, a slower increase at intermediate heights, and a rapid increase again at greater heights; transition heights are different for the three cases. It will be seen in § 4.1 that these general features of the daytime D region are present at middle latitudes during both summer and winter, and at solar maximum and minimum, although the electron concentrations over the same range of height are sometimes substantially different. It is to be noted, however, that cross-modulation measurements at Pennsylvania State College [21, 22] consistently show a minimum near 70 km in the height distribution of electron concentration and a maximum at lower heights, and similar features at rather lower heights seem to be required to reproduce VLF propagation data for summer at solar maximum over England [23].

The more limited data available for night-time hours reveal a greater degree of variability, as is indicated in radio-wave propagation data [24] and in the summaries of height distributions of electron concentration presented by Maeda [25] and Knight [26]. The latter results show that marked structure is often observed in the distributions above about 90 km. Furthermore, a distinct ledge sometimes occurs at lower heights [27].

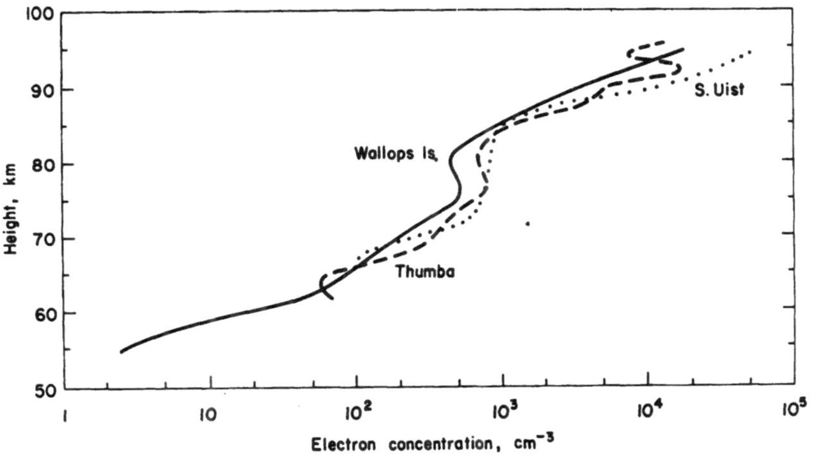

Fig. 1. Rocket-borne measurements of the height distributions of electron concentration: —— Wallops Island (38° N 76° W), 10 Sept. 1969, $\chi = 61°$ [18]; ····· South Uist (57° N 7° W), 19 June 1969, $\chi = 44°$ [19]; - - - Thumba (9° N 77° E), 19 Mar. 1970, $\chi = 48°$ [20].

2.2. Variations during Sunrise Period

Early observations of VLF and LF transmissions [8] revealed that pronounced changes occurred in the middle-latitude D region during the sunrise period, beginning when the solar zenith angle was near 98°; the corresponding variations of the height distributions of electron concentrations giving rise to these radio-wave effects have also been deduced [28]. These results and measurements using rocket-borne experiments by Mechtly and Smith [27] at Wallops Island and a cross-modulation experiment by Smith et al. [29] at Armidale, Australia, have shown that the major change consists of the development of a C layer between about 60 and 75 km, the electron concentrations near 80 km remaining at low values. The two sets of measurements, each referring to summer conditions, are shown in Fig. 2. It is seen from the rocket observations that an initial increase of electron concentrations occurring between about 72 km and 90 km near ground sunrise was followed by a decrease. It is to be noted, however, that the distribution shown for the solar zenith angle of 94° referred to a day in June 1965 whilst the other curves were for days in July 1964.

2.3. Variations with Solar Zenith Angle during Daytime

The partial-reflection and cross-modulation experiments are well suited to a study of the changes in electron concentrations during daytime and have been used for such a study by a number of workers [5, 29—32].

Fig. 3 shows the mean diurnal variations of electron concentrations measured by Coyne and Belrose [32] at particular heights over Ottawa for groups of days in May and September 1970, and for the 1969—1970 winter. The data for the summer and equinox months show a clear solar control. A definite asymmetry about noon is shown in the variation at 69 km for May and this was also evident in the corresponding data for June 1970. The measurements at different sites

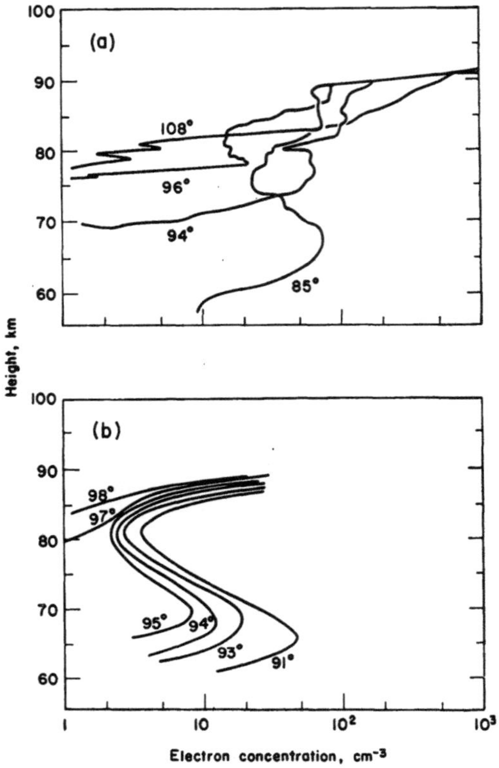

Fig. 2. Height distributions of electron concentration during the pre-sunrise period: (a) rocket-borne Faraday rotation and dc probe experiments at Wallops Island during July 1964 and June 1965 [28]; (b) cross-modulation experiment at Armidale, Australia, on 8 January 1965 [29].

have provided conflicting evidence concerning this diurnal asymmetry; thus cross-modulation and partial-reflection data for Crete (35°N 25°E) show an insignificant asymmetry during the autumns of 1964 and 1965 [33] but a marked asymmetry during May 1966 with afternoon values larger than those prior to noon [34]. In contrast, the morning values at Tsumeb (19°S 18°E) were reported [31] to be substantially larger than those in the afternoon. Finally, Aikin et al. [20] report that the rocket observations at Thumba also indciate a possible diurnal asymmetry with electron concentrations in the afternoon larger than in the morning.

3. Effects associated with Geomagnetic Storms

It is well-known from observations of the characteristics of VLF, LF and MF radio-wave propagation that pronounced D region effects occur at middle latitudes in association with certain geomagnetic storms [9, 24, 35, 36]. These studies have shown that in addition to changes occurring at the time of the storm, pronounced effects are observed for several days after the geomagnetic field has

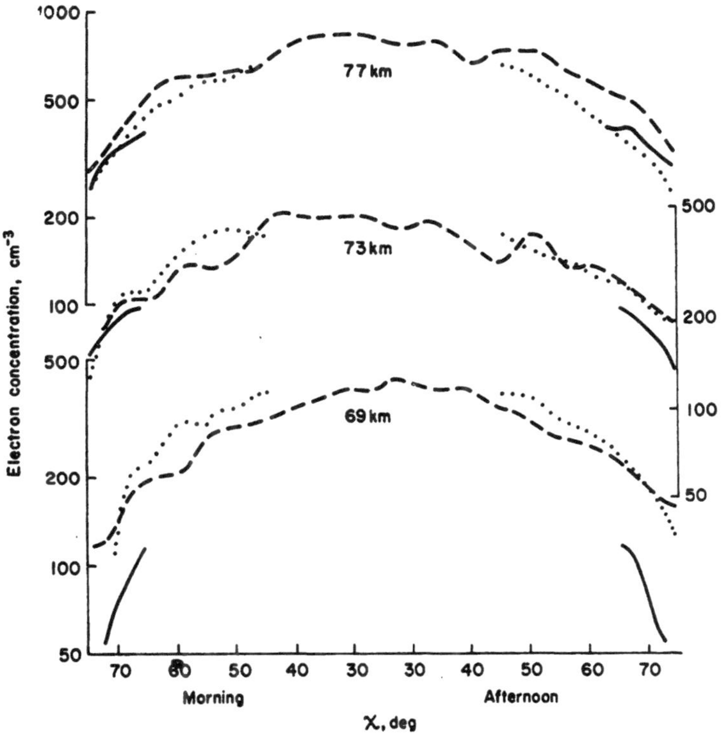

Fig. 3. The variations of electron concentrations at three heights measured with a partial-reflection experiment at Ottawa [32]: – – – mean of observations during seven days in May 1970; ····· mean of observations during seven days in September 1970; —— mean of observations during winter 1969/70.

lecovered. Fig. 4 shows recordings of the abnormal components of 16 kHz and 60 kHz transmissions from Rugby to Slough, i.e. the sky-waves polarized with the electric field vectors at right angles to the plane of propagation, at the time of and following the geomagnetic storm which occurred at 1418 UT on 8 March 1970 [37]. The variation of the phase of the 16 kHz wave, relative to the ground wave, and of the amplitude of the 60 kHz are shown; in each case the dotted curve represents the variation on normal days. Also shown is the variation of the planetary magnetic index *ap*. It is to be noted that the geomagnetic field had recovered by 10 March. A large advance in phase of the 16 kHz waves was observed during the nights of 8/9 and 9/10 March; the phase on the two subsequent nights showed an almost normal behaviour but anomalous conditions occurred again at sunset on 12 March. The phase change on the subsequent night corresponded to a reduction in reflection height of approximately 4.5 km from that for normal conditions. The recordings on the four later nights showed anomalous variations near sunrise and sunset. These anomalous phase variations at VLF during twilight periods are a characteristic feature of D region effects following geomagnetic storms [36]. It can be seen from Fig. 4 that the amplitude of the 60 kHz waves also shows a partial recovery on the nights of 10/11 and 11/12 March but severe after-storm effects on subsequent nights.

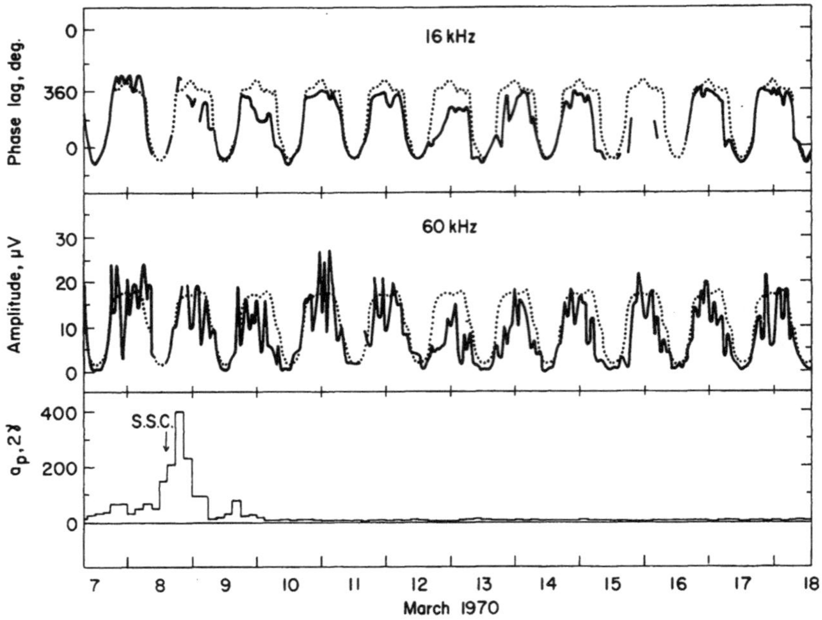

Fig. 4. Recordings of the abnormal components of 16 kHz and 60 kHz waves transmitted from Rugby to Slough at the time of and following the geomagnetic storm which occurred at 1418 UT on 8 March 1970 [37]: —— variations of the phase lag for 16 kHz, relative to the ground wave, and the amplitude of 60 kHz; ····· mean variations on normal days. The disturbance of the geomagnetic field is represented by the planetary index *ap*.

Lauter and Knuth [38] have emphasized that the effects at VLF and LF are most marked at times of large solar zenith angles but the studies of Belrose and Thomas [36] and Bourne and Hewitt [35] have indicated that effects at MF are also observed during daytime. Belrose and Thomas have suggested that the overall radio-wave effects might be understood if the anomalous ionization responsible was produced in the height range 80—90 km. No direct evidence is available but it is of interest to note that enhancements of airglow emissions originating at both upper and lower D region heights have been reported following geomagnetic storms [39, 40].

4. Seasonal and Latitudinal Variations

4.1. Seasonal

Although considerable information is available concerning the seasonal variation of MF and HF radio-wave absorption, its interpretation is complicated by changes in the relative contributions of deviative and non-deviative losses. Unambiguous information on the seasonal variations of D region electron concentrations is available only from measurements at a few middle-latitude locations. The results of partial-reflection measurements at Ottawa shown in Fig. 3 indicate very little change in concentrations between May and September 1970; although the concentrations at 69 km for the 1969/70 winter are shown to be substantially smaller, Coyne and Belrose [32] question the significance of the change because of the variability shown at low heights over Ottawa during winter months.

A clear indication of a seasonal variation of D region electron concentrations has been provided by rocket measurements at Wallops Island during solar minimum conditions by Mechtly and Smith [41]. Their results from six flights at values of χ near 60° showed that the electron concentrations between about 65 and 83 km were approximately four times as large on an equinox or summer day as on a winter day; little seasonal difference was shown at greater or smaller heights. In contrast, the corresponding results obtained from flights during solar maximum conditions, as reported by Mechtly et al. [42], revealed little clear seasonal variation in electron concentrations below 80 km but values observed between about 82 and 95 km on winter and spring days were larger than those for summer and autumn days. The two sets of results are illustrated by the height distributions for winter and summer days in Fig. 5.

The partial-reflection measurements for 1963—1967 at Christchurch, New Zealand, described by Gregory and Manson [43], when corrected to equivalent overhead sun conditions, showed that winter values of electron concentration for heights above 79 km were substantially larger than those for summer. However, conflicting evidence concerning the seasonal change at such heights was provided

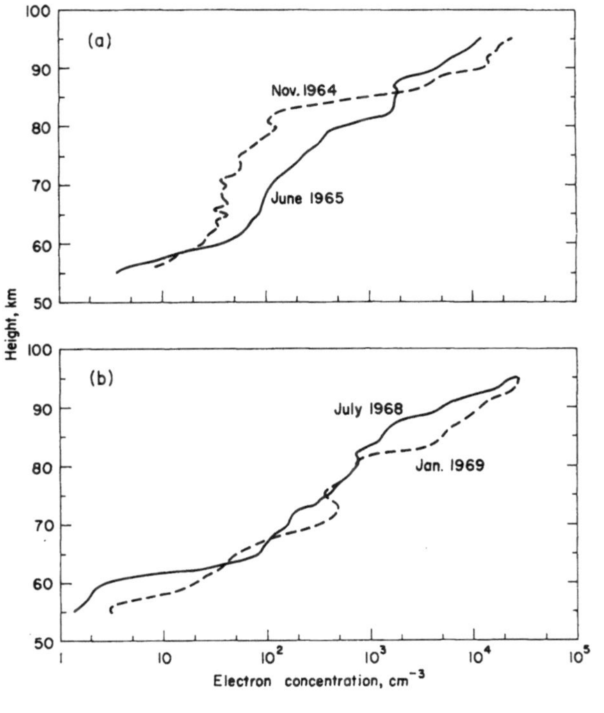

Fig. 5. Rocket-borne measurements of the height distributions of electron concentration for solar zenith angles near 60° at Wallops Island: (a) low solar activity [41]; (b) high solar activity [42].

by a comparison of partial-reflection results for winter at 35°S and for summer/autumn at 35°N during 1965 carried out by Gregory and Manson [30]. For values of χ near 60° the electron concentrations at about 80 km and above were found to be larger by a factor of 2 at the northern station.

4.2. Latitudinal

The direct measurements of electron concentrations available are insufficient to examine latitudinal variations in any detail. Thrane et al. [31] have presented data derived from partial-reflection and cross-modulation measurements at four locations ranging in latitude from 69°N to 19°S in the Europe-Africa sector; the mean height distributions for χ values near 50° in summer and equinox months showed similar shapes and values of electron concentration. More detailed information was provided by Mechtly et al. [44] from measurements during rocket flights carried out over the latitude range 13°S to 58°S near the West coast of S. America during March and April 1965. The four height distributions measured at values of χ between 58° and 67° all showed the same general features noted in § 2.1. Although the magnetic dip angle at the four launching sites varied from 0° to 58°, no evidence was found of a systematic latitudinal variation in electron concentrations at the lower heights, as might have been expected for ionization produced by cosmic rays. Instead, they found an anomalous latitudinal variation for electron concentrations in the height range 70—90 km: the values observed near the magnetic equator and at middle latitudes were smaller by a factor of two than those at intermediate latitudes. Such an equatorial anomaly in the latitudinal variation of D region electron concentrations, analogous to the well known behaviour of F region ionization, had been indicated previously by studies of the latitudinal variation of MF radio-wave absorption [45, 46]. A recent study by George [47] has shown that such an anomalous variation of absorption, with maxima near magnetic dip angles $\pm 20°$ and a minimum near the magnetic equator, occurred in all seasons during sunspot minimum and maximum years. Furthermore, the absorption was essentially constant at middle and high latitudes during summer and equinox months, but in winter showed an enhancement at latitudes corresponding to magnetic dip angles in the range 45—70°. This corresponds to the well-known anomalous absorption in winter months which is discussed in § 5.2.

5. Changes at Middle Latitudes in Winter

The variation of absorption for an equivalent frequency of 1 MHz at vertical incidence, as derived from observations of transmissions at 2.61 MHz propagating over a path of 295 km [11], is shown in Fig. 6 and illustrates three aspects of D region variations associated with winter: the sudden change from summer conditions occurring during October and November, and usually referred to as the "November effect"; the increased variability of absorption; and the anomalously large absorption relative to that expected from summer values on the basis of a $\cos^{0.75}\chi$ seasonal variation. It is known that all these aspects arise in part from changes in electron concentrations. It will be convenient to consider each in turn.

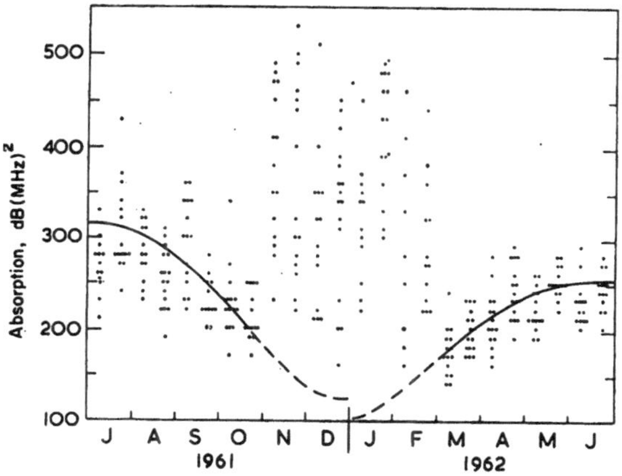

Fig. 6. Variation of the non-deviative absorption at noon for an effective frequency of 1 MHz at vertical incidence derived from measurements at 2.61 MHz over a path of 295 km [11]. The curves correspond to a $\cos^{0.75}\chi$ seasonal variation.

5.1. The November Effect

Early observations [48, 49] showed large changes in amplitude of LF signals transmitted across the Atlantic Ocean within a few days in November. The effect has been confirmed by several more recent studies of radio-wave propagation data and is illustrated in Fig. 7. This shows for the period August to December 1971 the variations of the amplitude near noon and midnight of the abnormal component of 16 kHz waves transmitted from Rugby to Slough. It is seen that a clear increase in amplitude occurs at noon, beginning at the end of September, but no such sudden change is observed at midnight.

In general, the change from summer to winter conditions is more easily identified at VLF and LF than at MF and HF, the greater variability in absorption at the higher frequencies serving to obscure it. It is of interest to note, however, that partial-reflection measurements by Gregory and Manson [43] at Christchurch, New Zealand, have revealed increases in electron concentrations at heights above 80 km between autumnal equinox and early winter but no such changes at lower heights. These workers have suggested that this effect in the D region is associated with the reversal in atmospheric circulation at heights between 20 and 100 km, towards the west in summer and towards the east in winter; the development of the eastward circulation at stratospheric heights in winter would enable perturbations at lower heights to propagate to the D region, as predicted theoretically [50]. However, the manner in which these perturbations lead to changes in electron concentrations has not been considered quantitatively.

5.2. Variability in Winter

Analyses of radio-wave absorption measurements at MF and HF by Appleton and Piggott [10] and of forward-scatter data at VHF by Bailey et al. [51] provided evidence of increased variability of the D region during winter. Partial-reflection

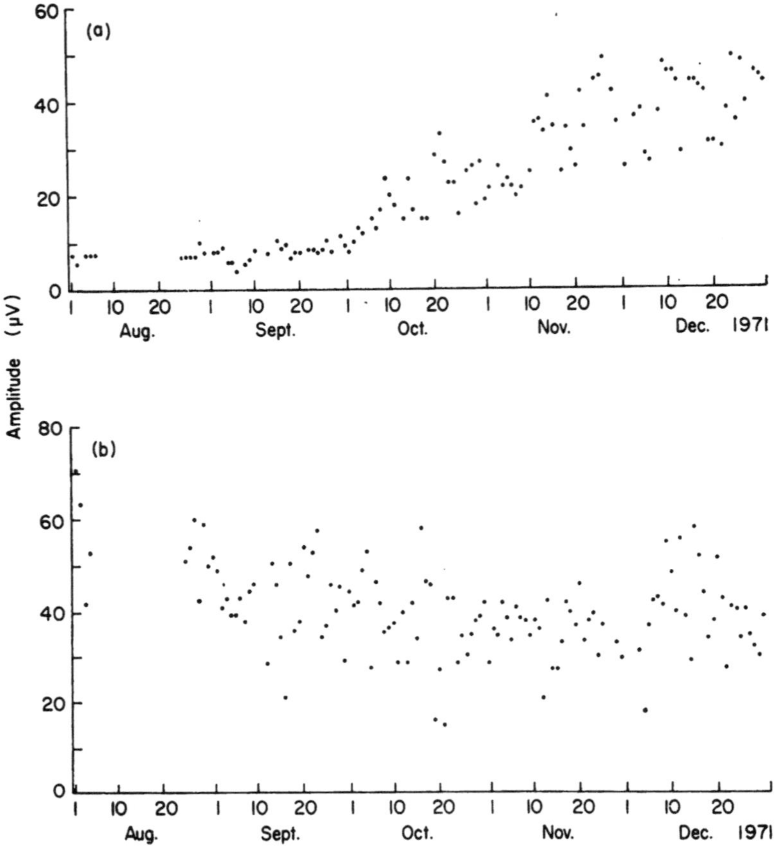

Fig. 7. Variations of the amplitude of the abnormal component of 16 kHz waves transmitted from Rugby to Slough during August—December 1971: (a) near noon, (b) near midnight.

measurements by Belrose et al. [52] and Gregory and Manson [43] have confirmed this finding. Furthermore, these measurements have been interpreted as indicating that a major part of the phenomenon arises from changes in electron concentration. Gregory and Manson have shown that the relative variability of electron concentrations is most marked at heights between 65 and 69 km, and Belrose et al. have also emphasized that the effect is most pronounced below 75 km. It is expected that such changes at low heights might contribute to the increased fluctuations of radio-wave propagation conditions in winter illustrated in Fig. 7(a), and also to the variability in absorption at MF shown in Fig. 6 since theoretical work [53] has demonstrated the importance of electron concentrations at low heights in determining the absorption at oblique incidence. The results of Gregory and Manson [43] also showed significant variations of electron concentrations between 80 and 88 km during winter and it might be expected that such changes could contribute to the variability of vertical incidence MF and HF radio-wave absorption.

A number of factors could contribute to the apparent increased variability at heights below 75 km in winter. Reid [54] has postulated that changes in positive-ion composition arising from variations in atmospheric temperature might be important. Changes in both positive and negative-ion composition resulting from variations in concentrations of minor neutral constituents, such as atomic oxygen or water, could also be important. At greater heights, above about 80 km, the changeability in the D region is associated with the degree of anomalous MF and HF radio-wave absorption.

5.3. Anomalous MF and HF Radio-wave Absorption during Winter

The anomalously large absorption of MF and HF radio waves during winter was first noticed by Appleton [55], and the main features of the phenomenon were later established by Dieminger [56], Appleton and Piggott [10], Beynon and Davies [57] and Thomas [58]. It is known to be limited to middle latitudes, the lower boundary being near 37–38°N in the North Atlantic [59] and 30–35°N in Japan [60]; the upper latitude boundary is obscured by the ionization effects of high-energy precipitated particles. Two types of anomalous days have been identified [61], namely those days on which the absorption is high but is consistent

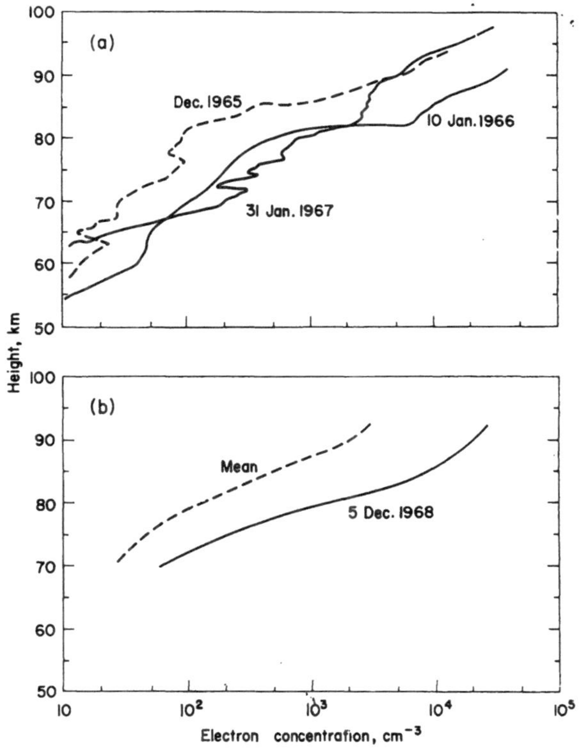

Fig. 8. Rocket-borne measurements of the height distributions of electron concentration on normal winter days (– – –) and days showing anomalously high MF radio-wave absorption (——): (a) Wallops Island, χ near 60° [63, 64]; (b) South Uist, $\chi = 80°$ [14].

with the average variation in winter, and those groups of days on which the absorption is excessively large.

Indirect evidence provided by radio-wave propagation data [62] indicated that the anomalously large absorption could be attributed to conditions in the upper D region. More recently, rocket-borne observations by Mechtly and Shirke [63] and Sechrist et al. [64] at Wallops Island and by Beynon and Williams [14] at S. Uist have revealed increased D region electron concentrations on anomalous days relative to normal days, the greatest increases being observed in the height range 80—86 km. These results are illustrated in Fig. 8. It has been proposed by Lastovicka [65] that such large enhancements of electron concentrations are probably responsible for the excessive absorption observed on groups of days, the average variation of high absorption throughout the winter being due to enhanced collisional frequencies at heights above 84—85 km [66]. However, Taubenheim [67] has since suggested that increases of both neutral pressure and electron concentration in the height range 85—105 km are required to explain the average winter anomaly. Furthermore, his studies indicate that short-term variations in pressure cause little change in radio-wave absorption since the change in collisional frequency is compensated by the accompanying change in electron concentration.

Mechanisms which could give rise to enhanced electron concentrations above about 80 km have been proposed by Maehlum [68], Armstrong et al. [69], Reid [54], and Geller and Sechrist [70]. A particular need is to distinguish between effects resulting from changes in positive or negative ion composition, caused by variations in concentrations of minor neutral constituents or temperature, and those arising from the production of ionization by high-energy particles.

6. Solar Cycle Variation

There is considerable information on the solar cycle variation of radio-wave absorption at LF, MF and HF [71, 72, 10]. In general, the absorption increases with increasing sunspot number, but recent work at MF [72] has indicated that the rate of increase varies with season, being most rapid in January and least in March. A study of the global morphology of MF absorption [47] has shown that the latitudinal variations are similar for years of high and low solar activities.

There is little direct information available for examining the change of the height distribution of electron concentration with the level of solar activity. The results of Mechtly and Smith [41] and Mechtly et al. [42] for Wallops Island shown in Fig. 5 indicate that there is little solar-cycle change between the distributions for the two summer days, except that below about 68 km the electron concentrations at high solar activity are smaller than those for low solar activity. However, comparison of the results for the winter days shows that the concentrations at high solar activity are substantially larger at all heights, but particularly near 80 km. A comparison of the median profile deduced from five distributions for all seasons at low solar activity with that from the corresponding five distributions for high solar activity [42] shows a similar change. Partial-reflection measurements at the higher latitude station Ottawa during 1962 [52] and 1970 [32] show that for March the mean distributions near noon were similar for the two periods but for June the concentrations in 1970 were larger by factors of two or more over the whole height range 60—80 km. Furthermore, Mitra and

Chakrabarty [73] have reported that at high latitudes the average concentrations for solar maximum are larger than those for solar minimum at all heights above 70 km, but the two sets of results are similar at lower heights.

7. Summary

The increased information on D region electron concentrations made available in recent years has served to establish certain features of the height distribution, and has also indicated some aspects of the temporal and spatial variations. There is, however, a continuing need for further measurements to provide a more general description of these variations. The important features of the observations made to date are summarized and some particular requirements for future studies are indicated as follows:

1. The steep gradient of electron concentration which occurs at a height between 80 and 90 km is a feature of the daytime D region at middle and low latitudes [18–20].
2. The pre-sunrise changes at middle latitudes during summer months consist of the development of a C layer at heights between 60 and 75 km, the electron concentrations at greater heights remaining at low values until after sunrise [27–29]. No measurements of the corresponding changes in winter or at low latitudes are available.
3. The diurnal variations of electron concentrations at particular heights show clear evidence of solar control. However, measurements at different locations show conflicting results concerning the asymmetry about noon [31–34].
4. Radio-wave propagation measurements have shown very pronounced D region changes associated with certain geomagnetic storms; the major effects persist for several days after the storm has ended [9, 35–38]. No direct measurements of electron concentrations have been made at such times.
5. The seasonal variation of electron concentrations for fixed values of solar zenith angle seems to vary with the epoch of the solar cycle. At solar minimum electron concentrations in the height range 65–83 km are larger in summer than winter [41, 30]; it has been reported that no seasonal change occurs at greater heights [41] although southern hemisphere values [43] have indicated that winter-time values are larger. For solar maximum conditions a limited number of rocket measurements have indicated that in the height range 82–95 km winter values are larger than summer, little change occurring at lower heights [42].
6. There is some evidence [44] that electron concentrations in the height range 70–90 km show an anomalous latitudinal variation similar to that found for MF absorption [45, 46] in which larger values were found at latitudes corresponding to magnetic dip angles $\pm 20°$ than at higher latitudes or near the magnetic equator.
7. The results of partial-reflection measurements have shown that the winter-time variability is most marked at heights below 75 km [43, 52]. In addition, the electron concentrations are enhanced on days of anomalously high MF/HF radio-wave absorption, especially between 80 and 86 km [14, 63, 64]. In view of the large variability in height distributions of electron concentrations in winter, more measurements are required.
8. Radio-wave absorption measurements indicate substantial changes of the D region during a solar cycle [10, 71, 72] but relatively little information is available

about the corresponding variation in the height distributions of electron concentration. A comparison of rocket measurements [42] indicates that the median concentrations at high solar activity are larger than those for low solar activity at all heights, but particularly near 80 km.

Acknowledgment

This paper is published by permission of the Director of the Appleton Laboratory of the Science Research Council.

References

[1] F. F. Gardner and J. L. Pawsey, J. Atmos. Terr. Phys. **3**, 321 (1953).
[2] J. S. Belrose and M. J. Burke, J. Geophys. Res. **69**, 2799 (1964).
[3] J. B. Gregory and A. H. Manson, J. Atmos. Terr. Phys. **31**, 683 (1969).
[4] J. A. Fejer, J. Atmos. Terr. Phys. **7**, 322 (1955).
[5] R. E. Barrington, E. V. Thrane and B. Bjelland, Canad. J. Phys. **41**, 271 (1963).
[6] R. A. Smith, in: Ground-based Radio Wave Propagation Studies of the Lower Ionosphere, DRTE, Ottawa, Canada, Vol. I, p. 235 (1967).
[7] E. A. Mechtly, S. A. Bowhill, L. G. Smith and H. W. Knoebel, J. Geophys. Res. **72**, 5239 (1967).
[8] R. N. Bracewell, K. G. Budden, J. A. Ratcliffe, T. W. Straker and K. Weekes, Proc. Instn Elect. Engrs **98**, 221 (1951).
[9] E. A. Lauter and K. Sprenger, Z. f. Met. **6**, 161 (1952).
[10] E. V. Appleton and W. R. Piggott, J. Atmos. Terr. Phys. **5**, 141 (1954).
[11] H. Schwentek, J. Atmos. Terr. Phys. **25**, 733 (1963).
[12] P. E. Krasnushkin, Dokl. Akad. Nauk SSSR, **139**, 67 (1961).
[13] D. G. Deeks, Proc. Roy. Soc. A **291**, 413 (1966).
[14] W. J. G. Beynon and E. R. Williams, J. Atmos. Terr. Phys. **32**, 1325 (1970).
[15] J. N. Rowe, A. J. Ferraro, H. S. Lee, R. W. Kreplin and A. P. Mitra, J. Atmos. Terr. Phys. **32**, 1609 (1970).
[16] E. A. Mechtly, C. F. Sechrist and L. G. Smith, J. Atmos. Terr. Phys. **34**, 641 (1972).
[17] R. H. Doherty, Radio Sci. **6**, 639 (1971).
[18] E. A. Mechtly and L. G. Smith, Radio Sci. **5**, 1407 (1970).
[19] F. D. G. Bennett, J. E. Hall and P. H. G. Dickinson, J. Atmos. Terr. Phys. **34**, 1321 (1972).
[20] A. C. Aikin, R. A. Goldberg, Y. V. Somayajulu and M. B. Avadhanulu, J. Atmos. Terr. Phys. **34**, 1483 (1972).
[21] H. S. Lee and A. J. Ferraro, J. Geophys. Res. **74**, 1184 (1969).
[22] A. R. Cole, A. J. Ferraro and H. S. Lee, Nature **222**, 761 (1969).
[23] W. C. Bain and M. D. Harrison, Proc. Instn Elect. Engrs **119**, 790 (1972).
[24] T. W. Straker, Proc. Instn Elect. Engrs **102 C**, 122 (1955).
[25] K. Maeda, J. Geomag. Geoelect. Kyoto **21**, 557 (1969).
[26] P. Knight, J. Atmos. Terr. Phys. **34**, 401 (1972).
[27] E. A. Mechtly and L. G. Smith, J. Atmos. Terr. Phys. **30**, 363 (1968).
[28] L. Thomas and M. D. Harrison, J. Atmos. Terr. Phys. **32**, 1 (1970).
[29] R. A. Smith, T. N. R. Coyne, R. G. Loch and I. A. Bourne, in: Ground-based Radio Wave Propagation Studies of the Lower Ionosphere, DRTE, Ottawa, Canada, Vol. I, p. 335 (1967).
[30] J. B. Gregory and A. H. Manson, J. Geophys. Res. **72**, 1073 (1967).
[31] E. V. Thrane, A. Haug, B. Bjelland, M. Anastassiades and E. Tsagakis, J. Atmos. Terr. Phys. **30**, 135 (1968).

[32] T. N. R. Coyne and J. S. Belrose, Radio Sci. **7**, 163 (1972).
[33] E. V. Thrane, J. Geophys. Res. **74**, 1311 (1969).
[34] A. Haug, E. V. Thrane, E. Tsagakis and M. Anastassiades, J. Atmos. Terr. Phys. **32**, 1865 (1970).
[35] I. A. Bourne and L. W. Hewitt, J. Atmos. Terr. Phys. **30**, 1381 (1968).
[36] J. S. Belrose and L. Thomas, J. Atmos. Terr. Phys. **30**, 1397 (1968).
[37] E. Hammond and M. D. Harrison, in: Rep. UAG-12, NOAA, Boulder, p. 237 (1971).
[38] E. A. Lauter and R. Knuth, J. Atmos. Terr. Phys. **29**, 411 (1967).
[39] G. M. Weill and J. Christophe-Glaume, Annls Géophys. **23**, 317 (1967).
[40] N. N. Shefov, Planet. Space Sci. **17**, 797 (1969).
[41] E. A. Mechtly and L. G. Smith, J. Atmos. Terr. Phys. **30**, 1555 (1968).
[42] E. A. Mechtly, S. A. Bowhill and L. G. Smith, J. Atmos. Terr. Phys. **34**, 1899 (1972).
[43] J. B. Gregory and A. H. Manson, J. Atmos. Terr. Phys. **31**, 703 (1969).
[44] E. A. Mechtly, M. Mukunda Rao, D. O. Skaperdas and L. G. Smith, Radio Sci. **4**, 517 (1969).
[45] M. D. Fligel, Ionosph. Res. **10**, 5 (1961).
[46] J. S. Shirke and G. W. Henry, Annls Géophys. **23**, 517 (1967).
[47] P. L. George, J. Atmos. Terr. Phys. **35**, 1893 (1971).
[48] H. J. Round, T. L. Eckersley, K. Tremellen and F. C. Lunnon, J. Instn Elect. Engrs **63**, 933 (1925).
[49] J. Hollingworth, J. Instn Elect. Engrs **64**, 579 (1926).
[50] J. G. Charney and P. G. Drazin, J. Geophys. Res. **66**, 83 (1961).
[51] D. K. Bailey, R. Bateman and R. C. Kirby, Proc. Inst. Radio Engrs **43**, 1181 (1955).
[52] J. S. Belrose, I. A. Bourne and L. W. Hewitt, in: Electron Density Profiles in Ionosphere and Exosphere, North-Holland Publ. Co., Amsterdam 1966 (p. 48).
[53] R. E. Evans and T. B. Jones, J. Atmos. Terr. Phys. **33**, 627 (1971).
[54] G. C. Reid, J. Geophys. Res. **75**, 2551 (1970).
[55] E. V. Appleton, Proc. Roy. Soc. A **162**, 451 (1937).
[56] W. Dieminger, J. Atmos. Terr. Phys. **2**, 340 (1952).
[57] W. J. G. Beynon and K. Davies, in: Proc. Conf. Phys. Ionosphere, Physical Society, London 1955 (p. 40).
[58] L. Thomas, J. Atmos. Terr. Phys. **23**, 301 (1962).
[59] B. Schaning, J. Atmos. Terr. Phys. **35**, 1003 (1973).
[60] N. Wakai, C. Ouchi and C. Nemoto, J. Radio Res. Lab. Japan **17**, 185 (1970).
[61] H. Schwentek, J. Atmos. Terr. Phys. **33**, 1647 (1971).
[62] E. A. Lauter and P. Tříska, Geomagn. and Aeron. **5**, 822 (1965).
[63] E. A. Mechtly and J. S. Shirke, J. Geophys. Res. **73**, 6243 (1968).
[64] C. F. Sechrist, E. A. Mechtly, J. S. Shirke and J. S. Theon, J. Atmos. Terr. Phys. **31**, 145 (1969).
[65] J. Lastovicka, J. Atmos. Terr. Phys. **34**, 1233 (1972).
[66] J. Taubenheim, J. Atmos. Terr. Phys. **33**, 1481 (1971).
[67] J. Taubenheim, Gerlands Beitr. Leipzig **81**, 396 (1972).
[68] B. Maehlum, J. Geophys. Res. **72**, 2287 (1967).
[69] R. J. Armstrong, F. Lied and E. V. Thrane, Physica Norvegica **4**, 157 (1970).
[70] M. A. Geller and C. F. Sechrist, J. Atmos. Terr. Phys. **33**, 1027 (1971).
[71] E. A. Lauter, Annls Géophys. **22**, 289 (1966).
[72] H. Schwentek, J. Atmos. Terr. Phys. **33**, 1839 (1971).
[73] A. P. Mitra and D. K. Chakrabarty, Space Research XI, 1013 (1971).

Lower Ionosphere Structure — Akademie-Verlag, Berlin 1974

COMPARISON OF ELECTRON DENSITY PROFILES IN THE LOWER IONOSPHERE AT EQUATOR AND MID-LATITUDES

Y. V. SOMAYAJULU, M. B. AVADHANULU and K. S. ZALPURI

National Physical Laboratory, New Delhi, India

The electron density profiles in the equatorial lower ionosphere obtained from rocket experiments are compared with the mid-latitude profiles obtained during similar conditions of solar activity and using the same techniques. For the comparison we have used median mid-latitude electron densities obtained by Mechtly et al. The comparison reveals that the electron density profiles in the altitude range 70—80 km for mid and equatorial latitudes agree very well both in magnitude and in general features. This is the altitude range where Lyman-alpha radiation acts as a dominant production source and no significant latitudinal dependence is expected to be observed. Thus the results are consistent with this concept. Significant departures are, however, observed below 70 km. In this altitude range the equatorial electron densities are found to be larger than the mid-latitude values in contradiction to the current concepts of production mechanisms. The differences are shown to be primarily due to the negative ion chemistry.

1. Introduction

In recent years considerable experimental effort has been made to determine the D region electron densities, ion densities and composition and their diurnal and seasonal variation by rockets and ground-based techniques. The most extensive data exist for mid-latitudes while for low latitudes the data obtained are rather limited. These limited data do, however, permit a comparison to be made of the equatorial D region profiles with the profiles obtained at mid-latitude under similar conditions of solar activity.

In the present study the equatorial electron density profiles measured by propagation techniques are utilized. For mid-latitudes, the average profiles given by Mechtly et al. [1] the profiles obtained by the wave interaction method by the Penn State group [2] and that of Bain and Harrison [3] are used. All these profiles are for periods of high solar activity and for near equinox. The details of the various electron density profiles [1, 3—5] are given in Table 1.

The electron density profiles, derived from rocket experiments conducted from Thumba, at the equator are shown in Fig. 1. These profiles cover a range of solar zenith angles from 6° to 53°. For comparison with mid-latitudes, the mean electron density profile suggested by Mechtly et al. [1] for similar conditions and the Bain and Harrison [3] profile for noon for $\chi \sim 30°$ are also shown in Fig. 1. The electron density profiles obtained from Penn State wave interaction technique for about the same period are also used and are given in Fig. 2.

Table 1

Details of electron density profiles shown in Fig. 1

S. No.	Ref.	Geographic location	Solar zenith angle	Remarks
1.	[4]	8° 30′ N 76° 54′ E (Thumba)	28°	*$R = 48$ 19 March 70 ISRO 17.05/NASA 14.424
2.	[4]	8° 30′ N 76° 54′ E (Thumba)	48°	$R = 48$ 19 March 70 ISRO 17.06/NASA 14.372
3.	[4]	8° 30′ N 76° 54′ E (Thumba)	53°	$R = 48$ 19 March 70 ISRO 17.04/NASA 14.425
4.	[5]	8° 30′ N 76° 54′ E (Thumba)	6°	$R = 115$ 27 March 70
5.	[1]	37° 50′ N 75° 29′ W (Wallops Is.)	60°	Median profile for solar maximum
6.	[3]	Mid-latitude	30°	Summer noontime model for solar maximum

* R = Sunspot number

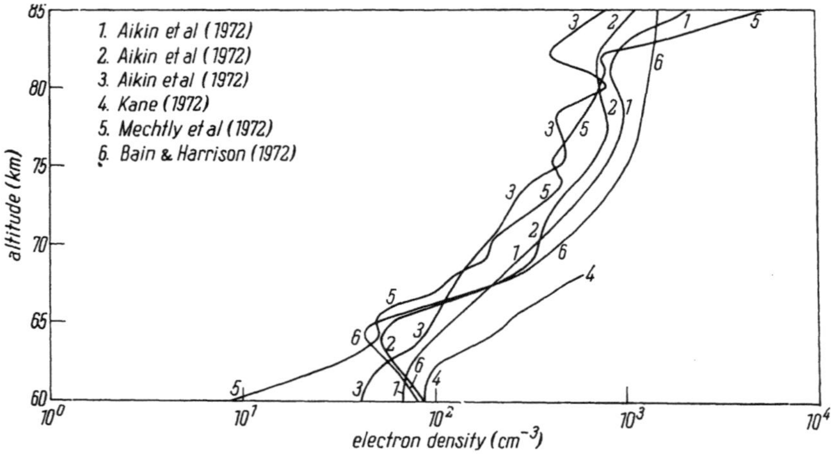

Fig. 1. Electron density profiles at equator and mid-latitudes for solar maximum conditions. Details are given in Table 1.

2. Electron Density in the 70–80 km Range

It is seen from the equatorial electron density profiles that the electron density increases with altitude in 70–80 km range, it shows a minimum or valley around 82 km. A comparison with mid-latitude profiles reveals that all these equatorial profiles are consistent both in general features and in the absolute values of the number densities with solar zenith angle control. This is the region which is believed to be dominated by Lyman-alpha production and the agreement between equatorial and mid-latitude profiles supports this conclusion.

3. Electron Density in the 60—70 km Altitude Range

The electron density profile in the 60—70 km range presents a more complex picture. It is noted that electron density decreases rapidly with decreasing altitude from 70 km to about 65 km, below which the density decreases more slowly or even shows an increase with decreasing altitude between 60 and 65 km. The slope between 60 and 65 km shows a diurnal variation, decreasing with time towards noon.

One remarkable feature is the asymmetrical behaviour observed in the afternoon. This is shown by the profile for the flight 14.372 for $\chi = 48°$ in the afternoon at 1530. This profile shows a valley with a minimum around 64 km; below that altitude the electron density increases with decreasing altitude down to 60 km. On the other hand the profile for $\chi = 53°$ in the forenoon does not display this behaviour. With the exception of this profile, all the other profiles relate to forenoon and give the impression that the minimum in electron density around 64 km flattens with decreasing solar zenith angle, presumably due either to detachment from negative ions or to ionization by Lyman-α radiation. In this region, a comparison with the mid-latitude profiles of Mechtly et al. [1] shows that mid-latitude values are very much lower than the corresponding equatorial values. As remarked earlier the electron density values as suggested by Mechtly et al. [1] may be too low to be realistic. On the other hand, the mid-latitude profile due to Bain and Harrison [3] for $\chi = 30°$ agrees very closely with 14.372 profile (for $\chi = 48°$). For further comparison the electron density profiles obtained from the wave interaction experiment of Penn State [2] around the period of rocket data are shown in Fig. 2. One consistent feature of these profiles is the presence of valley at or around 70 km altitude. The equatorial profiles indicate a change in slope or a minimum around 64 km. The discrepancy in the position of the valley in the results of the wave interaction experiment and those of Bain and Harrison [3] is not understood. If we take the results of the wave interaction experiment as representative of mid-latitudes then one of the main differences in the behaviour of the lower D region at mid-latitudes and the equator is height of the valley; it occurs at lower altitude at the equator.

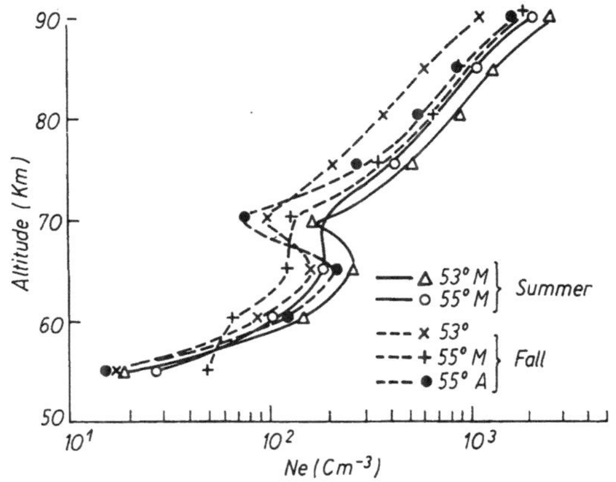

Fig. 2. Electron density profiles obtained from wave interaction experiment (After Rowe [2]).

4. Factors controlling the Electron Densities in the Altitude Range of 60—70 km

The equilibrium electron densities in the altitude range 60—70 km are controlled, apart from the two competing ionizing mechanisms, namely, solar Lyman-α radiation and galactic cosmic rays, by ion composition and ion chemistry.

The direct production processes involve the ionization of nitric oxide by solar Lyman-α radiation and all the neutral constitennts by galactic cosmic rays. In this altitude region the importance of the Lyman-α ionization decreases with decreasing altitude. On the other hand the ionization due to galactic cosmic rays is proportional to the total neutral atmospheric density and hence a decreasing function of altitude, while at 60 km level the cosmic-ray ionization is the dominant ionization mechanism. At altitudes close to 70 km solar Lyman-α radiation dominates as the ionization source. The changeover or transition from one to the other occurs in this altitude interval. The transition altitude varies in principle with latitude since, while the solar Lyman-α flux is independent of latitude for a given value of solar zenith angle, the cosmic-ray flux has a strong latitude dependence. Thus, one would expect a lower transition altitude at low latitudes compared with mid-latitudes.

The production rates due to the two mechanisms are computed for the solar zenith angles of 53°, 48° and 28° corresponding to the three rocket flights on 19 March 1970, using Meira's [6] values for nitric oxide. These are shown in Fig. 3. The cosmic-ray production rates are computed following Velinov [7]. It is clearly seen that at 60 km height the cosmic-ray production is the dominant one while at altitudes above 67 km, the solar Lyman-α is the dominant production mechanism. The cross over takes place for large zenith angles at around 65 km and at around

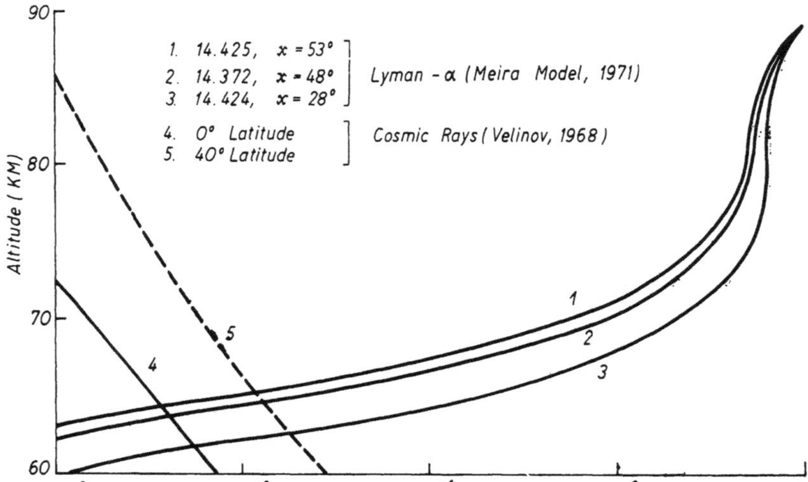

Fig. 3. Production rates due to Lyman-α and cosmic rays plotted against altitude for different zenith angles corresponding to the three rocket flights on the same day ($\chi = 28°$, $48°$ and $53°$).

62 km for low zenith angles. This change over forms a sort of valley or minimum around this altitude in the total production rate profile. It is seen that a change in slope in electron density profile occurs around 64 km. However, it is easy to see that this change in slope in electron density is not caused by the valley in the production rate profile. The electron density values based on the production rate profile, for example for $\chi = 28°$, have been computed using the relation

$$N_e = (q/\psi)^{1/2} \tag{1}$$

where N_e is the electron density, q is the production rate and ψ represents the loss coefficient given by

$$\psi = (1 + \lambda)(\alpha_D + \lambda\alpha_i) \tag{2}$$

where λ = ratio of negative ions to electron density, α_D = dissociative recombination coefficient for electrons and ions and α_i = ion-ion recombination coefficient.

Since the region below 80 km is dominated by water cluster ions $\alpha_D = 10^{-11}$ $m^3 s^{-1}$ is adopted [4]. An average λ-profile based on data available is used. These details are shown in Table 2.

Table 2

Altitude (km)	λ	$10^{11}\ \psi/m^3\ s^{-1}$	N_e/m^{-3} comp.	N_e/m^{-3} expt.
60.0	6.0	7.0	2.1×10^7	7.0×10^7
62.5	3.5	4.5	3.5×10^7	8.0×10^7
65.0	2.0	3.0	8.2×10^7	1.20×10^8
67.5	1.4	2.4	1.82×10^8	2.10×10^8
70.0	1.0	2.0	3.16×10^8	3.50×10^8

The computed electron densities show that the electron density profile due to cosmic rays and Lyman-α production would lead to a continuously decreasing electron density profile and would not account for a change in the slope around 64 km as observed. Therefore the cause of this change in the slope will have to be sought in the negative ion conversion to electrons due to detachment processes.

Current ideas on the negative ion chemistry of the D region are based on O_2^- acting as the precursor ion. The rapid attachment of electrons to molecular oxygen in a three-body reaction results in the formation of O_2^- [8] which then leads to the formation of other complex negative ions through two channels, one through $O_3^- \rightarrow CO_3^- \rightarrow NO_2^-$ and the other through $O_4^- \rightarrow CO^-_4 \rightarrow NO_3^-$.* Hydration of these ions can occur resulting in heavy cluster ions.

Several theoretical models of negative ion composition have been suggested [9—12] based on O_2^- as the precursor ion. The negative ion composition and the equilibrium electron concentration below 70 km, where large preponderance of negative ions is expected, are controlled by attachment and detachment processes and minor neutral constituents play an important part. For example, in the reactions mentioned above the concentration of neutral nitric oxide decides the relative preponderances of NO_2^- and NO_3^- because O_3^- or CO_4^- is the dominant

* Of these two, the latter reaction is considered faster and more effective at least in the altitude range of interest.

ion as in Reid's [9] model. On the other hand with a low concentration of NO, as in the Meira's [6] model, photo or collisional detachment would lead to a situation with CO_3^- and CO_4^- as the dominant ions with NO_2^- relatively unimportant. Rowe [2] observes that in order to reconcile the experimentally observed electron densities below 70 km obtained from the Penn State wave-interaction experiment, a much larger value of a weighted detachment rate is required than the normally used value of 5×10^{-3} s^{-1} for photo detachment. Thomas et al. [12] from a detailed study deduce a photo-detachment rate of 0.04, 0.04 and 0.03 for NO_2^-, NO_3^- and CO_3^- respectively. Thus these works point out the importance of detachment rate to account for the experimentally observed electron densities below 70 km.

There are a few mass-spectrometer measurements of negative ions available [13, 14]. Unfortunately these are either for night-time conditions or during an eclipse or a PCA event but none exists for daytime conditions. The significant feature of the results of these measurements is the detection of CO_3^-, CO_4^-, NO_2^- and NO_3^- as major negative ions. The results are conflicting regarding the presence of heavy cluster ions.

At altitudes close to 60 km the equilibrium electron density may be expressed as:

$$N_e = \frac{q + \gamma N^-}{\beta} \tag{3}$$

where N^- = total negative ion density, β = attachment coefficient for electrons.

If B is the lumped rate of reaction for the conversion of NO^+ ions to water clusters then β is given by

$$\beta = (1 + \lambda)\ B = \left(\frac{N_e + N^-}{N_e}\right) B. \tag{4}$$

Substituting the value of β in Eq. (3) we get

$$\gamma = \frac{\beta N_e - q}{N^-} = \frac{B(N_e + N^-) - q}{N^-}. \tag{5}$$

Mitra and Rowe [11] have shown that the value of $B = 10^{-2}$ s^{-1} and remains constant in the altitude range of interest. Using Eq. (5), γ is determined utilizing the experimentally determined value of N_e for flights 14.424 and 14.425 and taking $B = 10^{-2}$ s^{-1}. The profiles of γ thus derived are shown in Fig. 4. It is seen that for near noon time flight ($\chi = 28°$), the value of γ is of the order of 10^{-2} s^{-1} and is almost constant with increase of altitude. For morning time flight 14.425 the values at 67.5 and 70 km were considered less reliable owing to the possibility of large errors in the negative ion density due to their very low values which are determined from the difference between the concentration of electrons and total positive ions, and therefore these are neglected in this study. The rest of the values of γ for 14.425 for heights <67 km are in good agreement with the value obtained for 14.424 flight and also with the weighted mean of values of photodetachment rate derived from a theoretical study by Thomas et al. [12]. Thus the results of the present study appear to be consistent with the view that photodetachment is important in the daytime D region below 70 km and that the detachment rates are of the order of 10^{-2} s^{-1}.

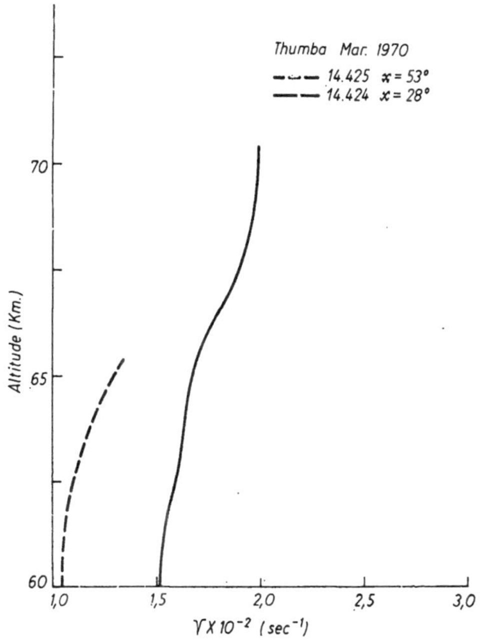

Fig. 4. Altitude variation of γ for flights 14.424 and 14.425.

Below 65 km where $q \ll \gamma N^-$, the electron density is given approximately by

$$N_e = \frac{\gamma}{\beta} N^-. \tag{6}$$

Since the negative ion density increases with decreasing altitude, the release of electrons from negative ions by detachment process would produce a change in the slope of electron density at altitudes below about 65 km. Thus the change in slope in the electron density profile between 60 and 70 km is to be attributed to change from the Lyman-α production control to γN^- control of the electron densities.

The profile as well as its slope in the 60—65 km range also shows a significant diurnal variation. From Eq. (6) we see that since β is a constant, a diurnal variation γN^- is implied. The magnitude and nature of γ which is a lumped detachment rate is not yet clear. For example a diurnal variation in the negative ion composition can cause γ to vary with time; alternatively, the diurnal variation of N^- may produce the diurnal variation of the features of the electron density below 65 km.

Indeed from the simultaneous measurement of electron and positive ion densities on two rocket flights in the forenoon of 19 March 1970 reported by Aikin et al. [4] it is observed that the negative ion densities as determined by the difference of the positive ion and electron densities decreased with time, especially in the lower altitude range. For instance for the morning flight (0830) the negative ion density at 60 km is about 8×10^8 m^{-3} (compared with $\simeq 1.3 \times 10^8$ at 1030). Keeping in view the limitations of the accuracy of measurement below 70 km, this may still be taken as indicative of a large supply of negative ions in the forenoon which is gradually depleted with time. Thus at night we have, so to speak, a "reservoir" of negative ions which acts as a source for electrons during the day for altitudes below 70 km.

If we take the above behaviour of the negative ion densities in the forenoon as representative of normal conditions, then with increasing time approaching noon and in the afternoon, the negative ion density at each altitude may reach a limiting value. This would then lead to an asymmetry with respect to noon in the negative ion densities and hence in the electron densities. In fact, in the q dominated region (above, say, 65 km), this asymmetry for large zenith angle may result in a valley in the profile because the electron densities decrease with increasing solar zenith angle while in the γN^- region the densities vary less steeply. Thus the valley in the electron density profile as observed on the flight 14.372 ($\chi = 48°$, time 1530) may have been due to this process, while in the forenoon on flight 14.425 ($\chi = 53°$, time 0830) the profile does not show any such valley. More data would be needed to confirm this point.

Acknowledgments

The authors wish to thank Dr A. P. Mitra for helpful discussions. One of the authors (K.S.Z.) is recipient of a fellowship from the Council of Scientific and Industrial Research, India, and gratefully acknowledges the financial support.

References

[1] E. A. Mechtly, S. A. Bowhill and L. G. Smith, J. Atmos. Terr. Phys. **34**, 1899 (1972).
[2] J. N. Rowe, Scient. Rep. No. 406, Penn. State Univ., Pennsylvania (1972).
[3] W. C. Bain and M. D. Harrison, Proc. IEE, **119**, 790 (1972).
[4] A. C. Aikin, R. A. Goldberg, Y. V. Somayajulu and M. B. Avadhanulu, J. Atmos. Terr. Phys. **34**, 1483 (1972).
[5] J. A. Kane, NASA, GSFC Rep. No. X-625-72-18 (1972).
[6] L. G. Meira, J. Geophys. Res. **76**, 202 (1971).
[7] P. Velinov, J. Atmos. Terr. Phys. **30**, 1891 (1968).
[8] L. M. Chanin, A. V. Phelps and A. M. Biondi, Phys. Rev. Lett. **2**, 344 (1959).
[9] G. C. Reid, J. Geophys. Res. **75**, 2551 (1970).
[10] E. E. Ferguson, Aeronomy Rep. No. 48, 217, Univ. of Illinois, Urbana, Illinois (1972).
[11] A. P. Mitra and J. N. Rowe, J. Atmos. Terr. Phys. **34**, 795 (1972).
[12] L. Thomas, P. M. Gondhalekar and M. R. Bowman, Nature **238**, 89 (1972).
[13] R. S. Narcisi, Aeronomy Rep. No. 48, 221, Univ. of Illinois, Urbana, Illinois (1972).
[14] F. Arnold and D. Krankowsky, J. Atmos. Terr. Phys. **33**, 1693 (1971).

Lower Ionosphere Structure — Akademie-Verlag, Berlin 1974

EQUATORIAL D REGION INVESTIGATIONS

J. S. Shirke and S. N. Pradhan

Physical Research Laboratory, Ahmedabad, India

A Nike-Apache rocket was launched from Thumba Equatorial Rocket Launching Station on 7 April 1972 at 1208 LT ($\chi = 3°$). The altitude distribution of electron concentration obtained from radio wave propagation experiment is discussed. An electron-neutral collision frequency model is determined. The electron density profile of this flight is compared with the profiles obtained by other investigators at low and mid latitude stations. The relative contributions are assessed of the solar Lyman α and solar X-ray fluxes in producing the long term variations in electron density profile. A downward displacement is noticed at the equator when compared with mid latitudes of a ledge in ionization situated near the mesopause. The reasons are discussed.

1. Introduction

A Nike-Apache rocket (ISRO 10.43) was launched from Thumba Equatorial Rocket Launching Station on 7 April 1972 at 1208 hours local time (solar zenith angle $\chi = 3°$). The rocket carried a Langmuir probe, a radio-frequency probe, an ultraviolet detector and a propagation experiment to study the aeronomic aspects of the equatorial lower ionosphere. This paper mainly deals with the results of the radio-wave propagation experiment. The electron number density profiles obtained by other investigators during the period 1965—1972 have been compared. The origin of the long term variation in electron density is investigated.

2. Experimental Technique

The propagation experiment consisted of a ground-based radio-wave transmitter operating at a frequency of 2.3 MHz. Two half-wave dipoles one oriented in the north-south direction and the other in the east-west were alternately used to obtain the ordinary and extraordinary modes of transmission respectively. A sensitive receiver with the antenna housed in a fibreglass nose-cone of the rocket monitored the field strength of the ground-based transmitter. The differential absorption was monitored as a function of altitude and used to determine the electron number density profile. For Thumba which is located very close to the dip equator the propagation is transverse to the magnetic field lines and hence the Sen and Wyller [1] generalized magneto-ionic formulation for the transversal case is applied for calculating the number densities of electrons.

3. Results and Discussion

The determination of the number density of electrons from the differential absorption data requires the knowledge of the electron-neutral collision frequency as a function of altitude. The collision frequency is assumed to be proportional to the pressure and can be expressed as:

$$\nu = kp$$

where k is a proportionality constant and p is the pressure in N m^{-2}. The value of the proportionality constant k varies with season and latitude from 6×10^5 to 9×10^5 [2].

Here we have tried to find the collision frequency model for low latitudes in the following manner. The electron number densities are accurately known at the levels of reflection of the ordinary and the extraordinary modes which in the present case are 96 km and 92.4 km respectively. In between the two levels the electron densities could be extrapolated linearly without committing a serious error. The measured differential absorption in this altitude region is used to find the collision frequencies as a function of altitude. Assuming the pressure profile to be known the constant of proportionality could be determined. The collision frequencies so obtained fit in with the relation:

$$\nu = 8.2 \times 10^5 p .$$

The atmospheric pressure profile measured at Thumba by an M-100 meteorological rocket on the previous night (private communication, Prof. P. R. Pisharoty) is used. The pressure profile was obtained only up to 80 km. Above 80 km level the values are extrapolated using expressions given by CIRA 1965. The electron number density profile has been derived assuming this collision frequency and the differential absorption data from flight 10.43. The results are shown in Fig. 1 by profile 6. The electron density profiles 1 to 5 are made at the dip equator by other investigators [3—7]. The profile 7 is made at Wallops Island, a mid-latitude station [8].

At low latitudes the radio wave absorption L as a function of solar zenith angle χ is expressed as

$$L = K \cos^n \chi$$

where K and n are constant.

For diurnal variation the value of the index n giving the solar zenith angle dependence on absorption is 0.95 as determined by Gnanalingam [9]. This value of the index has been used to normalize the electron density profile to solar zenith angle $\chi = 0°$ so that a meaningful comparison could be made between the profiles which are made at various zenith angle. All the profiles shown in Fig. 1 except three were made during the equinoctial months. The seasonal variations between the profiles are therefore automatically taken care of. Again these profiles were made during the period 1965—1972. The solar activity was at minimum in 1965 and reached its maximum in 1968, so the observed variations existing among the various profiles could be attributed to one of the long-term variations in solar activity. The most important source of ionization in the D region is the ionization of nitric oxide (NO) by Lyman-α radiation. Blamont and Vidal Madjar

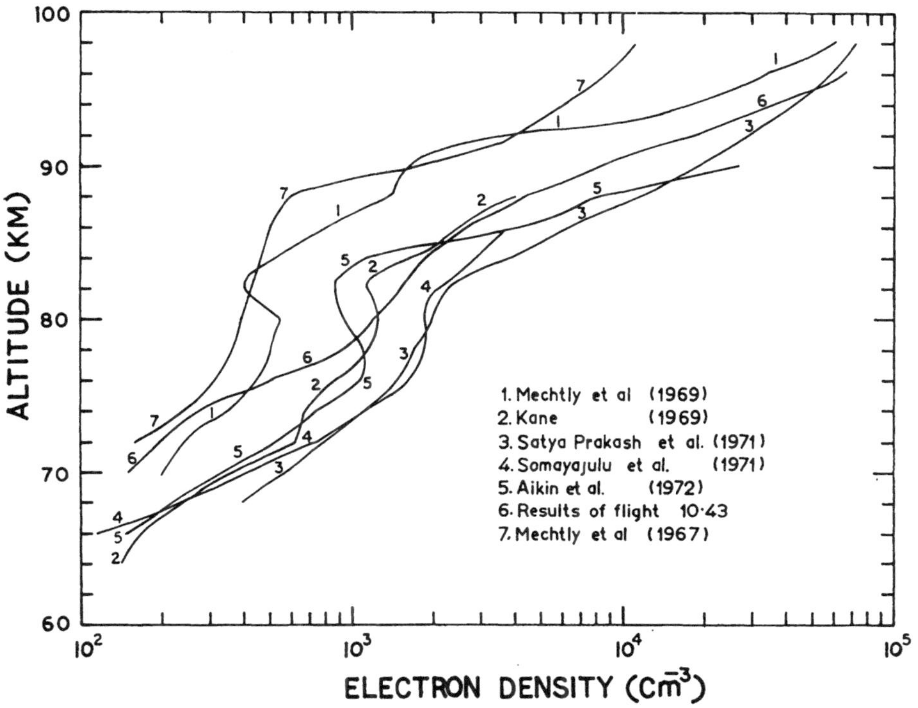

Fig. 1. Electron density profiles 1 to 6 are made at the dip equator and 7 at a mid-latitude station.

[10] have shown, though for a limited period, a cent per cent correlation between the intensity of solar Lyman-α radiation and the Zürich sunspot number. Their conclusion is based on the results of an experiment to monitor the solar Lyman-α on board the satellite OSO 5. This means that the Zürich sunspot number provides a very good index to study the variation in intensity of solar Lyman-α which is of consequence to D region ionization. The other important source of ionization in the D region is considered to be hard solar X-rays (< 10 Å). The solar X-rays are monitored by Naval Research Laboratory satellites and the flux values are published in *Solar Geophysical Data*. These are used in the following computations.

The correlation coefficient between Zürich sunspot number and electron number density is determined as a function of altitude and is shown in Fig. 2. The correlation coefficient between X-ray flux on 1—8 Å radiation and the electron number density as a function of altitude is also shown in Fig. 2. It is easy to see that the electron concentration in the altitude region 70—86 km is well correlated with the Zürich sunspot number, the average value being close to 0.8, whereas the correlation in the same region with X-ray flux is poor, the the average value being close to 0.3. The solar X-rays are known to produce short-term effects like solar flare when the electron density in the D region is enhanced. From absorption data at Colombo [9] it is found that absorption is better correlated with X-ray flux on a day-to-day basis. From Fig. 2 it appears that the long-term variation in the electron density is mainly due to variation in the intensity of the solar Lyman-α.

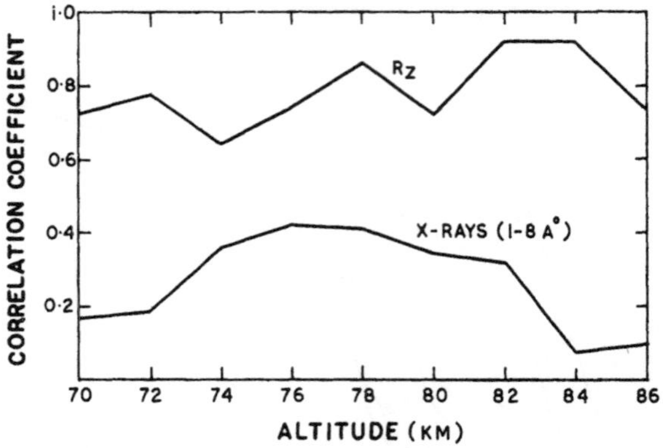

Fig. 2. Correlation coefficient of electron density with sunspot number R_Z and X-ray flux in 1–8 Å band.

A typical mid-latitude electron density profile made by Mechtly et al. [5] is also included in Fig. 1 so that a comparison could be made between low and middle latitudes. All the low latitude profiles show a sudden change in the slope of the electron density around 82 km. The mid-latitude profile shows a sudden increase around 85 km. This shows that the ledge in general is situated about 3 km lower at the equatorial station compared with the mid-latitude station.

From the meteorological observations at Thumba (private communication, Prof. P. R. Pisharoty) it has been found that near the equator the level of the mesopause is situated at a height lower by about 5 km compared with that at mid-latitude. Perhaps the lowering of the ledge in the D region at the equator is associated with this phenomenon in some way. It is also found that temperatures below the mesopause are lower by a few degrees than those at mid-latitude. The collision frequency model determined at higher altitude may not hold good below the mesopause. Very low electron density (profile 6) below 78 km may probably be due to an incorrect collision frequency model.

4. Conclusion

From the above discussion it can be said that the solar Lyman-α is the prime source for long-term variation in the ionization of the D region, though X-rays may produce solar flares, diurnal and seasonal effects. The difference in the altitude of the ledge in electron density at mid and low latitude is attributed to the different heights of the mesopause in the two zones.

Acknowledgments

Our thanks are due to Messrs. R. N. Mishra, S. R. Das and C. L. Pipalapure for their help in making the pay load and the transmitter. We would like to thank Mr. H. G. S. Murthy, Director TERLS and his staff for the kind co-operation extended by them during the flight. The financial assistance was provided by the Department of Space, Government of India.

References

[1] H. K. Sen and A. A. Wyller, J. Geophys. Res. **65**, 3931 (1960).
[2] E. A. Mechtly and L. G. Smith, J. Atmos. Terr. Phys. **30**, 1555 (1968).
[3] E. A. Mechtly, M. Mukund Rao, D. O. Skaperdas and L. G. Smith, Radio Sci. **4**, 517 (1969).
[4] J. A. Kane, NASA GSFC Rep. X-615-69-499 (1969).
[5] S. Prakash, B. H. Subbaraya and S. P. Gupta, J. Atmos. Terr. Phys. **33**, 129 (1971).
[6] Y. V. Somayajulu, M. B. Avadhanulu, K. S. Zalpuri and S. C. Garg, Space Research XI, 1131 (1971).
[7] A. C. Aikin, R. A. Goldberg and Y. V. Somayajulu, J. Atmos. Terr. Phys. **34**, 1483 (1972).
[8] E. A. Mechtly, S. A. Bowhill, L. G. Smith and H. W. Knoebel, J. Geophys. Res. **72**, 5239 (1967).
[9] S. Gnanalingam and P. A. J. Ratnasiri, Annls Géophys. **22**, 361 (1966).
[10] J. E. Blamont and A. Vidal Madjar, J. Geophys. Res. **76**, 4311 (1971).

D REGION DURING SOLAR ECLIPSES

D. K. Chakrabarty and A. P. Mitra

Radio Science Division, National Physical Laboratory, New Delhi, India

We present in this paper representative profiles of electron, positive and negative ion concentrations during solar eclipses and offer an ion-chemical scheme that satisfactorily explains the observed changes in these concentrations, for the solar eclipses of 7 March 1970, 12 November 1966 and 20 May 1966 occurring over Wallops Island, Brazil and Greece respectively. It is found that a mere change of the ionizing radiation during solar eclipses can explain the observed variations in all their aspects above 80 km. Below 80 km, however, a satisfactory explanation of only the positive ion density is possible. To explain the severe decrease in electron density and the increase in effective recombination coefficient below 80 km a gradual decrease in the detachment rate of electrons from heavy negative ions from a pre-eclipse value of $2 \times 10^{-2}\ s^{-1}$ to about zero at totality is necessary. The role of any decrease in atomic oxygen concentration is found to be minor. The consequent changes in ion composition between 60—90 km are given.

1. Introduction

To explain the observed features in the D region two theoretical approaches are possible. One is the use of a reaction scheme that includes all major positive and negative ion reactions. Although such an approach is desirable it has been shown by many authors [1, 2] that the currently available information on all the reactions is inadequate. An alternative approach is to use a simplified ion-chemical scheme based on channels that are well established and to simplify those that are still uncertain to include lumped parameters that are then empirically adjusted with observational results. Such an ion-chemical scheme (shown in Fig. 3) has been given by Mitra and Rowe [1]. This scheme has been found to explain major normal daytime features both qualitatively and quantitatively [2—4].

We first apply the Mitra and Rowe scheme in the identical condition used for normal time [4] with the change in electron production rates during eclipse. As the need arises the other parameters are also changed. The nature of these changes has been discussed separately for two specific heights, viz. 70 km where negative ions are important, and 80 km where they are not. Because of the absence of observational data below 70 km at totality, we limit our discussion to 70 km and above. Based on these changed parameters eclipse time models for the distributions of N_e, N^+, N^- and ion composition arc constructed.

Table 1

Eclipse Data

Date	Height	q_t/q_n	$(N_e)_t/(N_e)_n$						$(N^+)_t/(N^+)_n$	
Author			Mechtly 2nd cont.	Mechtly 3rd. cont.	Belrose 2nd cont.	Belrose 3rd. cont.				
7 Mar. 1970	70	1.1 — 2	1.1 — 2	9.0 — 3	—	—			—	—
	75	8.0 — 3	2.5 — 1	1.5 — 1	1.2 — 1	—			—	—
	80	1.6 — 2	1.7 — 1	1.2 — 1	1.8 — 1	6.9 — 2			—	—
	85	5.7 — 2	3.4 — 1	2.8 — 1	3.3 — 1	2.2 — 1			—	—
	90	6.5 — 2	3.6 — 1	3.6 — 1	3.9 — 1	3.1 — 1			—	—
Author				Haug	Kane (9% Vis.)				Willmore	
20 May 1966	70		7.9 — 3	—	2.8 — 2				3.0 — 1	
	75		8.0 — 3	8.0 — 2	4.8 — 2				2.1 — 1	
	80		1.5 — 2	4.5 — 1	1.1 — 2				3.7 — 1	
	85		5.6 — 2	—	7.9 — 2				—	
	90		6.0 — 2	—	2.5 — 1				—	
Author			Mechtly 2nd cont.	Mechtly 3rd cont.	Mester	Baker	Ulwick	Flood	Baker	Narcisi
12 Nov. 1966	70	6.3 — 3	2.0 — 2	—	8.3 — 3	1.3 — 2	—	—	4.0 — 1	—
	75	7.3 — 3	2.4 — 1	4.4 — 2	2.8 — 2	2.6 — 2	—	—	2.7 — 1	5.0 — 1
	80	1.1 — 2	4.1 — 1	1.6 — 1	3.4 — 2	5.4 — 2	5.0 — 2	1.5 — 1	—	4.5 — 1
	85	3.0 — 2	1.0 — 1	3.7 — 2	—	—	2.2 — 2	2.0 — 1	—	3.0 — 1
	90	5.5 — 2	5.5 — 1	2.4 — 1	—	—	1.4 — 1	3.6 — 1	—	3.0 — 1

Note: t, totality; n, normal time.
1.1 — 2 = 1.1×10^{-2}, 8.0 — 3 = 8.0×10^{-3} etc.

Table 1 (Contd.)

Date	Height	$H^+(H_2O)_n/(NO^+ + O_2^+)$ Normal	$H^+(H_2O)_n/(NO^+ + O_2^+)$ Totality	λ Normal	λ Totality	α_{eff} Normal	α_{eff} Totality	Z_0 level Normal	Z_0 level Totality
7 Mar. 1970	70	—	—	—	—	1.8 — 11	—	—	—
	75	—	—	—	—	1.5 — 11	1.0 — 10 (M)	—	—
	80	—	—	—	—	1.2 — 11	8.0 — 11 (B)	81 km	82.5 km
							6.5 — 11 (M)		
	85	—	—	—	—	3.5 — 12	3.5 — 12 (B)		
							1.5 — 12 (M)		
	90	—	—	—	—	5.0 — 13	7.0 — 13 (B)		
12 Nov. 1966	70	4.1 + 0 (N)	—	1.5 + 0	1.5 + 2	1.8 — 11	—		
	75	3.0 + 0	—	—	3.5 + 1	1.8 — 11	2.5 — 10 (M)		
	80	1.8 + 0	—	—	—	1.3 — 11	4.5 — 11 (M)	82.3 km	86.8 km
	85	1.5 — 2	2.0 + 2	—	—	3.9 — 12	4.2 — 11 (M)		
	90	—	7.0 — 3	—	—	6.0 — 13	1.2 — 12 (M)		
20 May 1966	70	—	—	1.5 + 0 (W)	7.5 + 0 (15%)	2.0 — 11	—		
	75	—	—	—	7.5 — 1 (15%)	1.9 — 11	1.3 — 10 (74 km L)		
	80	—	—	—	—	1.7 — 11	2.1 — 11 (L)	80.6 km	85.7 km
	85	—	—	—	—	4.4 — 12	—		
	90	—	—	—	—	5.0 — 13	—		

B, Belrose; M, Mechtly; N, Narcisi; W, Willmore; L, Landmark. $4.1 + 0 = 4.1 \times 10^0$; $1.5 - 2 = 1.5 \times 10^{-2}$; $1.5 + 2 = 1.5 \times 10^2$ etc.

2. Major Features of the D Region during Solar Eclipses

Several eclipse time studies are available [5—7]. The major features which have been identified as the eclipse progresses to totality and which need to be explained through the consideration of the ion-chemical scheme are: (i) a drastic decrease (by about two orders of magnitude) in electron density at 70 km; (ii) a comparatively small decrease of positive ion density (by a factor of about 4) at 70 km; (iii) increase in the ratio of $[H^+(H_2O)_n]/[NO^+ + O_2^+]$ by more than an order around 85 km; (iv) increase in the value of α_{eff} by more than an order below 80 km.

Table 1 gives all the available observed parameters for the three eclipses, most of them normalized with respect to the pre-eclipse value. Normalization avoids the systematic error of different techniques. It is to be noted from this table that at heights below 80 km variation of N^+ is much less than that of N_e. At 70 km when N_e decreases by a factor of 100, N^+ decreases by a factor of 4. This decrease of N_e and N^+ is less at higher altitudes. It is further to be noticed from this table that the variations of N_e and q are nearly the same below 80 km indicating a linear type of loss rate. At 90 km $(N_e)_t/(N_e)_n$ (t = totality, n = normal time) is approximately equal to square root of q_t/q_n, indicating thereby a square type of loss rate. In Table 1 is also given the values of $[H^+(H_2O)_n]/[NO^+ + O_2^+]$, λ, α_{eff} and the Z_0 level for pre-eclipse and totality conditions. Z_0 level goes up during totality time in all the three eclipses.

An eclipse totality time model for N_e, N^+ and ion composition based on the observations given in Table 1 is shown in Fig. 1.

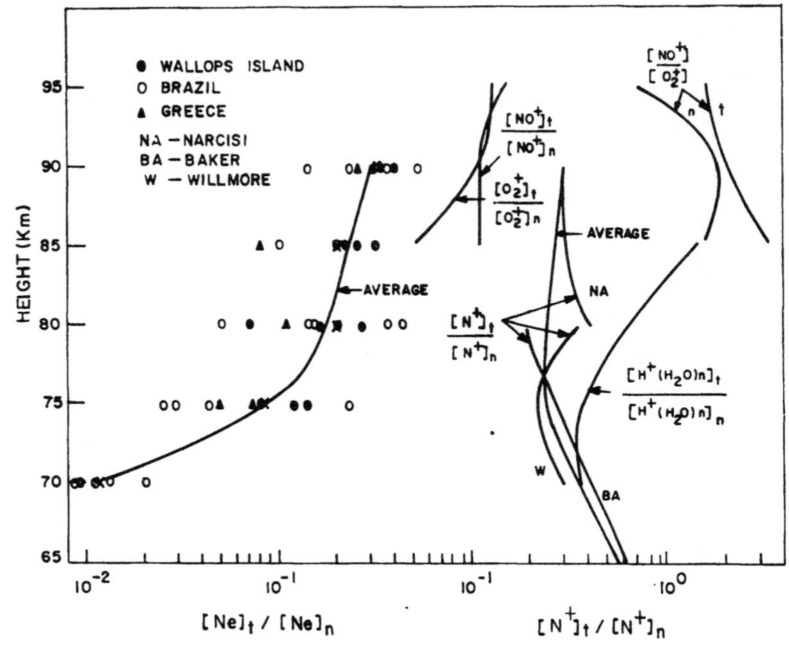

Fig. 1. Eclipse time models from observations.

3. Production Rates

The predominant sources of ionization in the D region at normal times are: (i) Lα ionizing NO, (ii) uv in the range of 1025—1118 Å ionizing $O_2(^1\Delta_g)$ and ionization of all the constituents, (iii) by X-rays less than 100 Å and (iv) by galactic cosmic rays. Another source of ionization which predominates during normal night time and may be important during the totality phase of eclipse is the electron production due to precipitating electrons.

Cosmic Rays. The ionization due to cosmic rays during eclipse has been assumed to be constant. For production rates due to cosmic rays the values adopted are from Velinov [8].

UV in 1025—1118 Å Range. For photoionization of $O_2(^1\Delta_g)$ by solar radiation in the range of 1025—1118 Å, we have used the values given by Huffman et al. [9]. The variation of this production rate during eclipse has been assumed to be proportional to the visible portion of the solar disk.

Lyman α. The calculation of rate of ionization of NO by Lα needs an NO density model and Lα flux. We have used the Meira [10] NO density model in this work and assumed no eclipse time variation of this minor constituent.

The pre-eclipse Lα flux values have been taken as 3.6 [11], 6.2 [12] and 4.2×10^{-3} Wm^{-2} [13, 14] for the eclipses of 20 May 1966, 12 November 1966 and 7 March 1970 respectively. At totality Lα flux values have been taken as 0.63% [12] and 0.58% [13] of the unobscured sun for the eclipses of 12 November 1966 and 7 March 1970. No measurement of Lα flux at totality is available for the eclipse of 20 May 1966. For this we have used the same percentage value as for the 12 November 1966 eclipse.

The variation of Lα flux between the first and the second contacts and between the third and the fourth contacts is assumed to be proportional to the visible portion of the solar disk; the variation ranges from the minimum value at totality to the maximum at the pre- or post-eclipse period.

X-rays. The calculation of the rate of electron production due to X-rays less than 100 Å requires solar flux, its spectrum and O_2 and N_2 densities.

Two X-ray observations are available for the eclipse of 20 May 1966. These are 6.0×10^{-7} and 5.0×10^{-7} Wm^{-2} [15] in the 0—8 Å band for the pre-eclipse condition. Also X-ray flux measurements by Landini et al. [16] when coupled with the X-ray photographs of the sun taken by Underwood and Muney [15] show that there were three X-ray spots on the sun at the time of the 20 May 1966 eclipse. The relative intensity of these spots was found to be in the ratio of 3:2:1. These have been used to calculate the production rate due to X-rays at different times during the 20 May 1966 eclipse.

Observations of X-ray fluxes are available for the eclipse of 7 March 1970 [17]. The pre-eclipse values in 0—8, 8—20 and 44—60 Å ranges are 2.6×10^{-6}, 3.5×10^{-5} and 1.65×10^{-4} Wm^{-2} respectively. These values are used to calculate the production rate due to X-rays for the eclipse of 7 March 1970.

No measurement of X-ray flux is available for the eclipse of 12 November 1966, except that given by Mariner 5 [18] for the pre-eclipse condition. We have used Mariner 5 data for the calculation of pre-eclipse time production rate due to X-rays for this eclipse. During the eclipse the production rate has been assumed to be proportional to the visible portion of the solar disk.

Measurements of residual X-ray fluxes are available for the eclipse of 7 March

1970 only. These are 2.7×10^{-8}, 7.5×10^{-6} and 2.6×10^{-5} Wm^{-2} for the 0–8, 8–20 and 44–60 Å ranges [17]. We have used these residual fluxes for all three eclipses.

The solar spectra have been built up from the values of X-ray fluxes mentioned above assuming exponential distribution with effective temperatures of 2×10^6, 2×10^6 and 0.5×10^6 °K for the 0–8, 8–20 and 44–60 Å wavelength regions respectively.

For the neutral atmosphere density we have used the model given by Groves [19]. No variation of O_2 and N_2 densities has been considered during eclipse.

Production Rate due to Precipitated Electrons. Ionization due to precipitated elec-

Table 2

Values of Lumped Parameters

Height (km)	B/s^{-1}	$\bar{X}$/s^{-1}	γ_2/s^{-1}	λ
60	1.6 − 2	2.4 + 1	5.0 − 1	7.5 + 0
65	1.6 − 2	1.0 + 1	1.0 − 1	2.1 + 0
70	1.6 − 2	1.5 + 0	2.0 − 2	1.1 + 0
75	1.6 − 2	2.5 − 1	2.0 − 2	5.0 − 1
80	1.5 − 2	8.0 − 2	2.0 − 2	—
85	1.0 − 3	5.0 − 2	2.0 − 2	—

Table 3a

Comparison of Observed and Theoretical Parameters at 70 km

Date		Observed			Theoretical I	Theoretical II
7 Mar. 1970	$(N_e)_e/(N_e)_n$	1.1 − 2 Mechtly 2nd cont.	9.0 − 3 Mechtly 3rd cont.		3.1 − 1	2.0 − 2
	$(N^+)_e/(N^+)_n$	—	—		3.3 − 1	7.3 − 1
	$(\alpha_{eff})_{tot}$/m^3 s^{-1}	—	—		9.8 − 12	5.4 − 10
	λ_{tot}	—	—		1.4 + 0	8.1 + 1
12 Nov. 1966	$(N_e)_e/(N_e)_n$	2.0 − 2 Mechtly 2nd cont.	8.3 − 3 Mester	1.3 − 2 Baker	2.6 − 1	2.7 − 2
	$(N^+)_e/(N^+)_n$	4.0 − 1 Baker	—	—	2.7 − 1	6.4 − 1
	$(\alpha_{eff})_{tot}$/m^3 s^{-1}	—	—	—	1.1 − 11	2.6 − 10
	λ_{tot}	1.7 + 2	1.7 + 2	1.1 + 2	1.4 + 0	5.5 + 1
20 May 1966	$(N_e)_e/(N_e)_n$	2.8 − 2 Kane	—	—	3.0 − 1	2.1 − 2
	$(N^+)_e/(N^+)_n$	3.0 − 1	—	—	3.1 − 1	6.9 − 1
	$(\alpha_{eff})_{tot}$/m^3 s^{-1}	—	—	—	1.0 − 11	5.2 − 10
	λ_{tot}	—	—	—	1.4 + 0	7.6 + 1

I, only q changed; II, q changed, γ_2 decreased to 0, $\alpha_i = 10^{-13}$
−2 means $\times 10^{-2}$; +2 means $\times 10^2$ etc.

Table 3b

Comparison of Observed and Theoretical Parameters at 80 km

Date		Observed			Theoretical I	Theoretical II
7 Mar. 1970	$(N_e)_e/(N_e)_n$	1.7 − 1 Mechtly 2nd cont.	1.2 − 1 Mechtly 3rd cont.	1.8 − 1 Belrose 2nd cont.	2.1 − 1	2.1 − 1
	$(N^+)_e/(N^+)_n$	—	—	—	2.1 − 1	2.2 − 1
	$(\alpha_{eff})_{tot}/m^3\ s^{-1}$	—	—	—	9.2 − 12	9.5 − 12
	λ_{tot}	—	—	—	4.4 − 3	4.2 − 2
12 Nov. 1966	$(N_e)_c/(N_e)_n$	4.1 − 1 Mechtly 2nd cont.	1.6 − 1 Mechtly 3rd cent.	1.5 − 1 Flood	2.0 − 1	1.9 − 1
	$(N^+)_e/(N^+)_n$	4.5 − 1 Narcisi	—	—	2.0 − 1	2.0 − 1
	$(\alpha_{eff})_{tot}/m^3\ s^{-1}$	—	—	—	9.1 − 12	9.3 − 12
	λ_{tot}	—	—	—	4.5 − 3	4.2 − 2
20 May 1966	$(N_e)_e/(N_e)_n$	4.5 − 1 Haug	1.1 − 1 Kane	—	2.2 − 1	2.1 − 1
	$(N^+)_e/(N^+)_n$	3.7 − 1	—	—	2.2 − 1	2.2 − 1
	$(\alpha_{eff})_{tot}/m^3\ s^{-1}$	—	—	—	9.3 − 12	9.5 − 12
	λ_{tot}	—	—	—	4.4 − 3	4.1 − 2

I, Only q is changed; II, q changed, γ_2 decreased to 0, $\alpha_i = 10^{-13}\ m^3\ s^{-1}$.
$-1 = \times 10^{-1}$ etc.

trons is a predominating source of electrons during night and need to be considered during eclipse time. The production rate due to this source has been taken from Manson and Merry [20].

4. Lumped Parameters B, $\bar{X}$ and γ_2

In the scheme of Mitra and Rowe three lumped parameters viz. B, $\bar{X}$ and γ_2 are involved in the following processes:

$$NO^+ \xrightarrow{B} H^+(H_2O)_n$$

$$O_2^- \xrightarrow{\bar{X}} X^- \xrightarrow{\gamma_2} N_e.$$

The values of B which have been used in the present work have been calculated from the available normal time positive ion composition measurements using the equation

$$B = \frac{[H^+(H_2O)_n] N_e \alpha_4}{[NO^+]}.$$

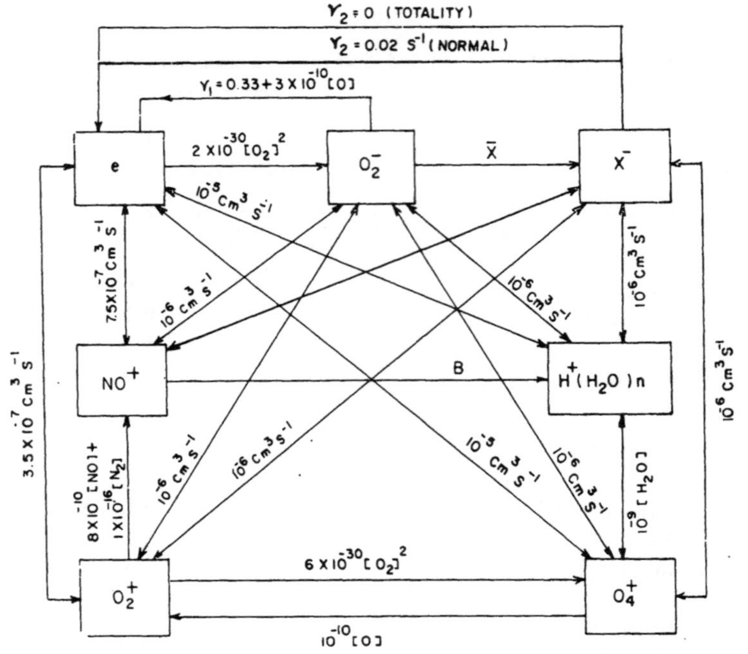

Fig. 2. Ion-chemical scheme for normal and totality conditions.

The estimated balues of B are given in Table 2. The values of $\overline{X}$ have been calculated using the relation

$$\overline{X}/\mathrm{s}^{-1} = 10^{-30}\,[O_2][M] + 4 \times 10^{-10}\,[O_3]$$

and are also shown in Table 2. B and $\overline{X}$ values are kept constant during the course of eclipse.

γ_2 is the detachment term introduced first by Mitra and Rowe [1] which includes the effect of collisional as well as photo detachment of electrons from heavy negative ions. Many authors [21, 22] have shown that a photo-detachment rate (of electrons from heavy negative ions) of value $10^{-2}\,\mathrm{s}^{-1}$ is necessary to reproduce a daytime electron density below 70 km close to observed values. In this work for pre-eclipse time we have varied γ_2 till it gives a value of λ in the 60—70 km range close to observed values. The value of γ_2 thus obtained by trial and error is also given in Table 2 with the corresponding λ values.

5. Ion Chemistry of D Region during Solar Eclipse

The continuity equations of four positive ions O_2^+, O_4^+, NO^+ and $H^+(H_2O)_n$ and two negative ions O_2^- and X^- framed on the basis of Mitra and Rowe scheme are solved simultaneously by a Fortran IV computer using the Runge-Kutta method for altitudes from 60 to 90 km in 5 km steps from the beginning to the end of the eclipse for all the three eclipses. Electron density has been calculated on the assumption of charge neutrality. The initial conditions were determined

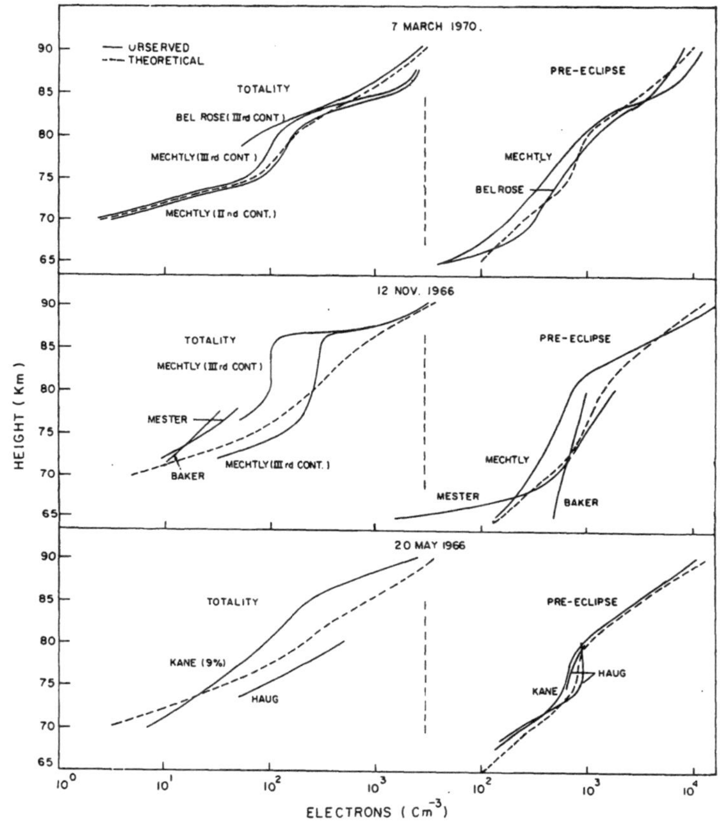

Fig. 3. Comparison of observed and computed electron density profiles for the eclipse of 7 March 1970, 12 Nov. 1966 and 20 May 1966.

by assuming the equilibrium condition. In this computation we have taken $\alpha_i = 10^{-13}$, $\alpha_1 = 7.5 \times 10^{-13}$, $\alpha_2 = 3.2 \times 10^{-13}$, $\alpha_3 = \alpha_4 = 10^{-11}\,m^3\,s^{-1}$, $k_1 = 2 \times 10^{-42}\,m^6\,s^{-1}$ and $k_2 = 3 \times 10^{-16}\,m^3\,s^{-1}$. ϱ_1 has been taken equal to $0.33\,s^{-1}$ for the pre-eclipse condition and varied according to the visible portion of the solar disk. The values of [O], $[O_3]$ and $[H_2O]$ have been taken from Shimazaki and Laird [23]. The computations are made under various conditions: Case I when only q is changed and other parameters are kept constant; Case II when q is changed and γ_2 is decreased slowly to zero; Case III when q is changed, γ_2 is made zero and [O] is decreased sinusoidally by a factor of 5; Case IV when q is changed, γ_2 is made zero, [O] is decreased by a factor of 5 and $\alpha_i = 10^{-12}\,m^3\,s^{-1}$. The results thus obtained are shown in Tables 3a and b for two height regions, 70 km where negative ions are important and 80 km where they are not.

In Table 3a, column I, the values are given when only q is changed, for 70 km. The decrease of positive ion concentration is close to observed values. The values of α_{eff} and λ are much less then that observed at totality. The decrease of N_e is also not close to observed values. To decrease the value of N_e further, γ_2 is decreased step by step. Finally γ_2 is varied according to the visible portion of the solar disk and made equal to zero at totality. The results thus

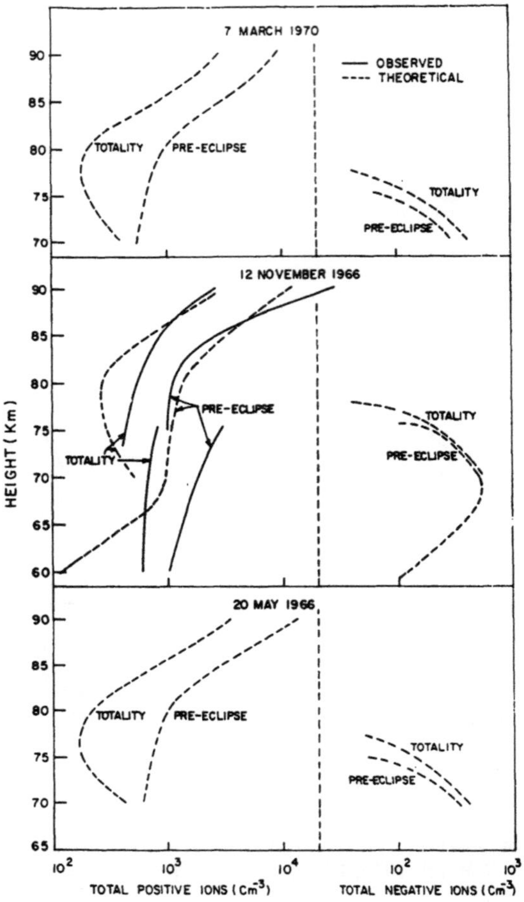

Fig. 4. Eclipse time models of N^+ and N^-.

obtained are shown in column II of Table 3a. The decrease of N_e now comes close to the observed values. The decrease of positive ion density is also reasonable. However, we find an increase in negative ion concentration during eclipse. This is not unexpected because if the detachment rate of electrons from heavy negative ions is lowered and the negative ions are not destroyed sufficiently quickly by recombination with positive ions, the concentration of negative ions will increase.

A change in the concentration of O is possible during eclipse particularly at 70 km. If [O] is decreased sinusoidally by a factor of 5, the changes in the values of parameters are found to be negligible except that of total positive ion concentration, where an increase in concentration is noticed at totality. This has become possible because the positive cluster ions formed are not destroyed sufficiently quickly. To offset this increase of positive ion concentration during eclipse, α_i has to be made equal to 10^{-12} m^3 s^{-1}.

For 80 km, the change in q is sufficient to explain all the observed parameters.

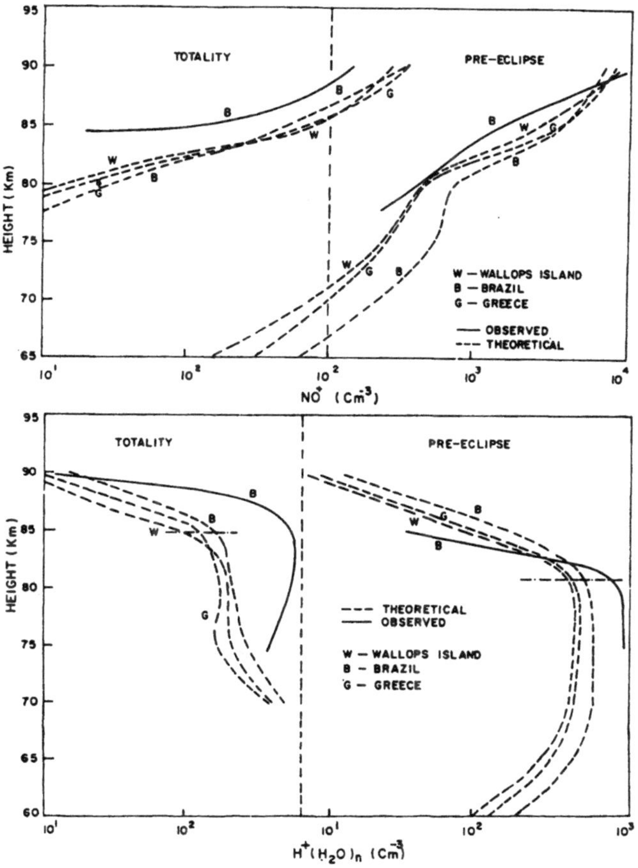

Fig. 5. Pre-eclipse and eclipse totality models of NO^+ and $H^+(H_2O)_n$.

No change in the values are noticed when γ_2, [O] and α_i are changed. This is evident from Table 3b.

From calculations of the variations of charged species during the course of the solar eclipse of 12 November 1966 (Brazil) we find that the percentage of $H^+(H_2O)_n$ of the total positive ion increases from about 75% at pre-eclipse value to about 99% at totality at 70 km. In the negative ion region while O_2^- decreases by about an order, the increase in concentration of X^- is marginal. At 80 km, it appears that the concentration of $H^+(H_2O)_n$ increases from about 48% at pre-eclipse value to about 99% at totality (of total positive ion). The ion-chemical scheme which holds good for pre-eclipse and eclipse totality conditions is shown in Fig. 2.

6. Eclipse Time Model of D Region

The theoretical profiles for the height distribution of N_e, N^+, N^-, $H^+(H_2O)_n$ and NO^+ obtained for all the three eclipses are shown in Figs. 3—5. Wherever available, the observed profiles are also shown. Agreement between theoretical and observed profiles is found to be good.

References

[1] A. P. MITRA and J. N. ROWE, J. Atmos. Terr. Phys. **34**, 795 (1972).
[2] D. K. CHAKRABARTY, Ph. D. thesis, University of Delhi, 1973.
[3] J. N. ROWE, Sci. Rep. No. 406, Penn. State Univ., Pennsylvania (1972).
[4] A. P. MITRA and D. K. CHAKRABARTY, Sci. Rep. No. 401, Penn. State Univ., Pennsylvania (1972) (p. 108).
[5] E. A. MECHTLY, C. F. SECHRIST, JR. and L. G. SMITH, J. Atmos. Terr. Phys. **34**, 641 (1972).
[6] J. S. BELROSE, D. B. ROSS and A. G. MCNAMARA, J. Atmos. Terr. Phys. **34**, 627 (1972).
[7] B. LANDMARK, A. HAUG, E. V. THRANE, J. E. HALL, A. P. WILLMORE, M. JESPERSON, B. M. PEDERSON, M. ANASTASSIADES, E. TSAGAKIS and J. A. KANE, J. Atmos. Terr. Phys. **32**, 1873 (1970).
[8] P. VELINOV, J. Atmos. Terr. Phys. **30**, 1891 (1968).
[9] R. E. HUFFMAN, D. E. PAULSON, I. C. LARRABEE and R. B. CAIRNS, J. Geophys. Res. **76**, 1028 (1971).
[10] L. G. MEIRA, JR., J. Geophys. Res. **76**, 203 (1971).
[11] J. E. HALL, J. Atmos. Terr. Phys. **32**, 1849 (1970).
[12] L. H. WEEKS and L. G. SMITH, Solar Phys. **20**, 59 (1971).
[13] L. G. SMITH, J. Atmos. Terr. Phys. **34**, 601 (1972).
[14] P. H. G. DICKINSON, J. Atmos. Terr. Phys. **34**, 621 (1972).
[15] J. H. UNDERWOOD and W. S. MUNEY, Solar Phys. **1**, 129 (1967).
[16] M. LANDINI, D. RUSSO and G. L. TAGLIAFERRI, Nature **211**, 394 (1966).
[17] C. A. ACCARDO, L. G. SMITH and G. PINTAL, J. Atmos. Terr. Phys. **34**, 613 (1972).
[18] A. MAVRETIC, Sci. Rep. No. 323, Penn. State Univ., Pennsylvania (1968).
[19] G. V. GROVES, AFCRL Sci. Rep. No. 70-0261 (1970).
[20] A. H. MANSON and M. W. J. MERRY, J. Atmos. Terr. Phys. **33**, 413 (1971).
[21] L. THOMAS, P. M. GONDHALEKAR and M. R. BOWMAN, Nature **238**, 89 (1972).
[22] R. P. TURCO and C. F. SECHRIST, JR., Radio Sci. **7**, 717 (1972).
[23] T. SHIMAZAKI and A. R. LAIRD, Radio Sci. **7**, 23 (1972).

ATOMIC HYDROGEN AND WATER VAPOUR IN THE LOWER ARCTIC THERMOSPHERE DURING A GEOMAGNETIC STORM AND PCA EVENT

G. M. Martynkevich and E. D. Byuro

USSR Academy of Sciences, Moscow, USSR

The radio-frequency mass spectrometer was launched on board the MP 12 rocket on 18 April 1972 from Heiss Island, to measure neutral composition during a geomagnetic storm with gradual commencement and a PCA event. The geophysical situation is described. High concentrations of H atoms and H_2O molecules were measured. It is interesting to note that the modulation percentage for the H_2O molecule ion current $I(H_2O^+)$ coincides essentially with those of other atmospheric constituents. The vertical distribution of the number density of the H atoms is calculated. Data from the flight performed on 18 April 1972 are compared with those obtained earlier. Some aspects of the measuring technique are discussed.

1. Introduction

A launching of an MP 12 rocket carrying a radio-frequency mass spectrometer [1] to study neutral composition and its variations was performed on 18 April 1972 at 0935 UT from Heiss Island (80°40′ N 58° E); apogee 190 km. A magnetic storm and a PCA event were in progress; the day was classed as a Polar Day.

2. Geophysical Situation

The lower thermosphere was stongly disturbed during the flight. The local ground data (Heiss Island), in particular, ionograms, riometer and geomagnetic micropulsation records, are indicative of this. The existence of disturbance is

Table 1

K index (Heiss Island)

day	Three hour intervals (UT)								
	0–3	3–6	6–9	9–12	12–15	15–18	18–21	21–24	ΣK
17 Apr. 1972	5	4	3	3	1	2	3	3	24
18 Apr. 1972	5	5	5	5	4	4	4	6	38
19 Apr. 1972	5	4	5	5	5	3	3	3	33
20 Apr. 1972	3	2	3	4	3	3	5	4	27

confirmed by the measurements of "soft" electron ($E \geqq 40$ keV) fluxes, which reached values of $(1-3.5) \times 10^7\ m^{-2}\,s^{-1}$. The fluxes of electrons were measured by Geiger counters [10] mounted on board the rocket. Global ground-based and satellite data [2—4] confirm also the conclusion that 18 April 1972 was one of the most disturbed days. For illustration Tables 1 and 2 give K indices for Heiss

Table 2

Kp index [3]

day	Three hour intervals									*Ci*	*Cp*	*Ap*
	1	2	3	4	5	6	7	8	ΣKp			
16 Apr. 1972	2^+	2^+	1	1	1^-	0^+	1^+	2	11^+	0.2	0.3	6
17 Apr. 1972	3^-	2^-	1^+	1	1	1	2^-	2	12^+	0.2	0.3	6
18 Apr. 1972	4^+	5^-	4^-	4^-	3^+	3	4^-	3^+	30^-	1.3	1.1	24
19 Apr. 1972	2^+	2^-	2^-	4^-	4^-	2^-	1	0^+	16	0.7	0.5	10

Island and Kp indices [3]. Hourly values of equatorial *Dst* variations [4] and fluxes of protons with energies $E \geqq 10$ MeV (satellite "Explorer" 41 [3] and $E = 5$ to 21 MeV (satellite ATS 1 [3]) as well other global data confirm the existence of the magnetic storm and PCA event.

3. Results of Measurements

The radio-frequency mass spectrometer operated well in the altitude range of 116—190—97 km. The following peaks were registered: 1(H), 12(C), 14(N), 15(?), 16(O), 17(OH), 18(H_2O), 26(?), 27(?), 28(N_2 + CO), 29(?), 30(NO), 31(?), 32(O_2), 34(?), 35(?), 36(?) and a group of poorly resolved peaks over the mass range of 38—47 amu. Earlier, as a rule, peaks with mass numbers 26, 27, 29, 31, 36, 39, 41—43, 45 and 46 had not been observed. It is possible that the observation of these peaks is due to the higher sensitivity of the mass spectrometer in the present flight. The peaks with mass numbers 34, 35 had been observed earlier [7] under disturbed conditions. A number of the peaks can be tentatively identified in the following manner: 15(NH), 29($^{14}N^{15}N + N_2H + COH$), 31(NOH), 34 ($^{16}O^{18}O + H_2O_2$), 35($H_3O_2$), 36($H_2O_2$), 45($CO_2H$), 46($NO_2$). The identification of other peaks seems difficult. The possibility of connecting the appearance of these unusual peaks with magnetic storm and PCA event cannot be excluded. We are inclined to accept the above point of view taking into account the following facts.

High number densities of atmospheric hydrogen atoms n_H were registered, the altitude dependence of n_H being shown in Fig. 1; calculated n_H densities are believed to be accurate within a factor of 2.5. It is easy to see that $n_H > 10^{16}\ m^{-3}$. A correction of $+\Delta n_H$ for the recombination of H and O atoms inside the mass analyser using the H_2O^+ peak reaches, for example, the value of $2 \times 10^{16}\ m^{-3}$ or less at an altitude of 119 km; at the level of 170 km and higher this correction can be neglected, since $n_{H_2O} \ll n_H$.

High concentrations of the H_2O molecules were also registered; this is clearly seen from the telemetry records of mass spectra (Fig. 2). The records were made on 18 April 1972 (PCA event, magnetic storm) and on 22 December 1970 (aurora)

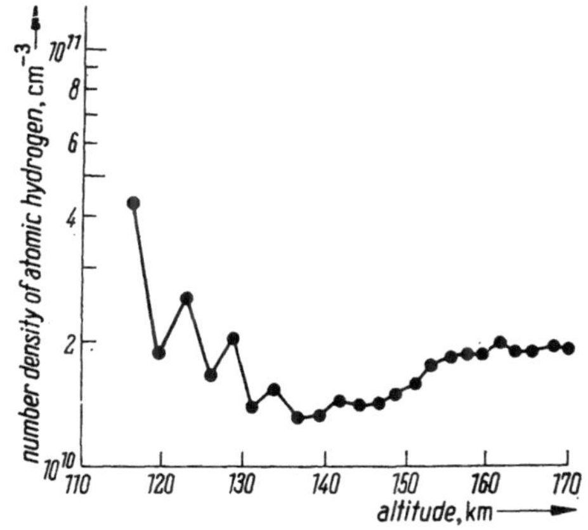

Fig. 1. Number density of atmospheric hydrogen atoms calculated according to H^+ peak as a function of altitude.

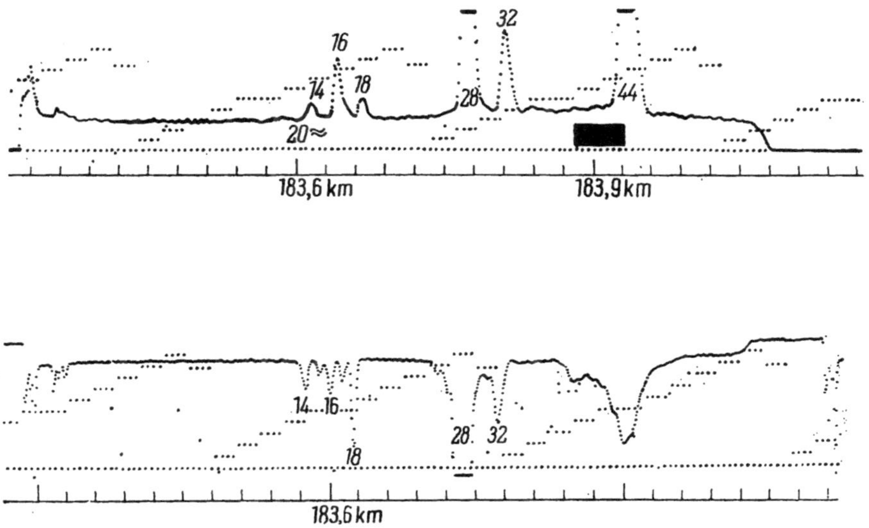

Fig. 2. Telemetry records of mass spectra obtained on 22 December 1970 (high sensitivity, aurora) and on 18 April 1972 (low sensitivity, magnetic storm, PCA event) at Heiss Island.

[7] on Heiss Island. H_2O molecule concentrations of the same order of magnitude were registered only on 25 July 1969 above p. Volgograd [7], also in a period of local disturbance. The percentage of modulation of the $I\,(H_2O^+)$ ion current during the flight on 8 April 1972 scarcely differs from that for $I\,(N_2^+)$ (Fig. 3) and the character of the altitude dependence of the $I\,(H_2O^+)$ ion current does not differ from that of any atmospheric component either. The character of the altitude

dependence of the I (H_2O^+) ion current is such that it excludes desorption as a possible cause of registration of the overwhelming majority of H_2O molecules which were present in the flight of 18 April 1972 and were partly of recombinational and partly of atmospheric origin.

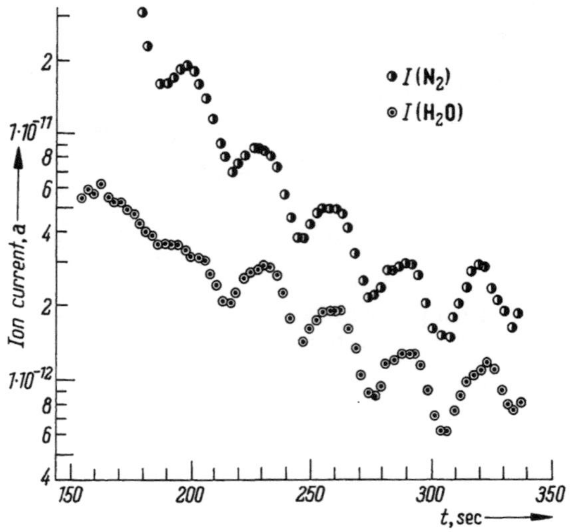

Fig. 3. Percentage modulation for I (N_2^+) ion currents during the launch on 18 April 1972 as a function of rocket flight time.

Preliminary calculations of the N_2 molecule number density and the thermosphere temperature over the N_2^+ peak near apogee leads to values $n_{N_2} = (7-12) \times 10^{14}\ m^{-3}$ and $T = 550$ K. These values do not exceed the limits compatible with the J 71 model [5], and are qualitatively consistent with slightly reduced values n_{N_2} measured by U. von Zahn also during a PCA event [6].

4. Discussion

The high concentrations of atmospheric hydrogen atoms obtained above are in good agreement with the results of our n_H measurements, performed earlier [7] under disturbed conditions and exceed n_H concentrations measured under relatively quiet conditions approximately by an order of magnitude, effective ionization cross section and mass discrimination having been taken into account, for example, in the flight of 24 November 1966, which is in complete agreement with paper [7]. More precise calculations of the upper limits of the concentrations of desorbed H_2 and H_2O molecules penetrating into analysers by diffusion at small Knudsen numbers (continuum flow) and by reflections from atmospheric particles at large Knudsen numbers allow us with complete confidence to neglect desorption effects on the measurements of hydrogen and H_2O molecules during the flight of 18 April 1972 and other flights. These calculations were carried out for the geometrical dimensions of the MP 12 rocket with the formulae derived in [9] and with a more precise geometrical factor in the region of molecular flow.

It is interesting and important to note a steady increase of atomic hydrogen number densities calculated from the data of spectral rocket and ground-based experiments (Lα, Lβ, Hα) [11—13]. In the initial stage a value of $n_H = 6.8 \times 10^{11}$ m^{-3} for the 110 km level was accepted [11], the data on the emission rates of scattered Lα radiation being traken into account. Later a value of $n_H = 4.5 \times 10^{13}$ m^{-3} at 100 km level was accepted [12] as the highest one. Recently C. S. Weller et al. [13] published the results of scattered Lα and Lβ radiation intensities measured aboard the same rocket and of simultaneous ground-based measurements of Hα emission rates. Study of these measurements led them to construct 11 models, in two of which (No. 7 and No. 11) a value of $n_H = 10^{14}$ m^{-3} at 100 km level is admitted. But this value cannot be considered as the highest limit from the data on Lα, Lβ and Hα emissions. Discrepancies between the data of Lα and Lβ emissions and Hα emission force us to consider the possibility of accepting a value of $n_H = 5 \times 10^{14}$ m^{-3} as the highest limit. This value does not differ significantly from the concentrations measured by our mass spectrometers in the 100—120 km range under quiet conditions and falling within the range $(6—30) \times 10^{14}$ m^{-3} [8].

In addition to the estimates of the H_2O^+ peak contribution to the concentration of hydrogen atoms, it is necessary to describe an estimation procedure for the effective recombination coefficient γ_{eff}^{H+O} of H and O atoms inside the analyser. This coefficient was determined as a ratio $\gamma_{eff}^{H+O} = n_{H_2O}^{meas}/n_{H_2O}^{calc}$ of the concentration of neutral H_2O molecules measured by mass spectrometer operating in the double "ions-neutrals" mode to the concentration of atmospheric neutral H_2O molecules $n_{H_2O}^{calc}$ calculated by the formula given in [15] from the data on the vertical distribution of H_2O^+ ions. The calculation of $n_{H_2O}^{calc}$ density as a function of altitude from the data of Arnold et al. [14] leads to the conclusion that γ_{eff}^{H+O} rises rapidly with increase of altitude and diminishes none the less rapidly with decrease of altitude. Thus at altitudes of 100—120 km γ_{eff}^{H+O} differs as insignificantly from unity as does the $+\Delta n_H$ correction from zero, while at 150 km level γ_{eff}^{H+O} can reach values of 30—100 and more. This means that at 100 km level and lower a recombination contribution is insignificant but at heights of 140—150 km H_2O molecules are almost entirely formed as a result of atmospheric H and O atoms recombination inside the mass analyser.

Acknowledgments

The authors express their gratitude to V. A. Troitskaya and her colleagues for a discussion of data on micropulsations and other geomagnetic data, and to V. G. Istomin and K. V. Grechnev for their assistance in producing a good ion current control scheme.

References

[1] V. A. Pavlenko, B. I. Zharkin, A. E. Rafalson and M. E. Stytsky, Kosm. Issled. **4**, 457 (1966).
[2] Kosm. Dannie No. 4, April 1972, Nauka, Moscow.
[3] Solar Geophysical Data (1972).
[4] M. Suguiura and D. J. Poros, Preprint X-645-72-355, GSFC, Greenbelt, September 1972.
[5] L. G. Jacchia, Smithson. Astrophys. Obs. Spec. Rep. No. 332 (1971).
[6] D. Offermann and U. von Zahn, Paper presented at COSPAR Meeting 1970.

[7] G. M. MARTYNKEVICH and E. D. BYURO, Space Research XIII, 261 (1973).
[8] E. D. BYURO, A. P. ZHUKOV, G. M. MARTYNKEVICH and E. A. SHVIDKOVSKY, Space Research IX, 501 (1969).
[9] A. I. IVANOVSKY and R. S. ZHANTUAROV, Trudy Central Aerolog. Observ. **56**, 97 (1956).
[10] V. F. TULINOV, Kosm. Issled. **5**, 240 (1967).
[11] V. G. KURT, Kosm. Issled. **4**, 111 (1966).
[12] B. A. TINSLEY, US Nat. Rep. to 15th General Assembly of IUGG, Moscow, August 1971, 510.
[13] C. S. WELLER, R. R. MEIER and B. A. TINSLEY, J. Geophys. Res. **76**, 7734 (1971).
[14] ESRO Report presented to XI COSPAR Meeting, Tokyo, Japan, 1968 (p. 26).
[15] G. M. MARTYNKEVICH, Space Research XII, 1311 (1972).

Lower Ionosphere Structure — Akademie-Verlag, Berlin 1974

SOFT ELECTRON FLUXES IN THE LOWER IONOSPHERE AT MEDIUM AND LOW LATITUDES

P. Rothwell, H. L. Collin, M. P. Gough and G. J. Jenkins

Physics Department, University of Southampton, Southampton, UK

The fluxes of electrons, with energies ranging from a few hundred eV to a few keV, have been measured with electrostatic analysers and channeltron from Petrel sounding rockets launched from the South Uist Hebridean range ($L = 3.5$) and the Thumba equatorial rocket range in the altitude range 100—150 km at various local times.

Electron fluxes of a few $\times 10^9$ particles $m^{-2} s^{-1} sr^{-1} keV^{-1}$ have been measured at South Uist near sunrise and sunset under a variety of magnetic conditions. At Thumba, electron fluxes of a few $\times 10^6$ particles $m^{-2} s^{-1} sr^{-1} keV^{-1}$ were measured near midnight. It is suggested that photoionization may be the predominant source of electrons in the keV energy range in the ionosphere, at medium and low latitudes.

Lower Ionosphere Structure — Akademie-Verlag, Berlin 1974

ENERGETIC ELECTRON PRECIPITATION AS A SOURCE OF IONIZATION IN THE NIGHT-TIME D REGION OVER THE MID-LATITUDE ROCKET RANGE, SOUTH UIST

M. P. Gough and H. L. Collin

Physics Department, University of Southampton, Southampton, UK

Two Petrel rockets were launched from South Uist, Scotland, one at a magnetically quiet time and the other during a magnetic storm. At the quiet time only low precipitation fluxes $\sim 4 \times 10^4$ electrons $m^{-2}\ s^{-1}\ sr^{-1} > 40$ keV were observed while during the storm these fluxes rose to as high as 3×10^9 electrons $m^{-2}\ s^{-1}\ sr^{-1} > 40$ keV. These precipitating energetic electron fluxes produced $< 25\%$ of the total night-time D region ionization at the magnetic quiet time but the storm time precipitation accounted for 99% of the total ionization.

A clear correlation is found between particle flux and magnetic activity Kp summed over 24 hours for 13 rockets launched from South Uist. This is used with the abundance of the different levels of activity to show that precipitating energetic electrons become a dominant source of the night-time D region on $15 \pm 11\%$ of the nights at solar minimum rising to $35 \pm 20\%$ of the nights at solar maximum.

When the magnetic field line passing through South Uist lies inside the plasmasphere, the electron precipitation continues up to $3^1/_2$ days after magnetic storms, in agreement with measured trapping lifetimes in the "electron slot" of the trapping region.

Direct Measurements in the Lower Ionosphere

Lower Ionosphere Structure — Akademie-Verlag, Berlin 1974

ION AND NEUTRAL COMPOSITION MEASUREMENTS IN THE LOWER IONOSPHERE

R. S. Narcisi

Air Force Cambridge Research Laboratories, L. G. Hanscom Field, Bedford, Mass., USA

The status of the quiescent and disturbed D and E regions as determined from rocket composition measurements is reviewed. Positive ion mass spectrometer measurements have been performed from low to high latitudes, providing information on the diurnal variation and on the effects of solar eclipses, PCA, aurora, sporadic E and meteor showers. Further, these measurements have contributed greatly to determining when and how transport processes are effective in redistributing the ionization. Less plentiful and of more recent origin, D region negative ion mass spectrometer measurements have raised entirely new problems; for example, the presence of extremely large negative ion clusters is unexplained. Measurements of those neutral species pertinent to ionospheric behavior are presented, including H_2O, NO and O concentrations deduced from ionospheric parameters. The experimental problem areas and the uncertainties and interpretive limitations of the measurements are discussed.

1. Introduction

This paper presents a review* of ion composition measurements in the lower ionosphere reported during the period 1967—1973. Rocket measurements conducted from low to high latitudes have provided information on the diurnal behavior of the D and E regions and the effects of solar eclipses, PCA, aurora, sporadic E, meteor showers and winds. The long sought D region negative ion composition has been measured.

Neutral species concentrations have been measured between 60 and 150 km by a variety of techniques. A comprehensive review of these measurements is not attempted. Rather, the minor neutral species which are presently believed to be of paramount importance in controlling ionospheric behavior are emphasized. The major neutral species above 80 km are presented, but in the form of newly developed models that represent a mean of the measurements.

---

* Updated version of part of a review by R. S. Narcisi in: Physics and Chemistry of Upper Atmospheres, B. M. McCormac (ed.), D. Reidel Publishing Company, Dordrecht, Holland (1973) p. 73.

2. The Lower Ionosphere

2.1. D Region

2.1.1. Positive Ions

Positive ion composition measurements have been made in the D region with rocket-borne mass spectrometers at equatorial, middle and high latitudes under a variety of conditions [1—10]. In all cases quadrupole mass spectrometers housed in cryogenically cooled pumping systems or in titanium getter pumps were utilized. Generally, the normal D region is found to be dominated by water-cluster ions, $H^+(H_2O)_n$, (Fig. 1) which decrease abruptly above about 82 km in

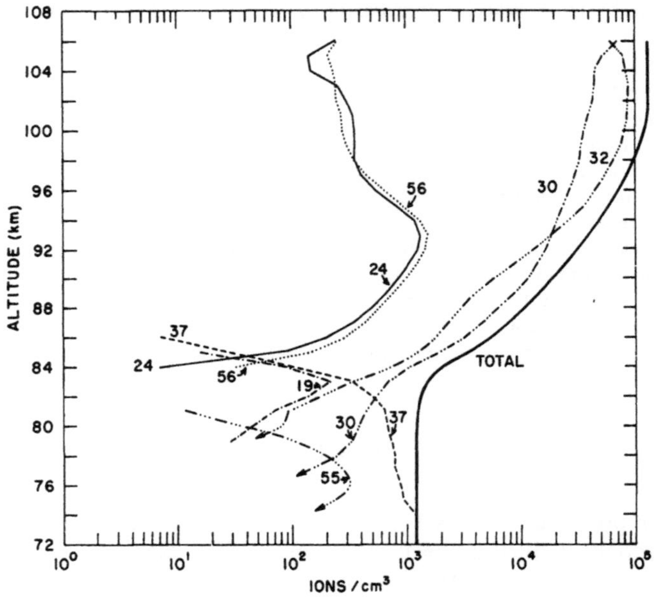

Fig. 1. Major positive ions in the D and E regions at a 20° solar zenith angle [7]. (Please note, units are in cm³ in the figures.)

the daytime and above about 86 km at twilight and at night. These altitudes also represent the transition levels above which NO^+ and O_2^+ dominate and metallic ions appear. This situation changes during high latitude ionospheric disturbances. Under conditions of auroral absorption (1.4—3.0 dB at 27 MHz) NO^+ and O_2^+ are found to be the major ions down to about 75 km [9]. Similarly, day, night and sunset rocket measurements during a PCA [8] with 30 MHz riometer absorptions of 3 dB, 0.4 dB and 0.7 dB, respectively, showed a suppression of the water-cluster ion cut-off level, and NO^+ and O_2^+ as the major ions above 77 km at night (Fig. 2), above 73 km in the daytime, and above 78 km at sunset, the latter two altitudes being lower altitude measurement limits. The PCA results have aided in clarifying the ion chemistry of the disturbed D region [11] and have permitted determinations of mesospheric nitric oxide, water vapor and atomic oxygen concentrations [12, 13].

In the quiescent, mid-latitude D region $H_5O_2^+$ is found to dominate the ion composition down to about 65 km with H_3O^+, $H_7O_3^+$ and $NO^+(H_2O)$ present in smaller amounts. The NO^+ ion is generally much less abundant than $H_5O_2^+$ in the daytime and very little NO^+ is found below 85 km at night and at twilight. Very rough variations with time of day between 75 and 85 km indicate an ion density of about $10^9\,m^{-3}$ near midday, about $10^8\,m^{-3}$ near midnight, a decay of about $(4-1) \times 10^8\,m^{-3}$ through sunset ($\chi = 88.5°-98.6°$) and about a threefold increase at sunrise ($\chi = 102.5°-90°$) [4]. D region ion concentrations are difficult to determine and subject to large errors; however, if the increase at sunrise ($\chi = 90°$)

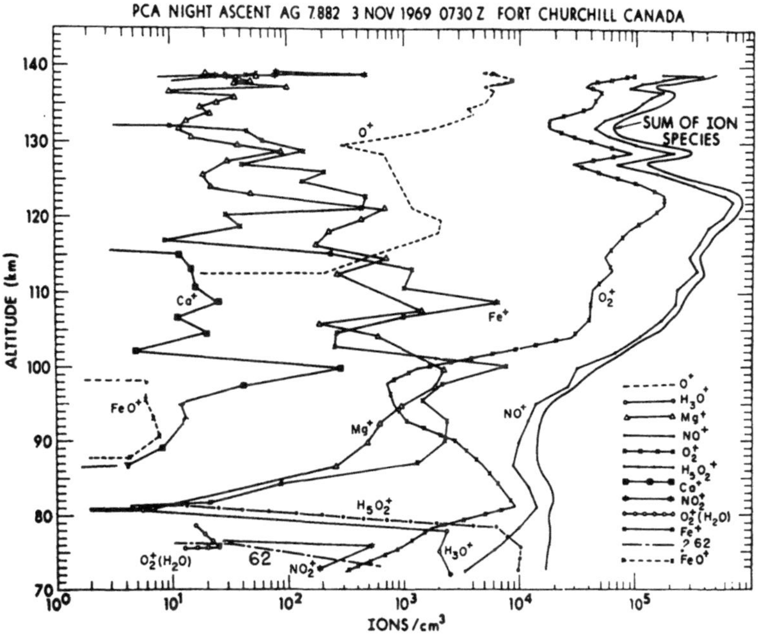

Fig. 2. Positive ion composition in the night-time D and E regions during a simultaneous PCA event and auroral event [8].

is correct, then water-cluster ions apparently can be produced by fairly low-energy solar radiation. Direct ionization of water conglomerates has been suggested as a possibility by Hunt [quoted in 4, 5].

Another serious problem associated with all the measurements is the possible fragmentation of weakly bound cluster ions that can occur through energetic collisions either as a result of the electric draw-in fields or by thermodynamic break-up at the increased shock-layer temperatures [4, 14]. In such reactions the ion loses one or more water molecules so that the measured composition may not be representative of the ambient. In any case, it seems clear that cluster-ions dominate. Subsonic measurements and experiments with the sampling electric field are presently being conducted to resolve these problems. Recent subsonic measurements showed an enhancement of $H_9O_4^+$ and a reduction in H_3O^+ between 76 and 79 km [15].

The present problem in the chemistry of the quiescent D region is to find a fast process which can convert NO^+ to water-cluster ions in the 70–86 km region. Rocket measurements during a solar eclipse indicate that the cluster-ion cut-off altitude rises from 82 km to 86 km from full sun to totality at the expense of a large decrease in NO^+ concentrations [7]. This was considered evidence for the conversion process and although the actual reaction could not be definitely determined, possible reactions with water conglomerates were suggested.

The complexity of the D region may be appreciated by the variety of ion species detected. The measured mass to charge ratios and possible identifications are: 19(H_3O^+), 21($H_3{}^{18}O^+$), 30(NO^+), 32(O_2^+, S^+), 34(S^+), 37($H_5O_2^+$), 39($H_5{}^{16}O^{18}O^+$), 41($Na^+ \cdot H_2O$), 46(NO_2^+), 48($NO^+ \cdot H_2O$, SO^+), 50($O_2^+ \cdot H_2O$, SO^+), 55($H_7O_3^+$), 60 ± 1 ($Na^+(H_2O)_2$), 63 ± 1($H_3O^+ \cdot CO_2$, HNO_3^+, $NO_2^+ \cdot H_2O$, SO_2^+, O_4^+), 66 ± 1($NO^+(H_2O)_2$), 73 ± 1($H_9O_4^+$), 80 ± 2($H_3O^+ \cdot CO_2 \cdot H_2O$, $HNO_3^+ \cdot H_2O$, SO_3^+, $H_xSO_3^+$), 91 ± 1 ($H_{11}O_5^+$), 96 ± 2(SO_4^+, $H_xSO_4^+$) and 109 ± 1($H_{13}O_6^+$). Atomic metal ions have been omitted. The ions 73, 91 and 109 amu were found to be enhanced in the cold high latitude summer mesopause [10], but all the above ion species except the 109 amu ion were also observed at mid-latitudes [16]. The sulfur ions and sulfur containing species appear to be restricted to below 86 km, falling off in concentration with the water-cluster ions. If the identification of these ions as sulfur is correct, it is a major problem to explain their presence in the D region [5]. The possibility that the sulfur ions may be fragments from ion clusters containing sulfur is being investigated in the laboratory.

2.1.2. Negative Ions

Negative ion measurements from seven rocket flights with cryogenically pumped quadrupole mass spectrometers have been reported. These include night-time quiescent measurements over Ft. Churchill, Canada [17], a night-time measurement from Andoya, Norway, during a weak auroral event [18], midday and midnight measurements in a PCA event from Ft. Churchill [19], and a measurement near totality of a solar eclipse from Wallops Island, Virginia [7]. Narcisi and coworkers found heavy ions with masses near 60, 62, 76, 78, 80, 98, 116, 134 and 152 amu in relatively large concentrations between 73 and 92 km and a layer of very heavy negative ion species centered near 88 km (Fig. 3). Between 90 and 92 km, the negative ion concentrations decrease by almost two orders of magnitude. These ions have been tentatively identified as $NO_3^-(H_2O)_n$, $n = 0-5$, with possible admixtures of $CO_3^-(H_2O)_n$, $n = 0-5$ and CO_4^-. Cluster-ion fragmentation due to the sampling electric field was detected and thus both the relative and absolute concentrations measured are questionable. The possible existence of very heavy ions greater than 150 amu was also evident (Fig. 3). Lighter ions with masses 16, 32, 35/37 and 46 amu, probably O^-, O_2^-, Cl^-, and NO_2^- are predominant above 90 km but in very small concentrations. With the exception of O_2^-, these species may be produced from contaminants or by the sampling method.

Drastic differences in the negative ion composition during a PCA were observed in comparison with quiet conditions. Near midday, O_2^- ions were the primary species and were found in large concentrations between 72 and 94 km. At night O^- was generally the major ion between 76 and 94 km; O_2^- was present below 80 km and became larger than O^- below 76 to 74 km. (The mass range of the PCA day spectrometer did not include 16 amu.) In both PCA flights the heavy negative ions measured in other flights were present below 85 km. The negative ion measurements are largely unexplained; details are discussed by Narcisi [11].

The measurements from the flight by Arnold et al. [18] in Fig. 3 are in sharp contrast to the measurements of Narcisi and co-workers. Large amounts of Cl^- ions (35 and 37) were measured below 80 km perhaps as a result of charge transfer with a chlorine contaminant. Additionally, CO_3^- (60) was the dominant ion along

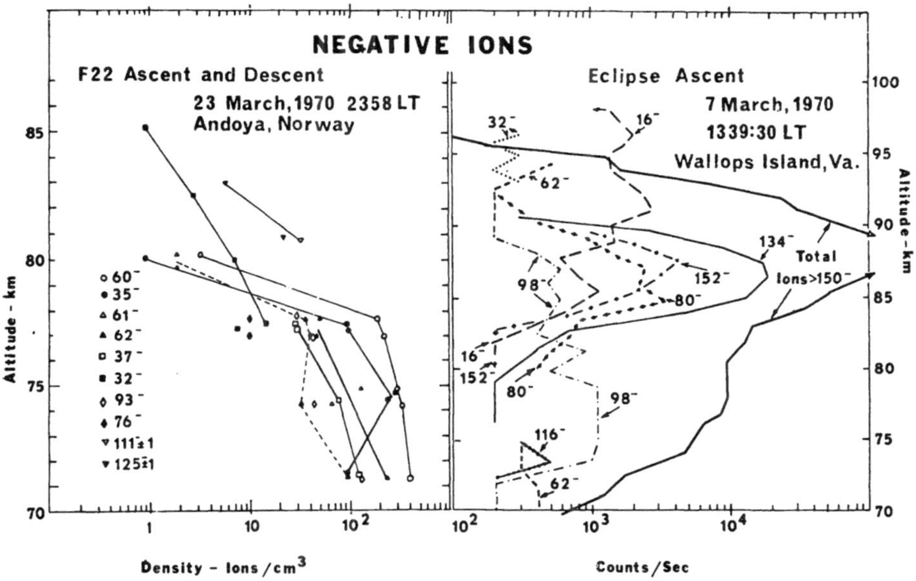

Fig. 3. Negative ion composition measurements in the D region. Left panel from Arnold et al. [18]. Right panel from Narcisi et al. [7].

with HCO_3^- (61) and NO_3^- (62) from 71—80 km. Mass 93 $\pm$ 1 amu ($CO_4^- \cdot H_2O$ or $NO_2^- \cdot HNO_2$) was found below 80 km instead of mass 98 $\pm$ 2 amu. Although O_2^- was similarly enhanced from 78—85 km owing to an energetic particle disturbance, the heavy cluster-ion mass numbers 111 $\pm$ 1($CO_4^-(H_2O)_2$) and 125 $\pm$ 1 ($NO_3^-(HNO_3)$) are completely different. The drop off in negative ion concentrations above 80 km may have resulted from an increasingly negative vehicle potential with altitude because no negative ions, including contaminants, were measured above 85 km. Arnold et al. also reported trace constituents of 68($O_2^-(H_2O)_2$) and 78($CO_3^-(H_2O)$). The uncertainties in the D region negative ion composition measurements point out the requirement for additional rocket and laboratory work.

2.2. E Region

2.2.1. Molecular Ions

Rocket measurements of the diurnal ion composition variations in the E region have been reported or reviewed by Keneshea et al. [20], Narcisi [5], Johnson [21] and Danilov [22]. NO^+ and O_2^+ generally dominate the E region ion composition (Figs. 1, 2, 4). In the daytime mid-latitude region, measurements of the NO^+/O_2^+ concentration ratio show excursions between 0.5 and 3 from 100—140 km. In twilight periods this ratio can become very large (Fig. 4) while at night it is about ten. A typical quiet night-time E region exhibits a bulge near 110 km of mainly NO^+ of about $(2-3) \times 10^9$ m^{-3}; however, NO^+ concentrations up to 10^{10} m^{-3} have also been observed near this altitude and were shown to be caused by atmospheric

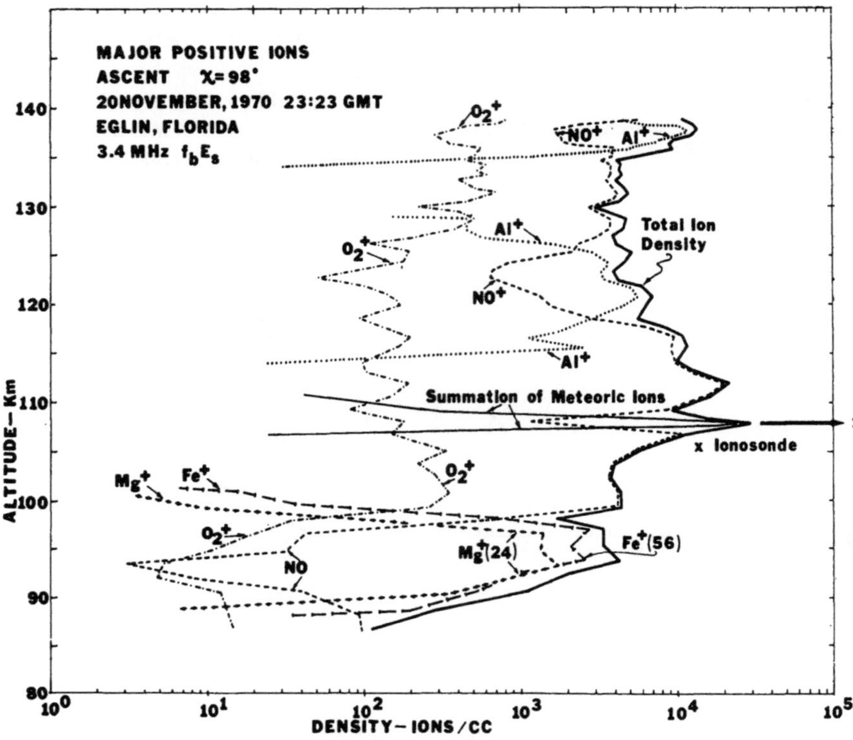

Fig. 4. Ion composition of the E region near sunset showing a blanketing sporadic E layer. The layer consisted of Si^+ (predominant), Mg^+, Fe^+ and smaller amounts of Na^+, Ni^+ and Ca^+. The Al^+ layers resulted from trimethyl aluminium released from an earlier rocket to measure neutral winds. Results are from Narcisi et al. [28].

ion transport processes [5], see Fig. 5. These measurements contributed to the development of a diurnal E region model [20].

E region measurements with quadrupole mass spectrometers have been made in auroras [8, 23—26]. In all cases the NO^+ density was greatly enhanced and, in most situations, the O^+ density was somewhat larger than expected (Fig. 2). Zipf et al. [25] measured both the ion and neutral composition in a bright auroral arc and found essentially no O_2^+ ions. This was consistent with their neutral measurements in that the NO density was greater than the O_2 density in the auroral arc. If correct, NO must somehow be increased by a very large amount in certain auroral arcs, but the processes by which this occurs are unknown.

2.2.2. Meteoric Ions and Sporadic E

A broad layer composed mainly of Mg^+ and Fe^+ with a peak ion density typically near 93 km is a permanent global feature of the ionosphere (Figs. 1, 2 and 4) and seems to be continuously maintained by ablating meteoroids [5]. The ion species measured in this layer include $^{23}(Na^+)$, $^{24-25-26}(Mg^+)$, $^{27}(Al^+)$, $^{39}(K^+)$, $^{40-42-44}(Ca^+)$, $^{52}(Cr^+)$, $^{54-56-57}(Fe^+)$, $^{59}(Co^+)$, $^{58-60}(Ni^+)$ and $^{72}(FeO^+)$ [3, 5, 8, 10, 27]. The relative abundances of these species are in quite good agreement with the relative elemental concentrations found in chondrite meteorites.

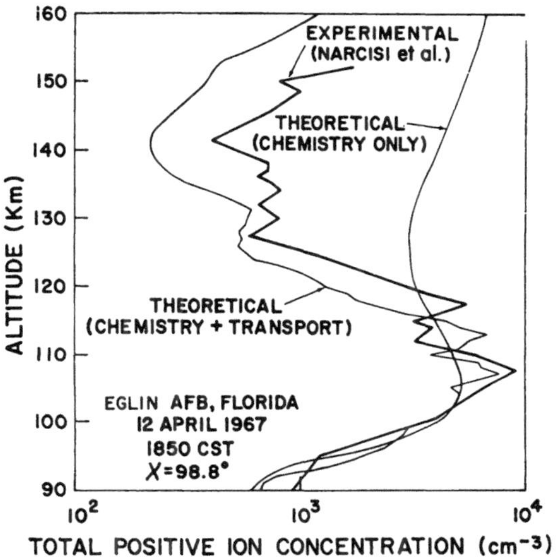

Fig. 5. Comparison of the measured total ion profile (mostly NO^+) with predicted profiles considering chemistry only and chemistry plus transport due to neutral winds calculated by Keneshea and MacLeod, see [5].

Thin meteoric ion layers are often found up to 140 km and also superimposed on the broad meteoric ion distribution (Fig. 2). Above 100 km, these layers may be composed of mainly Si^+, or Mg^+ or Fe^+ [5]. A silicon layer is typically located near 110 ± 5 km in the daytime and has been found to be enhanced on one occasion to produce a sporadic E layer, see Fig. 4. Intense mid-latitude sporadic E layers composed of meteoric ions have been observed in several other rocket flights [2, 29—31]. In twilight periods and at night, weak sporadic E layers may be composed mainly of either NO^+ or meteoric species. The molecular ion layer densities, however, are limited by electron recombination to a maximum ion density of about 2×10^{10} m^{-3} [2, 5, 32] as seen in Fig. 5. The meteoric species content in the E region was seen to be enhanced during a meteor shower [2], and the reduction of NO^+ and O_2^+ densities in a meteoric ion layer was observed many times in accord with theory (e.g. Fig. 4). Both upward and downward movements of thin meteoric layers have also been observed [5].

3. Mesosphere and Lower Thermosphere

3.1. Major Constituents

It is not within the scope of this paper to present a detailed survey of various composition measurements and their temporal and spatial variations. Instead, we shall rely upon recently developed mean models that represent a median of diurnal, seasonal and semi-annual variations for mid-latitude average conditions. Two such models by Champion and Schweinfurth [33] and Moe [34] are exhibited in Fig. 6 and show N_2, O_2, O, Ar and O_3 concentrations between 80 and 150 km. Where possible, Moe utilized experimental measurements to construct his median model. Unfortunately, the variations in the measurements are oftentimes a combination of experimental uncertainties and natural variations that are difficult

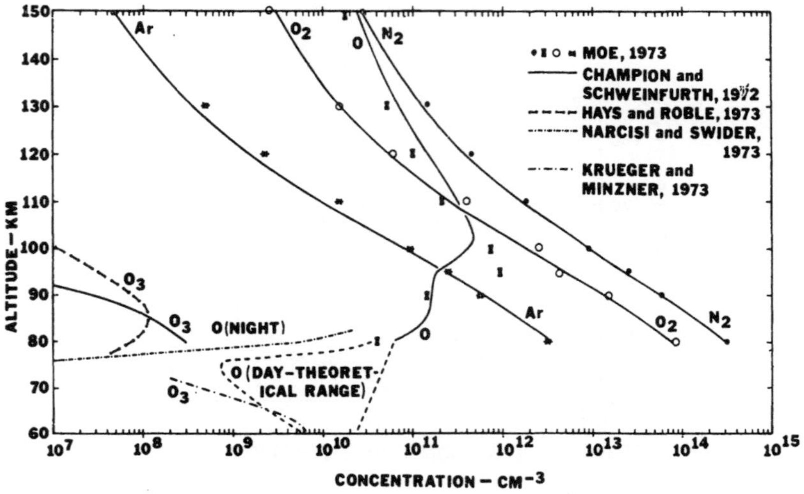

Fig. 6. The neutral composition between 60 and 150 km.

to separate. Measured deviations of factors of 2 and more about the mean profiles are not uncommon.

The O concentration continues to be a subject of considerable controversy. As can be seen in Fig. 6 the two models disagree most in [O]. Below 115 km the atomic oxygen concentration depends critically upon eddy diffusivity and its variation with altitude and time. Since the eddy diffusion coefficient is very difficult to determine it will require a large number of measurements to find the variability of O in this region. The atomic oxygen situation below 85 km is discussed in the next section.

The major species appear to be well mixed in the mesosphere according to their relative concentrations in surface air.

3.2. Minor Constituents

Trace species such as NO, H_2O, O_3, CO_2, mesospheric O, and N are most important for ionospheric processes as well as for the energy balance of the neutral atmosphere. Measurements of these gases are quite difficult to make and consequently few in number. It is now well recognized that NO plays a most critical role in lower ionospheric processes. Fig. 7 shows a compilation of the optical measurements of NO by Barth [35], Meira [36] and Tisone [37] along with the NO concentrations deduced from ionospheric measurements at mid-latitudes [20], and at high latitudes during a PCA [12] and two auroral events [24, 26]. Errors in all these profiles are about a factor of two and greater. Rusch and Barth [38] reported OGO 4 satellite measurements of NO gamma band emissions at twilight over a period of 14 months. They found NO tends to increase toward high latitudes, the increase being both altitude and latitude dependent. Increases of a factor of 2 in the mesosphere between 0° and 60° latitude in summer and winter were measured; however, little latitudinal variation was seen in fall and spring in the

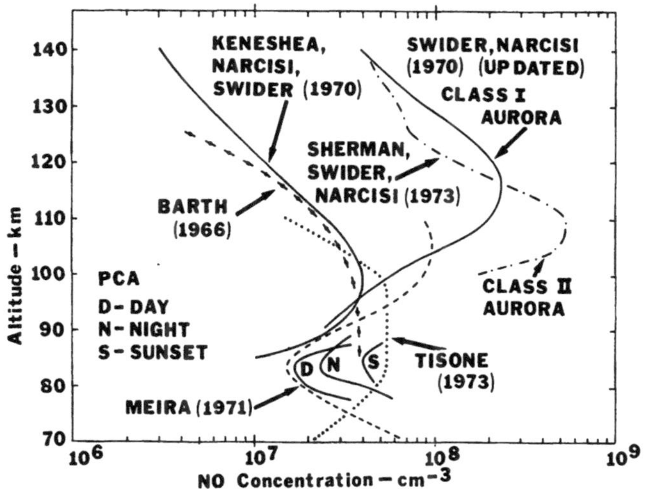

Fig. 7. Nitric oxide concentrations determined from optical and ionospheric measurements.

mesosphere and lower thermosphere. At high latitudes, 60° and greater, the NO content was highly variable ranging from mid-latitude levels to enhancements of factors of 3 to 6. The satellite measurements are in fairly good agreement with the results in Fig. 7. Offermann et al. [39] found NO to be about 6×10^{13} m^{-3} at 140 km in a mid-latitude night-time mass spectrometer measurement. The remarkable increases in NO concentrations by factors of 10^3—10^4 measured by Zipf et al. [25] in a bright auroral arc are, as noted earlier, unexplained.

Martell [40], analyzed air obtained with a cryogenic sampler and determined a water vapor mixing ratio of 3—5 ppm at the stratopause. No reliable measurements of H_2O have been made above this level. However, H_2O concentrations have been deduced by Narcisi and Swider [13] utilizing the ion chemistry and measured ionospheric parameters in the night-time D region during a PCA at Ft. Churchill, Canada. Fig. 8 shows this H_2O profile and although a factor of three error was assigned to the absolute values, the drop-off in concentration above 78 km is probably accurate. The H_2O mixing ratio was about 2 ppm from 70 to 78 km decreasing to about 0.5 ppm at 82 km. In comparison with the model predictions of George et al. [41] and Shimazaki and Laird [42], shown in Fig. 8, the H_2O profile indicates minimal turbulence in the mesosphere during this time.

From a survey of daytime and twilight measurements, Krueger and Minzner [43] have developed a mid-latitude ozone model that gives the annual mean ozone concentration and its variability at heights up to 72 km. The portion of this model between 60 and 72 km is shown in Fig. 6. The O_3 variabilities quoted range from about 35% up to 66 km and greater than 50% at higher altitudes. The mean of the Hays and Roble O_3 data above 77 km determined from OAO 2 stellar occultation [44] is also shown in Fig. 6.

Philbrick and Faucher [45] found CO_2 to be well mixed up to the turbopause (about 100 km) and Offermann and von Zahn [46] measured CO_2 at night to be

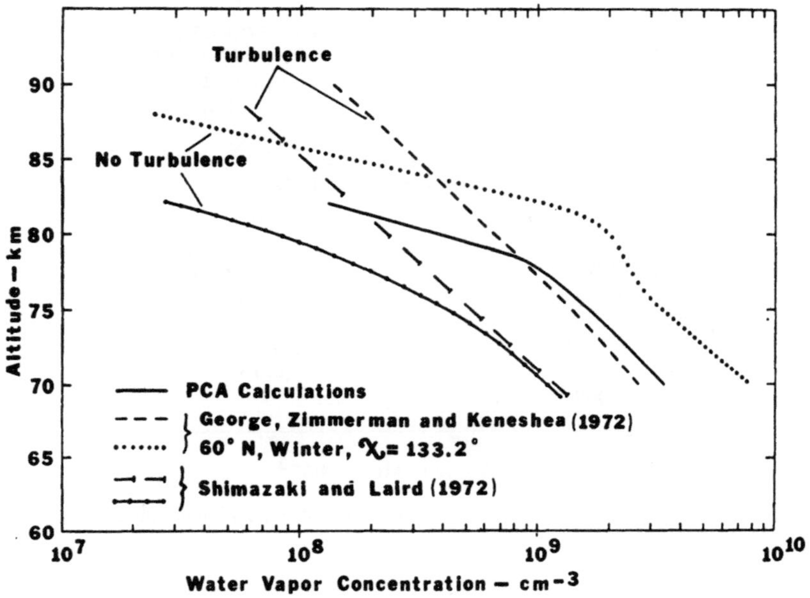

Fig. 8. Water vapor concentrations deduced from ion composition measurements during a PCA compared with model calculations.

roughly in diffusive equilibrium throughout the E region; both groups utilized mass spectrometers.

Atomic oxygen plays a most significant role in the D region in preventing O_2^+ hydration and in associative detachment reactions with negative ions. No measurements of O below 85 km are available. However, night-time atomic oxygen concentrations below 82 km were calculated in a similar manner as for H_2O and from the same PCA measurements [13]; these are shown in Fig. 6 and were quoted to have a factor of three error. The range of theoretical predictions from several models [41, 42, 47, 48] for daytime atomic oxygen concentrations below 80 km is also shown in Fig. 6.

Offermann et al. [39] determined atomic nitrogen to be about 4×10^{13} m^{-3} near 140 km in a mid-latitude night-time mass spectrometer measurement. Keneshea et al. [20] deduced a general upper limit for N of 10^{14} m^3 between 85 and 150 km from E region ion composition measurements.

References

[1] R. S. Narcisi, Space Research VII, 196 (1967).
[2] R. S. Narcisi, Space Research VIII, 360 (1968).
[3] R. A. Goldberg and L. J. Blumle, J. Geophys. Res. **75**, 133 (1970).
[4] R. S. Narcisi and W. Roth, Advan. Electron. Electron. Phys. **29**, 79 (1970).
[5] R. S. Narcisi, in: Physics of the Upper Atmosphere, Editrice Compositori, Publisher, Bologna 1971 (p. 12).
[6] R. A. Goldberg and A. C. Aikin, J. Geophys. Res. **76**, 8352 (1971).
[7] R. S. Narcisi, A. D. Bailey, L. E. Wlodyka and C. R. Philbrick, J. Atmos. Terr. Phys. **34**, 647 (1972).
[8] R. S. Narcisi, C. R. Philbrick, D. M. Thomas, A. D. Bailey, L. Wlodyka, D. Baker, G. Federico, R. Wlodyka and M. E. Gardner, in: COSPAR Symp. on Solar Particle Event of November 1969; AFCRL-72-0474, Special Reports No. 144 (1972).
[9] D. Krankowsky, F. Arnold, H. Wieder, J. Kissel and J. Zahringer, Radio Sci. **7**, 93 (1972).
[10] A. Johannessen and D. Krankowsky, J. Geophys. Res. **77**, 2888 (1972).
[11] R. S. Narcisi, in: Proc. COSPAR Symp. on Solar Particle Event of November 1969, AFCRL-72-0474, Special Reports No. 144 (1972).
[12] R. S. Narcisi, C. R. Philbrick, J. C. Ulwick and M. E. Gardner, J. Geophys. Res. **77**, 1332 (1972).
[13] R. S. Narcisi and W. Swider, EOS Trans. AGU **54**, 382 (1973).
[14] R. S. Narcisi, EOS Trans. AGU **51**, 366 (1970).
[15] R. A. Goldberg and A. C. Aikin, EOS Trans. AGU **53**, 1077 (1972).
[16] D. M. Thomas, R. S. Narcisi, A. D. Bailey and L. E. Wlodyka, EOS Trans. AGU **53**, 456 (1972).
[17] R. S. Narcisi, A. D. Bailey, L. Della Lucca, C. Sherman and D. M. Thomas, J. Atmos. Terr. Phys. **33**, 1147 (1971).
[18] F. Arnold, J. Kissel, D. Krankowsky, H. Wieder and J. Zahringer, J. Atmos. Terr. Phys. **33**, 1169 (1971).
[19] R. S. Narcisi, C. Sherman, C. R. Philbrick, D. M. Thomas, A. D. Bailey, L. E. Wlodyka, R. A. Wlodyka, D. Baker and G. Federico, in: Proc. COSPAR Symp. on Solar Particle Event of November 1969, AFCRL-72-0474, Special Reports No. 144 (1972).
[20] T. J. Keneshea, R. S. Narcisi and W. Swider, Jr., J. Geophys. Res. **75**, 845 (1970).
[21] C. Y. Johnson, Radio Sci. **7**, 99 (1972).
[22] A. D. Danilov, in: Aeronomy Rep. 48, COSPAR Symposium on D- and E-Region Ion Chemistry, University of Illinois, Urbana 1972 (p. 195).
[23] T. M. Donahue, E. C. Zipf, Jr. and T. D. Parkinson, Planet. Space Sci. **18**, 171 (1970).
[24] W. Swider, Jr. and R. S. Narcisi, Planet. Space Sci. **18**, 379 (1970).
[25] E. C. Zipf, W. I. Borst and T. M. Donahue, J. Geophys. Res. **75**, 6371 (1970).
[26] C. Sherman, W. Swider and R. S. Narcisi, EOS Trans. AGU **54**, 388 (1973).
[27] D. Krankowsky, F. Arnold, H. Wieder and J. Kissel, Int. J. Mass Spectrom. Ion Phys. **8**, 379 (1972).
[28] R. S. Narcisi, C. R. Philbrick, M. A. Macleod and N. W. Rosenberg, EOS Trans. AGU **53**, 462 (1972).
[29] R. S. Narcisi, A. D. Bailey and L. Della Lucca, Space Research VII, 123 (1967).
[30] J. M. Young, C. Y. Johnson and J. C. Holmes, J. Geophys. Res. **72**, 1473 (1967).
[31] A. C. Aikin and R. A. Goldberg, EOS Trans. AGU **53**, 1076 (1972).
[32] N. W. Rosenberg and S. P. Zimmerman, Radio Sci. **7**, 377 (1972).
[33] K. S. W. Champion and R. A. Schweinfurth, AFCRL-72-0579 (1972).
[34] K. Moe, J. Geophys. Res. **78**, 1633 (1973).
[35] C. A. Barth, Annls Géophys. **22**, 198 (1966).
[36] L. G. Meira, J. Geophys. Res. **76**, 202 (1971).

[37] G. C. Tisone, J. Geophys. Res. **78**, 746 (1973).
[38] D. W. Rusch and C. A. Barth, in: COSPAR Symp. on D- and E-Region Ion Chemistry, Aeronomy Rep. No. 48, University of Illinois, Urbana 1972 (p. 32).
[39] D. Offermann, K. Pelka and U. von Zahn, Int. J. Mass Spectrom. Ion Phys. **8**, 391 (1972).
[40] E. A. Martell, in: Physics and Chemistry of Upper Atmospheres, D. Reidel Publ. Co., Dordrecht, Holland 1973.
[41] J. D. George, S. P. Zimmerman and T. J. Keneshea, Space Research XII, 695 (1972).
[42] T. Shimazaki and A. R. Laird, Radio Sci. **7**, 23 (1972).
[43] A. Krueger and R. A. Minzner, NASA/GSFC X-651-73-22 (1973).
[44] P. B. Hays and R. G. Roble, Planet. Space Sci. **21**, 273 (1973).
[45] C. R. Philbrick and G. A. Faucher, Space Research XIII, 255 (1973).
[46] D. Offermann and U. von Zahn, J. Geophys. Res. **76**, 2520 (1971).
[47] E. Hesstvedt, Geophys. Norv. **27**, 1 (1968).
[48] M. R. Bowman, L. Thomas and J. E. Geisler, J. Atmos. Terr. Phys. **32**, 1661 (1970).

Lower Ionosphere Structure — Akademie-Verlag, Berlin 1974

POSITIVE IONS IN THE MESOSPHERE

L. C. Hale

Ionosphere Research Laboratory, Pennsylvania State University, University Park, Pa, USA

Positive ions have been measured using the parachute borne blunt probe, a subsonic system launched on meteorological rockets, which measures polar electrical conductivities in the 30 to 80 km altitude range. A central feature of the daytime positive conductivity profiles is a sharp break occurring between 60 and 70 km which indicates the transition from galactic cosmic ray to solar ultraviolet ionization, and, hence provides information about the nitric oxide density. In the 50–60 km region the positive ion conductivity is relatively constant with altitude and correlates well with atmospheric temperature down to 35 km.

Using temperature and pressure profiles and a reduced mobility, the conductivity profiles may be converted to density profiles. The average mid-latitude daytime profile at intermediate solar zenith angles showed a minimum at 63.5 km, and typical variations are less than ± 2 km, but ± 4 km is observed, indicating large variations in the NO density on these unusual days. On a fairly typical day for which electron density and Lyman α flux data were also available, the NO density was about 5×10^{13} m^{-3} in the 70–80 km altitude region, with λ of about 3 and α_d about 5×10^{-12} m^3 s^{-1} in the 65–80 km region. Analysis of ten shots on days when the nitric oxide did not vary drastically showed λ varying from about 1 to 10 and α_d from about $(2-10) \times 10^{-12}$ m^3 s^{-1}. α_d and λ varied cumulatively to give about a fivefold variation in the electron density which was reflected in A3 radio wave absorption measurements.

In the 50–80 km region dissociative recombination is apparently the principal ionization loss process, with λ relatively constant above 65 km. In the 50–60 km region positive ion density follows neutral atmospheric density, which requires a nearly altitude independent electron density in this range.

Data are also discussed for the night-time, low latitude (including solar eclipse) and high latitude (including PCA) cases.

1. Introduction

1.1. Scope of This Work

This paper is a summary of measurements, principally relating to positive ions, made using the parachute-borne blunt probe technique [1], and the implications of these measurements on aeronomical processes in the upper atmosphere. Most of the measurements and analysis relate to the mesosphere (about 50–80 km altitude), but the measurements have frequently penetrated the upper stratosphere; and these data are included, however, with less analysis of the aeronomical implications.

The paper will be divided into four parts:

Introduction, including a description of the measurement technique.

Mid-latitude daytime blunt probe measurements, which is the primary subject of the paper because many more data are available for this situation, and hence, the results have greater statistical significance.

Other blunt probe data, including night-time, low latitude (including solar eclipse) and high latitude (including PCA).

Summary and conclusions, primarily devoted to the aeronomical implications of the midlatitude daytime data, including values of λ (the negative ion to electron ratio), the nitric oxide density, the dissociative recombination coefficient, and the variability of these quantities inferred from the blunt probe measurements. The other measurements are summarized, and possible future experiments are discussed.

1.2. The Parachute Borne Blunt Probe

The parachute borne blunt probe is an instrument which was developed for subsonic measurement of polar electrical conductivities in the mesosphere and upper stratosphere. In the case of ions, the theory of operation is based on continuum principles stated by Hoult [2] and generalized by Sonin [3] and summarized by Hale, Hoult and Baker [1]. Comparisons between subsonic and supersonic operation [4] have shown that serious errors can be introduced by supersonic operation, and these errors become very large below 60 km.

In brief, the theory of ion collection is based on the fact that in the flow and environmental regimes satisfied by the blunt probe in the upper stratosphere and mesosphere, the ionic conductivity is not greatly perturbed by flow or electric field effects, and hence, ion collection can be expressed by a simple mobility mechanism:

$$dI = eN\mu E \, dA \tag{1}$$

where dI is the current collected by an element of probe surface, dA, E is the electric field at the surface, μ is the mobility, and N is the number density of the charged particles collected.

An illustration of the probe system is shown in Fig. 1. The probe geometry is an aluminum disk at the end of a cylindrical payload housing. The payload is suspended from a parachute so that the probe disk faces downward into the flow. The central portion of the disk, where the electric field is relatively uniform and calculable, is used as a collecting electrode and is held at the same potential as the rest of the disk, which then serves as a "guard" electrode. The disk is separated by an insulator, usually nylon, from the rest of the payload housing which is copper or aluminum and serves as a "return" electrode. A linearly swept potential is applied between the probe and return electrodes. This potential is typically of order of magnitude ± 5 to ± 10 V. The division of this applied potential between the probe and return electrode can be calculated [1, 5] and generally nearly all of the potential appears on the probe electrode, particularly in the daytime when photoelectrons from the return electrode tend to "lock" the potential of the return electrode to that of the surrounding atmosphere. The slopes of the I vs. U profiles are then related to dI/dU and then, using a geometrical factor, to the conductivities of the surrounding atmosphere. When the probe is negative these will be positive conductivities (σ_+), and when the probe is positive they are negative conductivities (σ_-). Fairly linear slopes are observed to represent both σ_+ and σ_- and at night, when there are generally no electrons present, are thought to be a good measure of the actual charged particle conductivities of positive and negative ions. In the daytime above about 40 km the negative conductivities

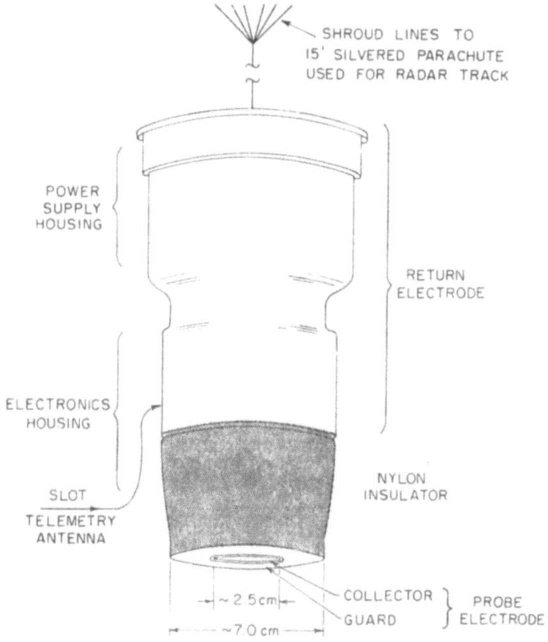

Fig. 1. Parachute borne blunt probe (Arcas rocket version).

greatly exceed the positive, and hence, are thought to be due to free electrons. The relatively linear *U-I* characteristic, indicating a mobility-like collection mechanism, led to a number of attempts to derive electron densities, but at present no good theory for doing this is known to exist. This difficulty is thought to be due to the non-linearity of electron drift velocity with electric field, and to the fact that the collection processes for electrons, as contrasted to ions, are determined at relatively great distances from the probe where the electric field is so weak that data on appropriate electron drift velocities are unavailable. Thus, the negative conductivity data from the blunt probe are used more as a relative than as an absolute indicator of electron densities.

It is generally recognized that outgassing, particularly of water vapor, from the surface of any instrument may seriously affect any direct measurement in the upper atmosphere, even if the instrument is carefully cleaned before launching. The following factors are thought to mitigate against this situation existing in the case of the parachute-borne blunt probe: 1. The instrument is deployed free of the nose cone and rocket motor. 2. The instrument has some time to outgas in a hard vacuum before data are taken. 3. The "open" geometry, in which the probe is in the flow upstream of the rest of the payload, so that outgassed particles will tend to be entrained in a direction away from the probe. 4. The fact that most environmental ions in the stratosphere are thought to be already hydrated to some degree, and additional hydration, if it occurred sufficiently close to the probe, would not affect the amount of current collected. 5. Most convincing, the fact that data obtained under similar geophysical conditions are quite reproducible. It is doubted whether outgassing effects would be equally reproducible.

In some of the data presented earlier [1, 4, 6—8] there were errors, including a factor of two error in calculating the appropriate electric field of the probe. The data presented here correct those errors.

2. Mid-Latitude Daytime Blunt Probe Measurements

2.1. Conductivity Measurements

Earlier work with the blunt probe included night-time, low latitude (including solar eclipse), and high latitude (including PCA) data, and these will be discussed later; more recently a number of measurements were made in the daytime at mid-latitudes, and these will be the primary subject of this paper since the larger number of measurements gives a better statistical picture than is available for the other situations. The data from ten rocket firings conducted over the period of October 1971 to February 1972 have been analyzed in some detail. These data were obtained at White Sands Missile Range (WSMR), New Mexico (latitude ~33° N, magnetic latitude ~42°), and at Wallops Island (WI), Virginia (latitude ~38° N, magnetic latitude ~50°), at various times and solar zenith angles (Table 1). The average positive conductivity of all these data is plotted in Fig. 2

Table 1
Parameters of ten WSMR and WI Rocket Launches

Date	LT (hrs)	Location	Solar Zenith Angle (°)	Transition Alt. (km)
22 January 1971	1300	WSMR	54	62
26 January 1971	1430	WSMR	62	63.5
1 February 1971	1315	WSMR	50	65
9 June 1971	0809	WSMR	53	63
28 July 1971	0705	WSMR	68	63
13 October 1971	1050	WSMR	44	64
6 January 1972	1337	WI	64	—
19 January 1972	1230	WSMR	53	62
31 January 1972	1305	WI	58	63.5
1 February 1972	1200	WSMR	51	63

along with data representing both the best agreement and the widest divergence from this average. The following features will be noted in the data: (i) a region from below 50 km to above 60 km in which the positive ion conductivity is relatively independent of altitude, (ii) a sharp break occurring in the altitude range 60—65 km.

The constancy of the conductivity with altitude in the 50—60 km region indicates that the conductivity is relatively independent of atmospheric density. A statistical analysis of these data by Cipriano [9] has shown that the positive conductivity correlates with atmospheric temperature, as measured with meteorological rocket "data-sondes", to a remarkable degree, with a correlation coefficient greater than 0.8 from 35 to 58 km (the highest altitude at which temperature data were always available). The temperature coefficient of the conductivity

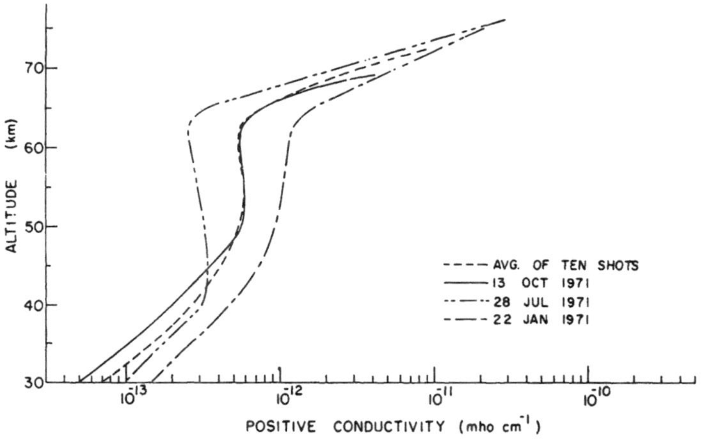

Fig. 2. Positive conductivity for ten shots of Table 1. Abscissa gives conductivity in mho cm^{-1}.

increased from 1.7%/K at 35 km to 3.7%/K at 45 km and 4.6%/K at 55 km. The positive conductivity variations in this region do not correlate well with negative conductivity, which is a relative indication of electron density.

The sharp break between 60 and 65 km can be interpreted as the transition altitude between galactic cosmic-ray ionization below and solar ionization above. Since all of these data were obtained when solar flare activity was relatively low, the solar ionization can be assumed to be due to the ionization of nitric oxide by solar Lyman α radiation [10]. The break in the curve can be used to determine the altitudes at which cosmic ray (Q_{CR}) and Lyman α ($Q_{L\alpha}$) ionization rates are equal, and this is about 63.5 km on the average for the ten shots. In principle this information can be used to calculate the nitric oxide density at the altitude where $Q_{L\alpha} = Q_{CR}$ if the incident Lyman α flux and O_2 density profile are known. Actually, this calculation is very sensitive to variations in atmospheric (O_2) density and can be performed more accurately if the Lyman α flux is measured directly, as will be done later for a particular day (31 January 1972, WI) on which such a measurement is available.

The ten shots of Table 1 showed a variation of less than ± 2 km in the transition altitude $Q_{L\alpha} = Q_{CR}$. Such a variation could be explained by the effect of large atmospheric density variations of the order $\pm 20\%$ on the absorption of Lyman α by O_2. However, other mid-latitude measurements for which the transition altitude $Q_{L\alpha} = Q_{CR}$ is available are:

11 December 1964	(1200 LT WSMR)	67.5 km
13 December 1966	(1200 LT WSMR)	62 km
2 September 1970	(0720 LT WSMR)	66.5 km
5 December 1972	(1314 LT WI)	59.5 km
16 January 1973	(1304 LT WI)	66 km

The large variation from the average transition altitude indicated here ($\sim \pm 4$ km) is more than double the variation that can be explained by the effect of $\pm 20\%$ variations in atmospheric density. It is necessary to admit the possibility of variations in the nitric oxide density at the transition altitude of at least an order of magnitude to explain this large variation in the height of the transition on the relatively unusual days, 11 December 1964, 2 September 1970, 5 December 1972 and 16 January 1973.

These latter two days are known to be days of low and high HF radio-wave absorption, respectively, at Wallops Island, Virginia, where these measurements were made. This shows an *inverse* rather than a direct correlation between high-frequency radio-wave absorption and the density of NO at the transition altitude.

2.2. Positive Ion Densities

The basic quantity measured by the blunt probe is conductivity σ_+, and the conductivity data alone provides useful synoptic information. In order to convert the positive ion conductivity data to positive ion density (N_+) data, assumptions must be made about the average mobility (μ_+) of the ions (since $\sigma_+ = N_+ e \mu_+$). Since the mobility depends upon atmospheric temperature and pressure, they must either be known or assumed, using an atmospheric model. In blunt probe experiments to date, the atmospheric conditions have not usually been measured above 60 km. The procedure used to convert the conductivity data to positive ion densities above 60 km has been to employ an appropriate model atmosphere and the relationship for mobility:

$$\mu_+ = \mu_0 \frac{P_0}{P} \frac{T}{T_0}$$

where μ_0 is an assumed reduced mobility ($\mu = \mu_0$ at STP) and P and T are pressure and temperature (P_0 and T_0 at STP).

In the calculations that follow, an assumed μ_0 of 1.8×10^{-2} m^2 V^{-1} s^{-1} and the appropriate CIRA 65 model atmosphere have been used. This value of μ_0 has been thought to be appropriate for small ions in air [12], and the use of a constant value of μ_0 provides a basis for the intercomparison of data. Fig. 3 shows the

Fig. 3. Average ion density for ten mid-latitude shots of Table 1.

positive ion density deduced for the average of the ten shots of Table 1. The following features will be noted: (i) The sharp break in the conductivity profile shows up as a deep minimum in the density profile, as might be expected for the transition from solar Lα ion production, which is attenuating very rapidly, and cosmic-ray production, which is increasing with decreasing altitude. (ii) The slope of the density profile in the 50—60 km region is closely proportional to atmospheric density. This has been interpreted [13] to indicate that there are substantial numbers of electrons (of order $(1-10) \times 10^7$ m^{-3}) with a density which is essentially constant with altitude in this range, this satisfying the lumped steady state positive ion conductivity equation:

$$Q = \alpha_d N_+ N_e + \alpha_i N_+ N_- \tag{2}$$

where Q is ion production rate, α_d is the dissociative recombination coefficient, and α_i the mutual neutralization coefficient.

Since Q is due to galactic cosmic rays (Q_{CR}) in this altitude range, and hence, proportional to atmospheric density ϱ, and if $N_+ \sim \varrho$ and $N_e \approx$ constant, then the equation is satisfied in this altitude range if the mutual neutralization term is smaller than the other two, which is possible since $\alpha_d \gg \alpha_i$. This situation is confirmed by the blunt probe negative conductivity data, which indicates an electron density that is essentially constant with altitude between 50 and 60 km, but which cuts off sharply at about 45 km. (iii) The positive ion density approaches a constant value below 40 km. This is consistent with $Q = Q_{CR} \sim \varrho$, and a three-body mutual neutralization coefficient proportional to density ϱ as the dominant ionization loss process.

2.3. Comparison with Electron Densities: λ Values

The blunt probe, as mentioned earlier, does not yet provide accurate absolute values of electron density. Thus, a comparison with electron density values is best made using electron densities determined using other techniques. A comparison of this type can be made for the data obtained on 31 January 1972 at Wallops Island, Virginia, where experimenters from the University of Illinois performed a combined probe and propagation experiment for the determination of electron densities on a Nike-Apache rocket launched 37 minutes prior to the blunt probe launch [14]. Fig. 4 shows the data from these shots for the comparison of electron and ion densities. It will be seen that the ion density exceeds the electron density in the 65—80 km range by a relatively constant factor of about 3 to 5, implying a relatively constant value of λ of about 2 to 4 in this altitude range. This is typical of other situations in which this comparison has been made [6].

This relatively constant λ implies that the operative attachment A and detachment D processes have a similar dependence on altitude in this range, which is at variance with current chemical models for this region, which hold that $A \sim \varrho^2$ (three-body attachment to form O_2^-) and that D, although somewhat uncertain, has a very different altitude dependence. One possibility which cannot be completely ruled out would be that the observed constant value of λ is in fact zero ($D = 0$). This would necessitate using a much higher value of reduced mobility μ_0 in converting conductivity data to density data and would have the effect of sliding the positive ion density profile to correspond approximately with the electron density profile in the 65—80 km region. The effective value of reduced mobility μ_0 required to do this is quite high, of order 5×10^{-4} m^2 V^{-1} s^{-1}. The consequences of maintaining $\lambda = 0$ to below 70 km are also difficult in terms of current theoretical models, and have the disadvantage of not so readily allowing the high effective electron loss rates that are frequently observed [15].

The broken line in Fig. 4 shows an electron density profile whose shape is more consistent with the positive ion data and the blunt probe negative conductivity data below 63.5 km. It has simply been normalized to the University of Illinois

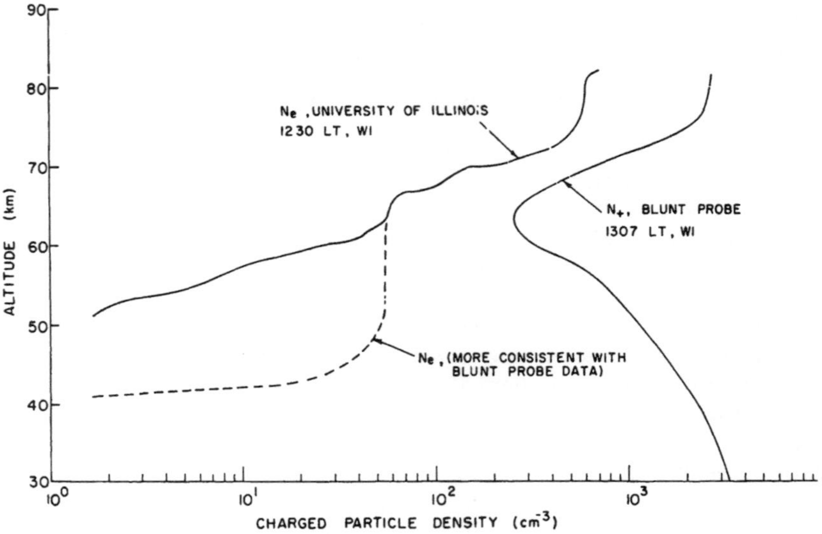

Fig. 4. N_+ and N_e profiles for 31 January 1972, Wallops Island, Virginia.

data at 63.5 km. In support of this profile, it must be pointed out that one would expect at least some break or ledge in the electron density data near the altitude of a deep minimum in positive ion density and ion production rate. This feature is borne out by the electron density profiles obtained from VLF and LF propagation studies by Bain and Harrison [16].

2.4. The Dissociative Recombination Coefficient

The knowledge of both positive ion and electron densities at the "break" or transition altitude where $Q_{CR} = Q_{L\alpha}$ enables the estimation (using Eq. (2)) of the dissociative recombination coefficient α_d at this altitude with more confidence than at higher altitudes, since the production rate at this altitude is simply twice the cosmic-ray value, which is probably better known than the production rate at higher altitudes, which is dependent on knowledge of Lyman α flux and NO density profiles. The mutual neutralization term in Eq. (2) at this altitude is small for any reasonable value of α_i ($\sim 10^{-13}$ m^3 s^{-1}) and the electron density is certainly in less dispute than at lower altitudes. Performing this calculation for the 31 January 1972 data of Fig. 3, the value of α_d becomes:

$$\alpha_d \approx 6 \times 10^{-12} \text{ m}^3 \text{ s}^{-1} \text{ (for Neher cosmic ray model) [17]}$$

$$\alpha_d \approx 3 \times 10^{-12} \text{ m}^3 \text{ s}^{-1} \text{ (for Velinov cosmic ray model) [18].}$$

These values are in the range expected for hydrated positive ions [19]. These values also provide evidence against the $\lambda = 0$ possibility mentioned previously, since reducing the positive ion density to the value of the electron density above 65 km would imply a corresponding increase in the estimated value of α_d to values greater than 10^{-11} m^3 s^{-1}, which are probably unacceptably large. In any case it may be concluded that the effective value of the electron loss coefficient, taken as $(1 + \lambda)\alpha_d$, was of order 2×10^{-11} m^3 s^{-1} on 31 January 1972 at the transition altitude of 63.5 km.

2.5. The Nitric Oxide Density Profile

The nitric oxide density can be calculated using Eq. (2) if the charged particle densities α_d and the Lyman α flux are known at all altitudes. On 31 January 1972 the Lyman α flux was measured as a function of altitude down to 64 km on the University of Illinois Nike-Apache rocket [20]. Using these values, and the charged particle densities of Fig. 3, and assuming constant α_d values of 6 and 3×10^{-12} m^3 s^{-1}, the NO profiles shown in Fig. 5 are generated.

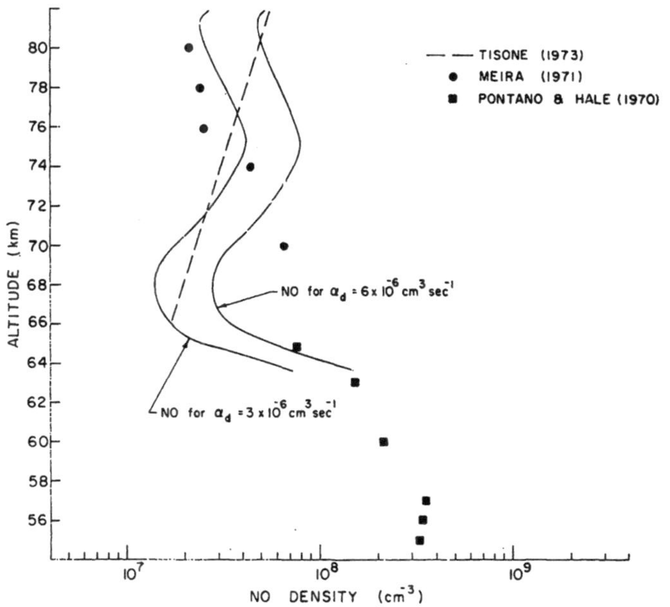

Fig. 5. NO profiles deduced for 31 January 1972, Wallops Island, Virginia.

These profiles show a minimum at about 70 km which is suggestive of the theoretical model of Brasseur and Nicolet [21] which includes photo-dissociation of NO and a "large" value of the eddy transport coefficient. The magnitude of NO values in the 70—80 km region are similar to those of Meira and the lower portion of the profiles match well with data obtained at lower altitudes by Pontano and Hale using a combined blunt probe and photoionization technique [7] (corrected here for the factor of two error in blunt probe theory mentioned earlier). A comparison is shown with the NO profiles deduced by Meira [22] and Tisone [23].

2.6. Negative Conductivity Measurements

As mentioned previously, the theory of the blunt probe for the collection of electrons is not yet well understood. Rational procedures may be considered for interpolating the shape of the profile over a limited altitude range, and this yields results consistent with the broken line profile of Fig. 4.

However, it is believed reasonable to expect that the blunt probe should provide a good relative measure of electron density at a given altitude. Using this supposition, Cipriano [9] found that negative conductivity variations measured in the cosmic-ray ionization region from below 50 km to about 63 km correlated well with variations in high-frequency radio-wave absorption taking place at much higher altitudes. This provided evidence that the absorption variability was due to changes in electron loss rate affecting a wide altitude range from 50 km to perhaps 100 km rather than to changes in ion production rate or collision frequency. However, two recent measurements on 5 December 1972 and 16 January 1973 do not fit well into this statistical picture. It will be noted that for the ten shots in Table 1 used for Cipriano's analysis there was relatively little variation in the height of transition from cosmic-ray to solar ionization, which is not true for the two later shots. Hence, Cipirano's conclusions may be regarded as correct for those days on which the transition altitude does not change greatly, but evidently are not applicable when this altitude changes by more than about 2 km. This is what would be expected if the change in transition altitude does in fact indicate a change in NO density, and leads to the conclusion that, at least, two physical mechanisms, involving both loss rate and NO variations, control the variability in high frequency radio wave absorption. The two unusual days, 5 December 1972 and 16 January 1973, showed an *inverse* correlation between radio wave absorption measured at Wallops Island and the implied variation in NO at the transition altitude. The negative conductivity on 31 January 1972 below the transition altitude was somewhat on the high side of the average of the data analyzed by Cipriano. If the effective recombination coefficient for electrons $[\sim(1 + \lambda)\alpha_d]$ is taken as about 2×10^{-11} m^3 s^{-1} for that day, then the observed range of variability of this quantity for the ten shots of Table 1 was from about $(4-10) \times 10^{-11}$ m^3 s^{-1}. The corresponding variations in high frequency radio wave absorption observed by the A3 technique was about 25 dB.

3. Other Blunt Probe Data

Other blunt probe data have been obtained using the blunt probe under a variety of conditions which are summarized in Fig. 6. The data are presented in terms of positive ion densities deduced from the conductivity data using a reduced mobility μ_0 of 1.8×10^2 m^2 V^{-1} s^{-1} and CIRA 65 model atmosphere. Also presented is the average daytime profile for comparison. (These profiles correct errors in previously published versions of some of these data [1, 5, 6].)

3.1. Mid-Latitude Night-time Profile

Most of the early blunt probe development work was done using night-time experiments. The data shown here from 45 to 65 km were obtained by two rockets at 2000 and 2200 LT on the night of 6 January 1966 at White Sands Missile Range, New Mexico [5]. The data also indicated zero electron density in this altitude

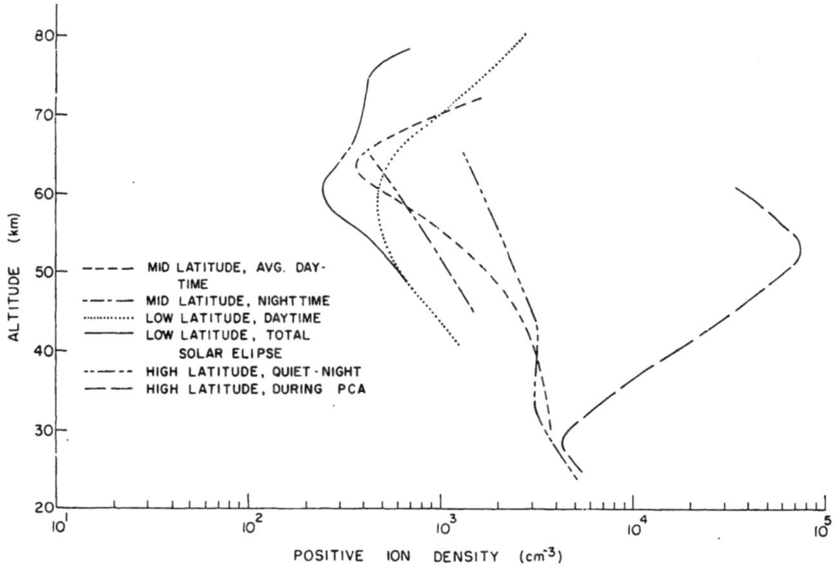

Fig. 6. Night-time, high and low latitude data compared with mid-latitude daytime.

range at night so that Eq. (2) reduces to:

$$N_+ = N_- = \left(\frac{Q_{\mathrm{CR}}}{\alpha_{\mathrm{i}}}\right)^{1/2}. \tag{3}$$

The data fit this model which for a constant α_{i} indicates that the charged particle densities should be proportional to the square root of atmospheric density, and this is observed. The value of the mutual neutralization coefficient α_{i} deduced from these data is:

$$\alpha_{\mathrm{i}} \approx 1.5 \times 10^{-13}\ \mathrm{m^3\ s^{-1}}\ \text{(Neher cosmic ray model) [17]}$$

$$\alpha_{\mathrm{i}} \approx 0.75 \times 10^{-13}\ \mathrm{m^3\ s^{-1}}\ \text{(Velinov cosmic ray model) [18].}$$

It was originally expected that the value of α_{i} determined at night could be used for a daytime model. However, if this were true the daytime values of ion density in the cosmic-ray ionization region would lie entirely below those measured at night, in accordance with Eq. (2), since recombination with electrons provides an additional loss process in the daytime. The fact that the night and day profiles cross indicates that the effective value of α_{i} for daytime conditions is much smaller and has been estimated to be less than $3 \times 10^{-14}\ \mathrm{m^3\ V^{-1}\ s^{-1}}$ [13]. A few data points obtained above 75 km at night indicate a deviation from the cosmic-ray model of Eq. (3) showing an enhancement of positive ion density of a factor of five on 6 January 1966 at 78 km. This indicates that additional sources of ionization can be important at night in the region above 75 km.

3.2. Low Latitude Data, including a Solar Eclipse

Blunt probe measurements were made in conjunction with the total solar eclipse of 12 November 1966 at Cassino, Rio Grande do Sul, Brazil (latitude ~32°S, magnetic latitude ~16.5°). The "low latitude daytime" profile of Fig. 6

was obtained about 90 minutes after totality at a solar zenith angle of 15° (this accounts for the different shape with respect to the mid-latitude data, which was all obtained at much higher zenith angles). This (corrected) data, when compared with measurements of electron density on the same rocket, gives a relatively constant λ of about 2 in the 70–80 km range. The large decrease in ion density at totality was previously interpreted to indicate a high value of the effective mutual neutralization coefficient α_i, of order 10^{-12} m^3 s^{-1}. However, since all other determinations of α_i from blunt probe data indicate a much lower α_i, it is now believed that this is probably not the correct interpretation. It seems more likely that the apparent large decrease in positive ion density was in reality due to a decrease in conductivity due to a decreased ion mobility at totality. This would happen if simpler ions proceeded to a more complex terminal state during totality.

There was no observable effect on ion densities of the solar eclipse below about 50 km, indicating that galactic cosmic-ray ionization dominates below this altitude. The ion densities are a factor of two to three lower than those observed at mid-latitudes, indicating that the ionization production rate was less than an order of magnitude less. This lends support to the latitude dependence of the cosmic-ray model of Neher.

3.3. High Latitude Data, including a PCA

High latitude data were obtained at Ft. Churchill, Manitoba, Canada [24] (latitude $\sim$58°N, magnetic latitude $\sim$70°). The "quiet" night data were obtained on 6 December 1967. The densities show an increase of about 2 to 3 over the mid-latitude (WSMR) night-time values, which implies an increase in ion production rate of 4 to 9 if α_i and atmospheric density are the same. This is greater than expected from the cosmic-ray models of Neher [17] or Webber [25] and could be due to the effect of temperature on either the recombination coefficient or the ion conductivity, considering the large temperature coefficient of positive conductivity previously mentioned.

The PCA profile shown was made at night (0530 LT) but three subsequent measurements during and after sunrise showed relatively little change in positive ion density, although they clearly showed the build-up of electron density. The density profile below 50 km can be fitted to calculated production rate profiles using satellite data on high energy proton and α-particle fluxes. The value of α_i consistent with this calculation is of order 10^{-14} m^3 s^{-1}. The detachment coefficient necessary to maintain an observed daytime electron density of about 500 at 50 km is about 0.1 s^{-1}.

4. Summary and Conclusions

4.1. Mid-Latitude Daytime Blunt Probe Data

The mid-latitude daytime positive ion conductivity profiles indicate a region of approximately constant conductivity from below 50 to above 60 km, and a sharp break occurring usually at about 63.5 km, but occasionally varying by as much as ± 4 km. The positive conductivity from 35 to 58 km correlates well with atmospheric temperature. The lack of correlation between positive conductivity and negative conductivity would seem to indicate that this variation with temperature is due more to changes in positive ion mobility than in positive ion density. The temperature coefficient of the (σ_+, T) relationship is much higher than can be explained by the classical dependence, and is probably due to a temperature dependence of the size of the ions, which for hydrated ions would indicate a difference in the degree of water vapor clustering.

The altitude of the sharp "break" observed in the positive conductivity profile may be interpreted under non-solar flare conditions as the altitude of transition from galactic cosmic-ray ionization to that produced by the ionization of nitric oxide by solar Lyman α radiation. This altitude is usually about 63.5 ± 2 km, and variations of this order could be explained by the effect of atmospheric density variation on the absorption of Lyman α by atmospheric O_2. However, occasional variations of up ± 4 km in the transition altitude indicate that the nitric oxide density can vary on occasion by an order of magnitude. Two such unusual days indicated an *inverse* relationship between the NO density at the transition altitude and high-frequency radio-wave absorption taking place at higher altitudes. This would tend to indicate that the effect of NO variability on radio-wave absorption is controlled by transport processes with smaller NO densities at the transition altitude accompanying larger values at higher altitudes, and vice versa.

Using an assumed reduced mobility and either measured or model atmospheric density profiles, positive ion density profiles may be calculated. These show a deep minimum at the transition altitude, and tend to be proportional to atmospheric density to below 50 km. These and other data support a model in this altitude range ($\sim$50—65 km) in which dissociative recombination is the principal ionization loss process, and the electron density is approximately constant with altitude. The ratio of attachment to detachment coefficient (and hence, λ) appears to increase linearly with atmospheric density in this altitude range. Below 40 km the ion density approaches a constant value, and the electron density becomes very small, which is consistent with galactic cosmic ray ionization and a three-body mutual neutralization process. (The transition to three-body ion-ion recombination indicates that the two-body and three-body values of α_i are equal at about 43 km which, using three-body data due to Sayers and a model atmosphere (CIRA 65), 26 fixes the two-body coefficient, α_{i2}, at about 2×10^{-14} m^3 s^{-1} [26], which is consistent with other estimates of the daytime α_i at higher altitudes using the blunt probe data.)

On 31 January 1972 a blunt probe experiment was performed in conjunction with other experiments as part of a coordinated group of experiments. The blunt probe conductivities and the transition altitude (63.5 km) were fairly typical, but somewhat above average. The use of electron density and Lyman α flux values kindly provided by University of Illinois experimenters [14, 20] has enabled a more accurate calculation of the dissociative recombination coefficient and λ values. λ appeared to be relatively constant with altitude above the transition altitude (63.5 km). This indicates that the ratio of attachment to detachment coefficients was about constant in this altitude range. The most probable value for λ is about 3. The dissociative recombination coefficient (α_d) was determined to be about 5×10^{-12} m^3 s^{-1} at the transition altitude, with some uncertainty due to a discrepancy in the available cosmic-ray flux data. If this value is assumed to remain constant above the transition altitude, a relatively constant (to $\pm 50\%$) nitric oxide density of about 5×10^{13} m^{-3} is implied in the 72—82 km region.

An analysis of the variability of blunt probe data for ten rockets in which the transition altitude and, hence, presumably the NO concentration, did not vary greatly indicated a variation in electron density below the transition altitude which correlated well with high-frequency radio-wave absorption and, hence, presumably electron density, at higher altitudes. If the electron loss coefficient above the transition altitude

$$\psi_e = (1 + \lambda)(\alpha_d + \lambda\alpha_i) \approx (1 + \lambda)\alpha_d \tag{4}$$

is taken as 2×10^{-11} m^3 s^{-1} for 31 January 1972, then these data imply a variability in this coefficient of $(4-10) \times 10^{-11}$ m^3 s^{-1}. It has been noted previously that variations in electron density are not correlated with variations in positive ion density [9]. Since the positive ion loss coefficient ($Q = \psi_+ N_+^2$) can be written as:

$$\psi_+ = \left(\frac{\alpha_d}{1 + \lambda}\right)$$

it is seen that the variation of the quantity $\psi_e = \alpha_d(1 + \lambda)$ must take place while the quantity $\psi_+ = \alpha_d/(1 + \lambda)$ remains relatively constant. For this to be true α_d and λ must increase together. Thus, if an increase in α_d were associated with a physical process such as an increased degree of hydration of ions due to increased water vapor, then this physical process must also increase λ so that $\alpha_d/(1 + \lambda)$ does not vary greatly. This would require either an increase in attachment or a decrease in detachment accompanying the increased dissociative recombination coefficient. The variations in electron density by a factor of five or so observed in the ten shots of Table 1 can thus be explained by the cumulative effects of variations in α_d and λ.

In summary, the "normal" mid-latitude daytime mesospheric ionization observations apparently can be described by a model based on the equation:

$$Q = Q_{L\alpha} + Q_{CR} = \alpha_d N_+ N_e = \alpha_d(1 + \lambda) N_e^2 = \alpha_d N_+^2/(1 + \lambda) \tag{5}$$

Above 65 km λ is approximately constant with altitude and the loss coefficient $\psi_e = \alpha_d(1 + \lambda)$ varies from about 4×10^{-12} to 1×10^{-10} m^3 s^{-1}, with α_d and λ varying together to give a cumulative effect. The range of α_d and λ values which satisfy this equation is possibly from ($\alpha_d \approx 2 \times 10^{-12}$ m^{-3} s^{-1}, $\lambda \approx 1$) to ($\alpha_d \approx 10 \times 10^{-12}$ m^3 s^{-1}, $\lambda \approx 10$). The nitric oxide density in the 70—80 km region is of order 5×10^{13} m^{-3}. Below the altitude of transition to cosmic ray ionization at about 63.5 km the positive ion density becomes approximately proportional to neutral atmosphere density and the electron density approximately constant to below 50 km. λ, which is approximately the ratio of attachment to detachment rates, is increasing linearly with atmospheric density from the value at the transition altitude. Ionization loss due to dissociative recombination dominates that due to mutual neutralization over the entire 50—80 km altitude range. On unusual days, as indicated by a large shift (2—4 km) in the transition altitude, large variations in the nitric oxide density occur at the transition altitude which are apparently inversely related to ionization density, and hence, to NO density at higher altitudes.

The model outlined above is, of course, invalid for solar flare or similar situations where it has been clearly shown that the loss coefficient for electrons may temporarily become much smaller, probably due to the creation of large numbers of simple ions with lower dissociative recombination coefficients [27, 28].

4.2. Other Blunt Probe Data

Night-time measurements at mid-latitudes indicated a mesospheric model in which $N_+ = N_-$, $N_e = 0$ and which involves cosmic-ray ionization below 75 km and a mutual neutralization coefficient of order $10^{-13}\,m^3\,s^{-1}$. An alternate explanation of these data might be that the terminal positive ions at night have very much lower mobilities than those observed in the daytime; this is suggested by the large decrease in positive ion conductivity during totality of a solar eclipse. In this case the ion densities might be a factor of three higher, and the mutual neutralization coefficient of order $10^{-14}\,m^3\,s^{-1}$.

In addition to the large decrease in ion conductivity in the upper part of the mesosphere during a solar eclipse at low geomagnetic latitude mentioned above, in the lower part of the mesosphere this and other low latitude data indicated an ionization density 2 to 3 times smaller than the mid-latitude values, suggesting a cosmic-ray flux no more than one order of magnitude lower than the mid-latitude values.

High latitude night-time conductivity measured at Ft. Churchill during quiet conditions was about two to three times the mid-latitude values, which was a greater difference than expected from current cosmic ray models. The effects of a PCA on positive ions penetrated to below 30 km, night and day, and were consistent below 50 km with proton and α-particle fluxes measured on satellites.

4.3. Future Experiments

This paper has summarized measurements of atmospheric electrical conductivity using the parachute borne blunt probe technique and the aeronomic implications of these data. It is believed that these data provide much information useful in the construction of an aeronomic model to explain mesospheric electrical structure and that this and other parachute borne systems will find wide application in direct atmospheric measurements in the mesosphere and stratosphere because the subsonic velocities do not introduce problems due to shock waves, yet the velocity prevents the build-up of pollution in the vicinity of the instrumentation which is a serious problem with balloons.

A fair sample of data has been obtained of mid-latitude daytime conductivities, and it is believed that this body of data will be most efficiently enhanced as an incidental by-product of other experiments aimed at more specific goals. It is of particular value to obtain data in conjunction with other experiments, as for example, the data of 31 January 1972, which was obtained at Wallops Island, Virginia, as part of a coordinated program to study the winter D region. (Other examples of such coordinated experiments in this program are 5 December 1972 and 16 January 1973 which are mentioned in this paper but for which the data have not yet been completely analyzed.) It is important that due to the large variability of many quantities in this region of the atmosphere, conjunctive experiments be performed at the same location and at nearly the same time. It is essential that density, Lyman α flux, and electron density be known. With these and the blunt probe data known, the dissociative recombination coefficient, λ and the NO profile can be calculated. (The transition altitude alone provides a relative measure of NO and can be used to identify days in which the NO distribution is unusual.) Conjunctive experiments to provide better data on ion mobilities, such as a subsonic Gerdien condenser, would be very useful. Other important experiments would include the direct measurement of NO and of possible detachment agents such as O and $O_2(^1\Delta_g)$. The possible correlation of ionization loss coefficient with water vapor gives a high priority to the development of some way of measuring water vapor. The correlation studies of Cipriano [9] between radio wave absorption and negative conductivity at lower altitudes indicated a good correlation with conductivities above 50 km, and it would be desirable to measure water vapor at least to this altitude. Because of extreme contamination problems associated with direct water vapor measurements, such measurements would probably have to be done by satellite or balloon borne remote sensing techniques.

Theoretical work to enable interpretation of blunt probe negative conductivity data in terms of electron densities is under way, and it would be hoped that eventually general agreement could be reached on the nature of the electron density profiles below 65 km. Ground-based electron density measurements performed in conjunction with the rocket measurements are very important, particularly in assessing the statistical validity of the rocket data.

The blunt probe data taken under situations other than mid-latitude daytime are so sparse as to have much less statistical significance. Certainly more nighttime and low and high latitude data are needed. Again, these can be most efficiently generated in conjunction with programs aimed at the study of specific geophysical phenomena.

It is intended to continue to develop the blunt probe technique to be compatible with current meteorological rocket and data acquisition systems such as the Loki-Dart system. This will provide an even lower cost technique for acquiring the synoptic data necessary to build a more complete model of mesospheric and stratospheric ionization phenomena.

Acknowledgments

The author wishes to thank Dr. John N. Rowe for helpful discussions and Mr. E. T. Chesworth and Mr. W. J. Burkhard for assistance with the aeronomical calculations. Thanks are due to Dr. L. G. Smith and Dr. E. A. Mechtly for providing Lyman α flux and electron density data from the University of Illinois rocket of 31 January 1972 at Wallops Island, Virginia.

This work was supported by the National Aeronautics and Space Administration Grant NGR 39-009-218 and by the US Army Research Office Grant DA-ARO-D-31-124-72-G158.

References

[1] L. C. HALE, D. P. HOULT and D. C. BAKER, Space Research VIII, 320 (1968).
[2] D. P. HOULT, J. Geophys. Res. **69**, 4617 (1964).
[3] A. A. SONIN, J. Geophys. Res. **72**, 4547 (1967).
[4] L. C. HALE, L. C. NICKELL, B. KENNEDY and T. A. POWELL, Radio Sci. **7**, 89 (1972).
[5] L. C. HALE, Space Research VII, 140 (1967).
[6] D. C. BAKER and L. C. HALE, Space Research X, 712 (1970).
[7] B. A. PONTANO and L. C. HALE, Space Research X, 208 (1970).
[8] J. D. MITCHELL, L. C. HALE, R. O. OLSEN, J. RANDHAWA and R. RUBIO, Radio Sci. **7**, 175 (1972).

[9] J. P. Cipriano, Ionosphere Research Laboratory Sci. Rep. No. 410, Pennsylvania State University 1973.
[10] M. Nicolet and A. C. Aikin, J. Geophys. Res. **65**, 1469 (1960).
[11] A. Dalgarno, in: Atomic and Molecular Processes, Academic Press, New York 1962.
[12] L. B. Loeb, in: Basic Processes of Gaseous Electronics, Univ. of California Press (1955).
[13] J. D. Mitchell and L. C. Hale, Space Research XIII, 471 (1973).
[14] E. A. Mechtly and L. G. Smith, private communication (1973).
[15] G. C. Reid, J. Geophys. Res. **75**, 2551 (1970).
[16] W. C. Bain and M. D. Harrison, Proc. IEE **119**, 790 (1972).
[17] R. V. Neher, J. Geophys. Res. **72**, 1527 (1967).
[18] P. Velinov, J. Atmos. Terr. Phys. **30**, 1891 (1968).
[19] M. A. Biondi, M. T. Leu and R. Johnson, COSPAR Symposium on D- and E-Region Ion Chemistry, Urbana, Illinois (1971).
[20] L. G. Smith, private communication (1973).
[21] G. Brasseur and M. Nicolet, Aeronomica Acta A, No. 113 (1973).
[22] L. G. Meira, J. Geophys. Res., **76**, 202 (1971).
[23] G. C. Tisone, J. Geophys. Res. **78**, 746 (1973).
[24] L. C. Hale, J. R. Mentzer and L. C. Nickell, in: Proc. COSPAR Symp. on the Solar Particle Event of Nov. 1969, AFCRL 72-0474 (p. 333).
[25] W. Webber, J. Geophys. Res. **67**, 5091 (1962).
[26] E. W. McDaniel, in: Collision Phenomena in Ionized Gases, Wiley, New York 1964.
[27] A. Haug and B. Landmark, J. Atmos. Terr. Phys. **32**, 405 (1970).
[28] A. P. Mitra and J. N. Rowe, J. Atmos. Terr. Phys. **34**, 795 (1972).

Lower Ionosphere Structure — Akademie-Verlag, Berlin 1974

NEW ASPECTS IN LOWER IONOSPHERE POSITIVE CLUSTER ION COMPOSITION: RESULTS FROM AN IMPROVED MASS SPECTROMETER PROBE EXPERIMENT

F. Arnold, K. H. Marien and D. Krankowsky

Max-Planck-Institut für Kernphysik, Heidelberg, FRG

In order to avoid errors due to fragmentation of cluster ions, a new mass spectrometer probe was designed with the following improvements: (i) a conically shaped sampling geometry to ensure a practically attached shock which allows most of the ions to be sampled under undisturbed flow conditions; (ii) a relatively small draw in potential of 5 volts; (iii) various operational modes of mass analyser and ion optical voltages depending on height in order to obtain optimum operating conditions; (iv) a pumping speed of the helium cryopump of more than 2000 l s^{-1}.

The first flight of the probe on 31 May 1972 from Andoya, Norway, reaching an altitude of 108 km, revealed drastic differences compared with previous results. 13 out of the 23 cluster ions found have not been observed in the ionosphere before. Most of them are not of the type $H^+(H_2O)_n$ and some do not contain water molecules. At all altitudes the cluster ions measured were heavier than in previous experiments.

Furthermore, the altitude range of measurements could be extended down to 35 km.

Lower Ionosphere Structure — Akademie-Verlag, Berlin 1974

CLUSTER IONS IN THE STRATOSPHERE: A MASS SPECTROMETRIC MEASUREMENT

D. Krankowsky and F. Arnold

Max-Planck-Institut für Kernphysik, Heidelberg, FRG

On the downleg portion of a rocket flight from Andoya, Norway, positive ion spectra were obtained between 55 and 35 km. At that time, a stable weak proton event caused enhanced ionization in this altitude range. In ten spectra, mass peaks were measured at mass numbers 37, 55, 60 ± 2 and 80 ± 2 amu. Masses 37 and 55 are identified as $H^+(H_2O)_2$ and $H^+(H_2O)_3$. A definite identification for masses 60 and 80 cannot be given. However, the relatively high abundance of these ions indicates characteristic changes in ion composition between the mesosphere and upper stratosphere, even if the measured results are affected by decomposition and discrimination during the measurement.

Lower Ionosphere Structure — Akademie-Verlag, Berlin 1974

ION COMPOSITION MEASUREMENTS AT ALTITUDES 100–180 km IN 1972

V. K. SEMENOV and A. D. DANILOV

Hydrometeorological Service, Moscow, USSR

The results of the upper atmosphere ion composition measurements by an RF mass spectrometer installed in MR 12 sounding rockets are presented. The rockets were launched at Volgograd in 1972: on 21 June at 1200 LT; on 23 June 1200 LT and on 1 December at 1202 and 1539 LT. The relative concentrations of the main ionospheric ions O^+, NO^+ and O_2^+ are shown and a comparison with similar results obtained by the same method as in 1971 is given.

1. Method

In 1972 the authors continued the investigation of the relative ion concentration variations and ionic layers in MR 12 geophysical rocket flights. The analogous results obtained in 1971 are published in [1].

The data about the flights in both 1971 and 1972 are presented in Table 1.

Table 1

Flight No.	Date	LT	Place	$\chi(^0)$	$P_{10}^{\odot}$
1	2 March 1971	13.33	Volgograd	60	126
2	1 October 1971	12.45	Volgograd	52	115
3	7 October 1971	12.30	Volgograd	54	115
4	26 December 1971	10.54	0°24′30″	30	120
5	21 June 1972	12.00	Volgograd	25	139
6	23 June 1972	12.00	Volgograd	25	131
7	1 December 1972	12.02	Volgograd	70	94
8	1 December 1972	15.39	Volgograd	86	94

$P_{10}^{\odot}$ is the Covington index.

A radio-frequency mass spectrometer MX 6407 with one analyser was used. One analyser is used instead of two (which are in the set-up of the device) for two reasons: the sensitivity of the second analyser (for the interval 1–4 amu) is not sufficient for the registration of the light ions in the height interval 100 to

180 km; also, for part of the experiments there was no spare space when the grouping of experiments in one rocket was made.

In Table 2 the main analytical characteristics of the devices used in the 1972 flights are presented. For the dynamical range two values are given: in the numerator, the absolute value for the given mass spectrometer; in the denominator, the value which was realized in the experiment. The following parameters were telemetered from the rocket to ground: ion current in two ranges of sensitivity (high and low), the sweeping voltage and the emission current of the ion source.

Table 2

Flight No.	amu range	Resolution $R/\Delta R$	Sensitivity (10^9 m^{-3})	Dynamical range	Sweep-time (s)
5	5—47	22	2	160/60	0.90
6	10—46	30	5	180/60	0.90
7	5—49	25	23	190/100	0.92
8	6—54	20	1.4	140/22	0.94

Though in this paper only the results on ion composition are considered, it is worth mentioning that in flights 5 and 6 the neutral composition was measured simultaneously by other authors with the same mass spectrometer.

For that purpose the mass spectrometer was switched cyclically by means of a special device: ion spectra, neutral spectra, ion spectra etc. In flight 8 at altitudes 98—120 km during the ascent the mass spectrometer operated in the neutral regime; at 120 km it was switched to ionic and only ion composition measurements were made till the end of the flight. The main results of the measurements are presented in Table 3 and Figs. 1 and 2. Something must be said about the treating of the initial data. The short sweep-time of the device allows small details in the ion concentration distribution to be registered, so narrow layers of ions

Table 3

Flight No.	h (km)	Height resolution ΔH (km)	Mass number registered M (amu)		Comments
5	103—178	min —1.0	16, 30, 32	←	O^+, NO^+, O_2^+
		max —0.1	18, 19, 28, 42	←	contamination
6	103—174	min —2.2	16, 30, 32	←	O^+, NO^+, O_2^+
		max —0.4	23, 24, 25, 26	←	in one layer
			18, 19, 28	←	contamination
7	103—183	min —1.1	16, 30, 32	←	O^+, NO^+, O_2^+
		max —0.3	24, 28	←	in one layer
			18, 19, 42	←	contamination
8	120—183	min —0.8	16, 30, 32	←	O^+, NO^+, O_2^+
		max —0.2	13, 14	←	results of charge transfer
			18, 19, 44, 46, 48	←	contamination

may be detected, but the short sweep-time leads to the lowering of the sensitivity of the device.

While building up the behaviour of the relative ion concentrations, analysis of the ion peaks detected was made. The peaks associated with the rocket concentration were separated. For the atmospheric ions an averaging over each 5 points were made to improve the reliability of the data presented.

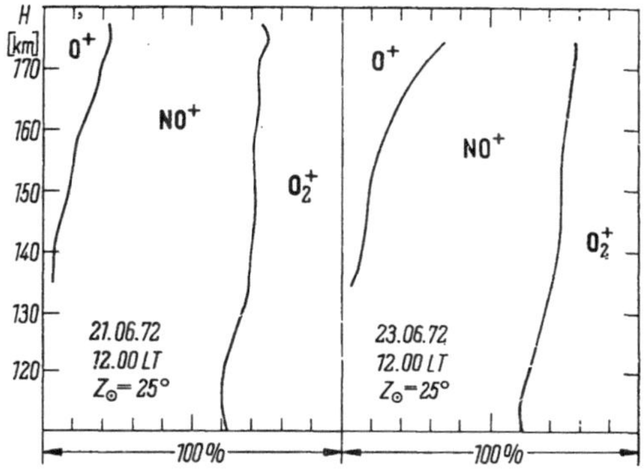

Fig. 1.

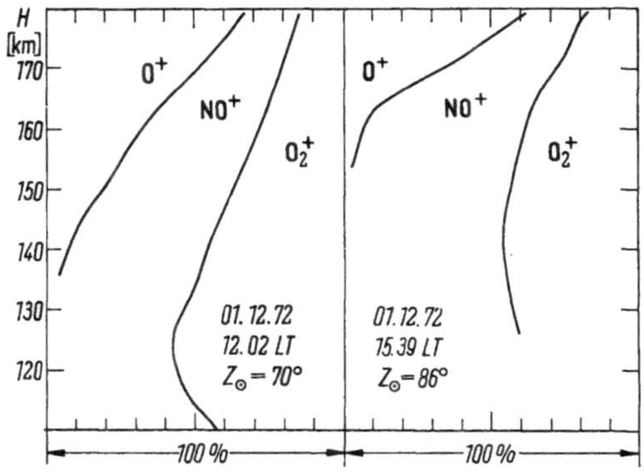

Fig. 2.

The absolute profiles for each ion were analysed, and the error for each ion and for different heights was estimated; then the relative ion composition was calculated on the basis of points taken from the altitude profiles for each ion and separated by about 10 seconds from each other.

The discrimination of light ions relative to heavy ions in the analytical system of the mass spectrometer was taken into account. The values of the discrimination coefficient 16^+ (O^+) relative to 32^+ (O_2^+) attained a factor of 2; that is why the real sensitivity of the device in the mass range 5—12 amu was rather low.

The relative ion composition in flight 8 at altitudes above 160 km was corrected with the help of the electron concentrations measured in that flight [2].

In obtaining the data on ion layers no averaging was done: all the detected values of the ion peaks were used, the data obtained during both ascent and descent were used, the shape of the peaks was analysed, and comparison with the corresponding neutral spectra, when available, was made.

The error of the relative ion concentrations measurements is about 10—15% and is due mainly to the light ions because of the uncertainty in the discrimination coefficient. The error in absolute determination of the altitude is ± 1.5 km.

2. Discussion

Layers of the meteoric ions were detected in two flights, 6 and 7. The layer of ions 23^+, 24^+, 25^+ and 26^+ in flight 6 was situated at 108 km; it had half-width less than 1 km and concentration $\sim 2\%$ of that of electrons. The main contribution to the layer was given by the 24^+ ions. Ion peaks for 23, 25 and 26 amu were detected near the limit of the sensitivity, the amplitudes of 23^+ and 26^+ being of the order of 1/4 of that of peak 24 amu. The layer was detected only for one spectrum, because the device had the lowest height resolution in comparison with all the others used.

In flight 7 a layer of the ions analogous to the layer detected in flight 1 and described in [1] was detected. The layer was situated at 130 km, its half-width was ~ 2.5 km. This layer was detected during both ascent and descent. The relative concentration of the 28^+ ions was 5 times higher than in [1] and was as high as 5% of the electron concentration. Also, near the limit of the sensitivity 24^+ ions were detected in the layer. It is interesting to note that in flight 8, which was made 3.5 hours after flight 7, the above-mentioned layer was not detected. The mass spectrometer in flight 8 was switched to the ionic regime at the altitude of the layer, so that the layer could not have been detected if it had moved down after flight 7. It is reasonable to identify ions 28^+ with Si^+ (as was done in [1]), and ions 24^+ with Mg^+ (flight 7). In flight 6 ions 23^+ are identified with Na^+ and ions 24^+, 25^+ and 26^+ may be considered as ions of magnesium isotopes [3]. Their cosmic abundances are about 7:1:1 which is close to that detected in the experiment.

The data of the 1971 and 1972 flights were used for the investigation of the diurnal variations of the ion composition. As one may seen from Table 1 all the flights (except flight 4 from the ship *Prof. Zubov*) were carried out in the same place and by the same method. The change of the solar radio emission flux was relatively small, $94-140 \times 10^{-22}$ W m^{-2} Hz^{-1}, so the data of the flights seem to be fairly comparable.

The variations of the relative ion concentrations with solar zenith angle for the altitude 160 km is shown in Fig. 3. Points, crosses and triangles correspond to experimental values of $[O^+]/[e]$, $[NO^+]/[e]$ and $[O_2^+]/[e]$. The solid lines give the trend of their variations. Though some scattering of the data is seen, one can see that the results of all experiments are in good agreement with each other, O^+ ions having the best pronounced dependence on the solar zenith angle χ (which is labelled $Z_\odot$ in the figures).

It is interesting to compare these results with the empirical ion composition model [4]. This is based on 12—15 rocket-borne mass spectrometer measurements for each altitude. The variation of the corresponding relative ion concentrations according to the model [4] is shown in Fig. 3 by the broken lines. As is seen, the results of this work (provided at the same place by the same method) confirm the conclusions from the model (based on the comparison of the different data); the confirmation of the conclusion about the diminishing of the O^+ concentration while moving from $\chi = 60°$ to $\chi \approx 30°$ seems especially interesting. In [4] this conclusion was obtained for the absolute O^+ concentrations. Now it is evident that this effect is due to the decrease in the relative concentrations, because the electron concentration shows no such effects.

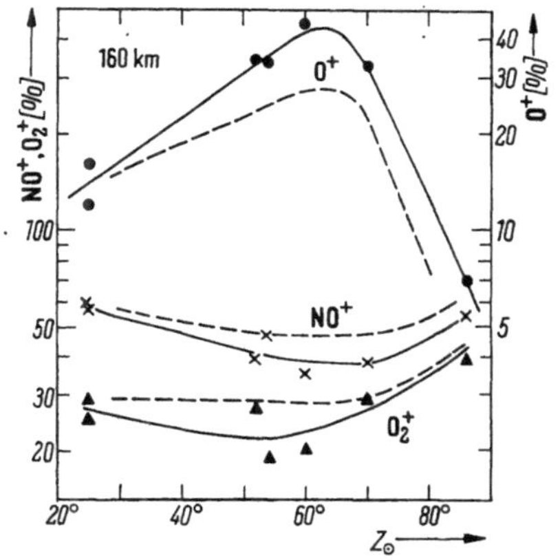

Fig. 3.

References

[1] A. D. Danilov and V. K. Semenov, Space Research XIII, 493 (1973).
[2] A. D. Danilov et al., paper presented at the XIV COSPAR Symp., Constance 1973.
[3] A. D. Danilov et al., Space Research X, 742 (1970).
[4] A. D. Danilov and M. N. Vlasov, Trudi IPG (Proc. Inst. Appl. Geophys.) **19**, 12 (1973).

Lower Ionosphere Structure — Akademie-Verlag, Berlin 1974

METEORIC IONS IN THE D AND E REGIONS

A. D. Danilov

Hydrometeorological Service, Moscow, USSR

The analysis of the experimental mass-spectrometer data leads to some conclusions on the behaviour of meteoric ion layers at altitudes 80—120 km. The dependence of the half-width of Fe^+ and Mg^+ layers on the altitude is found: the $[Mg^+]/[NO^+]$ ratio decreases with increase in altitude. The dependence on solar zenith angle for this ratio is found at 92—93 km. Some conclusions are drawn about the distribution of the magnesium neutral atoms.

The fact of the existence at altitudes 80—120 km of narrow layers of meteoric ions is widely known. In this work an attempt is made to collect and to analyse the results of all the mass-spectrometer measurements in which such layers were observed. The list of the flights used is presented in Table 1. In most of the flights only ions Fe^+ and Hg^+ were detected, so only for these two ions is it worth trying to find some common laws of their distribution. That is the purpose of the work.

The important factor which characterizes the metal ion layers is their width. We will define $\Delta h_{1/5}$ as the width of the layer at the level where the concentration is equal to 0.2 of that at the maximum of the layer. Fig. 1 shows the behaviour of the value of $\Delta h_{1/5}$ with altitude according to the experimental data. Two conclusions may be drawn on the basis of Fig. 1. The first is that the width of the layers decreases with altitude. Though the scattering of the data in Fig. 1 is rather high, the common tendency of decreasing $\Delta h_{1/5}$ with increasing h is seen. For daytime and twilight experiments the effect is especially pronounced. The value $\Delta h_{1/5}$ for $h < 100$ km, averaged over 24 experimental points, is equal to 9.1 km. For $h > 100$ km the same averaging over 17 points gives $\Delta h_{1/5} = 2.8$ km. Thus, from 90—95 km to 110—120 km the width of the layers decreases to about one third.

The observed fact of the diminution of the width of the layers agrees with the hypothesis of their formation under the action of the wind shear mechanism. Actually, the vertical component of the ion drift, induced by the horizontal wind system, is much less at 90 km (see for example [15, 16]) than at 120—130 km, where the optimal conditions for the wind shear exist.

The second conclusion which may be drawn from Fig. 1 is that the distribution of the layers with altitude is irregular. One third of all the layers are situated in the narrow interval 91—94 km, and almost half fall within the interval 91 to 97 km.

The variation with altitude of the $[Mg^+]/[NO^+]$ ratio, measured at layer maximum, is shown in Fig. 2. Despite the scattering of the points the tendency for

Table 1

No.	Author	Date	LT	Place	$\chi(°)$	$P_{10}^{\odot}$	h/km	M^+	$[M^+]/m^{-3}$
1	Istomin [1]	02 Aug. 1958	morn.	Middle latitudes	54	250	105	Fe^+	1.2×10^{10}
								Si^+	7.5×10^9
2	Istomin [1]	15 Jun. 1960	morn.	Middle latitudes	75	166	105	Mg^+	1.4×10^{10}
								Ca^+	5.4×10^8
							101	Fe^+	1.5×10^{10}
							120	Mg^+	
								Si^+	
3	Narcisi [2]	31 Oct. 1963	12.00	Eglin	44	82	95	Mg^+	5×10^9
								Na^+	9×10^8
								Ca^+	1.6×10^8
4	Narcisi	11 Mar. 1964	12.00	Eglin		75	95	Fe^+ (?)	
5	Narcisi [3]	06 Mar. 1965	01.32	Churchill	—	74	100	Mg^+	
6	Narcisi [4]	17 Oct. 1965	23.20	Eglin	—	—	89	Fe^+	
								Mg^+	
								Ca^+	
							93	Fe^+	
							95	Mg^+	
							98	Fe^+	
7	Young et al. [5]	26 Jan. 1966	18.29	White Sands	103	85	106	Mg^+	1×10^{11}
								Fe^+	
								Na^+	
								Ca^+	
								Al^+	
								Si^+	
8	Zloodko [6] and Klyueva	12 Dec. 1966	21.39	Middle latitudes	133	158	100	Mg^+	6.5×10^8
							106	Mg^+	5.4×10^8
								Si^+	2×10^9
9	Zloodko [6]	20 Dec. 1966	19.00	Middle latitudes	117	108	120	Si^+	4×10^9
10	Narcisi et al. [7]	05 Nov. 1966	11.55	Cassino	17	99	93	Mg^+	
								Fe^+	
11	Narcisi et al. [7]	12 Nov. 1966	11.54*	Cassino	20	120	93	Mg^+	
								Fe^+	
							104	Fe^+	
							105	Mg^+	
12	Narcisi et al. [7]	12 Nov. 1966	12.09**	Cassino	20	129	92	Fe^+	
								Mg^+	
13	Narcisi et al. [8]	12 Apr. 1967	18.01	Eglin	89	130	93	Mg^+	1×10^9
								Fe^+	4×10^9
							109	Si^+	1.2×10^9
								Mg^+	1×10^8
							113	Fe^+	5×10^7

Table 1 (Cont.)

No.	Author	Date	LT	Place	$\chi(°)$	$P_{10}^{\odot}$	h/km	M^+	$[M^+]/m^{-3}$
14	Narcisi et al. [8]	12 Apr. 1967	18.49	Eglin	99	130	93	Fe^+	7×10^8
								Mg^+	3×10^8
							106	Fe^+	2.7×10^8
							117	Si^+	3×10^8
15	Narcisi et al. [8]	18 Apr. 1967	04.20	Eglin	103	129	100	Fe^+	1×10^8
							112	Mg^+	3×10^8
								Fe^+	1×10^8
							118	Mg^+	1.3×10^8
							125	Fe^+	1.7×10^7
							128	Fe^+	6
16	Narcisi et al. [8]	18 Apr. 1967	05.21	Eglin	90	129	92	Fe^+	1.9×10^8
								Mg^+	1.5×10^8
							110	Fe^+	1×10^{10}
							111	Mg^+	6×10^9
							117	Mg^+	9×10^8
								Fe^+	1×10^8
17	Narcisi et al. [8]	04 Dec. 1967	16.20	Churchill	96	126	92	Fe^+	1.9×10^9
								Mg^+	8×10^8
18	Zloodko [9]	21 Mar. 1969	23.58	Heiss Is.		231	104	Mg^+	3×10^9
19	Zloodko [10]	18 Apr. 1969	17.00	Volgograd	73	147	121	Si^+	5×10^{10}
								Mg^+	8×10^9
20	Zloodko [10]	20 Apr. 1969	17.04***	Volgograd	73	149	100	Mg^+	2.8×10^9
21	Zloodko [11]	13 Mar. 1970	02.32	Heiss Is.	111	198	120	Mg^+	10^{10}
								Na^+	4×10^9
								Si^+	1.7×10^9
22	Zloodko [11]	Jul. 1970		Middle latitudes	107		120	Mg^+	1×10^9
23	Goldberg [12]	09 Mar. 1970	19.38	Thumba	110	164	98	Mg^+	9×10^8
24	Goldberg [12]	10 Mar. 1970	01.08	Thumba	162	158	93	Mg^+	2×10^9
							95	Al^+	1.6×10^8
25	Goldberg [12]	19 Mar. 1970	08.27	Thumba	53	130	92	Mg^+	3.5×10^8
							118	Mg^+	2.5×10^7
26	Philbrick et al. [13]	20 Nov. 1970		Eglin	98		95	Mg^+	1.8×10^9
							96	Fe^+	3×10^9
							122	Al^+	5.6×10^9
27	Krankowsky [14]	08 Aug. 1970		Andoya	55		90	Fe^+	
								Mg^+	

* Solar eclipse, 80%. ** Solar eclipse, total. *** Geomagnetic disturbance.

decreasing [Mg+]/[NO+] with altitude is seen; at night-time (solar zenith angle $\chi > 100°$) the tendency is especially pronounced. The same picture may be seen for the [Fe+]/[NO+] ratio, and the night-time values of [Mg+]/[NO+] and [Fe+]/

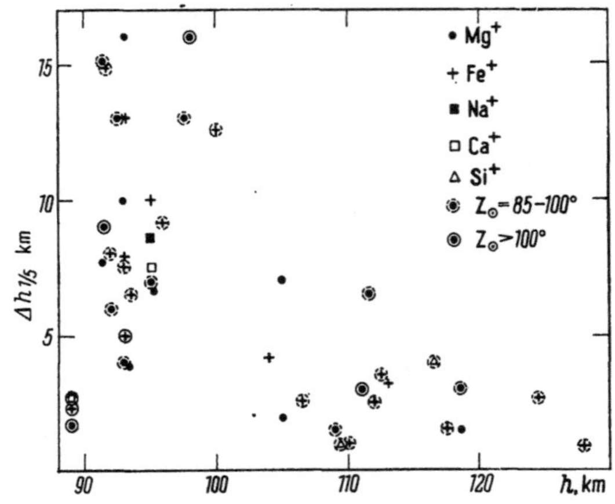

Fig. 1.

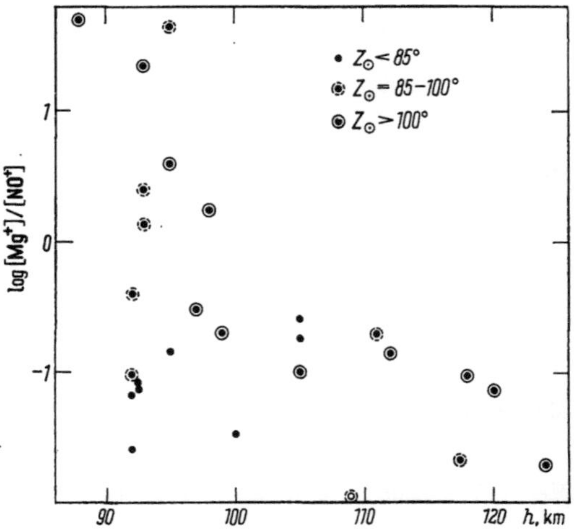

Fig. 2.

[NO+], measured at layer maximum, are presented in Fig. 3. Probably, part of the scattering of the data in Fig. 2 may be explained by the diurnal variations of the [Mg+]/[NO+] ratio. The number of experimental points is not sufficient to obtain the diurnal variations at all altitudes considered; the results of the data analysis for the interval 92—93 km are shown in Fig. 4. Even though the number of experimental data is small, one may confirm the increase of the [Mg+]/[NO+] ratio with increase of solar zenith angle χ (which is labelled $Z_\odot$ in the figures).

On the basis of the experimental data presented one may try to draw some conclusions about the processes governing the behaviour of the Mg^+ ions, and also about the distribution of the corresponding neutral atoms.

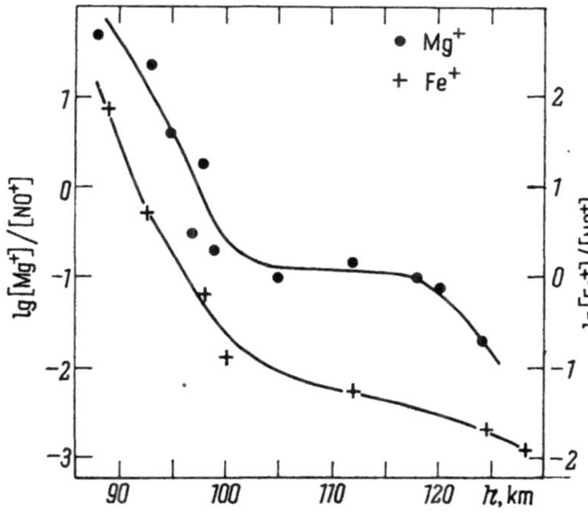

Fig. 3.

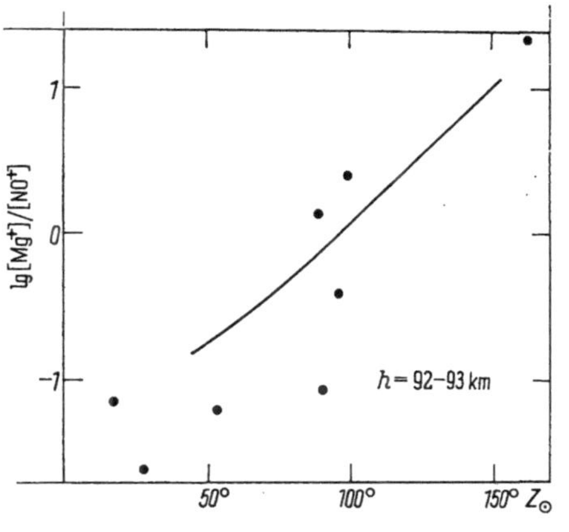

Fig. 4.

The photochemistry of the metal ions has been considered in [17—20], but there are still many uncertainties connected with the values of the rate constants for some reactions. The list of the processes with Mg^+ involved is presented in Table 2. The scheme is practically identical with that of Swider [18] and Anderson and Barth [19]. The neutral model used is given in Table 3. The scheme of particle transformations is shown in Fig. 5. How atoms Mg^+ will be distributed in this scheme depends on the ratio of rates of reactions (2) and (5), i.e. (with the values of γ_2 and α_5 accepted) on $[O_3]/[O_2][N_2]$. At 90—100 km we have $[O_3]\gamma_2 \gg \alpha_5[O_2]\,[N^2]$

Table 2

Eq. No.	Reaction	Rate constant	Comments and references
(1)	$Mg + h\nu \rightarrow Mg^+ + e$	$j_{Mg} = 4 \times 10^{-7}\ s^{-1}$	After Swider [18]
(2)	$Mg^+ + O_3 \rightarrow MgO^+ + O_2$	$\gamma_2 = 2.3 \times 10^{-16}\ m^3\ s^{-1}$	Experimental [21]
(3)	$MgO^+ + e \rightarrow Mg + O$	$\alpha_3^* = 1 \times 10^{-10}\ m^3\ s^{-1}$	After Swider [18]
(4)	$MgO^+ + O \rightarrow Mg^+ + O_2$	$\gamma_4 = 1 \times 10^{-16}\ m^3\ s^{-1}$	Experimental [21]
(5)	$Mg^+ + O_2 + N_2 \rightarrow MgO_2^+ + N_2$	$\alpha_5 = 3 \times 10^{-42}\ m^6\ s^{-1}$	After Swider [18]
(6)	$MgO_2^+ + e \rightarrow Mg + O_2$	$\alpha_6^* = 3 \times 10^{-10}\ m^3\ s^{-1}$	After Swider [18]
(7)	$MgO_2^+ + O \rightarrow MgO^+ + O_2$	$\gamma_7 = 10^{-16}\ m^3\ s^{-1}$	After Swider [18]
(8)	$Mg + O_2^+ \rightarrow MgO^+ + O$	$\gamma_8 < 5 \times 10^{-16}\ m^3\ s^{-1}$	After [19]*
(9)	$Mg + NO^+ \rightarrow MgO^+ + N$	$\gamma_9 < 5 \times 10^{-16}\ m^3\ s^{-1}$	After [19]*
(10)	$Mg + NO^+ \rightarrow Mg^+ + NO$	$\gamma_{10} = 10^{-16}\ m^3\ s^{-1}$	Accepted
(11)	$Mg + O_2^+ \rightarrow Mg^+ + O_2$	$\gamma_{11} \approx 0.1\ \gamma_{10}$	After [17]

* ... not taken into account.

Table 3

h/km	$[O_3]/m^{-3}$ [22] night	day	$[O]/m^{-3}$ [22, 23]	$[O_2]/m^{-3}$ [23]	$[M]/m^{-3}$ [23]	$[e]/m^{-3}$ [24]
90	5×10^{14}	5×10^{13}	2×10^{17}	1.6×10^{19}	7.5×10^{19}	2×10^{8}
100	10^{13}	5×10^{11}	2×10^{17}	1.7×10^{18}	6.9×10^{18}	1.9×10^{9}
110	2×10^{9}	2×10^{9}	1×10^{17}	1.9×10^{17}	1.6×10^{18}	2.0×10^{9}
120	—	--	4×10^{16}	3.2×10^{16}	4.6×10^{17}	1.8×10^{9}
130	—	—	6×10^{15}	1.0×10^{16}	1.5×10^{17}	1.6×10^{9}

and Mg^+ ions are mainly transformed into MgO^+ ions. Photochemical equilibrium gives in this case two equations:

$$[Mg^+][O_3]\gamma_2 = [MgO^+][O]\gamma_4 \tag{12}$$

and

$$[MgO^+][e]\alpha_3^* = (j_{Mg} + [NO^+]\gamma_{11})[Mg]. \tag{13}$$

When $[NO^+]$ is higher than $4 \times 10^3\ cm^{-3}$ the role of the direct ionization is smaller than that of reaction (11), moreover the j_{Mg} value is to be diminished because of the absorption of the radiation, so we can consider reaction (11) as a main source of the Mg^+ ions. Then (12) and (13) give the expression for $[Mg^+]/[NO^+]$:

$$[Mg^+]/[NO^+] = \frac{[Mg][O]\gamma_{11}\gamma_4}{[e][O_3]\gamma_2\alpha_3^*}. \tag{14}$$

Using this expression we may to fing the diurnal variation of $[Mg^+]/[NO^+]$ at 93 km. The data used are presented in Table 4. The electron concentration at these heights is rather changeable; the uncertainity of the ozone concentration is also very great, so the data obtained are purely illustrative and are used only to show the character of the $[Mg^+]/[NO^+]$ variations with $Z_\odot$. As is seen from

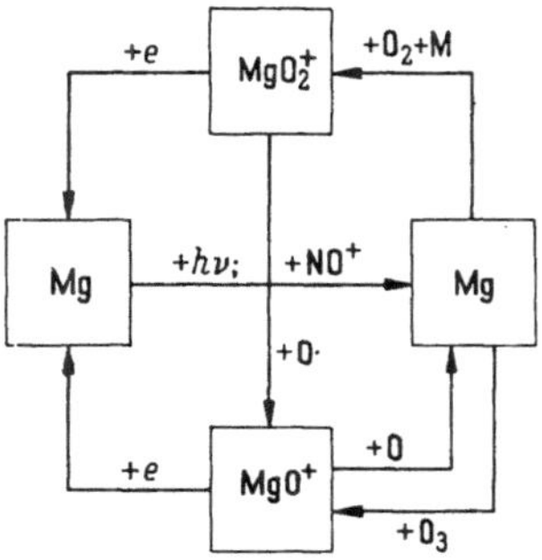

Fig. 5.

Table 4

Conditions	$[O_3]/m^{-3}$	$[e]/m^{-3}$	$[O]/m^{-3}$	$[Mg]/m^{-3}$
Night	2×10^{14}	5×10^{8}	2×10^{17}	10^{10}
$Z_\odot = 90°$	1×10^{14}	1×10^{10}	2×10^{17}	10^{10}
$Z_\odot = 50°$	4×10^{13}	1×10^{11}	2×10^{17}	10^{10}

Fig. 4, where the results of the calculations are shown by a curve, the theoretical change of the $[Mg^+]/[NO^+]$ ratio is very similar to the experimental. It was suggested in the calculations that neutral magnesium concentration does not vary from day to night and it was found that to agree both values of $[Mg^+]/[NO^+]$ it is necessary to put $[Mg] \approx 10^{10}\ m^{-3}$.

Let us consider now the vertical distribution of [Mg]. The expression (14) gives the connection between $[Mg^+]/[NO^+]$ and [Mg] at 90–100 km. Above 110 km the main process of Mg^+ loss became reaction (5). In that case the expression for $[Mg^+]/[NO^+]$ (when $j_{Mg} \ll [NO^+]\gamma_{11}$) is:

$$[Mg^+]/[NO^+] = \frac{[Mg][O]\gamma_{11}\gamma_7}{[O_2][N_2]\alpha_5\alpha_6{}^*}. \tag{15}$$

At 110 km the rates of reactions (2) and (5) are almost equal, so it is difficult to write a sufficiently simple expression for $[Mg^+]/[NO^+]$.

The expressions 14) and (15) give the values of $[Mg^+]/[NO^+]$ in photochemical equilibrium conditions. But the experimental data used for Mg^+ concentrations at layer peaks are the results of the simultaneous action of both photochemistry and wind shear. Probably, rather roughly one may write:

$$[Mg^+]_{lay} = [Mg^+]_{ph}\left(1 + \frac{l}{\Delta h_{1/2}}\right) \tag{16}$$

where $[Mg^+]_{ph}$ is the photochemical concentration, $[Mg^+]_{lay}$ the observed concentration at layer peak, $\Delta h_{1/2}$ the half-width of the layer and l the characteristic dimensions of the region where the collection of the ions takes place. The latter value is defined by $\tau_{Mg^+}\overline{V}$ (τ_{Mg^+} is the photochemical life time of the Mg^+ ions and $\overline{V}$ the average velocity of the vertical drift) when τ_{Mg^+} is not very high. When τ_{Mg^+} is high l has to be limited by the structure of the vertical drift profile.

At 90—100 km τ_{Mg^+} is less than 500 seconds for the photochemical theory used. With $\overline{V} \approx 1\ \mathrm{m\ s^{-1}}$ and $\Delta h_{1/2} \approx 3-5$ km we obtain that at these altitudes the term in brackets in (16) is about unity. At 110—120 km the τ_{Mg^+} value becomes rather high ($\tau_{Mg^+}\overline{V} \approx$ hundreds km), so we used $l = 20$ km, which roughly corresponds to the mean distance between the convergence and divergence points in the experimental ion drift profiles in [15].

Now the expression (15) is written as follows:

$$([\mathrm{Mg}^+]/[\mathrm{NO}^+])_{\mathrm{lay}} = \left(1 + \frac{l}{\Delta h_{1/2}}\right) \frac{[\mathrm{Mg}][\mathrm{O}]\gamma_{11}\gamma_7}{[\mathrm{O}_2][\mathrm{N}_2]\alpha_5\alpha_6{}^*}. \tag{17}$$

Using (15), (17) and the experimental profiles of the $[Mg^+]/[NO^+]$ (see Fig. 3), we obtain the following Mg concentrations at night:

Height/km	90	100	110	120
$[Mg]/m^{-3}$	4×10^{10}	10^8	10^4	10^2

These estimates of the Mg concentration, though obtained by the scheme in which some uncertainties are involved, permit us nevertheless to assess the quantity and the distribution of the neutral magnesium in the upper atmosphere. The photochemistry of iron ions is even less known than that of magnesium ions. There are some ideas that similar processes may govern the behaviour both of Mg^+ and Fe^+ ions [18]. If so, it follows from Fig. 3 that neutral iron atoms exist in the upper atmosphere in about the same quantities as magnesium atoms.

References

[1] V. G. Istomin, Iskysstvennie Sputniky Zemli **11**, 98 (1961).
[2] R. S. Narcisi, Aeronomy Rep. of Illinois University **10**, 307 (1965).
[3] R. S. Narcisi, Symposium D'Aeronomie, Paris, **1966** (p. 154).
[4] R. S. Narcisi, Space Research VII, 123 (1967).
[5] J. M. Young et al., J. Geophys. Res. **72**, 1473 (1967).
[6] A. D. Zhlood'ko and N. M. Klyueva, Trudy IEM **16**, 3 (1970).
[7] R. S. Narcisi et al., Aeronomy Rep. of Illinois University **32**, 450 (1969).
[8] R. S. Narcisi et al., Aeronomy Rep. of Illinois University **32**, 355 (1969).
[9] A. D. Zhloodko, Space Research XI, 1051 (1971).
[10] A. D. Zhloodko, in: "Issledovanie ionosphere v period povishennoi solnetchnoi activnosty", Leningrad 1970 (p. 44).
[11] A. D. Zhloodko, Trudy IEM **20** (1972).
[12] R. A. Goldberg, Aeronomy Rep. of Illinois University **48**, 160 (1972).
[13] C. R. Philbrick et al., Space Research XIII, 441 (1973).
[14] A. Johannessen and D. Krankowsky, J. Geophys. Res. **77**, 2888 (1972).
[15] K. Fujitaka et al. Planet. Space Sci. **33**, 680 (1971).
[16] E. Constantinides and J. F. Bedinger, J. Atmos. Terr. Phys. **33**, 461 (1971).
[17] R. S. Narcisi, Space Research VIII, 360 (1968).
[18] W. Swider, Planet. Space Sci. **17**, 1233 (1969).
[19] J. C. Anderson and C. A. Barth, J. Geophys. Res. **76**, 3723 (1971).
[20] W. Swider, Nature **217**, 438 (1968).
[21] E. E. Ferguson and F. C. Fehsenfeld, J. Geophys. Res. **73**, 6215 (1968).
[22] T. Shimazaki and A. R. Laird, J. Geophys. Res. **75**, 3221 (1970).
[23] A. D. Danilov, Chemistry of the Ionosphere, Plenum Press, New York 1970.
[24] K. Rawer and S. Ramakrishnan, Rep. to URSI Meeting, Warsaw 1972.

Lower Ionosphere Structure — Akademie-Verlag, Berlin 1974

MASS SPECTROMETRIC INVESTIGATIONS OF THE NIGHT-TIME POLAR IONOSPHERIC STRUCTURE

A. D. Zhlood'ko

Hydrometeorological Service, Moscow, USSR

Mass spectrometric investigations of ion composition of the polar ionospheric E and F1 regions by means of MP-12 rocket-borne radio-frequency mass spectrometers were carried out at Heiss Island in 1970—1972. Results of three rocket experiments show that from 100 to 170 km the predominant ions were NO^+, O_2^+ and O^+, the concentration of which in some cases increases as the *Kp* index increases. Besides the main atmospheric ions, the impurity ions 23^+, 24^+, 25^+ (Na^+, Mg^+) were registered, of which isotopes 24^+ were predominant.

The peculiarity of the night polar ionosphere is closely related to the predominant role of corpuscular currents in the ion structure formation. As in the 250 to 100 km height region the low energy particle currents are absorbed (as a result of which atmospheric heating is observed) and interact identically with the atmospheric gases O_2 and N_2, so in the night polar ionosphere a difference in the height distribution of the predominant ions compared with that in the mid-latitude ionosphere can be expected.

Since the intensity of corpuscular currents is closely related to the state of planetary geomagnetic activity, it is interesting to consider the problem of the correlation of the polar ionospheric ion structure with the *Kp* index.

The analysis of experimental data obtained by means of satellites [1] showed that the electron currents increase approximately by a factor of five per unit of the increase in the *Kp* index. Hence in the case when the corpuscular currents are absorbed at lower heights, a positive correlation of *Kp* with the concentration of the predominant ions NO^+, O_2^+, O^+ sould be observed in the lower ionospheric polar region. This is true for those polar regions where the inflow of soft electrons is mostly observed, e.g. for the auroral zone. Due to such a height dependence of the spectral characteristics of corpuscular currents [2] and the spatial inhomogeneity of the currents one can expect marked local variations in the night polar ionospheric ion composition.

Let us consider the results of mass spectrometric ion composition measurements of the near-polar ionosphere; these results were obtained from the rocket sounding data on Heiss Island in 1970—1972 during the decay of the 11-year cycle of solar activity. Ion composition measurements were made by means of MP-12 rocket-borne radio-frequency mass spectrometers of the MX 6407 type. Ion concentration was estimated on the basis of the known sensitivity of the mass spectrometer MX 6407, which was determined approximately with an accuracy of factor 2 in simultaneous n_e and n_i measurements during earlier rocket experiments.

The rocket No. X 101 was launched at 0007 LT on 28 November 1970. The solar zenith angle χ was 119.9° (labelled $Z_\odot$ in the figures), the planar index $Kp = 3$, $\Sigma Kp = 19$. According to the data obtained by the observatory magnetometric service on Heiss Island, the local geomagnetic activity index $K = 6$. Rocket sounding was carried out at the end of the period characterized by the negative bay disturbance of the geomagnetic field H component and by the increase in Z component. The maximum disturbance was registered two hours

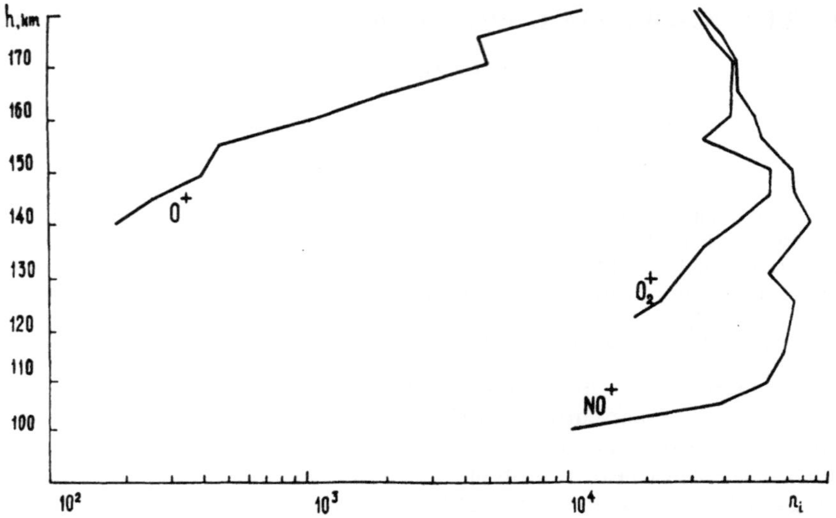

Fig. 1. Concentration profiles of predominant ions in the night polar ionosphere on 28 November 1970, 0007 LT, $Kp = 3$.

before the rocket launch. Totally overcast sky during the rocket launch did not allow us to determine whether aurorae were naturally occurring. In this period diffusive reflections from the Es layer at 100 km were observed on the ionosonde ionograms.

Fig. 1 shows the measurement results of the three main ions $16^+(O^+)$, $30^+(NO^+)$, $32^+(O_2^+)$ characterizing the lower ionospheric structure. In the region of the occurrence of sporadic Es layer at 110 km the marked peak of 24^+ ions and minor peaks of other impurity ions 23^+, 26^+ were observed (see [3]). Impurity ions were not registered in the 110—143 km height range but at greater heights they were certainly observed up to the trajectory peak. In this flight the ions 17^+, 18^+, 19^+, 20^+, 21^+, 28^+, 43^+, 45^+, 46^+ were also registered, the origin of which (except ions 28^+) seems to be related to contamination by the rocket.

The rocket No. X 105 was launched at 2400 LT on 23 January 1971. The solar zenith angle χ was 118.2°, $Kp = 2$, $\Sigma Kp = 16$, $K = 5$. The geomagnetic activity in this launch was similar to that on 28 November 1970: an hour before the rocket launch the hourly negative bay of the geomagnetic field H component, coinciding with simultaneous growth of the Z component, was observed. During the rocket flight the following ions were registered by means of a mass spectrometer: 14^+, 16^+, 28^+, 30^+, 32^+, 46^+. Measurement results of the main atmospheric ions are shown in Fig. 2. Ions 14^+ and 28^+ were observed at the upper limit of sensitivity of the instrument.

Rocket No. X 128 was launched at 0344 LT on 5 March 1972. During the rocket launch $\chi = 101.7°$, $Kp = 1$, $\Sigma Kp = 15$, $K = 3$. According to observations carried out on Heiss Island the geomagnetic field during the rocket launch was an undisturbed one. Pulsation values of the H, Z and D characteristics of the geomagnetic field were not greater than the statistical mean fluctuations in the

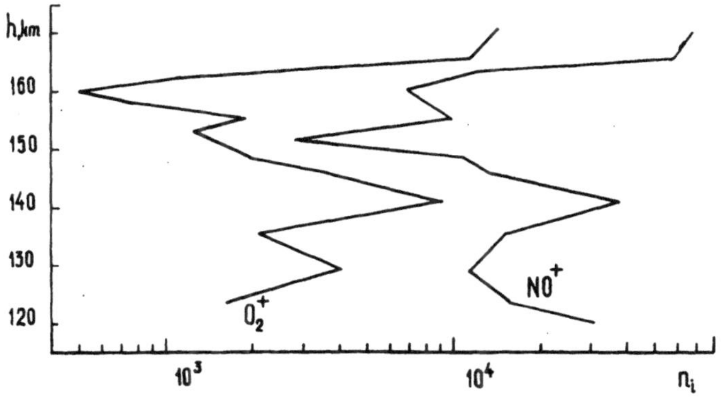

Fig. 2. Concentration profiles of predominant ions in the night polar ionosphere on 22 November 1971, 2400 LT, $Kp = 2$.

quiet period. Fig. 3 shows the main ion concentration profiles plotted according to the experimental mass spectrometer data. Besides the ions mentioned, during the rocket flight at the height of the appearance of the sporadic Es layer from 105 to 110 km the ions 24^+ were also registered, their concentration being low.

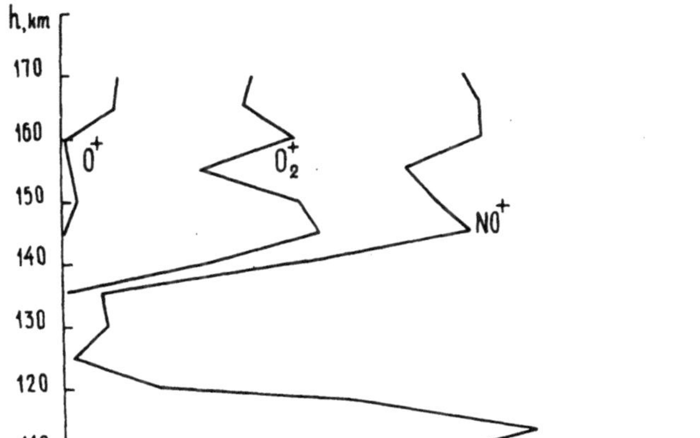

Fig. 3. Concentration profiles of predominant ions in the night polar ionosphere on 5 March 1972, 0344 LT, $Kp = 1$.

Rseults of mass spectrometric measurements of ion concentration carried out at different levels of geomagnetic activity make it possible to study the correlation of n_i with Kp in the polar region. Considering only the measurement results in 1970—1972 one can see that in the Es layer in the altitude range 120—130 km the concentration of ions O_2^+ and NO^+ increases as Kp increases. At the same time such a clear dependence in the F1 region ($H > 130$ km) is no longer observed. Furthermore, if we consider previous measurements on the night ion composition on Heiss Island [4, 5], it should be noted that the high molecular ion concentrations in the polar E region can be observed also at low Kp values. Thus, in the night experiment on 2 November 1967 in the E layer maximum at 130 km the following values were obtained: $n_{O_2^+} \cong n_{NO^+} \cong 1 \times 10^{11}$ m^{-3} at $Kp = 1$, and in the experiment on 13 February 1970 in the E layer maximum at 120 km $n_{O_2^+} \cong 4.5 \times 10^{10}$ m^{-3} and $n_{NO^+} \cong 1.1 \times 10^{11}$ m^{-3} at $Kp = 1$. One can see that these values of predominant ion concentrations are comparable with the results obtained on 27 November 1970 at $Kp = 3$.

Therefore we must conclude that in the lower ionospheric polar region in some cases positive correlation of n_i with Kp is observed, but it is characterized by a low correlation coefficient. The situation is no better when n_i is correlated with the local K index. Hence to establish the correlation of the polar ionospheric ion composition with the parameters of outer ionization sources, a more physically meaningful parameter must be used and not the Kp index. It is likely that the apparent time that elapses after the onset of geomagnetic disturbance is of great importance to the polar ionosphere. However it is not yet clear during what time the equilibrium values of ion concentration are established since the inflow of corpuscular currents can result in noticeable variation of atmospheric temperature.

References

[1] B. J. O'Brien, J. Geophys. Res. **69**, 13 (1964).
[2] D. A. Curnett, J. Geophys. Res **70**, 2485 (1965).
[3] A. D. Zhlood'ko, V. N. Lebedinets and V. B. Shushkova, Space Research XIV (1974).
[4] A. D. Zhlood'ko and N. M. Klyueva, Space Research X, 746 (1970).
[5] A. D. Zhlood'ko, Kosm. Issled. **11**, 1323 (1973).

Lower Ionosphere Structure — Akademie-Verlag, Berlin 1974

SOME FEATURES OF THE EQUATORIAL D REGION AS REVEALED BY THE LANGMUIR PROBE EXPERIMENTS CONDUCTED AT THUMBA

S. Prakash, B. H. Subbaraya and S. P. Gupta

Physical Research Laboratory, Ahmedabad, India

Langmuir probes have been used at Thumba for lower ionospheric investigations. When due care is taken to maintain the cleanliness, not only of the sensor, but also of the rocket body which acts as the return electrode, the probe has been found to give consistent results from flight to flight. In this paper some results on the equatorial D region obtained at Thumba by using Langmuir probes are presented and discussed.

1. Introduction

Langmuir probes have been used extensively for ionospheric investigations on both rockets and satellites. When used on rockets for lower ionospheric studies, it has been found that absolute electron densities cannot be obtained from the classical Langmuir probe. The measured probe currents need to be calibrated in terms of electron densities and the calibration factor could in general depend on various parameters of the probe as well as of the medium [1]. At heights below about 95 km classical probe theory is not applicable owing to the existence of a high collision regime. Also at these heights the ambient properties vary rapidly with height. Hence the calibration factor may also change rapidly with height. Hence for D region studies the probe has been flown with accompanying radio-frequency propagation experiments and the probe currents calibrated at frequent height intervals [2].

2. Langmuir Probe Experiments at Thumba

Several Langmuir probes have been flown at Thumba over the past few years for the investigation of the lower equatorial ionosphere. The probing electrode is usually a stainless steel ogive situated at the nose tip of the rocket and insulated from the rest of the rocket body. A guard electrode is used to maintain the field near the sensor normal to its surface and to minimize leakages. The probe system is specially designed to minimize leakage effects, internal capacitances etc. [3]. With the nose tip geometry, effects due to rocket spin, precession etc. as well as wake effects are found to be a minimum. The probe assembly is well cleaned and kept in a hot oven to prevent moisture from condensing on its surface until the probe has to be fitted to the rocket nose-cone. During the pre-launch period the probe is continuously flushed with dry nitrogen until a few minutes from launch. Due care is taken to maintain the cleanliness not only of the probing sensor, but also of the rocket body which acts as the return electrode.

When such precautions are taken it has been found that the Langmuir probe can give consistent results from flight to flight. When flight time ionograms obtained at Thumba and Kodaikanal are used to convert the probe currents into electron densities for several daytime flights, it is found that at the E region peak the conversion factor remains nearly constant (within $\pm 10\%$) for all flights on which the probe and the return electrode had identical configurations. On several flights the Langmuir probe was accompanied by an onboard resonance probe which could measure absolute electron densities down to a height of 90 km. On these flights the variation with height of this conversion factor could be studied. These have also given consistent results and the details will be published separately. For D region studies the probe results have been compared with simultaneous propagation measurements. The two have been found to agree within reasonable limits except in the 80 km region where the discrepancies could be due to uncertainties in the collision frequencies used. The purpose of the present paper is to present some results on the equatorial D region obtained from the Langmuir probe data. Particular attention is paid to three daytime flights which had identical probe and rocket configurations and carried the same experiments. The three flights were conducted at similar solar zenith angles but on days with widely differing solar X-ray emission intensities. Geomagnetically all the three days were classified as quiet days.

Flight 1 was a very quiet day with almost no spots on the sun. The X-ray fluxes from the sun were extraordinarily low especially around the flight time, in all the wavelength bands monitored by the satellite Solrad 9. Flight 2 was a quiet day but the solar X-ray fluxes were larger than on flight 1. Flight 3 was

Table 1

Flight	Solar zenith angle	Zürich sunspot No.	Solar X-ray flux/10^{-3} W m^{-2} 0.5–3 Å	1–8 Å	8–20 Å	Remarks
Nike Apache 10.44 13 Oct 1972 1259 IST	24°18′	14	$<1.1 \times 10^{-5}$	2.6×10^{-4}	2.8×10^{-3}	at $T - 5$ minutes from Solrad-9
			1.3×10^{-6}	1.3×10^{-5}	6.6×10^{-3}	at $T + 6$ minutes from Solrad-10
Nike Apache 10.38 28 Jan 1971 1110 IST	29°37′	99	$<1.1 \times 10^{-5}$	1.5×10^{-3}	2.1×10^{-2}	Solrad-9
Nike Apache 20.07 29 Aug. 1968 1415 IST	33°00′	64	3.6×10^{-4}	10.8×10^{-3}	4.0×10^{-2}	Solrad-9

on a day with very much enhanced solar X-ray fluxes. The flight was during a short-wave fade-out as reported by the Kodaikanal Astrophysical Observatory. At the flight time the solar X-ray emission in the 8—20 Å band was near its peak. The enhancement in the 0—8 Å band had passed its peak and was in the declining phase. Even though the solar X-ray fluxes were enhanced in all the emission bands the enhancements did not reach the flare levels. Table 1 gives a summary of the X-ray data for the three flights.

3. Results

The electron density profiles obtained on the three flights are shown in Fig. 1. The following features are noted.

1. The presence of a ledge in the electron density profiles around the height region of 60 km on two of the three profiles shown in the figure. This appears to be a common feature of the noon-time D region at Thumba. It is generally seen on the daytime profiles whenever the probe is sensitive enough to detect electron densities of the order 10^7 m^{-3} (10 cm^{-3}) or less.

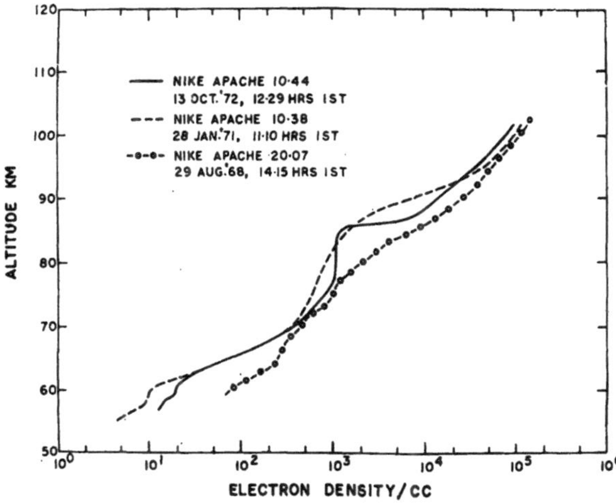

Fig. 1. Electron density profiles obtained from the Langmuir probe experiments on flights 1, 10.44; 2, 10.38; and 3, 20.07. Details of the flights are given in Table 1.

2. On flights 1 and 2 the profiles nearly overlap in the height range 63—75 km. On flight 3 the electron densities are larger in the height region 60—65 km. Above 65 km the profile gradually tends to merge with the other two profiles.

3. In the height range 75 to about 90 km both the profile shapes and the absolute electron densities are different for the three flights. On flight 3 the electron densities were larger than on flights 1 and 2 throughout the height region and differences are most marked in the 80—85 km region. Flight 1 shows a steep gradient at 86 km, the electron density increasing by a factor of 5 in half a kilometre. On flight 2 the gradient is less steep whereas flight 3 shows a relatively monotonic increase in the height range.

In order to facilitate a discussion of these results ionization production rates were calculated as a function of height for the three flights (Fig. 2). Solar X-ray fluxes as measured by satellites Solrad 8 and 9 were used together with standard atmospheric models for calculating ion production due to X-rays. For flight 3,

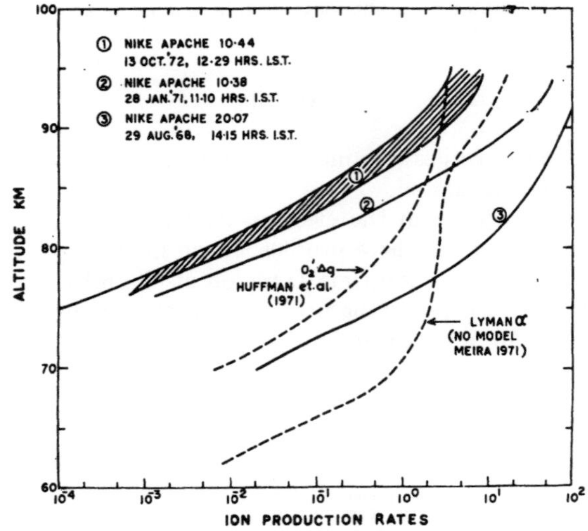

Fig. 2. Total ion production rates for the three flights of Fig. 1. Cosmic ray ionization due to Velinov is also included.

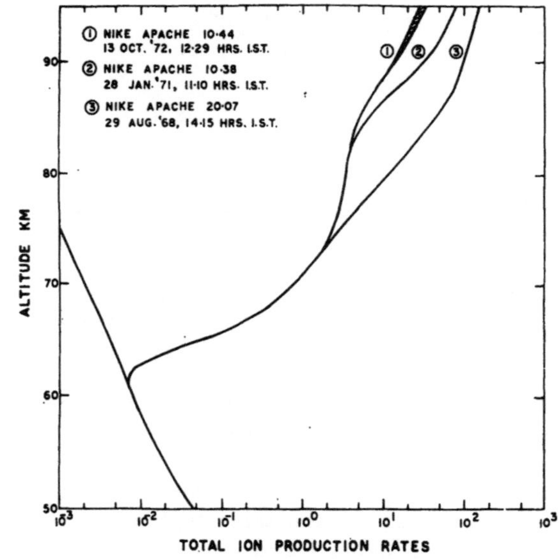

Fig. 3. Ion production rates due to X-rays (1—10 Å) for the three flights are compared with the production rates due to solar Lyman α and ionization of $O_2'\Delta_g$ by solar ultraviolet. The two curves for X-ray production for flight 1 are for the X-ray fluxes at $T - 5$ minutes and $T + 6$ minutes respectively.

when there was disturbance on the sun, detailed information regarding the solar X-ray spectrum was available from the OSO 4 data. Ionization production rates due to cosmic rays are taken from Velinov [4]. Ionization due to solar Lyman alpha is calculated for a flux of 5×10^{-3} W m^{-2}, at the top and used the recent nitric oxide concentrations of Meira [5]. Ionization of $O_2{}'\Delta_g$ according to the revised estimates of Huffman et al. [6] are also included. These are shown in Fig. 3. The electron density profiles of Fig. 1 will now be discussed in the context of these production rates.

4. Discussion

Ionization production in the 60 km height region is primarily due to cosmic rays and this does not show any ledge corresponding to those observed in Fig. 1. The ledges of Fig. 1 are therefore not directly due to primary ionization production. Ionization loss at these heights is essentially by formation of negative ions by attachment to ambient molecules. Electrons can be removed from these negative ions either by collisional or photo detachment. The recent work of Thomas et al. [1] shows that inclusion of photo detachment from negative ions can affect the electron density profile at heights below about 70 km. The effect is substantial below 65 km and an electron density of about 10^7 m^{-3} at 60 km can be explained on their model. The first two profiles of Fig. 1 do show electron densities of this order at 60 km and one may conclude that these are in agreement with the photo detachment model of Thomas et al. However, it may be noted that measurements made at middle latitudes show that negative ions at these heights are predominantly heavy water clusters [8] and the model of Thomas et al. does not consider these water-cluster ions.

Above 60 km ionization of nitric oxide by solar Lyman alpha begins to compete with cosmic-ray ionization, becomes the major source of ionization at 65 km and remains so up to about 75 km for all the flights. Above 75 km the high degree of variability of the solar X-ray emission makes the ion production function vary from flight to flight. The electron density profiles of Fig. 1 follow the production functions of Fig. 3 in a general fashion but not in detail. This could be due to (i) variations in solar Lyman alpha flux with solar activity (for which there is some evidence now), (ii) variations in nitric oxide concentration at these heights, (iii) variations in neutral atmospheric properties and (iv) changes in ion chemistry. All of them are possible and it is not possible quantitatively to assess their relative contributions. The high degree of variability in the profiles above 75 km and the near constancy below indicate that the two regions are probably characterized by altogether different ion chemistry.

It was mentioned earlier that flight 1 was on an extraordinarily quiet day. The solar emission in the soft X-ray bands was so low that the entire lower ionosphere below 95 km (Fig. 2) was controlled by ionization due to ultraviolet radiation from the sun, especially Lyman alpha ionization of nitric oxide. Fig. 4 shows the ion production rates as well as electron densities of flight 10.44 plotted as a function of height. The two curves follow one another closely up to 86 km. The marked gradient at 86 km in the electron density profile is a unique feature of this flight. On no other daytime flight has such a dramatic change been seen. After this change the two curves again follow one another up to about 95 km. Electron density profiles obtained from probe and propagation experiments at middle latitudes very often exhibit a change in slope in this heigt region and this is usually attributed to changes in positive ion composition [9]. The lower heights are found to be dominated by water-cluster ions and other heavy ions whereas above this height the ions are mostly NO^+ and O_2^+ [10]. The heavy ions are assumed to have much larger recombination rates. Ion composition measurements made at Thumba [11] also show the dominance of water clusters and heavy ions at lower heights and a transition to molecular ions at 85 km. Hence the observed

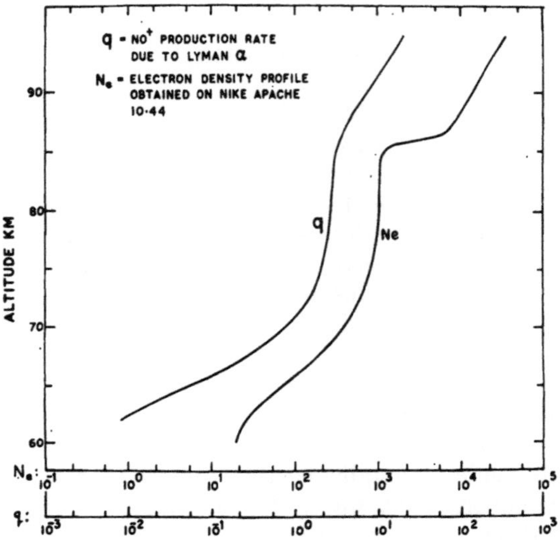

Fig. 4. Electron density profile for flight 1 compared with the ion production function.

change at 86 km could be attributed to such a change in ion composition. From Fig. 4 the height region below 86 km can be divided into two height intervals in each of which the effective loss rate is nearly constant. In the height range of 63—68 km an effective loss rate of 10^{-11} m^{-3} s^{-1} applies. In the height range 70—80 km a recombination coefficient of 5×10^{-12} m^{-3} s^{-1} is applicable. Recent laboratory measurements of recombination rates for the water-cluster ions [12] have given values 1.3, 2.7 and 4.6×10^{-12} m^3 s^{-1} respectively for mass 19, 37 and 55 respectively whereas the values for the much heavier ions approach 10^{-11} m^3 s^{-1}. The present results therefore indicate the predominance of the heavier water clusters at lower heights.

Acknowledgments

The authors would like to express their thanks to the staff of TERLS for their co-operative efforts in making the rocket experiments successful. Solar X-ray data for the flights was supplied by Dr. D. M. Horan of Naval Research Laboratory, USA. Detailed X-ray spectrum information for flight 3 was provided by Prof. A. P. Willmore of University College, London, from the OSO 4 data. Mr. I. L. Kripalani has, over the years, assisted us in payload construction. Mr. R. I. Patel has done much of the computations involved. These are gratefully appreciated.

References

[1] L. G. Smith, in: Small Rocket Instrumentation Techniques, North Holland Publ. Co., Amsterdam 1969 (p. 1).

[2] E. A. Mechtly and L. G. Smith, J. Atmos. Terr. Phys. **30**, 1555 (1968).

[3] S. Prakash and B. H. Subbaraya, Rev. Scient. Instrum. **38**, 1132 (1967).

[4] P. Velinov, J. Atmos. Terr. Phys. **30**, 1891 (1968).

[5] L. G. Meira, J. Geophys. Res. **76**, 202 (1971).

[6] R. E. Huffman, D. E. Paulsen, J. C. Larrabee and R. B. Cairns, J. Geophys. Res. **76**, 1028 (1971).

[7] L. Thomas, P. M. Gondhalekar and M. R. Bowman, J. Atmos. Terr. Phys. **35**, 397 (1973).

[8] R. S. Narcisi, A. D. Bailey, L. E. Wlodyka and C. R. Philbrick, J. Atmos. Terr. Phys. **34**, 647 (1972).

[9] G. C. Reid, J. Geophys. Res. **75**, 2551 (1970).

[10] R. S. Narcisi and A. D. Bailey, J. Geophys. Res. **70**, 3687 (1965).

[11] R. A. Goldberg, J. Geophys. Res. **76**, 8352 (1971).

[12] M. A. Biondi, M. T. Leu and R. Johnson, paper presented at COSPAR Symp. on D and E region ion chemistry, Seattle 1971.

Lower Ionosphere Structure — Akademie-Verlag, Berlin 1974

ROCKET MEASUREMENTS WITH VARIABLE FREQUENCY IMPEDANCE PROBE AT THE MAGNETIC EQUATOR

K. M. Kotadia[a], N. K. Parikh[a] and K. G. Jani[b]*

[a]Physics Department, Gujarat University, Ahmedabad, India
[b]Arbeitsgruppe für Physikalische Weltraumforschung, Freiburg im Breisgau, FRG

Two rockets with variable frequency impedance probe, Langmuir probe, cylindrical probe and magnetic aspect sensor were flown at Thumba in August 1972. The nose-cone itself was used as sensor antenna of the impedance probe. Results obtained by the dc Langmuir probe are compared with those obtained by the variable frequency RF impedance probe. Here the results of a morning flight are reported.

The Indian-made Centaure rocket rose to an altitude of 150 km. The electron density profile deduced from the impedance probe measurements from 95 km to 150 km altitude after due correction for the effect of the earth's magnetic field seemed to be consistent with that deduced from Langmuir probe measurements and also with the ground-based ionosonde measurements. No valley was noticeable above the E-layer peak, although several cyclic undulations were observed in the height range 120—135 km. The profile recorded during the descent of the rocket gave values of electron densities lower by a factor of 1.4—1.6 than during the ascent in the height range 105—140 km. The cyclic variations are attributed to nutational motion of the rocket.

1. Introduction

Special interest is being shown in the investigation of the ionosphere over the earth's magnetic equator because of the situation of a strong electrojet current system in the E region of the ionosphere. Thumba (8°33′N 76°56′E; $I = 0.6°$S) happens to be just at the magnetic equator and many investigators from different countries have conducted a variety of rocket experiments. The experiment with the variable frequency impedance probe was performed for the first time at Thumba using a 1-metre long nose-cone as sensor antenna. The aim of the experiment was to find the electron density distribution in the height range 95 to 150 km. The technique and genesis of the impedance probe have been described by Jacobs and Rawer [1] and Bross, Jacobs and Rawer [2]; but they used a whip or rod antenna projecting out of the nose-cone. It was thought worth while to test the feasibility of using the nose-cone itself as sensor and to see how the measurements of electron density by impedance probe compare with those by dc Langmuir probe which has been flown several times before by Satya Prakash and his group [3, 4].

* On leave from Physics Department, Gujarat University, Ahmedabad.

2. Experimental and Analytical Procedure

An Indian-made Centaure rocket was launched on 12 August 1972 at $T = 0745$ IST (UT + $5^1/_2$ hr), Flight No. 05.14. The payload consisted of impedance probe (weight 2.2 kg), Langmuir probe, cylindrical probe and magnetic aspect sensor. The details of internal assembly and rocket will be given in a separate paper. The initial setting of the rocket was at an elevation of 85° and bearing 270°, i.e. westwards. The rocket was thought to have fallen in the sea at a distance of 213 km at $T + 384$ s. The ground radar tracked the rocket up to a height of 82 km and the rest of the trajectory was determined by computer. According to the computed trajectory, the apogee was found to be at 149.56 km at $T + 199$ s. When the sensor enters the ionized medium, the simple neutral air capacitance between the nose-cone and the rocket body transforms into a kind of resonant circuit shown in Fig. 1, the values of components in the equivalent resonant

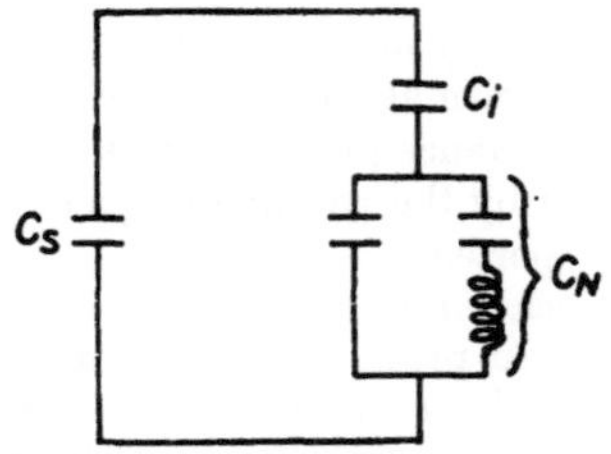

Fig. 1. Equivalent circuit of the impedance probe sensor.

circuit depending on the electron/ion density. As the frequency of the constant-voltage oscillator sweeps every half second from 1 MHz to 8 MHz, a current minimum corresponding to parallel resonance and maximum corresponding to series resonance (Fig. 2) are observed which after suitable detection and amplification are transmitted to the ground telemetry station. The telemetry record thus gives the variation of probe current against the oscillator frequency. The frequency corresponding to the major current minimum is taken in the calculation of plasma frequency at different heights.

Morning time was chosen to give the conditions when the normal E and E2 layers are in the formation stage and the equatorial electrojet is just beginning to build up. Another flight was performed at 1532 hr IST on 17 August 1972 but the results of this flight have not yet been studied.

The relation used for the calculation of plasma frequency f_N is

$$f_N^2 = \frac{1}{2}\left[\frac{f^2 - f_B^2}{f^2 - f_B^2 \sin^2\theta}\left\{2f^2 - f_B^2\sin^2\theta + [(2K^2f^2 - f_B^2\sin^2\theta)^2 + 4K^2f^4(1 - K^2)]^{1/2}\right\}\right] \quad (1)$$

where f = parallel resonance frequency, f_B = gyromagnetic frequency, θ = inclination of rocket axis with the magnetic field,

$$K = \frac{C_s C_i}{C_0(C_s + C_i)} \simeq \frac{C_s}{C_0} \quad \text{when} \quad C_i \gg C_s$$

C_i = ion sheath capacitance, C_s = shunt capacitance due to internal geometry, C_0 = effective free space or neutral air capacitance between the nose-cone and rocket body, C_N = effective capacitance of sensor in plasma (Fig. 1). In the present set-up, C_s is found to be 30 pF, C_0 19.11 pF and therefore $K \simeq C_s/C_0 = 1.57$. Eq. (1) is a simplified form of those given by Pyati and Weil [5] and Balmain [6, 7] obtained by omitting certain logarithmic terms and by assuming that the electron density to be measured is larger than 6×10^{10} m^{-3}. If ion sheath capacitance C_i and shunt capacitance C_s are neglected, the relation for plasma frequency reduces to

$$f_N^2 \approx f^2 \frac{f^2 - f_B^2}{f^2 - f_B^2 \sin^2 \theta}. \tag{2}$$

Calculations show, for example at 110 km, that the value of f_N obtained by Eq. (1) is approximately 1.5 times that obtained by Eq. (2), indicating that the effect of shunt and ion sheath capacitances is quite significant and this may change due to the precession and nutation of the rocket.

3. Results

It may be noted from Fig. 2 that there is one major current minimum corresponding to the parallel resonance frequency of the effective circuit. Sometimes another minimum but a minor one is seen at a lower frequency, and a maximum occurs at a higher frequency than that of the major minimum. It is found that the current maximum following the major minimum happens to be at a frequency twice the value of the gyromagnetic frequency f_B. However, the cause of the lower minimum is not understood, but we believe that it has something to

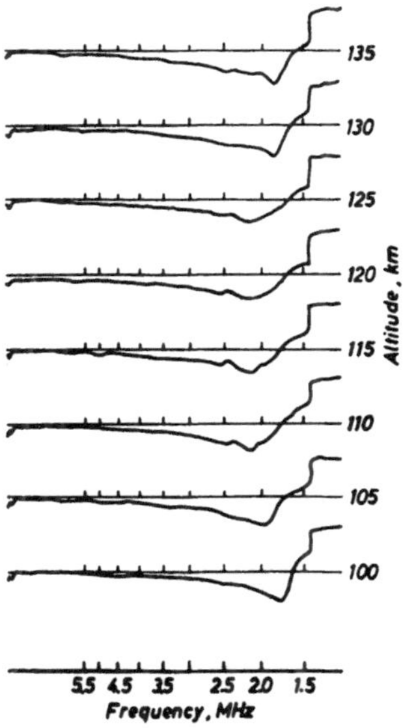

Fig. 2. A sequence of telemetry records of impedance probe measurements at different heights. Ascent on flight No. 05.14, 12 August 1972, 0745 IST, from Thumba.

do with the magnetic field and its inclination with the rocket axis and the ion sheath. Further, the depth of the current minimum increases a little at greater heights which is quite understandable because the resistance component due to electron collisions is less at greater altitudes.

4. Electron Density Profile

The electron density profile obtained by the impedance probe measurements of the parallel resonance frequency in the ionized medium is shown in Fig. 3. Data reduction is based upon Eq. (1). It is seen that the electron density rises to a maximum of 1.35×10^{11} m^{-3} following almost a parabolic distribution law. It then remains almost constant up to 118 km above which altitude cyclic undulations of apparent amplitude about 5×10^{10} m^{-3} and period of 3 seconds

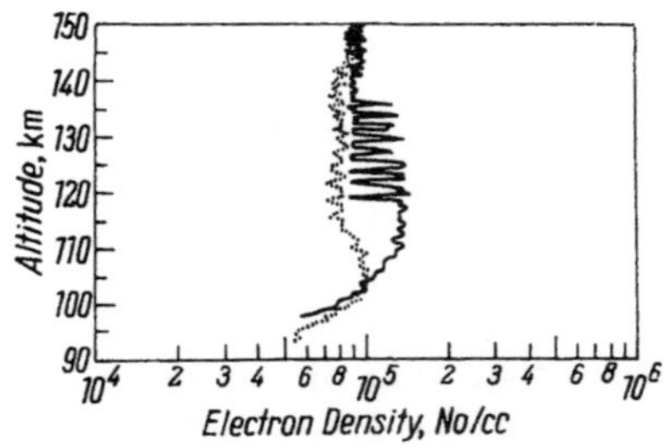

Fig. 3. Electron density distribution profile deduced from impedance probe measurements during ascent and descent. Flight No. 05.14, 12 August 1972, 0745 IST. —— ascent; ····· descent.

wavelength $\simeq 3$ km) are observed up to 135 km altitude. The peaks of these undulations become broader as the rocket goes up and finally they die out. The descent starts at 149.4 km. During descent undulations are also observed of almost the same period as during the ascent but with amplitude very much reduced, about 8×10^9 m^{-3} or nearly one-sixth of the amplitude observed during ascent. Also the electron density measured by the impedance probe during its downward journey is less than that measured during the upward journey by a factor of about 1.4—1.6 in the height range 105—140 km. This ascent/descent difference in electron density is believed to be due to the wake effect of the heavier mass heading during the downward journey of the rocket. Such wake effect is also observed in many flights with other probes. As regards the cyclic variations of electron density, it may be said that the rocket has three motions, spin, precession and nutation (or wobbling) while precessing. According to the rocket-range data, the precession period was about 35—40 s and the nutation period was about 2—5 s. It therefore appears that the 3 s period undulations in electron density might have been caused by the nutational motion of the vehicle, which probably changes the orientation of the sensor with respect to the magnetic field and also changes the conditions of the surrounding plasma. The densities measured in the lower E region during ascent and descent are not much different. The low-frequency (or high period) changes in the electron density due to the precessional motion of the rocket will not be detected. Rao et al. [8] observed the variations in electron density from capacitance probe measurements having a period of about 8 s (wavelength $\simeq 8$ km) in the height range 120—140 km and they attributed these to gravity waves. However, we believe that gravity waves of wavelength less than 10 km are hard to conceive of, although acoustic waves of such low periodicity can occur, but the rocket travels at a speed much faster than that of sound, so it is most probably the motion of the rocket which is responsible for the periodic variations in a given height range.

5. Comparison with Langmuir Probe Results

The electron density distribution profiles deduced from Langmuir probe (LP) and impedance probe (IP) measurements are shown in Fig. 4 only for the ascent flight. Some of the distinct features of these profiles may be summarized as

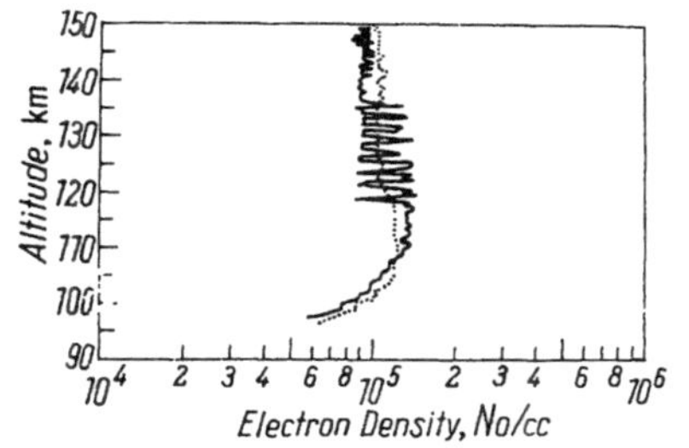

Fig. 4. Comparison of electron density distribution profiles measured by RF impedance probe (——) and dc Langmuir probe (·····). Flight No. 05.14, 12 August 1972, 0745 IST.

follows: (i) In the lower E region, LP gives values of electron density about 1.5×10^{10} m^{-3} larger than those measured by IP. Near the maximum of the E layer, IP gives larger values than LP by almost the same amount. Once again LP gives larger electron density in the height range 135—150 km. (ii) The E layer peak density as measured by IP is observed at 110 km while the LP finds it at 105 km altitude. (iii) The cyclic undulations observed by IP in the 118—135 km range are not found in the LP results. However, the LP electron density values follow a profile which passes through the average of the undulations observed by IP.

Thus if the undulations seen in the IP results can be attributed to a rocket nutation effect we have almost consistent measurements of electron density profile by the two probes. The Langmuir probe employs a conical sensor at the tip of the nose-cone and is very small in size (base diameter 38 mm, length 96 mm) compared with that of the impedance probe sensor (length 1 m, half cone angle = 5.5°, equivalent cylinder diameter 133 mm and length 1000 mm). The tip sensor does not experience variations in density due to the motion of the rocket since the configuration is such that the effective receiving area is almost independent of the rocket orientation.

If the rocket had spin motion only the standard evaluation of IP data would not show the cyclic variations because the axial symmetry would remain unchanged (angle θ constant). However, when the rocket has additional motions like precession and nutation, the standard evaluation is not necessarily correct for the nose-cone sensor because of its large surface area and the large surrounding parts of the rocket structure. These influences may be the reason why our kind of sensor shows rather large aspect effects due to the geometry of the system, particularly variations of the angle θ between configuration axis and magnetic field, and the plasma ion sheath and impedance parameters.

6. Conclusions

This paper gives the results of a rocket experiment with a variable frequency RF impedance probe performed at Thumba (mag. dip = 0.6° S) on 12 August 1972 at 0745 IST.

(i) The electron density distribution as measured by the impedance probe shows a maximum at 110 km altitude. Distribution below this peak level follows a parabolic law.

(ii) Cyclic undulations of apparent amplitude 5×10^{10} m^{-3} and period of about 3 s are observed by IP sensor in the height range 118—135 km. These are possibly due to the nutational motion of the space vehicle and aspect sensitivity of our kind of IP sensor. This phenomenon certainly needs further study and may even give access to improved information about certain plasma parameters. It seems to show that the simplified reduction equation (1) is not always applicable.

(iii) The measured electron density during the downward journey of the rocket is nearly 1.6 times lower than that measured during the upward journey in the height range 105—140 km. This perhaps results on account of the wake effect of the heavier mass heading during descent.

(iv) The electron density distribution profile measured by the RF IP is consistent with that measured by the dc Langmuir probe except for the cyclic undulations. This difference may be due to the large inequality in the sizes of the sensors of the two probes (small size of LP sensor).

(v) The peak electron density measured by the IP agrees with that found from the ionogram.

Acknowledgments

The authors are very grateful to Prof. K. Rawer and Mr. R. Kist of Arbeitsgruppe für physikalische Weltraumforschung, Freiburg (FRG) for their kind cooperation in this experiment. We also acknowledge with deep gratitude the services rendered by the authorities of the Thumba Equatorial Rocket Launching Station. The results of the Langmuir probe were kindly made available by Prof. Satya Prakash and Dr. S. P. Gupta. Mr. B. M. Patel, a senior research fellow of the Indian Council of Scientific and Industrial Research helped in the testing and integration of the payload. We are very grateful to the University Grants Commission for their financial support to our space research project on rocket-borne experiments in the ionosphere. Finally we are also very grateful to Research Engineer Mr. W. Ott of Arbeitsgruppe für physikalische Weltraumforschung, Freiburg, for the fabrication of the payload and technical help throughout the experiment.

References

[1] K. G. Jacobs and K. Rawer, Space Research V, 706 (1965).
[2] P. Bross, K. G. Jacobs and K. Rawer, Space Research VII, 467 (1967).
[3] S. Prakash, B. H. Subbaraya and S. P. Gupta, Space Research IX, 237 (1969).
[4] S. Prakash, S. P. Gupta and B. H. Subbaraya, Planet. Space Sci. **18**, 1307 (1970).
[5] V. P. Pyati and H. Weil, Radio Sci. **66D**, 291 (1965).
[6] K. G. Balmain, IEEE Trans. Antennas and Propagation, **AP-17**, 389 (1964).
[7] K. G. Balmain, IEEE Trans. Antennas and Propagation, **AP-12**, 605 (1964).
[8] U. R. Rao, S. S. Degaonker, B. Dhar and R. M. Patel, Nature **224**, 676 (1969).

Lower Ionosphere Structure — Akademie-Verlag, Berlin 1974

ROCKET MEASUREMENTS OF ELECTRON DENSITY AND ELECTRON TEMPERATURE AT SUNSET

YU. K. CHASOVITIN, V. G. KHRYUKIN, N. M. KLYUEVA and G. P. KOMRAKOV

Hydrometeorological Service, Moscow, USSR

The N_e and T_e profiles for the 85—175 km height region obtained at Volgograd during five rocket launches in 1970—1971 are presented. The N_e profiles are characterized by irregular structure and the occurrence of Es layer. It is shown that the N_e height distribution in the Es layers is quite well approximated by a symmetrical layer model and in some cases for such approximation a parabolic model can be used. From the N_e variation at sunset it is estimated that α' values are equal to $(3-6) \times 10^{-13}$ m^3 s^{-1} at 100 km and $(1-2) \times 10^{-13}$ m^3 s^{-1} at 140 km. The experimental T_e profiles were compared with the model profiles of neutral gas temperature.

In 1970—1971 at Volgograd by means of an MR-12 rocket [1] five experiments were carried out with simultaneous measurements of electron density and electron temperature in the ionosphere at sunset. Electron density N_e was measured by a high-frequency impedance probe and electron temperature T_e by a cylindrical Langmuir probe. Measuring accuracy for N_e was about 15% and for $T_e \pm 100\%$. A detailled description of apparatus and technique used is given in [2, 3]. The basic information on these experiments is given in Table 1. In spite of the fact

Table 1
Rocket Experiments on N_e and T_e Measurement

Launch No.	Date	Time LMT	χ	Kp	Max. height of rocket ascent	Notes
1	01 July 70	21.02	99°	3+	148 km	
2	03 July 70	19.47	90°	1	166 km	
3	06 July 70	20.44	98°	1	148 km	
4	25 Sept. 71	17.17	86°	4	176 km	
5	01 Oct. 71	17.17	87°	3−	171 km	only N_e

that the experiments were carried out on different days, the analysis of data obtained makes it possible to follow the general regularities of the variation of N_e and T_e at sunset.

Experimental profiles of electron density obtained in the 85—175 km height region are shown in Fig. 1a (figures on curves correspond to numbers of launches in Table 1). From these profiles one can clearly see the transition from twilight

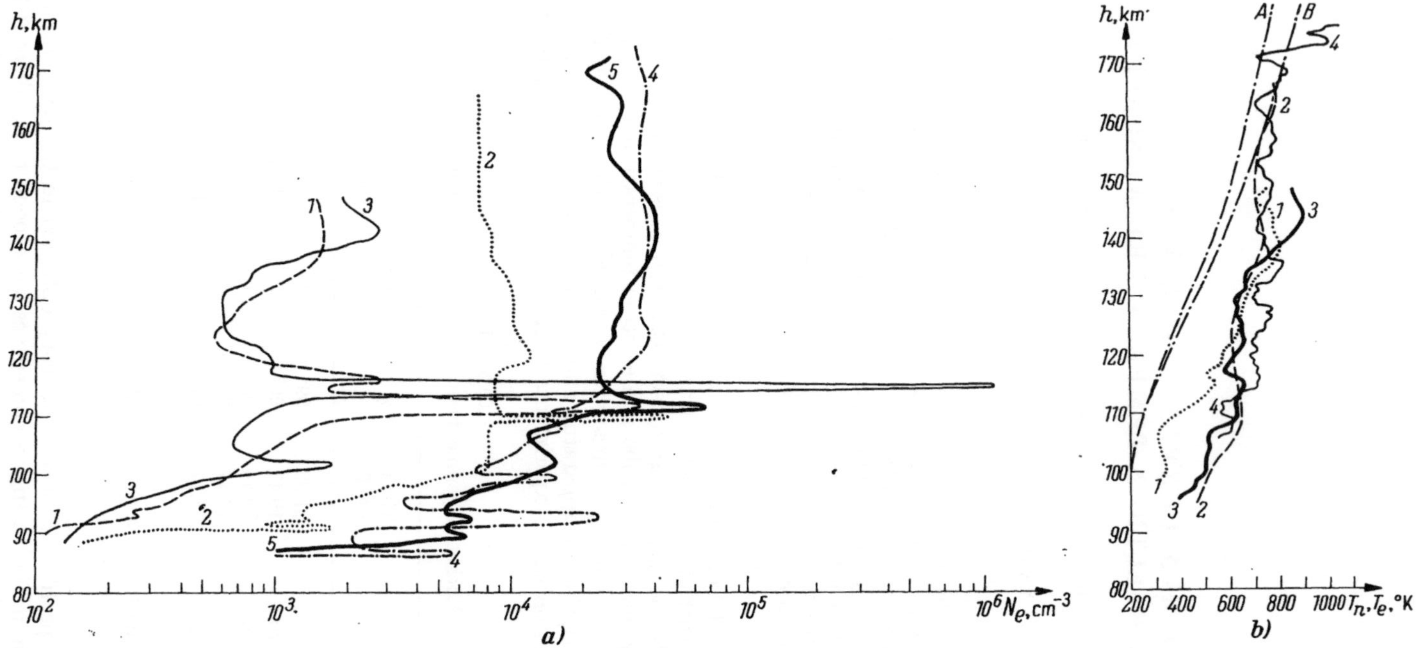

Fig. 1. Profiles of electron density (a) and electron temperature (b) at sunset, Volgograd: 1 July 1970, 2102 LMT (1); 3 July 1970, 1947 LMT (2); 6 July 1970, 2044 LMT (3); 25 Sept. 1971, 1717 LMT (4); 1 Oct. 1971, 1717 LMT (5). Model profiles T_n [9] (b) for launch No. 4 (A), for launches Nos. 1—3 (B).

N_e profiles, the form of which is close to the daytime ones (curves 4, 5) to typical night-time N_e profiles (curves 1, 3). The first are characterized by a sharp N_e gradient in the 85—90 km height region, by more or less monotonic increase of N_e at greater altitudes and by a weakly expressed maximum in the E region. Low N_e values of about 10^9 m^{-3}, the well-defined E region and minimum above it, an occurrence of intensive sporadic formations are characteristic of the latter ones [4, 5]. The profile, obtained in launch No. 2, is some intermediate type of N_e height dependence.

The N_e dependence on the solar zenith angle χ is clearly seen from Fig. 1a. It is interesting to compare the experimentally observed N_e dependence on χ at sunset with some theoretical model. As such a model the N_e dependences on χ

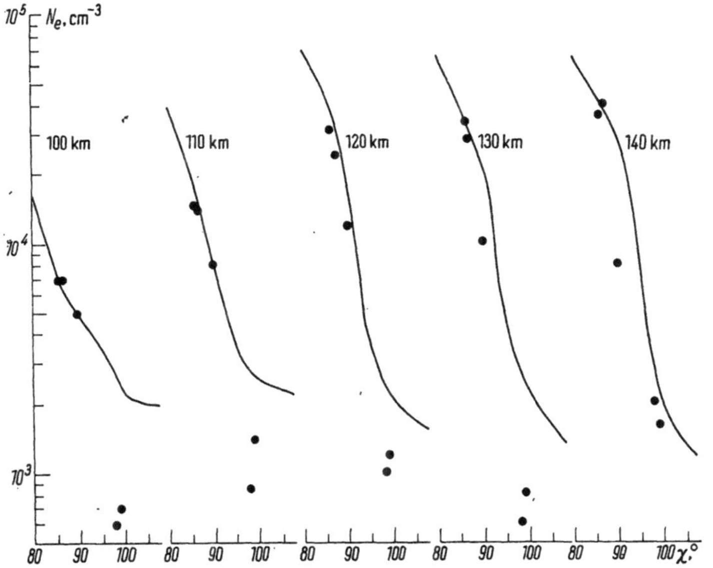

Fig. 2. Theoretical $N_e(\chi)$ dependences for different heights [6] (solid lines) and experimental N_e values for different solar zenith angles (points).

calculated in [6] for different altitudes were used. In these calculations the night ionization source was taken into account; it was assumed that this was the scattered H Ly α and H Ly β solar radiations the intensities of which were 1% and 0.4% of their daytime values respectively. It was suggested that disappearance of charged particles is defined by the dissociative recombination process with the coefficient $\alpha^* = (2—4) \times 10^{-13}$ $(300\ K/T_e)$ $m^3\ s^{-1}$ depending on the type of ions.

In Fig. 2 the theoretical $N_e(\chi)$ dependences, taken from [6], are shown by solid lines for different altitudes, the corresponding experimental N_e values for the launches considered are given by points (some smoothed N_e values were used without taking into account sharp maxima which seem to be caused by dynamic processes). It is clear from this figure that the experimental N_e values (except some points at $\chi \approx 100°$) and their variations with increase in χ are in satisfactory agreement with theoretical curves; this confirms the correctness of the theoretical pre-conditions used. This makes it possible to estimate the value of the effective recombination coefficient corresponding to the experimental data obtained. Assuming that in the calculations the electron temperature was taken to be equal to the neutral gas temperature, one can obtain for α' values varying from $(3-6) \times 10^{-13}$ m^3 s^{-1} at 100 km to $(1-2) \times 10^{-13}$ m^3 s^{-1} at 140 km. These values are not in contrast with present data on the effective recombination coefficient in the ionospheric E layer [4, 7]; those differences of experimental and theoretical N_e values which are observed at $\chi \approx 100°$ in the 100—130 km height region (Fig. 2) are probably associated with a somewhat overestimated value of the night-time ionization source used in calculations.

In all experiments well-defined electron density maxima were observed in the 90—120 km height interval; they can be identified with the Es layers recorded on the ionograms taken during the rocket flights. As is seen from Fig. 1a, these layers of increased electron density had both different thickness and different maximum N_e values. The greatest N_e value (1.1×10^{12} m^{-3}) at Es layer maximum was registered in the experiment carried out on 6 July 1970. On the ionogram taken on this flight there were reflections from the Es layer up to a frequency of 9.3 MHz.

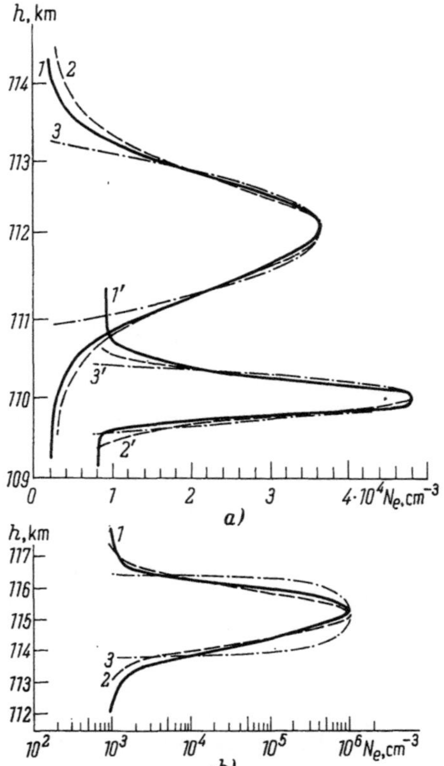

Fig. 3. Electron density height distribution in the Es layers: (a) launches Nos. 1 and 2 (curves 1 and 1′ respectively); (b) launch No. 3 (curve 1); 2,2′ is the symmetrical layer model, 3,3′ is a parabolic model.

For some maxima the detailed electron density distributions were calculated. These distributions are shown in Fig. 3a for launches Nos. 1 and 2 (curves 1 and 1′ respectively) and in Fig. 3b for launch No. 3 (curve 1). It is seen that these layers are practically symmetrical with respect to the maximum; this makes it possible to approximate the experimental N_e profiles in the Es layers by theoretical models. For such models the models of parabolic and symmetric layers were taken [8].

In Fig. 3a curves 2 and 2′ correspond to the symmetrical layer model with thickness of 1640 m and 500 m at 0.5 $N_{e(max)}$; curves 3 and 3′ correspond to the parabolic model with a half-thickness of 1170 m and 420 m at the N_e background. In Fig. 3b curve 2 corresponds to the symmetrical layer model with thickness of 650 m, curve 3 corresponds to the parabolic model with a half-thickness of 1300 m. From comparison of experimental and calculated N_e profiles it is seen that for launches Nos. 1 and 3 the experimental profiles (Fig. 3a, curve 1; Fig. 3b, curve 1) are in good agreement with the symmetrical layer model; both approximations can be used for the Es layer observed in launch No. 2 (Fig. 3a, curve 1′).

Experimental profiles of electron temperature in the 95—175 km height region are shown in Fig. 1b (figures on curves correspond to numbers of launches in Table 1). As is seen from Fig. 1b, in all experiments T_e increased with height practically monotonically. From this figure it is also seen that at sunset the electron temperature decreases as the solar zenith angle increases so that already at $\chi = 86°$ it is falls practically to the night-time values. This is perhaps due to peculiarities of variations of the heat input to the electron gas at sunset.

In Fig. 1b the profiles of neutral gas temperature T_n are also given; these profiles were taken from the Jacchia 71 model [9] for corresponding levels of solar and geomagnetic activity (curve A is for launch No. 4, curve B is for launches Nos. 1—3). As is seen from comparison of the T_e and T_n profiles, electron temperature approaches T_n in the 150—170 km height region but is greater than T_n below 150 km. The latter is typical of the great majority of probe measurements of T_e [10, 11]. To explain this the hypothesis of the occurrence of an additional heat source for electron gas at these heights is usually used [10, 11]; however, for final decision on this question further theoretical and experimental investigations as well as improvement of the T_e measurement procedure in this region must be made.

References

[1] A. D. Zhlood'ko et al., Meterol. i Hydrol. No. 3, 72 (1970).
[2] G. P. Komrakov et al., Kosm. Issled. **8**, 278 (1970).
[3] Yu. B. Burakov et al., in: Proc. Inst. Exp. Met. No. 16, Hydrometeoizdat, Moscow, 1970 (p. 47).
[4] G. S. Ivanov-Kholodny and G. M. Nikolsky, The Sun and Ionosphere, Nauka, Moscow 1969.
[5] K. Maeda, J. Geomagn. Geoelect. **21**, 557 (1969).
[6] T. J. Keneshea et al., J. Geophys. Res. **75**, 845 (1970).
[7] A. D. Danilov and M. N. Vlasov, Photochemistry of Ionized and Excited Particles in the Lower Ionosphere, Hydrometeoizdat, Leningrad 1973.
[8] V. L. Ginzburg, Electromagnetic Wave Propagation in the Plasma, Nauka, Moscow 1967.
[9] L. G. Jacchia, Smithson. Astrophys. Obs. Spec. Rep. No. 332 (1971).
[10] A. P. Willmore, Space Sci. Revs. **11**, 607 (1970).
[11] P. M. Banks, Proc. IEEE **57**, 258 (1969).

Lower Ionosphere Structure — Akademie-Verlag, Berlin 1974

POSITIVE ION COMPOSITION IN THE LOWER IONOSPHERE DURING THE GEMINID METEOR SHOWER AND THE OCCURRENCE OF A WINTER ANOMALY

P. A. Zbinden, M. A. Hidalgo*, P. Eberhardt and J. Geiss

Physikalisches Institut, University of Bern, Bern, Switzerland

The positive ion composition in the altitude region from 68 to 152 km was measured with a cryopumped mass spectrometer on a Skylark rocket launched from Salto di Quirra, Sardinia (39.6 °N 9.40 °E), on 14 December 1971 at 1311 CET. The measured D region ion profiles deviate from typical undisturbed mid-latitude ion compositions. Water-cluster ions are dominant only below 77 km. Above this altitude NO^+ dominates and its profile reaches at approximately 80 km an ion density close to 10^{10} m^{-3}. These particular features are probably associated with the winter anomaly present on the launch day. In addition to the typical constituents of the water-cluster region, ions of masses 23, 39, 41 and 59 were measured. There is some evidence that these ions or at least a substantial fraction of them are Na^+, NaO^+, $Na^+ . H_2O$ and $Na^+ . (H_2O)_2$. In the region around 80 km the isotopic ratio $n(34^+)/n(32^+)$ is about a factor 10 greater than in the E region, and indicates the presence of sulphur ions. Profiles of Na^+, Mg^+, Al^+, Si^+, K^+, Ca^+, Cr^+, Fe^+ and Ni^+ were recorded in the E region and traces of some other metals were tentatively identified. In the limits of error the isotopic compositions of the major metals were found to be the same as on earth.

1. Introduction

The objective of the experiment was to study the ion composition of the mid-latitude lower ionosphere, and in particular the metal ion composition at the time of the Geminid meteor shower and with the presence of a sporadic Es layer. Our mass spectrometer was specially designed to give a good mass and altitude resolution so that elemental and isotopic abundances of the metallic and other ions could be determined over the whole mass range. This was achieved by using a single focusing magnetic ion mass spectrometer fitted with a double collector system. The mass spectrometer was provided at its entrance with a high speed cryopump filled with liquid helium.

The mass spectrometer was launched from Salto di Quirra, Sardinia (39.6 °N 9.40 °E) on 14 December 1971 at 1311 CET. At launch time the solar zenith angle was 64°. According to the ionograms taken with an ionosonde located at 20 km approximately downrange, a sporadic Es layer with blanketing and top frequencies of 3.2 MHz and 5 MHz respectively was present at about 110 km [1]. Non-horizontal stratifications of varying intensity, probably due to gravity

* ESRO Fellow on leave of absence from the University of Zaragoza, Facultad de Ciencias, Laboratorio de Termologia, Zaragoza, Spain.

waves, were seen in the ionograms during the whole day [1]. The transmitter-receiver link (2.83 MHz) between Aranjuez and Balerma (Spain) (midpoint coordinates 38.38 °N 3.24 °W), operated by the Max Planck Institute for Aeronomy, Lindau, FRG, measured on 14 December 1971 a high absorption with $L_M \simeq 48$ dB [2]. The geomagnetic activity was low $(\Sigma Kp = 5^+)$ and the solar flux was 122 in the usual units (10^{-22} W m^{-2} Hz^{-1}).

2. Instrumentation

The mass spectrometer is a modification of a previously flown dual mode instrument [3]. The mass spectrum from 11 to 73 AMU is scanned in 1.1 s. The mass analyser has two exit slits for simultaneous recording of two masses with a mass ratio of 0.85. Thus most ions are measured twice in each scan interval. The time between these two sequential measurements corresponds to a height difference of approximately 100 m in the region around 95 km. The 1% mass resolution of the analyser is approximately 60. The cryopump has a pumping speed of 1 m^3 s^{-1} at an internal pressure of 10^{-4} torr. With a 2 mm diameter inlet orifice the mass resolution was not affected by external pressures up to 0.5 torr. The holding time of the cryopump is 90 minutes without gas load, with an external pressure of 1 torr it decreases to approximately 50 minutes. Details pertaining to the mass spectrometer calibration are given in [4].

The mass spectrometer was part of the ESRO sounding rocket payload S-82. In addition, the payload contained a double plasma density and temperature probe giving also an indication of the electric field [5], an ejectable parachute package including a blunt ion density probe for a total ion density measurement in the D region [6], and a set of microphone detectors to measure the micrometeorite particle flux [7]. The instrument was opened to the ambient medium at 68 km and the first ion appeared at 68.5 km. The resolution remained unaffected by the ambient gas. The peak altitude reached was 152 km and on the downleg data were obtained until 61 km. The half angle of precession was 4.9°.

3. E Region Results

Fig. 1 represents the upleg ion density profiles of the major positive E region ions. As usually NO^+ and O_2^+ are dominant in the E region. O^+ and N^+ appear as minor constituents. Over the altitude range 105–130 km the ratio NO^+/O_2^+ is seen to oscillate around a value of 1. The wave-like structure of these profiles above 105 km could be an artifact of the rocket precession as the modulation in the upleg and downleg data (not shown) is anticorrelated (no correction was applied for a possible dependence of the mass spectrometer sensitivity on the direction of the incoming ions). At lower heights the NO^+ profile shows an unusual high density at approximately 80 km. This will be discussed in the next section.

The following metal ions were identified on the basis of their isotopic or elemental abundances: ^{23}Na, ^{24}Mg, ^{25}Mg, ^{26}Mg, ^{27}Al, ^{28}Si, ^{29}Si, ^{39}K, ^{40}Ca, ^{41}K, (^{45}Sc), ^{52}Cr, ^{54}Fe, ^{56}Fe, ^{57}Fe, ^{58}Fe, ^{58}Ni, ^{59}Co, ^{60}Ni, (^{61}Ni). The total metal ion profile (see Fig. 1) shows the existence of two distinct layers at 95 and 119 km altitude which were also seen on the downleg. At 95 km the metal ion density is very high and the molecular ions NO^+ and O_2^+ are depleted and show a deep minimum, the metal ions becoming dominant. No sporadic Es layer is observed in our profiles.

Fig. 2 shows the upleg density profiles of the major metallic ions. In the upper layer at 119 km Si^+ is the most abundant metal ion. At lower heights it disappears and Fe^+ and Mg^+ dominate at 95 km. In the upper layer at 119 km altitude a stratification according to mass is apparent. The concentration of the heavier ions peaks at slightly higher altitudes than that of the lighter ones. This could be interpreted as the result of an ion downward transport.

The measured isotopic ratios for Mg^+, Si^+, Fe^+ and the main isotopes of Ni^+ are in agreement with the terrestrial values. For Mg^+, the measured isotopic ratios $^{25}Mg^+/^{24}Mg^+$ and $^{26}Mg^+/^{24}Mg^+$ were 0.124 ± 0.004 and 0.139 ± 0.008 respectively

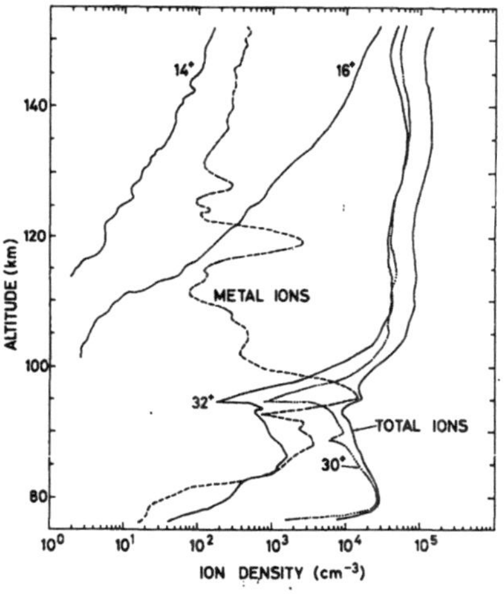

Fig. 1. Upleg density profiles of the major positive E region ions.

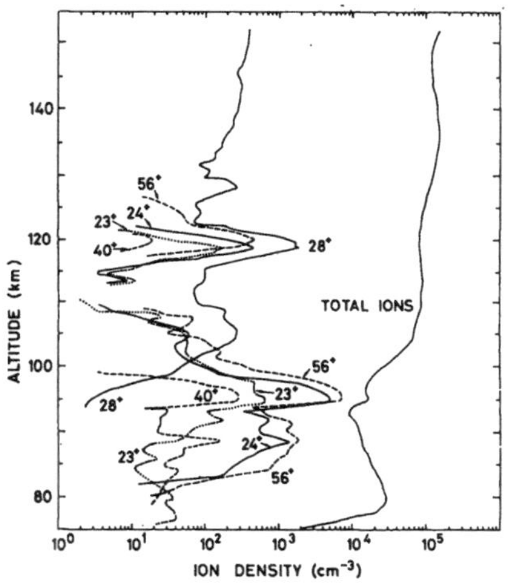

Fig. 2. Upleg density profiles of the major metallic ions.

(weighted averages over total altitude range). If a substantial fraction of the observed metal ions originated from the Geminid meteor shower then the isotopic composition of these meteoritic metals is identical with those in chondrites and in the earth's crust. In addition to the metal ions shown in Fig. 2 also layers of 39^+, 43^+, 44^+ and 45^+ ions were seen at the 119 km altitude level. On the basis of the isotopic as well as the relative abundances compared with Mg^+, the following tentative identifications were made: $39^+ = K^+$; $43^+ = AlO^+$; $44^+ = SiO^+$. The 45^+ concentration is three orders of magnitude higher than what would be expected from solar system abundances if this ion is identified as Sc^+. Goldberg and Aikin [8] have found a similar excess during the β-Taurid meteor shower. In the lower metal layer at 95 km mass 41^+ is also detected (see Fig. 4). The measured ratio $41^+/39^+$ is close to the terrestrial $^{41}K/^{39}K$ ratio. Mass 59^+ peaks also at 95 km and its abundance relative to Mg^+ is similar to the solar system value for ^{59}Co.

4. D Region Results

Fig. 3 shows the measured profiles of the more representative mesospheric ions. Our profiles show some important departures from the quiescent daytime mid-latitude D region. It is seen that mass 37^+ ($H^+.(H_2O)_2$) is the most abundant ion until 77 km where the ion mass 30^+ (NO^+) becomes dominant. The NO^+ ion shows a very sharp gradient at 76 km with an increase of more than two orders of magnitude in less than 3 km. Mass 48^+ also shows a similar sharp gradient. This close relationship suggests the identification of 48^+ as $NO^+.H_2O$. The profiles of $NO^+.H_2O$ and $H^+.(H_2O)_2$ have maximum concentrations at the same altitude of 77 km. Above, they decrease with similar trends, disappearing at 85—86 km. The topside profile of $H^+.(H_2O)_2$ shows a pronounced drop off at 86 km, approximately 3 km higher than normally observed in the quiescent daytime mid-latitude D region [9].

During downleg, when the spectrometer sampled in the wake, the measured profiles were essentially identical with those of Fig. 3. This confirms the authenticity

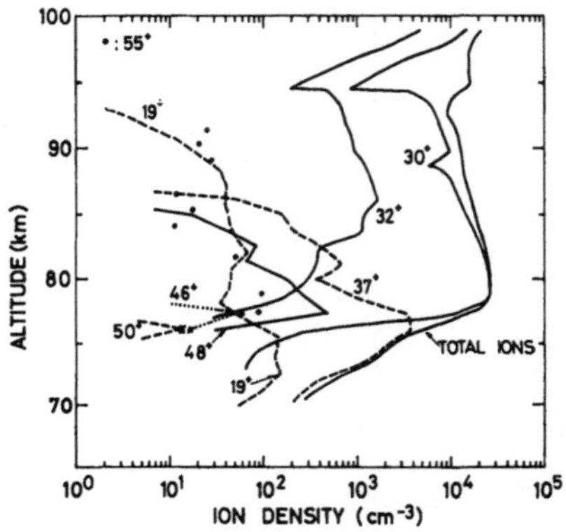

Fig. 3. Upleg density profiles of the typical D region ions.

of the features shown, and in particular it proves that the overabundance of NO^+ relative to the rest of the D region ion constituents is not an artifact due to sampling effects or data reduction.

We mentioned above that the day of launch was reported as an anomalous winter day. Measurements of the mesospheric electron concentration under such conditions [10—12] show large enhancements of the electron density in the region between 65—85 km altitude. Enhancements up to a factor of 50 are observed at approximately 83 km. Having in mind charge neutrality, and the virtual absence of negative ions above 75 km (at least during normal conditions), the comparison of the above quoted electron profiles with our total profile clearly shows that our measurements reflect the presence of a winter anomaly over Sardinia during the launch day. Therefore, we consider our profiles to be representative of the ion composition of the mid-latitude ionosphere during an anomalous winter day.

Minor mesospheric ions are also depicted in Fig. 3. $19^+(H^+.H_2O)$ and $48^+(NO^+.H_2O)$ are correlated. The 19^+ concentration decreases with increasing 48^+ densities. This suggests that NO^+ may be involved in the formation of water-cluster ions at this altitude. Ion mass $55^+(H^+.(H_2O)_3$ or possibly $NaO_2^+)$ is present in some dispersed spectra and is indicated by dots. 46^+ (possibly NO_2^+) and $50^+(O_2.(H_2O))$ were also detected.

As shown in Fig. 4 the ions 41^+ and 59^+ are found in the D region correlated with 37^+ and 48^+ (see Fig. 3 for 48^+). We therefore suggest their identification in this region as $Na^+.H_2O$ and $Na^+.(H_2O)_2$, and not as $^{41}K^+$ and $^{59}Co^+$ as at higher altitudes. This identification of 41^+ was already proposed by Narcisi [9]. 39^+ could be the main potassium isotope, but then its abundance relative to the other metals would be about a factor 20 greater than in chondrites. Although this might be due

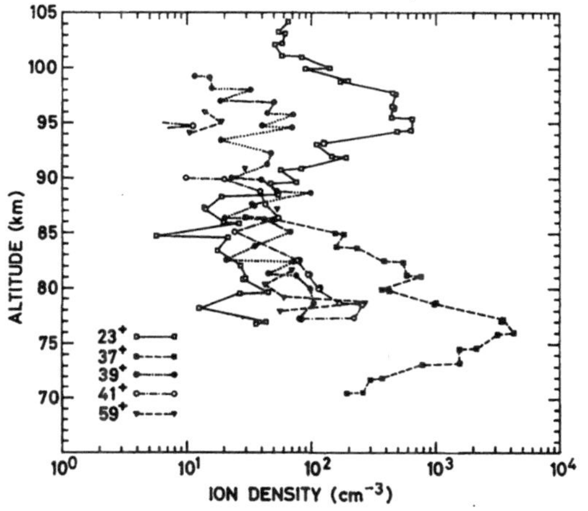

Fig. 4. Upleg D region profiles of 23^+, 37^+, 39^+, 41^+ and 59^+.

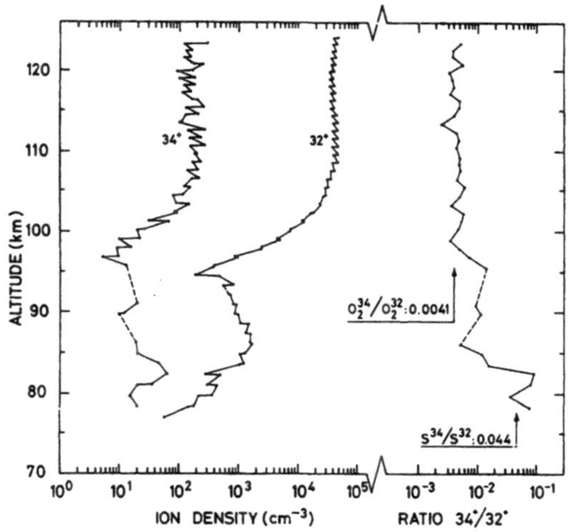

Fig. 5. Density profiles for ions 32 + and 34^+ and their ratio.

to a particular chemical behaviour, the correlation and disappearance with the group of Na^+ and its derivates (23^+, 41^+, 59^+) suggests the identification of 39^+ as NaO^+.

Ion mass 32^+ interpreted as O_2^+ in the E region, presents a marked break in its profile at 84 km. An analysis of the ratio $34^+/32^+$ is shown in Fig. 5. Each point on the density profiles corresponds to a value measured with one collector. The points in the ratio curve refer to a ratio $34^+/32^+$ averaged over the two channels of one spectrum. It can be realized, in spite of the incertitudes involved in the evaluation of mass peaks close to instrumental detection limits, that the $34^+/32^+$ abundance ratio departs at 83 km from the isotopic O_2 ratio, suggesting the presence of sulphur ions in the range from 77 to 83 km. With this identification in mind a possible contribution of SO^+ to 48^+ cannot be excluded.

Acknowledgments

We would like to thank Mr. H. Wyniger, H. Hofstetter, R. Liniger, J. Stettler, A. Juza, S. Schürch and L. André for their help in building and calibrating the mass spectrometer and associated electronics. We are grateful to Drs. E. Kopp and H. Balsiger and Mr. J. Fischer and L. Weber for their assistance and help during the different phases of the launch. We are indebted to Dr. W. Becker, Max Planck Institute, Lindau FRG, for the evaluation and comments on the ionograms, and to Dr. H. U. Widdel of the same institution for providing us with information on radio absorption and making available his unpublished results. One of us (M. A. H.) is indebted to ESRO for a fellowship.

Preparation and launching of the rocket was performed by ESRO. Our work was supported by the Swiss National Science Foundation (grants NF 5094, 2.73.68, 2.213.69, 2.405.70, 2.592.71 and 2.768.72).

References

[1] W. BECKER, private communication.
[2] G. ROSE, J. WEBER, H. U. WIDDEL and P. GALDON, this volume, p. 331.
[3] E. KOPP, P. EBERHARDT and J. GEISS, J. Planet. Space Sci. **21**, 227 (1973).
[4] P. ZBINDEN, Ph. D. Thesis, 1971, University of Bern (Switzerland).
[5] A. PEDERSEN, U. FAHLESON and C. FÄLTHAMMAR, Space Research XII, 1369 (1972).
[6] R. BERKELJON, G. ENGLERT, R. JAESCHKE, D. KÖHN, A. PEDERSEN and R. SCHEPER, Tech. Note ESRO TN-102 (ESTEC) (1970).
[7] B. A. LINDBLAD, G. ARINDER and T. WIESEL, paper presented at the XV COSPAR meeting, Madrid (1972).
[8] R. A. GOLDBERG and A. C. AIKIN, Goddard Space Flight Center Rep. X-625-72-402 (1972).
[9] R. NARCISI, in: Physics of the Upper Atmosphere, Bologna, 1971 (p. 12).
[10] S. A. BOWHILL, E. A. MECHTLY, C. F. SECHRIST and L. G. SMITH, Space Research VII, 246 (1967).
[11] C. F. SECHRIST, E. A. MECHTLY, J. S. SHIRKE and J. S. THEON, J. Atmos. Terr. Phys. **31**, 145 (1969).
[12] L. THOMAS, J. Atmos. Terr. Phys. **30**, 1211 (1968).

Note added in proof: After submission of this paper the data were re-evaluated and the absolute ion densities in the D region were adjusted to fit the total ion density profile measured with the parachute blunt probe (R. Jaeschke and A. Pedersen, private communication). This leads to a different absolute ion density profile below 92 km, without the pronounced bulge at 80 km. Absolute ion densities given in this paper for altitudes below 92 km must thus be considered as preliminary. The relative ion abundances were not significantly changed by this re-evaluation of our data.

Lower Ionosphere Structure — Akademie-Verlag, Berlin 1974

NEUTRAL TEMPERATURE MEASUREMENTS BETWEEN 85 AND 100 KM BY LASER SOUNDING FROM THE GROUND

G. MEGIE, J. E. BLAMONT and M. L. CHANIN

Service d'Aéronomie du CNRS, Verrières le Buisson, France

Observations of the night-time atmospheric sodium layer by laser scattering lead to a measurement of the neutral temperature at the base of the thermosphere between 85 and 100 km. It will permit a survey of temperature profiles as a function of time with high resolution either in altitude or time.

We present here preliminary results which have been obtained with a sodium lidar experiment set up at the Haute-Provence Observatory (44 °N 5 °E). Two types of studies have been developed, using that technique: vertical density

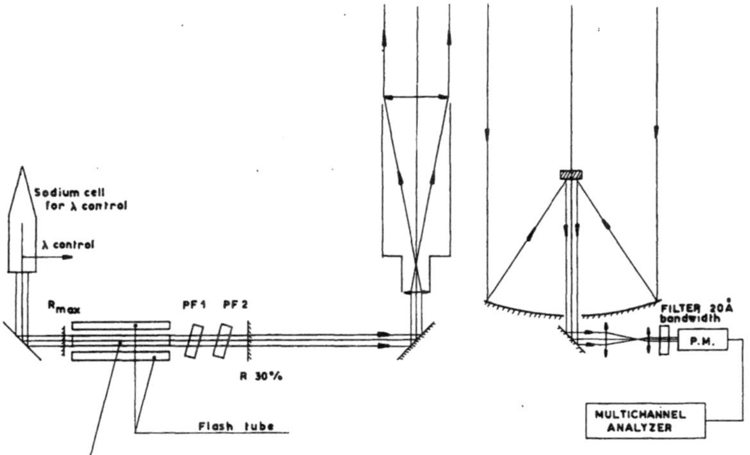

Fig. 1. Experimental set-up.

distribution of sodium atoms which shows stratification related to the dynamics of the atmosphere at the mesopause level (characteristics described in [1—3]); temperature measurements made with the absorption technique previously used by Blamont and Chanin on artificial sodium clouds [4].

Fig. 1 shows the experimental set-up. The transmitter part of the lidar is a flashlamp pumped dye laser. It is built with a bi-elliptical reflector cavity, with the dye cell on the common axis. Two Fabry Perot interferometers are inserted inside the cavity to reduce the linewidth of the emission to 0.1 Å. The energy emitted is 0.3 J with a repetition rate of 0.5 Hz. We use a part of the output signal going out on the side of the high reflectivity mirror to control automatically the wavelength of the emission with a sodium resonance cell. The laser pulse is then sent directly upward in the atmosphere through an afocal system to reduce

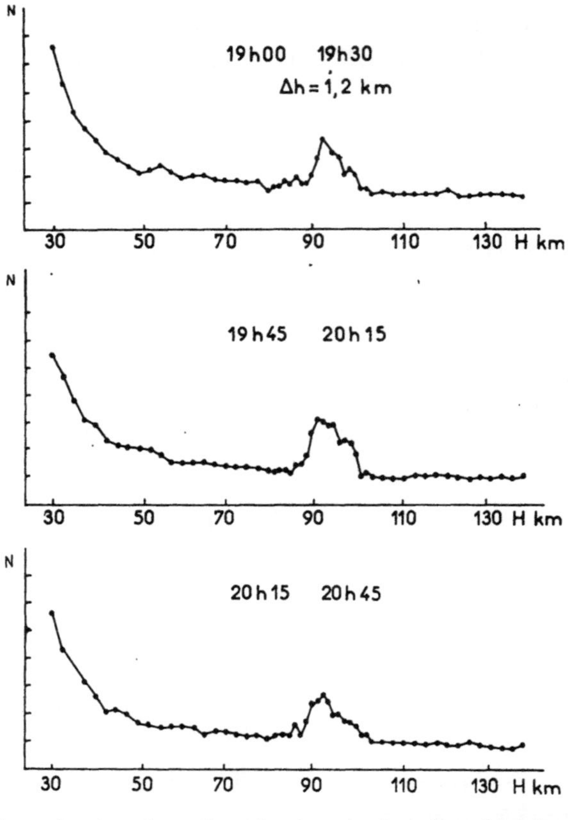

Fig. 2. Typical echoes showing the sodium density as a function of height; 28 October 1972.

the beam divergence. The backscattered signal is received on a 80 cm diameter telescope and the altitude discrimination is made by the use of a multichannel analyser whose resolution can be reduced to 600 metres. Fig. 2 shows some typical echoes obtained in less than half an hour for three series of 600 laser bursts.

The sodium atoms which backscatter the laser pulses are in the ground state and therefore at this altitude in thermal equilibrium with the ambient; thus to measure the temperature of the sodium is to measure the temperature of the atmosphere. This temperature is directly deduced from the line-shape of the D_2 line which is backscattered by the sodium atoms.

The principal advantage of the absorption method is to transform a line-shape measurement into an intensity measurement. Instead of making this measurement by interferometer methods, the backscattered signal is partially absorbed through an absorption cell containing sodium vapour at a known temperature T_a. The resonance line emitted by the atmospheric sodium has a Doppler width

$$(\Delta\nu_D)_n = \frac{\nu_0}{c}\left(\frac{2RT_n}{M}\right)^{1/2}$$

where T_n is the ambient temperature. The absorption coefficient in the cell has a width

$$(\Delta\nu_D)_a = \frac{\nu_0}{c}\left(\frac{2RT_a}{M}\right)^{1/2}.$$

If the intensity entering the cell is I_0 and the intensity after traversing the cell is I, the reduction factor is

$$R = \frac{I}{I_0}.$$

This quantity can be shown to be a function only of T_n/T_a and τ_a the optical thickness of the absorption cell. To measure τ_a we use sodium emission provided by a scattering cell at a known temperature T_α so that R is now a function of

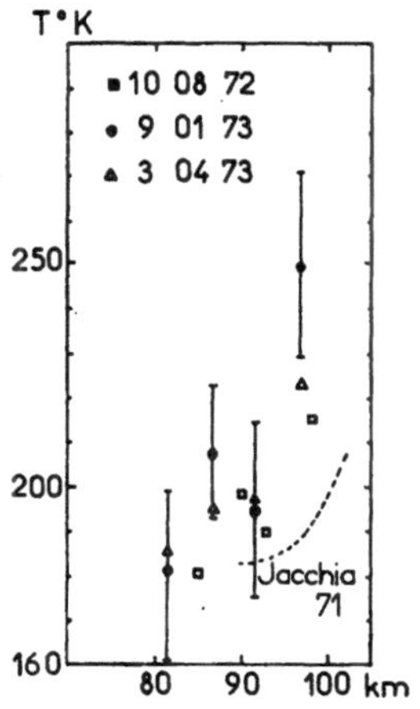

Fig. 3. Temperature as a function of altitude.

three temperatures T_n, T_a, T_α. Knowing T_a and T_α we can easily deduce T_n, the temperature of the atmosphere.

Fig. 3 shows three temperature profiles obtained at different periods of the year. These profiles have been integrated over periods of at least 5 hours with an altitude resolution of 5 km. The overall error is then $\pm 25\,°K$. The main features which can be deduced from these preliminary studies are: a positive gradient between 85 and 100 km showing a systematic discrepancy with the 1971 Jacchia model; in the winter profile (9 Jan. 1973) an anomaly at 87 km which may be correlated with some dynamic effects at this altitude, such as propagating gravity waves.

This method will now be improved as far as the receiver and the dye laser are concerned and systematic measurements are scheduled to start at the same place in the very near future.

References

[1] M. C. W. SANDFORD and A. J. GIBSON, J. Atmos. Terr. Phys. **32**, 1423 (1970).
[2] C. J. SCHULER et al., Appl. Opt. **10**, 1689 (1971).
[3] J. E. BLAMONT, M. L. CHANIN and G. MEGIE, Annls Géophys. **28**, 833 (1972).
[4] J. E. BLAMONT et al., Phys. Rev. Lett. **6**, 403 (1961).

Winter Anomaly and Flare Effects

Lower Ionosphere Structure — Akademie-Verlag, Berlin 1974

SOME RESULTS OBTAINED FROM THE EUROPEAN COOPERATION CONCERNING STUDIES OF THE WINTER ANOMALY IN IONOSPHERIC ABSORPTION

H. Schwentek

Max-Planck-Institut für Aeronomie, Institut für Ionosphärenphysik, Lindau/Harz, FRG

In winters 1970/71 and 1971/72 an international effort was made to measure ionospheric absorption L at a constant solar zenith angle χ: $L(\cos\chi = 0.2)$ at 16 mid-latitude stations between Spain (4° W) and Japan (141° E), at latitudes between 36° N and 54° N, in order to prepare a basis for a comparative study. As a typical feature of the behaviour of absorption in winter, it was found that at nearly all stations three types of values occurred more or less sporadically: Some values as low as in summer, values following the same particular trend in each winter, and groups of excessive values. It was confirmed that both amplitude and duration of the anomaly decreased with latitude. According to measurements on board a ship, the transit zone where the winter anomaly disappeared was centred at about 39° N. A comparison between the measurements made on board a ship and at three fixed stations around showed a somewhat complicated correlation depending on time. The correlation coefficient of the data of the different stations, in general, decreases rapidly with distance between the stations (in the range from 0 to about 500 km). At greater distances, up to 1000 km, the correlation is small but always significant. A correlation coefficient characterizing the day-to-day variation of absorption only shows a tendency to decrease exponentially with distance reaching at about 2000 km the level of significance (95%).

In winter 1971/72 a wave-like structure with a period of about 6 to 7 days was found at Central European stations, and with a period of 25 days at the Japanese stations. In winter 1970/71, in Central Europe only, periods appeared between 19 and 25 days.

1. Introduction

In winters 1970/71 and 1971/72 an international effort was made to measure ionospheric absorption at frequencies between 1.5 and 3 MHz simultaneously at 16 mid-latitude stations from Spain (4° W) to Japan (141° E), at latitudes between 36° and 54° N (Fig. 1). The data obtained from these measurements have been compiled and distributed amongst the participants. The names of these are found under § 6; frequencies and methods used are given in these distributed data sets. The purpose of this paper is to present some results and preliminary conclusions.

The proposal was to measure daily during winter the diurnal variation of absorption and to determine the typical parameters which describe the "character of the day": Subsolar absorption L_0, exponent n, and absorption at a constant solar zenith angle $\chi = 78.5°$, that is L ($\cos\chi = 0.2$). This value from these typical data has proved to be very suitable; plotting these parameters as a function of day in winter, a special pattern appears. Fig. 2 shows two examples [1]. Three

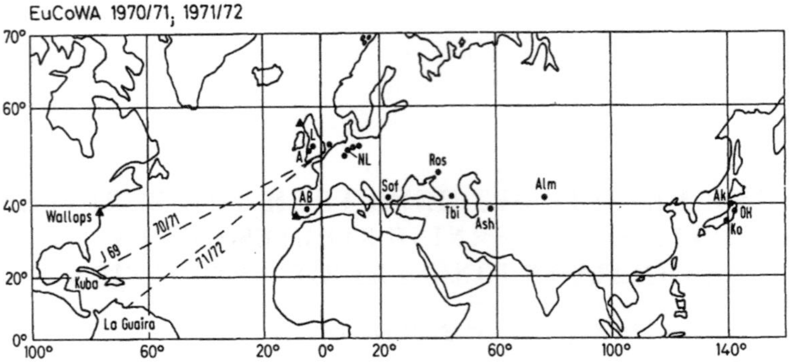

Fig. 1. Map showing the position of A1 stations or A3 circuits (reflection points), respectively, at which absorption was measured in winters 1970/71 and 1971/72 in the framework of a European Cooperation. It should be noted that two groups of stations exist, one group covering a rather small area at about 52 °N from 10 °W to 10 °E, and one group forming a chain extended from 10 °W to 141 °E at latitudes around 40 °N.

--- routes on which absorption was measured on board ship; ▲ rocket launching sites (Wallops Is., S. Uist in the Hebrides, and Arenosillo in Spain).

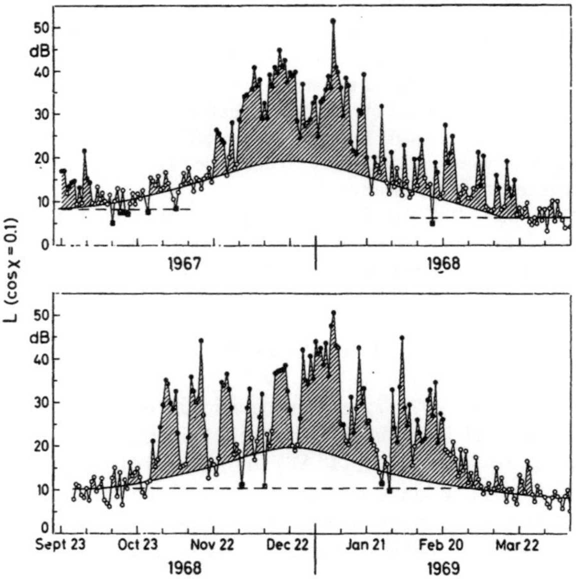

Fig. 2. Two examples of typical patterns showing the different behaviour of absorption at a constant solar zenith angle ($\cos\chi = 0.1$) during winter; A3 circuit Norddeich-Lindau (2.61 MHz; 296 km). The data L ($\cos\chi = 0.1$) were obtained from the diurnal variation of absorption. ■ data as low as in summer; ○ "normal anomaly" (data following a regular trend which occurs in each winter) ● excessive absorption.

types of data may be distinguished: (i) some data at the same level as in summer, (ii) a regular increase of absorption beginning after the September equinox with a maximum around 21 December, and decreasing until the March equinox, (iii) data of excessive absorption which are superimposed more or less sporadically on the trend described under (ii).

Now the question was: Is this pattern typical for all stations; and if it were typical, what is the correlation between the shapes of these patterns; furthermore, how can these patterns be understood and explained? Finally, the southern boundary of the appearance of the winter anomaly and its variation were of particular interest.

2. Presentation of Measured Data

The first result of a comparative study of the variation of absorption during winter was that the typical pattern shown in Fig. 2 appeared at all stations, more or less pronounced. In order to demonstrate this, Fig. 3 shows some selected

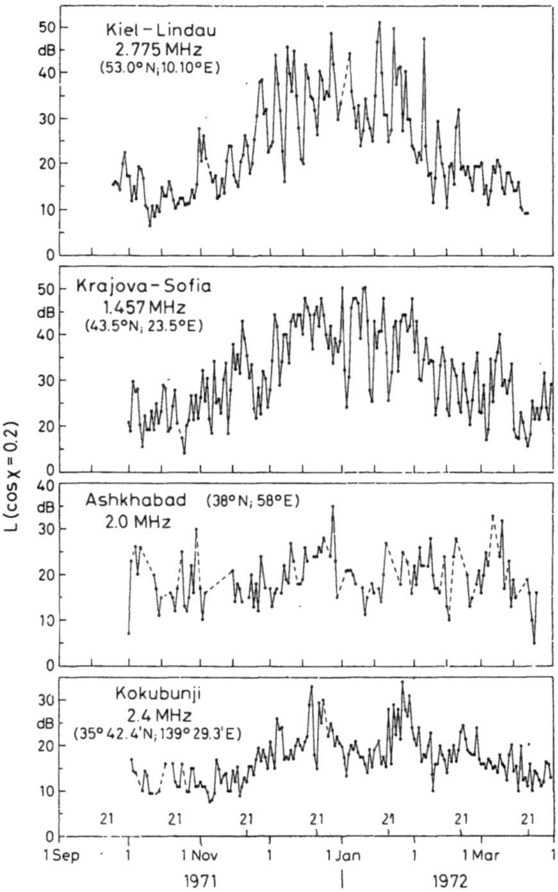

Fig. 3. The behaviour of absorption in winter 1971/72 at some selected stations at different geographical positions.

examples. Of course, since distances and frequencies vary, these differences must be kept in mind when interpreting A 3 data: e.g. effects tend to be stronger on lower than on higher frequencies, and for larger than for shorter distances. At the latitude of 53° N (circuit Kiel-Lindau; 2.77 MHz; 300 km) the anomaly was strong with the maximum mean amplitude in late December. But the amplitude at 43.5° N was also very remarkable, due also to the operating frequency which

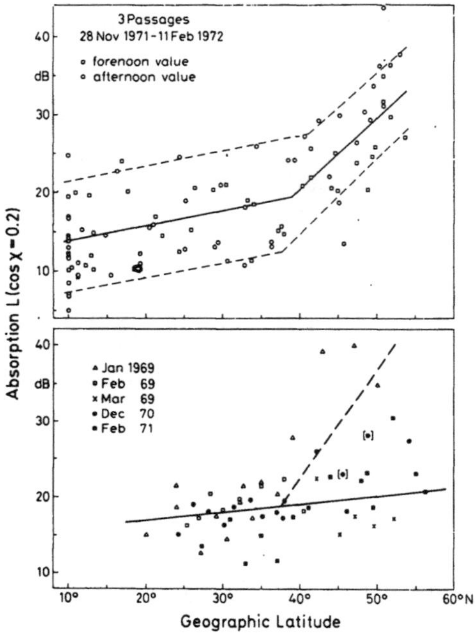

Fig. 4. The southern boundary of the winter anomaly in ionospheric absorption obtained from measurements on board ship during three different winters [4—6]: top, 1.7 MHz; below, 2.11 MHz.

was as low as 1.5 MHz. At lower latitudes, about 38° N and 36° N, the anomaly tends to vanish, although it is clearly detectable, at least, at Kokubunji. These results support some similar ones which were obtained earlier in Europe [2] and in Japan [3].

By measurements on board ship it was possible to show more precisely that, and where, a real boundary of the winter-anomaly zone exists. This is demonstrated by Fig. 4 which shows results of all measurements of this kind carried out up to now. Two points should be noted: first, something like a southern boundary of the appearance of the winter anomaly can be defined at about 39° N; secondly, a larger scatter in the points, also south of 39° N. It should be emphasized here that, despite the appearance of a boundary at about 39° N, *continuous* measurements carried out at stations at latitudes less than 39° still showed a considerable amount of winter anomaly in the winter pattern of absorption values L (cos χ = const), for instance at Kokubunji (36° N) (Fig. 3), and in Spain (38.4° N) (Fig. 8). Therefore, preferably we should speak of a *zone* where the winter anomaly vanishes. This zone seems to be in the range of latitudes between 40° and 30° N. It would be most interesting to study absorption and simultaneously parameters characterizing the mesosphere, the lower thermosphere, and the lower ionosphere just in that zone by use of a dense network. It should be mentioned that the Nimbus 3 measurements of thermal radiances emitted from the stratosphere (in the centre of the 15 μm CO_2 band) also indicated a remarkable change in the behaviour at latitudes between 30° and 40° N [7].

3. On Correlations between Absorption Data measured at Different Stations

A question of main interest was how the winter-absorption patterns obtained from the various stations (Fig. 3) are correlated with each other. As a measure, the correlation coefficient of absorption data was evaluated for each possible pair of stations. These coefficients, as a function of distance between the stations, or reflection points, respectively, are shown in Fig. 5 for distances less than

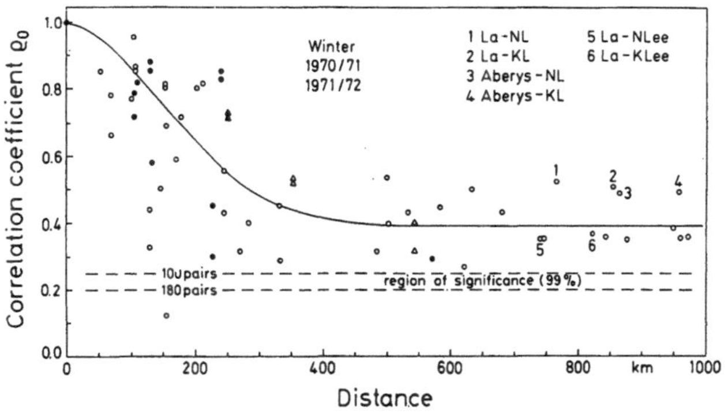

Fig. 5. Correlation coefficient ϱ_0 of absorption data obtained from pairs of stations or circuits, plotted as a function of distance, for winters 1970/71 and 1971/72. The plots for the two single winters are rather similar, and, therefore, were combined in this figure. The curve is drawn tentatively, roughly linking running median values. The region of significance above which the probability of a correlation is more than 99% is indicated based on 100 and 180 pairs of values; in most cases the number of pairs was between 120 and 170.

● same frequency, ○ different frequencies, stations in Central Europe; ▲ △ stations in Japan.

1000 km. As a result, the correlation coefficient decreases rapidly with distance, and reaches a certain level for distances greater than about 500 km; all correlations are significant with one exception only. The data were obtained mainly from stations in Central Europe, with six values from Japan. The Japanese values, marked by triangles, fit well in the data field. Obviously, the scatter in the points is large; it will be due also to the different methods which were used, as A 1 and A 3, where particularly the A 3 circuits were rather different in frequency, distance (thus equivalent frequency), and relation to the earth's magnetic field. For example, the correlation coefficient is considerably higher for La-NL and La-KL than for La-NLee and La-KLee, although neither the distances between the stations nor the relations of the circuits to the earth's magnetic field differ much. The main differences are in the equivalent frequencies f_{eq} of the circuits, and in the distances d

between transmitter and receiver (Table 1). This means that the correlation decreases also with height of reflection of the waves at the E layer thus supporting the view that the excessive winter anomaly is produced well above the mesopause.

Table 1

Circuit	NL	KL	NLee	KLee
d (km)	296	300	419	502
f_{eq} (MHz)	1.45	1.55	1.03	0.93
ϱ_0 (La)	0.52	0.50	0.35	0.37

A more detailed interpretation of Fig. 5 would be necessary but cannot be given now.

The "constant level" which appears at distances of about 500 km is probably due to the fact that winter conditions existed at all stations, or in other words, it might be due, at least partly, to the "normal anomaly" (Fig. 2). This view is supported by a result which was obtained from an analysis using for correlation

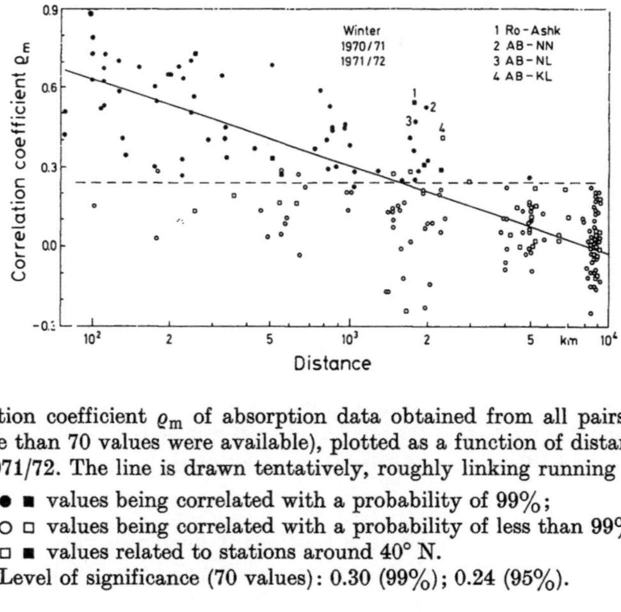

Fig. 6. Correlation coefficient ϱ_m of absorption data obtained from all pairs of stations or circuits (if more than 70 values were available), plotted as a function of distance, for winters 1970/71 and 1971/72. The line is drawn tentatively, roughly linking running median values.

● ■ values being correlated with a probability of 99%;
○ □ values being correlated with a probability of less than 99%;
□ ■ values related to stations around 40° N.
Level of significance (70 values): 0.30 (99%); 0.24 (95%).

the absorption values reduced with respect to running mean values (over ± 15 days), thus eliminating the "normal anomaly" and emphasizing the day-to-day variability of the data. In Fig. 6 the result of this approach is shown. Only those pairs of stations were used from which more than 70 values were available for comparison; the corresponding level of significance for a probability of 95% is shown as a broken (horizontal) line. We conclude that the correlation coefficient of the day-to-day variability of absorption shows a tendency to decrease exponentially with distance; it falls below the level of significance at about 2000 km.

In the light of the results outlined above it is interesting to study Fig. 7 where the variation of absorption is shown obtained at three fixed stations and on board a ship moving in the same region (see Fig. 1) during two time intervals in winter 1971/72.

During passage 1 a similar trend is found at the three stations and the ship up to and including 6 December. From 6 to 7 December there is a first discrepancy, since absorption at NL still decreased considerably, while at the other stations absorption increased. From 7 to 8 December a remarkable change occurred; at NL, La, and AB absorption increased, at NL and La excessively; the ship, how-

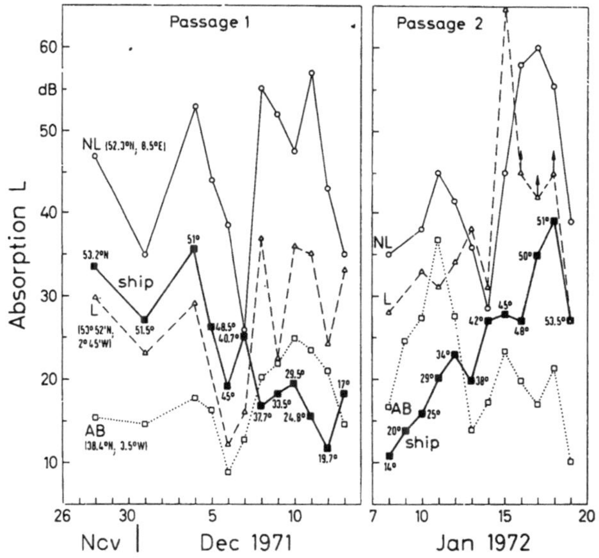

Fig. 7. Absorption L measured simultaneously at three fixed European stations and on board a ship moving in the region of those stations (mv *Hanau* of Hapag-Lloyd). NL circuit Norddeich-Lindau (2.61 MHz; 296 km; A3; equivalent frequency 1.5 MHz; L at $\cos\chi = 0.2$); ship (1.73 MHz; A1; L at $\cos\chi = 0.2$); L Lancaster, England (2.14 MHz; A1; L at noon); AB circuit Aranjuez-Balerma, Spain (2.83 MHz; 374 km; A3; equivalent frequency 1.34 MHz; L at $\cos\chi = 0.2$).

L-NL: $\varrho_0 = 0.52$, $D = 768$ km; L-AB: $\varrho_0 = 0.37$, $D = 1684$ km; NL-AB: $\varrho_0 = 0.74$, $D = 1831$ km.

ever, seems just to have moved out of the zone where the large anomaly occurred, for it moved from 40.75 °N 22.35 °W to 37 °N 30.5 °W. Then, the behaviour at NL, La, and AB is similar from 7 to 12 December. Although the variation at AB and at the ship is also similar from 8 to 12 December, this seems to be accidental, because the distance of the ship from the circuit AB was about 2000 km at 38° N, and, according to Fig. 6, a correlation should no longer exist. Moreover, from 7 to 8 December, the trends at AB and the ship were just opposite. Summarizing these results it may be said that correlations between absorption data obtained at various stations are rather complicated. The impression is that notions such as 'climate' or 'weather' as used for the troposphere seem to be transferable to the behaviour of the region around and, particularly, above the mesopause. Therefore, a comprehensive statistical analysis gives results comparable to climate, the study of day-to-day variations to weather.

4. Concerning Wave-like Structures in the Absorption Data

In winter 1971/72, in the data obtained from the circuits Norddeich—Lindau and Aranjuez—Balerma an obvious wave-like structure can be detected. The data of both circuits are plotted in Fig. 8 and linked tentatively by quasi-periodic curves for demonstration. The period, which was found by an auto-correlation

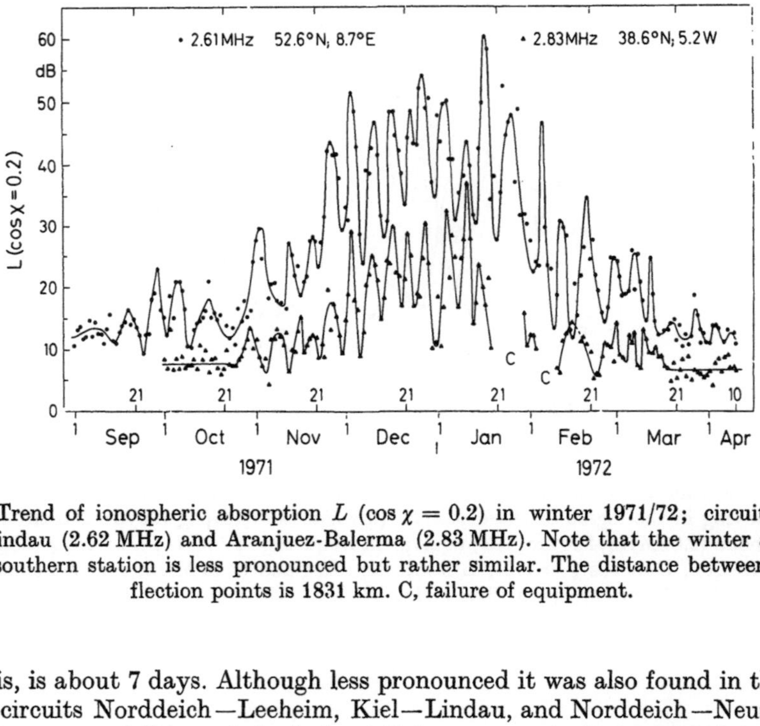

Fig. 8. Trend of ionospheric absorption L ($\cos \chi = 0.2$) in winter 1971/72; circuits Norddeich-Lindau (2.62 MHz) and Aranjuez-Balerma (2.83 MHz). Note that the winter anomaly at the southern station is less pronounced but rather similar. The distance between the reflection points is 1831 km. C, failure of equipment.

analysis, is about 7 days. Although less pronounced it was also found in the data of the circuits Norddeich—Leeheim, Kiel—Lindau, and Norddeich—Neustrelitz. The occurrence of this period of about 7 days seems to be restricted to winter 1971/72, for an analysis of the Norddeich—Lindau data for the whole sunspot cycle 1961 to 1972 showed that there was no winter during that cycle in which a 7-day period occurred so clearly.

In winter 1970/71, another period, however, was found in the data of some European stations, and in winter 1971/72 also in the data of the Japanese stations. In Europe, periods of 19, 20, 22 and 25 days were found, in Japan, clear periods of 25 days only. In winter 1970/71 in Japan no periods could be detected. Taking all the results together, it may be said that wave-like structures or periods can appear in absorption data but as an irregular phenomenon only.

5. Conclusions and Remarks

1. Although many results from absorption measurements have already been obtained up to the present, monitoring of ionospheric absorption in the frequency range from 1.5 to 3 MHz can still be used as an effective diagnostic tool, (a) for determining certain conditions in the upper D and lower E region above the station, or at the point of reflection, as a basis for coordinated measurements of various associated parameters at those heights by rockets and other methods, (b) for the study of large scale processes occurring just above the mesopause, at least in areas of about 1000 km in diameter, (c) for the study of the hemispheric or even global morphology of absorption.

2. The most reliable measurement is made by recording the diurnal variation of absorption from which all the characteristic parameters can be derived; of these the exponent n in $L = L_0 \cos^n \chi$ seems to play a very important role. Thus, the "absorption pattern" for each winter and each station can be determined. These patterns can be used also for studies of ionospheric-mesospheric-stratospheric relationships. Although a typical pattern was found to appear at each station the variety in their shapes is large, not only from station to station but also from winter to winter, depending on latitude, operating frequency and method of measurement. The most suitable parameter for comparison of different circuits or stations is L ($\cos \chi = 0.2$) derived from the diurnal variation of absorption, because it shows the highest correlation.

3. It was confirmed that the winter-anomaly pattern depends strongly on latitude. The amount and duration of the winter anomaly decrease with latitude and vanish in the zone between 40 and 30° N.

4. The absorption pattern depends on the method used. Particularly, using A 3, the distance between transmitter and receiver plays an important role, in other words, the equivalent frequency, or the height of reflection. The shorter the distance the higher the correlation coefficient.

5. The correlation between absorption at various stations is rather complicated. The correlation coefficient of the data decreases rapidly with distance up to 500 km, and is roughly constant and significant up to 1000 km. The correlation coefficient of absorption data related to running mean values (± 15 days) shows a tendency to decrease exponentially with distance, reaching the level of significance at about 2000 km.

6. Periods of 7, 19, 20, 22 and 25 days could be identified in the absorption data, but their occurrence is rather irregular.

7. Concerning methods, method A1 seems to be most useful for synoptic work, especially if a standardized frequency is used. A1 gives not only the amplitudes but also the height h' of reflection and the shape of the pulses as an indicator of the conditions at the level of reflection. Method A3 seems to be most suitable for long-term measurements and a study of the influence of the solar cycle. The comparison, however, of results obtained from circuits different with regard to distance, frequency, and relation to the earth's magnetic field can give valuable hints concerning the causes of differences in absorption; and this is important also for practical purposes.

8. It should be noted that particle precipitation due to solar flares can disturb the seasonal variation of absorption, for instance, that of L ($\cos \chi = 0.2$), in the course of the whole year. Such a disturbance can be detected most clearly in

summer when the scatter from day to day in the absorption data is small (Fig. 9). In winter, when the scatter is large, particle precipitation can naturally be also one of the influences determining the actual absorption [8], but it is difficult to detect and separate it; but it is certainly not the cause of the real winter anomaly [9].

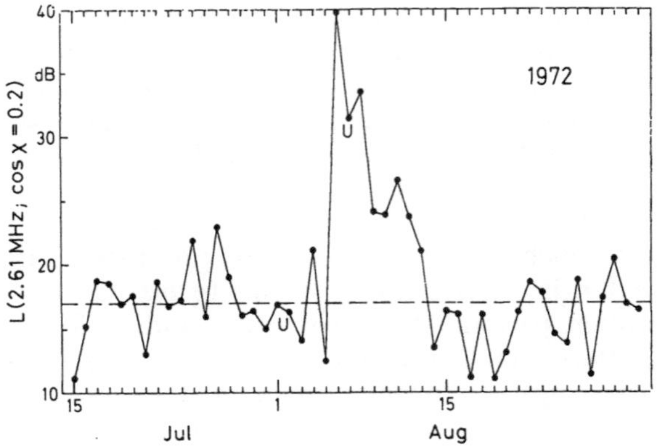

Fig. 9. Excessive absorption $L(\cos\chi = 0.2)$ in summer due to particle precipitation during the strong solar events in August 1972. The data were obtained from the diurnal variation of absorption where the short-wave fade-outs caused by solar X-rays were excluded. U, unreliable.

In conclusion it may be mentioned that 10 years ago a preliminary approach had already been made to obtain a survey of the behaviour of absorption in the northern hemisphere using measurements of absorption (A1) made at four stations and many supplementary f_{min} data during the IGY [10].

6. Workers in the European Cooperation

B. Beckmann, R. Eyfrig, Fernmeldetechnisches Zentralamt der Deutschen Bundespost, D 61 Darmstadt, FRG
G. Entzian, Observatorium für Ionosphärenforschung, DDR 2565 Kühlungsborn, GDR
J. K. Hargreaves, University of Lancaster, UK
U. Hense, Rundfunk- und Fernsehtechnisches Zentralamt, DDR 1199 Berlin-Adlershof, GDR
T. B. Jones, R. E. Evans, University of Leicester, UK
G. Nestorov, Geophysical Institute, Sofia, Bulgaria
Z. Ts. Rapoport, IZMIRAN, USSR
Yu. N. Faer, A. M. Svechimikov, Rostov on Don, USSR
K. I. Tukhashvili, Tbilisi, USSR
M. Shirmammedov, Ashkhabad, USSR
M. P. Rudina, E. N. Rybina, Alma Ata, USSR
G. Rose, H. Schwentek, J. Weber, H. U. Widdel, Max-Planck-Institut für Aeronomie, D 3411 Lindau, FRG
K. Sinno, M. Higashimura, Radio Research Laboratories, Tokyo 184, Japan

P. Tříska, Geophysical Institute of Czechoslovak Academy of Sciences, Prague 4, Czechoslovakia
N. Wakai, Ch. Ouchi, Ch. Nemoto, Radio Research Laboratories, Tokyo 184, Japan
E. R. Williams, University College of Wales, Aberystwyth, UK

References

[1] H. Schwentek, J. Atmos. Terr. Phys. **33**, 1647 (1971).
[2] W. Dieminger et al., J. Atmos. Terr. Phys. **28**, 317 (1966).
[3] N. Wakai et al., J. Radio Res. Lab. Japan **17**, 185 (1970).
[4] E. A. Lauter and B. Schäning, J. Atmos. Terr. Phys. **32**, 1619 (1970).
[5] B. Schäning, J. Atmos. Terr. Phys. **35**, 1003 (1973).
[6] W. Elling, K. H. Geisweid and H. Schwentek, this volume, p. 313.
[7] S. Fritz and S. D. Soules, Mon. Weath. Rev. **100**, 582 (1972).
[8] J. K. Hargreaves, J. Atmos. Terr. Phys. **35**, 291 (1973).
[9] H. Schwentek, J. Atmos. Terr. Phys. **33**, 1839 (1971).
[10] L. Thomas, J. Atmos. Terr. Phys. **23**, 301 (1962).

A NOTE CONCERNING DURATION AND INTENSITY OF THE WINTER ANOMALY IN IONOSPHERIC ABSORPTION

Z. Rapoport

IZMIRAN, USSR

Comments on the duration and intensity of the winter anomaly in ionospheric absorption arising from discussion on the preceding paper on the winter anomaly presented by H. Schwentek.

We consider the dependence of the following two characteristics of the winter anomaly: duration D (in days) iand intensity I (in dB) at geomagnetic latitude Φ. Let us clear up those characteristics. By using a graph of absorption data $L(t)$ at constant sun's zenith angle, day by day, we determine the approximate position of the curve characterizing the normal winter anomaly [1]. The duration

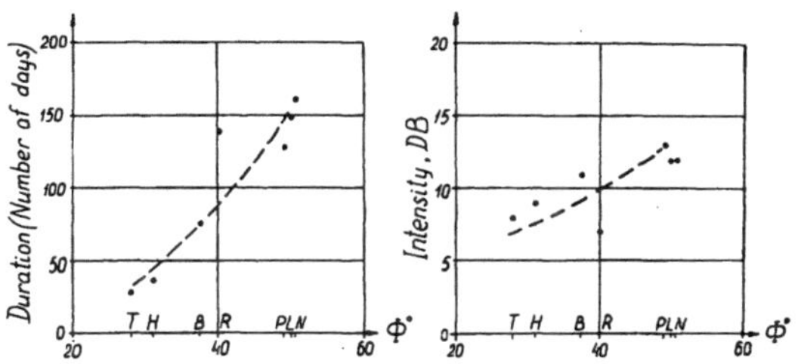

Fig. 1.

D of the winter anomaly is easily determined from this curve, i.e. the number of days per year when the anomaly is observed, as well as its intensity I which is the maximum value of the difference between the absorption level in winter (near 22 December) and the normal level in summer.

Our data are not complete and in some cases it is difficult to plot the above curve. The following data could be used: Tokyo (Kokubunji), Okamazaki—Hiraiso, Aranjuez—Balerma, Rostov-on-Don, Kiel—Panska Ves, Norddeich—Lindau and Norddeich—Neustrelitz. Dependences $D(\Phi)$ and $I(\Phi)$ are presented in Fig. 1. These show a distinct tendency to increase with increasing geomagnetic latitude. Along with this fact it should be noted that there is a non-negligible scattering of the points on the graphs. This could be connected with the effect of the longitude of the observation sites.

In Fig. 2 we give the graphs $D(I^0)$ and $I(I^0)$ where I^0 is the magnetic inclination in degrees. Naturally, the tendency mentioned above is distinctly pronounced on these graphs as well. We would like to indicate one more peculiarity of the graphs $L(t)$. The scatter is different for different stations, with a tendency to

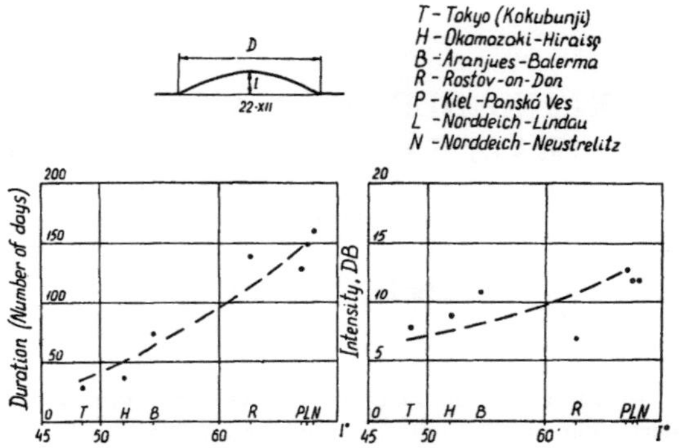

Fig. 2.

increase with the latitude. Maximum scatter occurs for Lancaster, the station of highest latitude ($\Phi = 53°$).

The increase in scatter of absorption values as well as of D and I with increasing geomagnetic latitude (or inclination) is, in our view, evidence of a definite influence of corpuscular emission on the winter absorption anomaly.

References

[1] H. SCHWENTEK, J. Atmos. Terr. Phys. **33**, 1647 (1971).

Lower Ionosphere Structure — Akademie-Verlag, Berlin 1974

SPATIAL EXTENT OF THE WINTER ANOMALY IN ABSORPTION

N. WAKAI, C. OUCHI and C. NEMOTO

Hiraiso Branch, Radio Research Laboratories, Ministry of Posts and Telecommunications, Nakaminato-shi, Japan

The simultaneous observation of the ionospheric absorption of MF waves at oblique incidence was made at three observatories in Japan in winter from 1971 to 1972. The apexes of reflection from the E region for four propagation paths are situated at 44°, 41°, 38° and 35° geographic latitude. The spatial extent of the winter anomaly is derived from correlation analysis for different combinations of the daily absorption value at noon in the four circuits covering about 1000 km in meridional distance. The results of measurement suggest that the winter anomaly is weakened towards the lower latitudes and has a spatial extent of the order of 300 km with a correlation coefficient of 0.7.

1. Introduction

The winter anomaly in absorption noticed at mid-latitude stations is characterized phenomenologically by the following: (i) the absorption is enhanced in winter above the level expected from the seasonal trend of the solar zenith angle, (ii) the excessive absorption appears for periods of several days, and (iii) it attenuates towards the lower latitudes [1], disappearing around 30° [2].

The spatial extent of the phenomenon was first reported by Thomas [3] who concluded from American and European data for the IGY and IGC, that the anomaly occurred over an area of the order of 10^6 km^2. Observation of the ionospheric absorption by means of the reception of Loran-A signals at 1850 kHz from Okamazaki transmitting station was made at Hiraiso in the IQSY [4]. Thereafter this project was revived in the IASY (1969—1971) with the aim of obtaining the solar cycle dependence of the ionospheric absorption by comparing with the data in the IQSY. Accordingly, the same technique of measurement as in the IQSY was employed. Another purpose of the project is to study the spatial extent of the winter anomaly in absorption.

From these points of view, the observing program was intensified year by year by adding three transmission circuits located in different latitudes. The fundamental characteristics of the ionospheric absorption including the solar cycle dependence will be published in a separate paper [5]. Therefore, this paper discusses just the spatial extent of the winter anomaly based on the data obtained in winter from 1971 to 1972 in four transmission circuits with apexes of the E reflection mode covering about 1000 km in meridional distance.

2. Method of Absorption Measurement

The Loran-A pulse (40 μs in width) signal at 1850 kHz is transmitted from two stations of relatively high radiation power (usually 160 kW peak) with the same pulse repetition frequency. The absorption measurement by the reception of Loran-A signals at a distance where the sky-wave component of 1-hop-E mode is predominant is definitely advantageous, combining the merits of the A1 (pulse

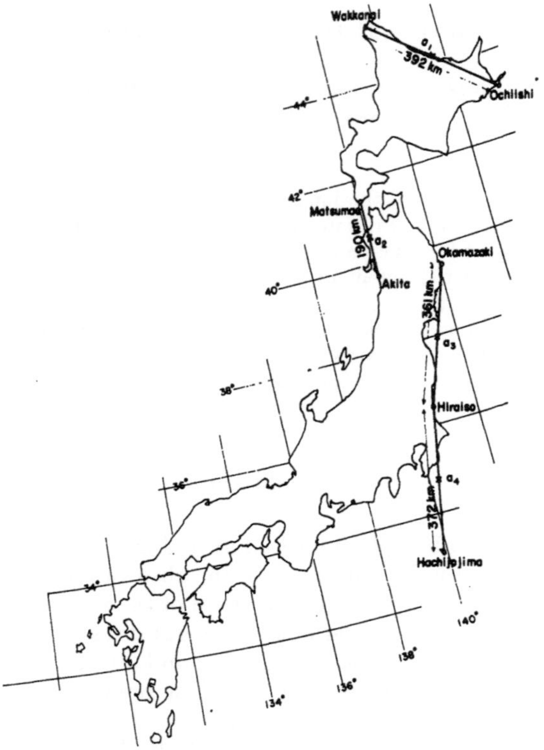

Fig. 1. Four propagational paths and their midpoints for the simultaneous observation of absorption.

vertical) and the A3 (cw oblique) methods [6]. Those techniques, used in the IQSY at Hiraiso [4], were also followed during the IASY.

The absorption measurement at Hiraiso (36°22′ N 140°37.5′ E) was started in December 1968 by receiving the signals from Okamazaki (39°25′ N 142°02′ E) transmitting station 361 km north, and then intensified by adding in December 1969 the recording of signal from Hachijojima (33°05.3′ N 139°49′ E) 372 km south of Hiraiso. Akita Radio Wave Observatory (39°44′ N 140°08′ E) joined in this program in November 1970, receiving Matsumae (41°25.3′ N 140°04.8′ E) 190 km distant. Wakkanai Observatory (45°24′ N 141°41′ E) cooperated in winter 1971/72 in receiving Ochiishi (43°11.2′ N 145°31.4′ E) 392 km distant. Although the photographic recording of the A-display on the oscilloscope by exposure for

30 seconds is made every 30 minutes for 24 hours at Hiraiso, it is limited to three shots at Akita and one shot at Wakkanai around noon. The geographic situation of the four propagation paths is shown in Fig. 1, together with the midpoints of the paths, a_1 to a_4. It is to be mentioned that the midpoints of the paths which correspond to the reflection point of predominant 1-hop-E mode at the ionosphere cover about 1000 km in meridional distance. Calibration of the field intensity was made once a day by means of a standard signal generator.

The ionospheric absorption in the daytime is of primary concern for this project. In order to derive the ionospheric absorption from the field intensity measured, however, the unabsorbed field intensity based on the intensity in the night-time has to be determined by more or less statistical processing. In this report, the highest peak of the level in the monthly occurrence frequency of night-time field intensity is referred to as the unabsorbed field intensity which probably arises from the Es reflection suffering from the least ionospheric absorption [6]. Accordingly the absorption can be obtained by subtracting the field intensity measured at a given time from the unabsorbed intensity thus calculated monthly.

This is the case in two circuits centred at Hiraiso (Okamazaki-Hiraiso, and Hachijojima-Hiraiso). On the other hand, no measurements were made at night on a routine basis in two other circuits (Matsumae-Akita and Ochiishi-Wakkanai). Therefore complementary measurement at night was made in August 1971 for the former and in June 1972 for the latter, when the sporadic E layer was active, in order to determine the unabsorbed field intensity.

3. Results of Measurement

Fig. 2 shows the day-to-day variation in absorption at noon in six months from 1971 to 1972 for the four circuits in order of decreasing latitude of apexes towards the bottom of the figure. The ordinate gives the ionospheric absorption normalized with respect to the seasonal trend, i.e. to the level expected from the solar zenith angle. The similarity of the day-to-day variation of absorption along the four paths is obvious in the figure, particularly when we notice several groups of high absorption days. Furthermore, it is seen that the excessive absorption shaded in the figure is much more pronounced in higher latitudes with earlier onset and later finish, although the beginning and end could not be determined definitely. The mean values of absorption for the four paths down to south exceed the reference level (0 dB in the figure) by 5.6 dB, 5.1 dB, 2.3 dB and 0.8 dB, respectively. In other words, this mean deviation would be a measure of the severity of the anomaly.

Thomas [3] reported that the winter anomaly in absorption covers at least an area of the order of 10^6 km^2, based on the correlation analysis of the A1 and $f_{\min}$ data during winters in IGY and IGC. He examined the spatial correlation of absorption for the twenty days of highest absorption in each month in terms of the Rawer correlation number [7].

In this paper, the correlation coefficient between values of the daily absorption in different pairs of propagation paths is calculated from the data for six months as shown in Fig. 2. Six coefficients thus calculated are indicated in Table 1 and shown in Fig. 3 against the separation of the respective apexes. If we define the spatial extent of the anomaly as the distance equivalent to a correlation coefficient of 0.7, it ranges about 300 km. Open circles in Fig. 3 represent the correlation coefficient calculated for data in summer (May—August 1971) in the same treatment of data (no observation at Wakkanai in this period). No appreciable change is found in correlation coefficients with the distance.

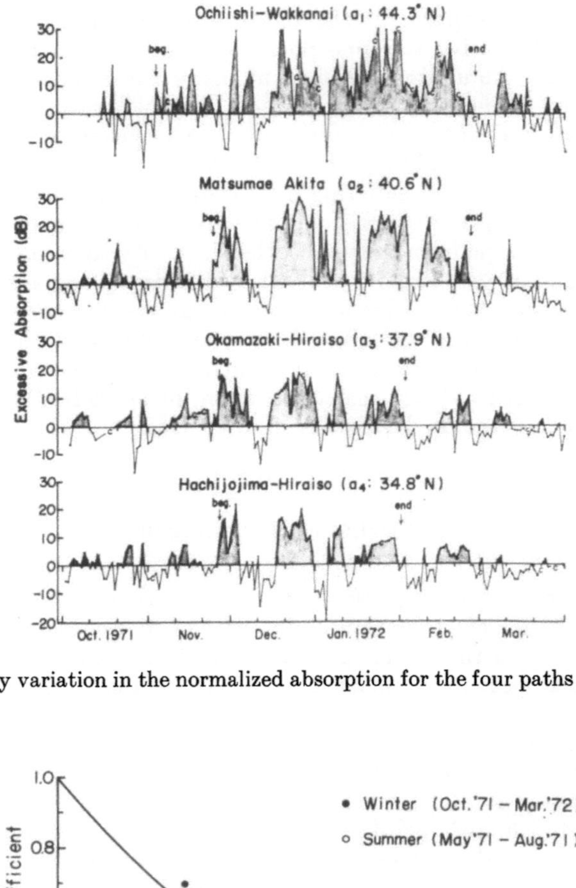

Fig. 2. Day-to-day variation in the normalized absorption for the four paths in winter 1971/72.

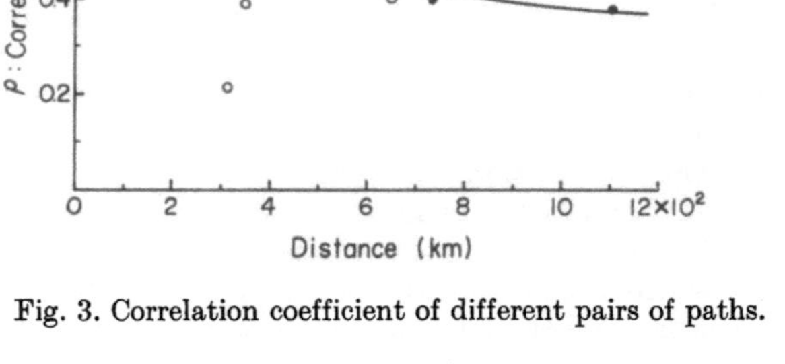

Fig. 3. Correlation coefficient of different pairs of paths.

Table 1
Correlation Coefficient of Different Pairs of Paths

Combination	Separation of apexes (km)	Correlation coefficient
(Ochiishi-Wakkanai) and (Matsumae-Akita)	504 ($a_1 - a_2$)	0.43
(Ochiishi-Wakkanai) and (Okamazaki-Hiraiso)	738 ($a_1 - a_3$)	0.40
(Ochiishi-Wakkanai) and (Hachijojima-Hiraiso)	1102 ($a_1 - a_4$)	0.38
(Matsumae-Akita) and (Okamazaki-Hiraiso)	316 ($a_2 - a_3$)	0.61
(Matsumae-Akita) and (Hachijojima-Hiraiso)	650 ($a_2 - a_4$)	0.54
(Okamazaki-Hiraiso) and (Hachijojima-Hiraiso)	352 ($a_3 - a_4$)	0.70

4. Concluding Remarks

A spatial correlation of the order of 0.4 seems to exist independently of any separation of apexes. This kind of fluctuation forming the background level of the additional absorption due to the winter anomaly may be ascribed to the solar radiation, which is inherent throughout the year. The spatial extent of the winter anomaly in lower mid-latitudes (e.g. Japanese area) would be of the order of several hundreds of kilometres (300 km for $\varrho = 0.7$). A larger scale of anomaly could be appropriate to higher mid-latitudes (around 50° north) as revealed by Thomas [3].

Acknowledgments

The authors wish to express their thanks to Mr. T. Kogaku of Akita Radio Wave Observatory and to Mr. I. Yamazaki of Wakkanai Radio Wave Observatory for their efforts given to the cooperative observation of the ionospheric absorption.

References

[1] W. Dieminger et al., J. Atmos. Terr. Phys. **28**, 317 (1966).
[2] N. Wakai et al., J. Radio Res. Lab. Japan **17**, 185 (1970).
[3] L. Thomas, J. Atmos. Terr. Phys. **23**, 301 (1961).
[4] N. Wakai et al., J. Radio Res. Lab. Japan **16**, 1 (1969).
[5] C. Ouchi et al., J. Radio Res. Lab. Japan, in press (1974).
[6] IQSY Instruction Manual, No. 4, Ionosphere, 49 (1963).
[7] W. R. Piggott and K. Rawer, URSI Handbook for Ionogram Interpretation and Reduction, Elsevier Publ. Co., Amsterdam 1961.

SHIPBORNE OBSERVATIONS OF THE SOUTHERN BOUNDARY OF THE WINTER ANOMALY IN IONOSPHERIC ABSORPTION

W. Elling, H. K. Geisweid and H. Schwentek

Max-Planck-Institut für Aeronomie, Institut für Ionosphärenphysik, Lindau/Harz, FRG

In the period 28 November 1971 to 12 February 1972, during three passages of the North Atlantic, the diurnal variation of ionospheric absorption L was measured by method A1 on 1.7 and 5.0 MHz as a function of geographic and geomagnetic parameters. The measurements were carried out on board a cargo vessel moving between Hamburg (53° N 10° E) and La Guaira (11° N 67° W).

The diurnal variation of absorption showed large modifications from day to day. The behaviour of L ($\cos \chi = 0.2$) was somewhat different at each passage but showed a common trend: The winter anomaly vanished in a transit zone between 34° and 44° N centred at about 39° N; the same was true for the behaviour of L ($\cos \chi = 1$). The exponent n in $L = L_0 \cos^n \chi$ can be presented as a function both of geographic latitude and of geomagnetic dip angle. An influence of the magnetic dip angle on the southern boundary of the winter-anomaly zone seems to be possible. The results of the measurements made on 5 MHz, in general, supported those obtained on 1.7 MHz.

1. Introduction

One of the approaches to determine the southern boundary of the winter-anomaly zone are measurements on board a cargo ship moving on a suitable route. Another approach is to compare the winter-anomaly patterns, plots of the day-to-day variation of ionospheric absorption, obtained at selected stations by methods A1 or A3. By comparing the patterns of several mid-latitude stations (circuit Norddeich—Lindau at 52.6° N, Freiburg at 48° N, circuit S. Marinella—Capo San Lorenzo at 41° N, circuit Okamazaki—Hiraiso at 38° N, and Kokubunji at 36° N) Wakai, Ouchi and Nemoto [1] showed that in 1964/65 (sunspot minimum) in Japan a southern boundary of the occurrence of the winter anomaly in ionospheric absorption existed at about 37° N. This result was supported by Lauter and Schäning [2] and Schäning [3] who made measurements on board a ship crossing the North Atlantic in winters 1969/70 and 1970/71. In the framework of the European Cooperation on studies of the winter anomaly in ionospheric absorption, in the midwinter period 28 November 1971 to 12 February 1972, absorption was measured again on board ship (mv *Hanau* of Hapag-Lloyd) during three passages of the North Atlantic (Fig. 1) (method A1; 1.7 and 5.0 MHz).

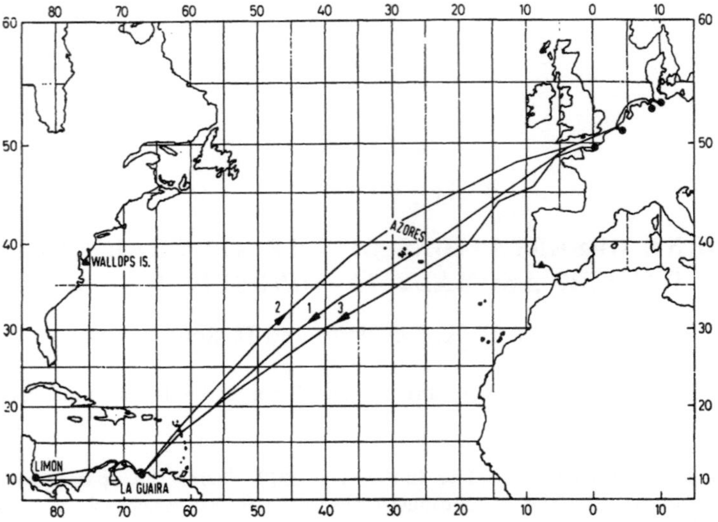

Fig. 1. Map showing the three routes, 1, 2, 3, of mv *Hanau* on which measurements were carried out in December 1971, January and February 1972.
Hamburg (53° N 10° E), La Guaira (11° N 67° W). ● harbour; ▲ rocket launching site.

2. Equipment

The measurements were carried out applying a new, mobile A1 equipment built up in solid state technique (with the exception of the power stages and modulator of the transmitter where tubes were used; pulse power 30 kW at 200 μs). Always 60 of the received pulses were superimposed photographically on one frame $24 \times 24\ \text{mm}^2$ during the daytime at 2-minute intervals, during the night at 10-minute intervals. Thus, a very quick data processing was possible enabling continuous control of the calibration constant of the equipment including the antenna, and information on the variation of absorption during the passage.

3. Results

Ionospheric absorption of short waves is a rather complicated function of several parameters. For instance, the diurnal variation L can be described by

$$L(f, t + \tau, \varphi, I) = L_0(R, \varphi) \cos^n \chi(t), \tag{1}$$

with $n = n(f, \varphi, I)$, where f is the operating frequency, t time, φ geographic latitude, I magnetic dip angle, R sunspot number, τ delay time, specifying the asymmetry of the diurnal variation, χ solar zenith angle. The behaviour of the characteristic parameters L ($\cos \chi = 0.2$), $L_0 = L$ ($\cos \chi = 1$), and n, as a function of geographic latitude φ and magnetic dip angle I according to Eq. (1) will be presented here. Figs. 2 and 3 show absorption at a constant solar zenith angle

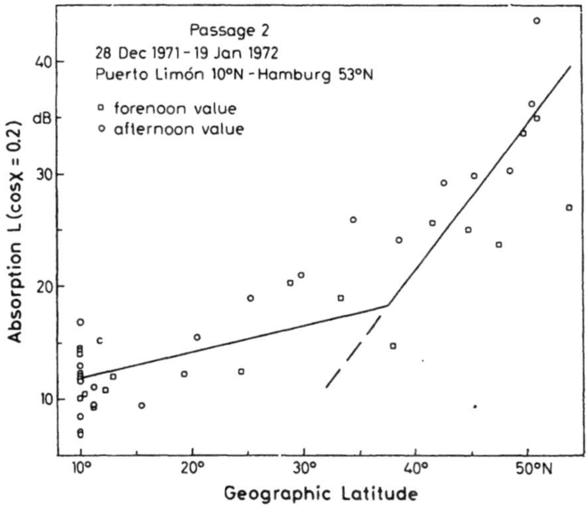

Fig. 2. Absorption L (1.7 MHz; $\cos \chi = 0.2$) as a function of geographic latitude. The lines are drawn tentatively to indicate the trends.

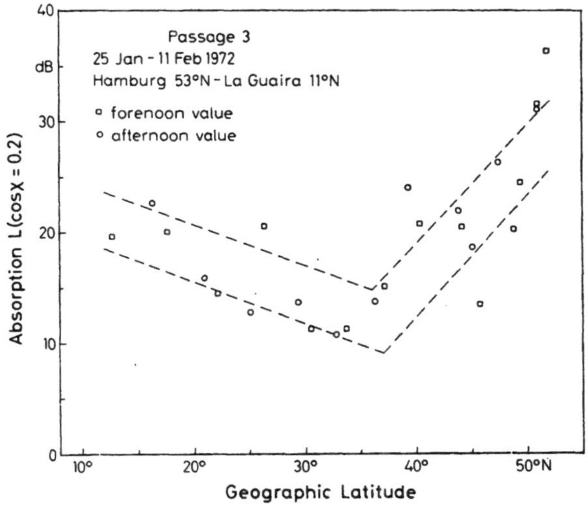

Fig. 3. Absorption L (1.7 MHz; $\cos \chi = 0.2$) as a function of geographic latitude. The lines are drawn tentatively to indicate the trends.

$\chi = 78.5°$, $L(\cos \chi = 0.2)$, as a function of geographic latitude. It reveals that the behaviour of absorption was different during the two passages although a common feature can be noted: in both cases L decreased from 53° to about 36° where something like a boundary of the winter-anomaly zone appeared. Fig. 4 shows subsolar absorption as a function of geographic latitude. The lines are linking running median values; the intersection is at 39°. In Fig. 5 the averages of absorption obtained from the forenoon and afternoon branches of the diurnal variation are

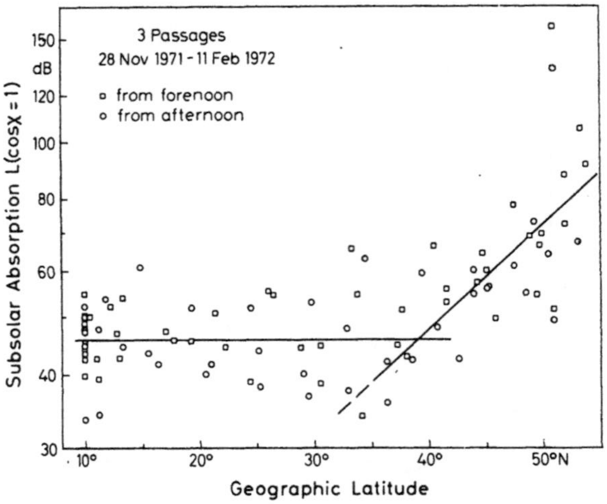

Fig. 4. Subsolar absorption L (1.7 MHz; $\cos \chi = 1$) as a function of geographic latitude obtained from measurement of the diurnal variation of absorption during the three passages of the ship.

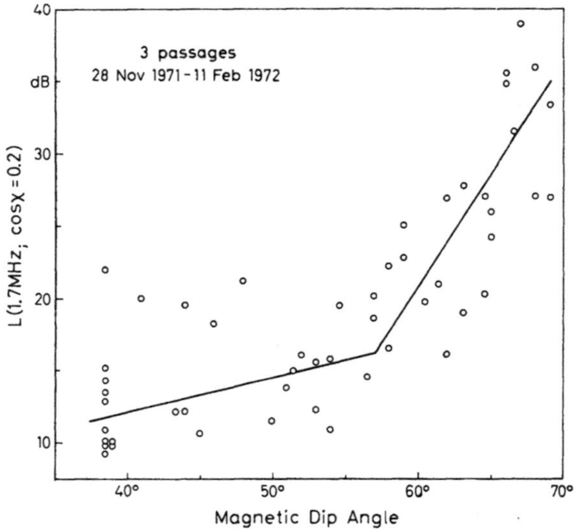

Fig. 5. Absorption L (1.7 MHz; $\cos \chi = 0.2$) as a function of the geomagnetic dip angle; the dip angle 57° corresponds to a geographic latitude of 38° N (see the route of the ship, Fig. 1).

shown as a function of geomagnetic dip angle. Also in this presentation a southern boundary can be clearly determined. It should be noted that high absorption occurred also on some days at low latitudes (10° to 20° N). But whether the cause for this high absorption is the same as for high absorption in the winter-anomaly zone should be considered as an open question. Finally, the exponent n was studied as a function of geographic latitude and geomagnetic dip angle I (Fig. 6). Although

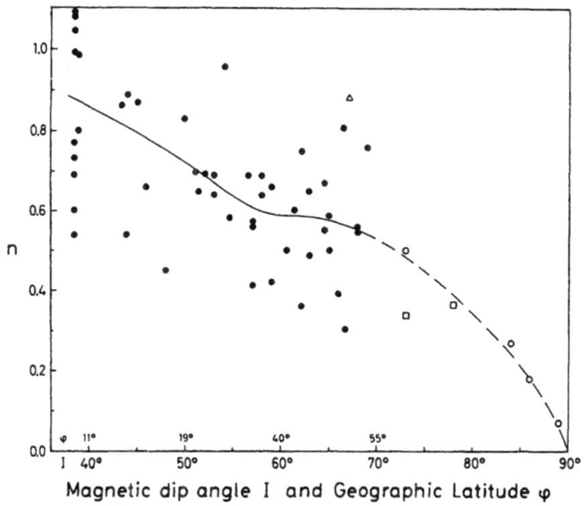

Fig. 6. Exponent n (mean of forenoon and afternoon value) as a function of the geomagnetic dip angle. Some mean values of n obtained at various locations situated at dip angles greater than 70° are included to demonstrate more clearly the trend towards the geomagnetic pole. The triangle indicates the median value of the exponent obtained from A3 measurements on 2.61 MHz over 300 km at Lindau for all those days on which measurements on board the ship were carried out. The data represented as open circles [4] and squares [5] are listed in the following table.

Location	Geomagnetic dip angle	Exponent n	Symbol
Resolute Bay (Canada)	89°	0.07	circle
Baker Lake (Canada)	86°	0.18	circle
Churchill (Canada)	83°	0.27 Equinox	circle
		Winter 0.35	not marked
Prince Rupert (Canada)	73°	0.50	circle
Kjeller (Norway)	72.3°	0.60	not marked
Kjeller (Norway)	72.3°	0.34	square (from seasonal variation)
Tromsö (Norway)	77.9°	0.36	square (from seasonal variation)

the scatter in the points is large, the (solid) curve linking the running median values shows a definite trend indicating vaguely the boundary of the winter-anomaly zone. We do not consider this as a reliable result, but only as a stimulation to study this problem in more detail. If a real geophysical background exists, then the geomagnetic influence on the position of the boundary of the winter-anomaly zone must also be taken into account.

It should be noted that the delay time τ increases sharply with decreasing latitude. At low latitudes, the diurnal variation of absorption becomes rather asymmetric with delay times up to 3 hours; after midday absorption remained almost constant for one or two hours.

4. Conclusions

Taking together all available measurements concerning the southern boundary of the winter-anomaly zone it is evident that there is certainly no fixed definite limit but a transit zone between about 40° N to 30° N. This must be emphasized since the variation of absorption at fixed stations evidently shows the winter-anomaly pattern also at latitudes down to 35°.

And a last but important remark: Since the parameters describing the winter anomaly depend strongly on sunspot cycle [6], it should be expected that also the "southern boundary", varying within the transit zone, will very probably depend on it also.

References

[1] N. Wakai, C. Ouchi and C. Nemoto, J. Rad. Res. Labs. Japan **17**, 185 (1970).
[2] E. A. Lauter and B. Schäning, J. Atmos. Terr. Phys. **32**, 1619 (1970).
[3] B. Schäning, J. Atmos. Terr. Phys. **35**, 1003 (1973).
[4] K. Davies, National Bureau of Standards Monogr. No. 80 (1965).
[5] H. Schwentek, Nachrtechn. Z. **21**, 32 (1968).
[6] J. Röttger and H. Schwentek, J. Atmos. Terr. Phys. **36**, 363 (1974).

For a complete description of the results and the equipment outlined above see: W. Barke, W. Elling, K. H. Geisweid, G. Heimesaat, A. Loidl, C. Römer, H. Schwentek and R. Zellermann, The southern boundary region of the winter anomaly in ionospheric absorption in winter 1971/72 observed on board the cargo vessel mv *Hanau* of Hapag-Lloyd moving between 10° and 55° N, Mitteilungen aus dem Max-Planck-Institut f. Aeronomie, Nr. 51. Springer-Verlag, Berlin, Heidelberg, New York 1974.

Lower Ionosphere Structure — Akademie-Verlag, Berlin 1974

THE USE OF LF A3 ABSORPTION MEASUREMENTS IN STUDYING THE WINTER ANOMALY

J. Laštovička and P. Tříska

Geophysical Institute, Czechoslovak Academy of Sciences, Prague, Czechoslovakia

The lower boundary of the winter anomaly of ionospheric absorption is estimated to be at 84—85 km, as follows from the analysis of changes of the seasonal variation of absorption with the solar zenith angle. Some evidence is given for a dual origin of the winter anomaly: "background" winter enhancement of absorption is due to a considerably enhanced collision frequency, whereas days or groups of days of excessive absorption, which are added to the "background" winter anomaly, are due to enhancements of the electron density.

1. Introduction

This paper is based mainly on the A3 absorption measurements made at the Průhonice Observatory on 272 kHz (reflection point 49°34′ N 16°03′ E, $L = 2.13$). These absorption data are used (i) for estimating the lower boundary of the winter anomaly height interval and (ii) for studying a dual origin of the winter anomaly of ionospheric absorption.

2. Lower Boundary of the Winter Anomaly

Fig. 1 shows the mean seasonal variation of absorption on 272 kHz at three constant values of the solar zenith angle (sunset, $\chi = 80-85°$, $\chi = 75-80°$) for the period 1966—1968. The bottom curve (sunset) displays a pronounced winter increase of the absorption. The middle curve also displays some winter increase, but comparatively small. The top curve displays no winter increase with respect to summer values. Using various $N(h)$ profiles and the reflection height determination based on a simplified form of Booker's quartic, we can calculate the reflection height for all three curves in Fig. 1. These computations are described in [1]. It can be concluded that reflections from altitudes of about 83—84 km (top curve) do not display the winter anomaly, which is distinctly present, however, at reflections from altitudes of about 86—87 km. It enables us to estimate the lower boundary of the winter anomaly to be at about 84—85 km, which is in good agreement with [2].

Fig. 2 shows the mean seasonal variation of the sunset absorption measured on four frequencies at the Kühlungsborn and Průhonice Observatories. The

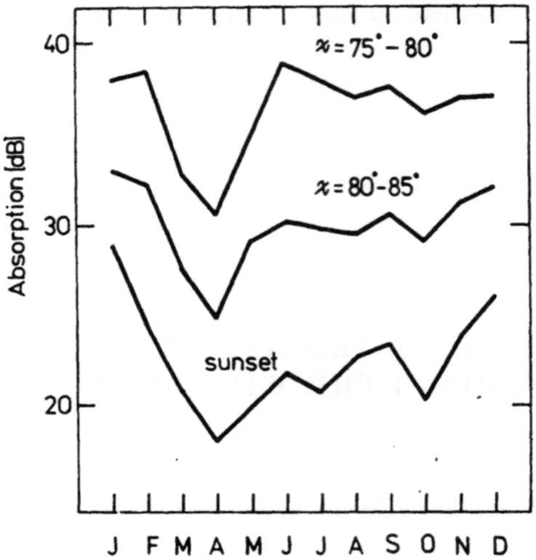

Fig. 1. Mean seasonal variation of absorption at 272 kHz for three fixed values of the solar zenith angle χ, 1966—1968.

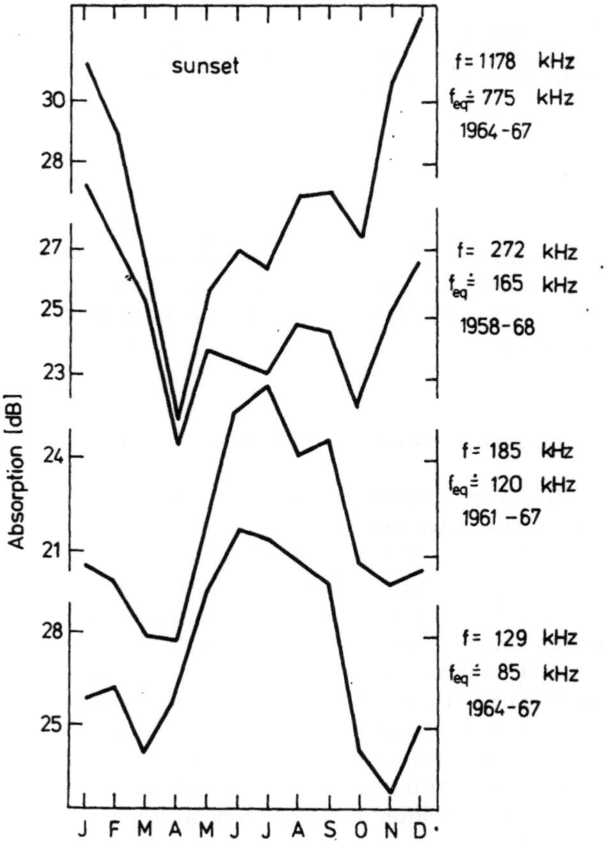

Fig. 2. Mean seasonal variation of the sunset absorption. The frequency of 272 kHz was received at the Průhonice Observatory; the other three frequencies were received at the Kühlungsborn Observatory (GDR).

frequencies and equivalent frequencies are also given. We can see that the magnitude of the winter anomaly decreases with decreasing frequency and is replaced by the "summer" anomaly at the lowest frequencies. The lower boundary of the winter anomaly of about 85 km is consistent with these data.

3. Two Conceptions of the Winter Anomaly

There are two conceptions of the winter anomaly. Firstly, it can be considered as the winter-time occurrence of days, or groups of days, of excessive absorption, or secondly, as an anomaly in the seasonal variation of absorption. Both these components of the winter anomaly can be clearly detected from long series of

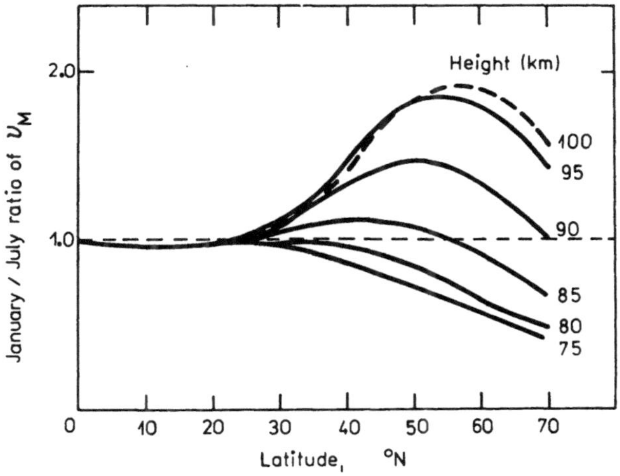

Fig. 3. Ratio of January—July values of the "monoenergetic" collision frequency, ν_M, as a function of northern geographic latitude, for different fixed heights (after Taubenheim [8]).

absorption measurements (e.g. [3]. The days of excessive absorption are accompanied by high enhancement of the electron density in the whole D region (e.g. [4]). On the other hand, no significant role of systematic changes of the collision frequency can be expected, as follows from both measurements [5] and theoretical calculations [6]. Now there is a general belief that the days of excessive absorption are caused by high electron density enhancements (e.g. [7]).

The winter anomaly regarded as an anomaly in the seasonal variation of absorption (e.g. Fig. 1) is more difficult to explain. Using the LF-absorption data, we can give some evidence of the dominating role of the collision frequency changes in the winter anomaly. The collision frequency below about 95—100 km may be considered to be proportional to total air pressure. Taubenheim [8] used Groves' [9] data on the seasonal variation of pressure and obtained the seasonal variation of the collision frequency between 75 and 100 km shown in Fig. 3.

We shall try to give some evidence supporting the collision frequency origin of the winter anomaly.

(i) The zero level of the seasonal variation of the collision frequency for 272 kHz is located at a height of about 83 km, so that it seems to be quite realistic to expect the lower boundary of the winter anomaly to be at about 84—85 km. The electron density in days of excessive absorption is enhanced in the whole D region with a maximum relative enhancement observed at about 82 km [10]. The "normal" $N(h)$ profiles under conditions of quiet sun give summer values higher than or roughly equal to winter values [11], i.e. they do not display any winter anomaly.

(ii) The seasonal variation of the collision frequency and its height dependence make it possible to explain both the winter anomaly observed at higher frequencies and the "summer anomaly" at lower frequencies shown in Fig. 2. Taking into account the fact that in summer the "normal" electron densities under active sun conditions are according to Mechtly et al. [11] equal to or lower than those in winter, we cannot explain the high summer anomaly observed at lower frequencies by changes of the electron density. It is rather difficult also under conditions of quiet sun.

(iii) The minimum monthly values of the sunset absorption at 272 kHz are systematically higher in winter than in summer by about 10—15%. This can be fully explained by collision frequency changes (reflection height of about 89 to 90 km), but rather difficult to explain on the basis of the above mentioned summer-winter differences in the electron density [11], mainly for the year 1966.

(iv) The latitude dependence, mainly the low-latitude boundary, of the seasonal variation of the collision frequency (Fig. 3) agrees well with the latitude dependence of the winter anomaly.

In conclusion it can be considered that the winter anomaly of the ionospheric absorption consists of two parts. The following rough pattern of the winter anomaly can be given. There is the "background" increase of absorption mainly due to a considerably enhanced collision frequency, which results in an anomaly in the seasonal variation. Additional high increases of absorption are due to high enhancements of the electron density and result in days, or groups of days, of excessive absorption.

In our opinion, in the future it will be necessary to study closely not only the winter-time increase of absorption, but also deep equinoctial minima of absorption, mainly the spring minimum.

References

[1] J. Laštovička, J. Atmos. Terr. Phys. **34**, 1233 (1972).
[2] E. A. Lauter and P. Nitzche, J. Atmos. Terr. Phys. **29**, 533 (1967).
[3] H. Schwentek, J. Atmos. Terr. Phys. **33**, 1647 (1971).
[4] M. A. Geller and C. F. Sechrist, Jr., J. Atmos. Terr. Phys. **33**, 1027 (1971).
[5] J. N. Rowe, A. J. Ferraro, H. S. Lee and A. P. Mitra, in: Aeronomy Rep. No. 32, University of Illinois, Urbana 1969 (p. 341).
[6] J. Taubenheim, Gerlands Beitr. Geophys. **81**, 396 (1972).
[7] L. Thomas, J. Atmos. Terr. Phys. **33**, 157 (1971).
[8] J. Taubenheim, J. Atmos. Terr. Phys. **33**, 1481 (1971).
[9] G. V. Groves, Air Force Surveys in Geophysics, No. 218, AFCRL, Bedford, Massachusetts (1970).
[10] E. A. Mechtly and J. S. Shirke, J. Geophys. Res. **73**, 6243 (1968).
[11] E. A. Mechtly, S. A. Bowhill and L. G. Smith, J. Atmos. Terr. Phys. **34**, 1899 (1972).

SOLAR LYMAN ALPHA CONTROL OF THE A3 IONOSPHERIC ABSORPTION ON 2775 kHz

T. Laštovička and P. Tříska

Geophysical Institute, Czechoslovak Akademy of Sciences, Prague, Czechoslovakia

The A3 absorption data measured at the Panská Ves Observatory on 2775 kHz are compared with the satellite X-ray and Lyman-α data for summer 1969. At the solar zenith angle of 60° the Lyman-α control of absorption was observed under undisturbed and even moderately disturbed conditions. It is shown that the role of X-rays in absorption decreases considerably with increasing solar zenith angle between 60° and 80°.

The object of this paper is to compare the radio-wave absorption data with satellite X-ray and Lyman-α data. It can be one of the possible approaches to the determination of a mutual role of these two most important D region ionization sources.

We will compare the ionospheric absorption data measured by the A3 method on 2775 kHz at the Panská Ves Observatory (reflection point 52°27′ N 12°27′ E; distance 517 km; equivalent frequency of about 1 MHz) with X-ray data in the band 1—8 Å measured by the Solrad 9 satellite [1] and Lyman-α data measured by the OSO 5 satellite [2]. The comparison for the period 1 June—31 August 1969 and mainly for the solar zenith angle χ of 60° is presented in this paper.

Fig. 1 shows the dependence of the absorption L on the magnitude of the X-ray flux for $\chi = 60°$. We can see that for the X-ray flux lower than about 1×10^{-6} $\mathrm{W\,m^{-2}}$ ($= 10^{-3}\,\mathrm{erg\,cm^{-2}\,s^{-1}}$) the absorption is essentially independent of X-rays. In the whole remaining part of X-ray flux values, in Fig. 1 up to about 5×10^{-3} $\mathrm{erg\,cm^{-2}\,s^{-1}}$, the absorption increases slightly with increasing X-ray flux, but it does not seem to be controlled by X-rays.

Fig. 2 shows the dependence of absorption on the Lyman-α intensity. In spite of a great scatter of values, the absorption clearly rises with increasing Lyman-α intensity. A much lower increase of the Lyman-α than of the X-ray flux is needed to result in the same increase of absorption. Also the correlation with absorption seems to be better for Lyman-α than for X-rays.

Fig. 3 shows the time course of absorption, X-ray flux, Lyman-α intensity and solar radio noise measured at $\lambda = 10.7$ cm. It illustrates why the solar radio noise data cannot always be used instead of the Lyman-α radiation data. We can see the existence of a large day-to-day variability of absorption and particularly extreme decreases of absorption that are not caused by changes in ionizing radiation. The envelope of the absorption curve follows closely the Lyman-α curve contrary to a rather different behaviour of the X-ray curve. The general impression from Fig. 3 also supports a greater role of Lyman-α. It can be concluded that the data

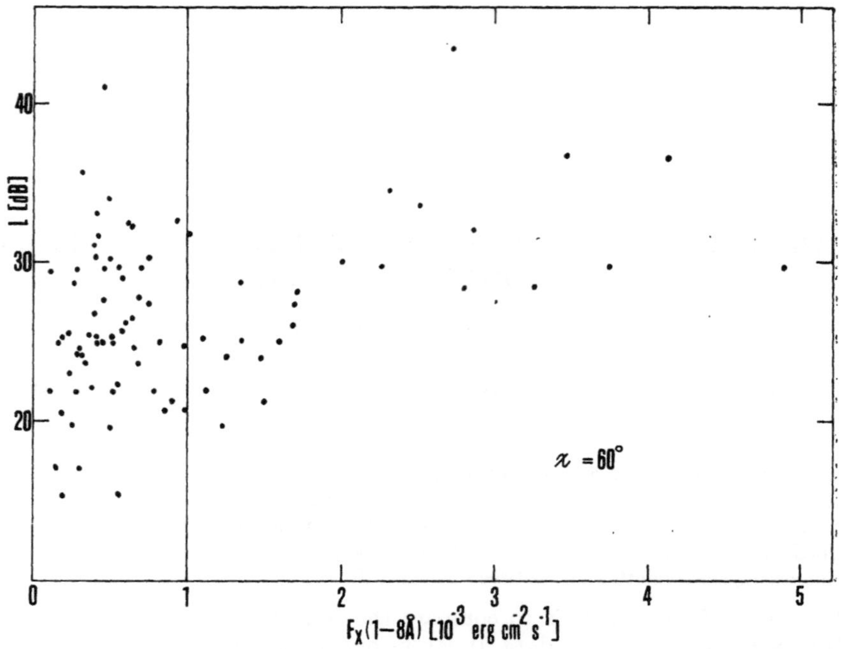

Fig. 1. Dependence of the absorption L on the X-ray (1–8 Å) flux at $\chi = 60°$, June–August 1969.

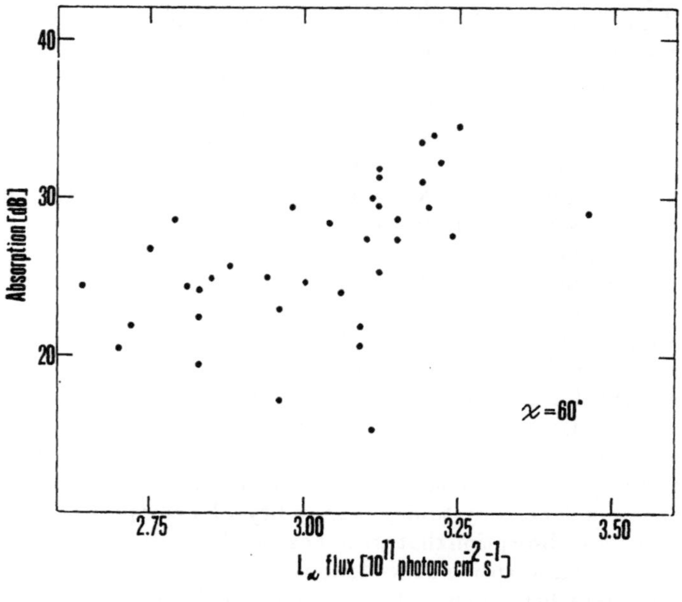

Fig. 2. Dependence of the absorption on the Lyman-α intensity at $\chi = 60°$, June–August 1969.

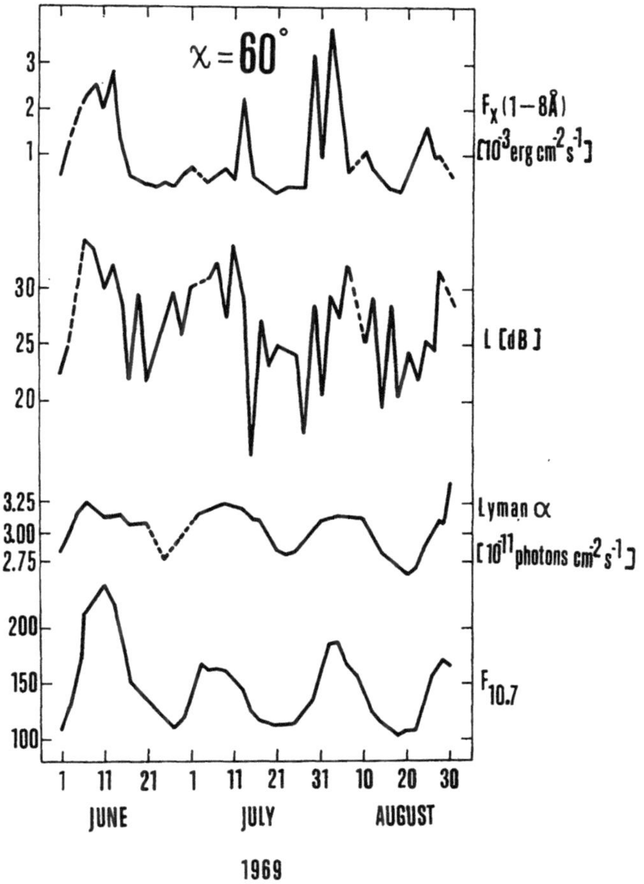

Fig. 3. Time course of the X-ray (1—8 Å) flux, of absorption, of the Lyman-α intensity and of the solar radio noise intensity at wavelength 10.7 cm (Covington index), June—August 1969.

contained in Figs. 1—3 confirm the idea of a less significant role of X-rays and support the idea of the Lyman-α control of the absorption studied.

In order to confirm this conclusion, some correlation coefficients were calculated. They are shown in Table 1. The absorption correlates considerably better with

Table 1

Correlation Coefficients describing the Correlation between Absorption and X-rays ($\varrho_{L,X}$), Absorption and Lyman-α ($\varrho_{L,\alpha}$) and X-rays and Lyman-α ($\varrho_{X,\alpha}$) for all data and for two subgroups: data for F_X (1—8 Å) $\gtrless 1 \times 10^{-6}$ Wm^{-2} (=10^{-3} erg cm^{-2} s^{-1}).

	All data	$F_X < 10^{-6}$ Wm^{-2}	$F_X > 10^{-6}$ Wm^{-2}
$\varrho_{L,X}$	0.38	0.08	0.46
$\varrho_{L,\alpha}$	0.52	0.40	0.78
$\varrho_{X,\alpha}$	0.27	0.27	0.32

Lyman-α ($\varrho_{L,\alpha}$) than with X-rays ($\varrho_{L,X}$). There is nearly no correlation between absorption and X-rays for the X-ray flux lower than 1×10^{-6} W m^{-2} (= 10^{-3} erg cm^{-2} s^{-1}), which is in accordance with the result of Fig. 1. The correlation between X-rays and Lyman-α ($\varrho_{X,\alpha}$) is sufficiently low to enable us to distinguish the effects of X-rays and Lyman-α in radio-wave absorption by the statistical method used. The dependence of absorption on X-rays and Lyman-α was calculated by Laštovička [3] using the least squares method. Assuming linear or quadratic dependence, the best fitted equation is:

$$L\,(\mathrm{dB}) = 0.23 F_X^2 + 0.57 F_X + 11.3 F_{L\alpha} - 8.92$$

F_X in 10^{-6} W m^{-2} (= 10^{-3} erg cm^{-2} s^{-1}), the photonflux $F_{L\alpha}$ in 10^{15} m^{-2}s^{-1}. This equation completely confirms the conclusions based on Figs. 1—3 and Table 1.

It can be concluded that at the solar zenith angle of 60° the radio-wave absorption is dominated by Lyman-α radiation over the whole range of the X-ray flux values studied, i.e. up to about 5×10^{-6} W m^{-2}. The role of X-rays is negligible for the X-ray flux lower than about 1×10^{-6} W m^{-2} and increases with increasing

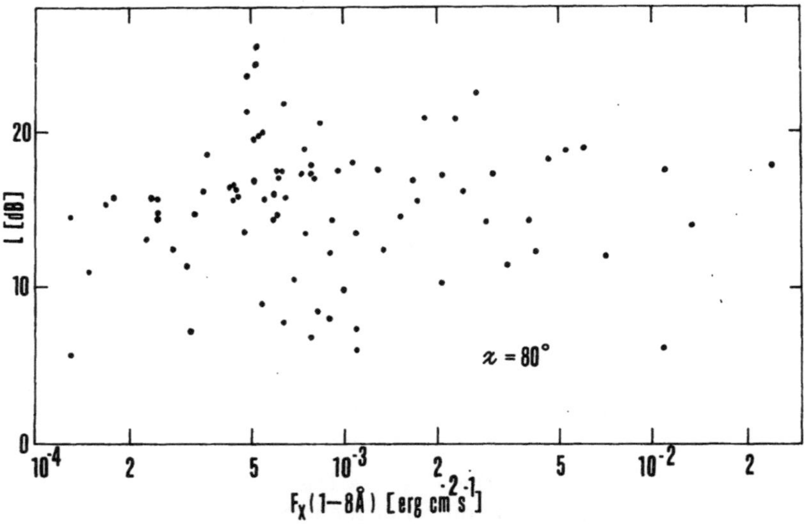

Fig. 4. Dependence of the absorption L on the X-ray (1—8 Å) flux at $\chi = 80°$, June—August 1969.

X-ray flux. The absorption is significantly affected by some factors different from ionizing radiation, which cause irregular changes of absorption shown in Fig. 3.

Fig. 4 shows the situation at higher solar zenith angles: for $\chi = 80°$ the X-ray influence vanishes almost completely and even for X-ray flux as high as about 20×10^{-6} W m^{-2} it is questionable whether one should allow for any X-ray influence on absorption. The dependence of the X-ray influence on absorption on the solar zenith angle is presented in a simplified form in Fig. 5. We can see a gradual shift of the "no-influence" boundary and simultaneously a considerable decrease of the role of X-rays with increasing solar zenith angle. The same trend, i. e. a decreasing role of X-rays with increasing solar zenith angle is preserved towards the lower solar zenith angles, at least down to 40°.

It appears that the results obtained can be useful for several ionospheric implications. For example, we tried to calculate the height profiles of radio-wave absorption and of ionization rates of various ionization sources. From these height

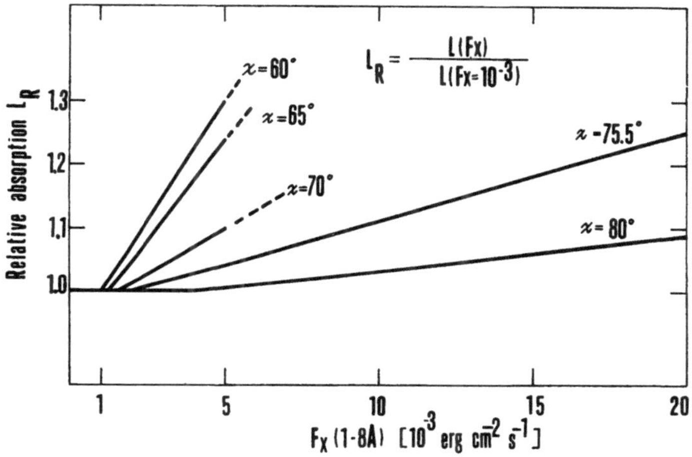

Fig. 5. An approximate representation of changes in the X-ray influence on absorption with changes of the solar zenith angle χ.

profiles we can obtain the ratio of the roles of X-rays and Lyman-α in absorption. The preliminary results show that the best agreement between calculated and observed ratios of X-ray and Lyman-α influence on absorption is obtained for the nitric oxide densities higher than Meira's [4] NO density by a factor of about 2. The agreement with Meira's data is fairly good, taking into consideration various approximations used in the calculations.

All the results presented in this paper were obtained under summer 1969 conditions. They are likely to change slightly for other seasons and other levels of the solar activity.

Acknowledgment

The authors would like to thank Dr. A. Vidal Madjar of Laboratoire de Physique Stellaire et Planétaire, CNRS, France for making available unpublished satellite Lyman-α data.

References

[1] Solar-Geophysical Data No. 304, Part II; No. 305, Part II; No. 306, Part II, U.S. Department of Commerce, Boulder, Colorado, USA (1969, 1970).
[2] A. Vidal Madjar, private communication (1971).
[3] J. Laštovička, J. Atmos. Terr. Phys. **35**, 815 (1973).
[4] L. G. Meira, J. Geophys. Res. **76**, 202 (1971).

VLF PHASE VARIATIONS ASSOCIATED WITH HF ANOMALOUS WINTER ABSORPTIONS

K. SINNO and T. ISHII

Radio Research Laboratories, Tokyo, Japan

Phase and intensity measurements of NPG, 18.6 kHz transmitted from Jim Creek (48°12′ N 121°55′ W) in USA were carried out at Inubo Radio Wave Observatory, RRL (35°42′ N 140°52′ E) near Tokyo during August 1971—May 1972.

A good coincidence in time was found between VLF irregular phase advances appearing in winter and HF absorption of field intensity of WWV. This may be reasonably explained by the fact that VLF phase advancement depends on the lowering of reflection height in the course of anomalous winter absorption in the HF band.

Further, a significant relation was found in this period between wind velocity in the stratosphere, observed by Japanese meteorological rockets, and the winter anomaly.

A full paper will be published in *J. Radio Res. Labs Japan.*

Lower Ionosphere Structure — Akademie-Verlag, Berlin 1974

EXPERIMENTAL RESULTS OF RADIO WAVE ABSORPTION MEASUREMENTS IN SOUTHWEST EUROPE

G. Rose[a], J. Weber[a], H. U. Widdel[a] and P. Galdon[b]

[a]Max-Planck-Institut für Aeronomie, Institut für Ionosphärenphysik, Lindau/Harz, FRG
[b]Observatorio del Ebro, Roquetas (Tarragona), Spain

Some results of ground-based radio wave A3 absorption measurements performed in Spain are presented and compared with rocket-borne *in situ* measurements of the D region. Because the transmission paths are located close to the southern boundary of the winter anomaly some dynamic features of this effect can be observed more clearly. Periods of very low absorption imbedded in the winter anomaly were observed quite regularly, caused by movements of its southern boundary. With a net of three observation sites, the drift velocity of a large scale winter anomaly front movement was determined independent of *in situ* rocket measurements. Pressure variations derived from foil cloud fall rates correlate positively with absorption. Integral absorption determined from *in situ* guard-ring Langmuir-type probe measurements and simultaneously measured pressure values agreed quite well with the absorption actually observed. The differential absorption profiles show that, under winter anomalous conditions, absorption is increased below 90 km together with deviative components near the reflector. In the examples shown, 50% of the total amount of absorption was located below and above about 90 km in both winter anomalous and "normal" cases.

1. Introduction

In 1967 a close cooperation between the Observatorio del Ebro, Tortosa, the Instituto Nacional de Técnica Aeroespacial (INTA) and the MPI für Aeronomie in Lindau/Harz started to achieve both ground-based and rocket-borne measurements in the ionospheric D region in connection with the problems of winter anomaly of radio-wave absorption.

2. Experimental Arrangements

For the ground-based experiments the well known A3 method was used. A continuous wave transmitter was installed in Aranjuez (40.04° N 3.62° W) equipped with a tuned vertical quarter-wave antenna with ground net radials. The operating frequency is 2830 kHz, the power output 800 watts. At distances between 350 and 425 km from the transmitting site up to three simultaneously operated receiving stations were installed.

At the beginning, two of the transmission paths were orientated as close to the geomagnetically north-south direction as the geographic conditions permitted;

one station was located in Alceda (43.19° N 3.90° W) near Santander in the north, the other in Balerma (36.72° N 2.86° W) near Almeria in the south of Spain. A third receiving station was installed at the Observatorio del Ebro in Tortosa (40.83° N 0.50° E) located approximately east of the transmitter site. The receiving station in the north of Spain was closed down after more than one year of operation in 1968. It was moved later on, 1969, to the rocket launching site El Arenosillo of INTA (37.1° N 6.73° W) close to Huelva in the south west of Spain. Fig. 1 shows the geography of the transmission paths.

At all the receiving stations the same type of equipment is used: five $\lambda/2$ long-wire antennas favouring the ordinary wave $1 \times$ E hop during the day and,

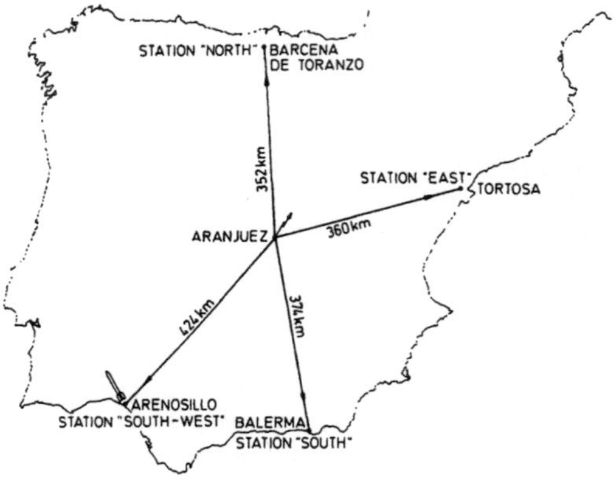

Fig. 1. The geography of the different A3 propagation paths; three of them have been operated at the same time up to now.

if present, the ordinary $1 \times$ Es layer reflections during the night, whereas possible other propagation modes are widely reduced. Night-time Es-layer propagation is used for calibration purposes.

Rhode und Schwarz field-strength meters (type HFH) and constant signal generators are used together with standard attenuators which are mounted at the receiver input sides for calibration purposes and to match the relevant HF antenna outputs to the optimum recording ranges of the equipments, which are normally calibrated once a day to obtain a stability of the order of ± 0.2 dB.

3. Purpose of the Measurements

Since the existence of winter anomaly (WA) of radio-wave absorption in geographical latitudes around 40° N had already been proved [1], one of the main objects of our ground-based absorption measurements in Spain was now the detailed study of winter-anomalous absorption and its variation from year to year close to its southern boundary [2] to gather the background for an optimal timing of rocket-borne *in situ* measurements, which are then supplemented by the ground-based data. In this connection, the question of the best characterizing parameter, which allows comparison of absorption measurements in different places reliably, was also investigated. Further, with the aid of different transmission paths, operated at the same time, drifts of large-scale absorption patterns can be analysed and compared with the results of rocket-borne neutral wind measurements in D region heights in order to verify (and localize) a possible substantial source of anomalous absorption.

4. Results

From our ground-based absorption measurements it can safely be concluded that anomalous winter absorption occurred during all winters in southern Spain (e.g. on the transmission path Aranjuez—Balerma $\phi_m = 38.38°$ N, $\lambda_m = 3.23°$ W) from 1967/68 to the last winter 1972/73, but with rather different durations and intensities in the different years. This is clearly demonstrated by Fig. 2 which presents the mean noon absorption values $\bar{L}_M$ obtained for the intervals between 11 h and 13 h GMT for each day of the different winters.

The homogeneity of winter anomalous events over Spain could be estimated by comparing the absorption parameters from the different transmission paths. A rather close ($r = 0.88$) correlation was found between the noon absorption values $\bar{L}_M$ obtained for the winter months of the transmission paths to the north and to the south of Spain, their midpoints being separated by 360 km. If the daily variation of absorption L is approximated by $L = L_0 \cos^n \chi$, the correlation between the exponents n of these lines was found to be $r = 0.64$, the correlation coefficient between the subsolar absorption values 0.73, but between the ratios $L_0/(n + 1)$, 0.90. In all these cases solar zenith angles smaller than 80° only were taken into account in the calculation of L_0 and n. The ratio $L_0/(n + 1)$ is the integral $\int_0^1 L(\cos \chi)\, d(\cos \chi)$ and therefore a parameter which characterizes the absorption of a whole day but with elimination of the dependence of absorption on the position of the sun and on the different durations of sun irradiation at different places. After normalizing to vertical incidence and standard frequency conditions, if considered necessary, $L_0/(n + 1)$ can readily be used for a comparison of observations which were obtained on different frequencies and at places of different geographical latitudes and longitudes.

Fig. 3 shows the seasonal variations of the parameters n, L_0, $L_0/(n + 1)$ and that of a parameter r for the transmission path Aranjuez—Balerma in the south of Spain. (r is the correlation coefficient between log (cos χ) and log L which indicates how well a $\cos^n \chi$ law is fitted for different days.) As can be seen from Fig. 3, WA as observed by our ground-based radio-wave absorption measurements in southern Spain is characterized by rather high winter values of subsolar (L_0) and integrated absorption $L_0/(n + 1)$. The rather strong fluctuations of n from day to day are superimposed by a distinct slow annual (and semi-annual) variation. (It is interesting to see that the big solar event, which started at the beginning of August 1972, looks relatively small in this presentation when compared with the effect of WA absorption.)

Before the ground-based absorption measurements are compared with some of the results of our rocket-borne experiments, the following result should be mentioned. One of the salient features which we observed in our absorption measurements was periods of days imbedded into the winter anomaly which displayed a very low noon absorption (see Fig. 2, filled areas). In most cases, the lowest absorption of the whole year was measured then. After such periods, winter

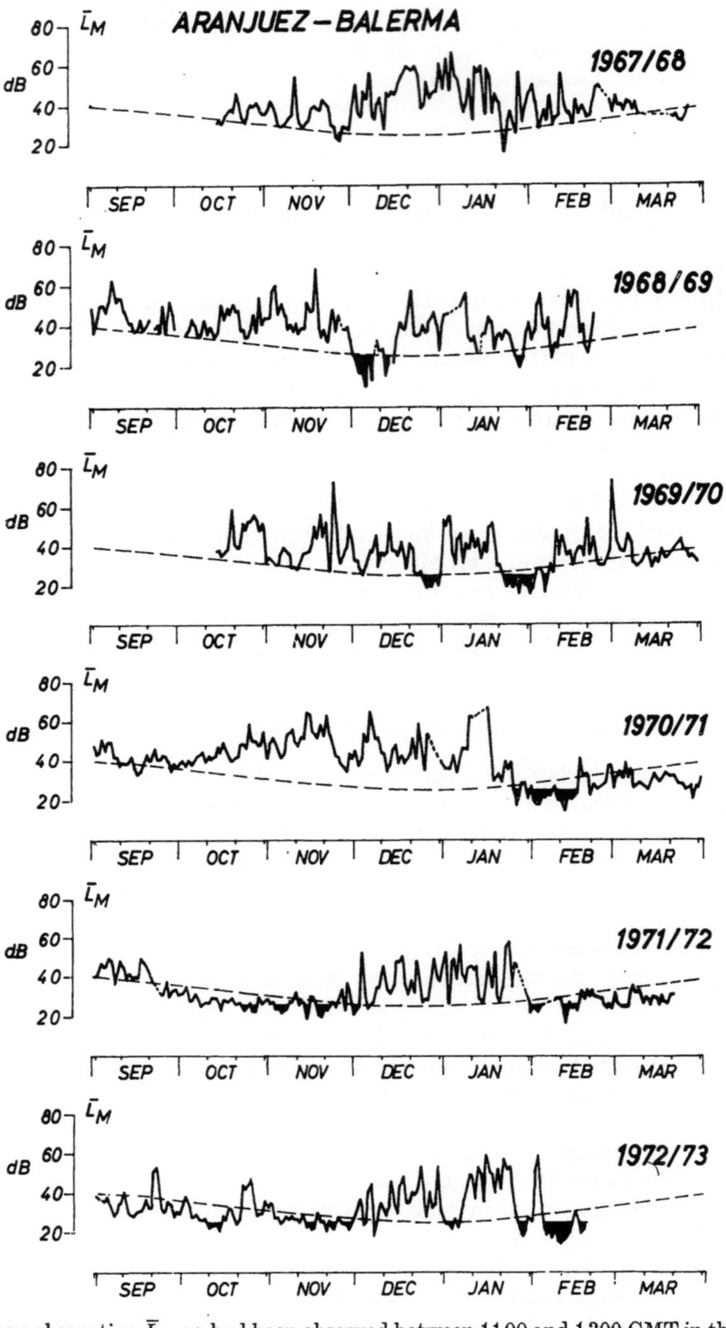

Fig. 2. Noon absorption $\bar{L}_M$ as had been observed between 1100 and 1300 GMT in the different winters in Spain.

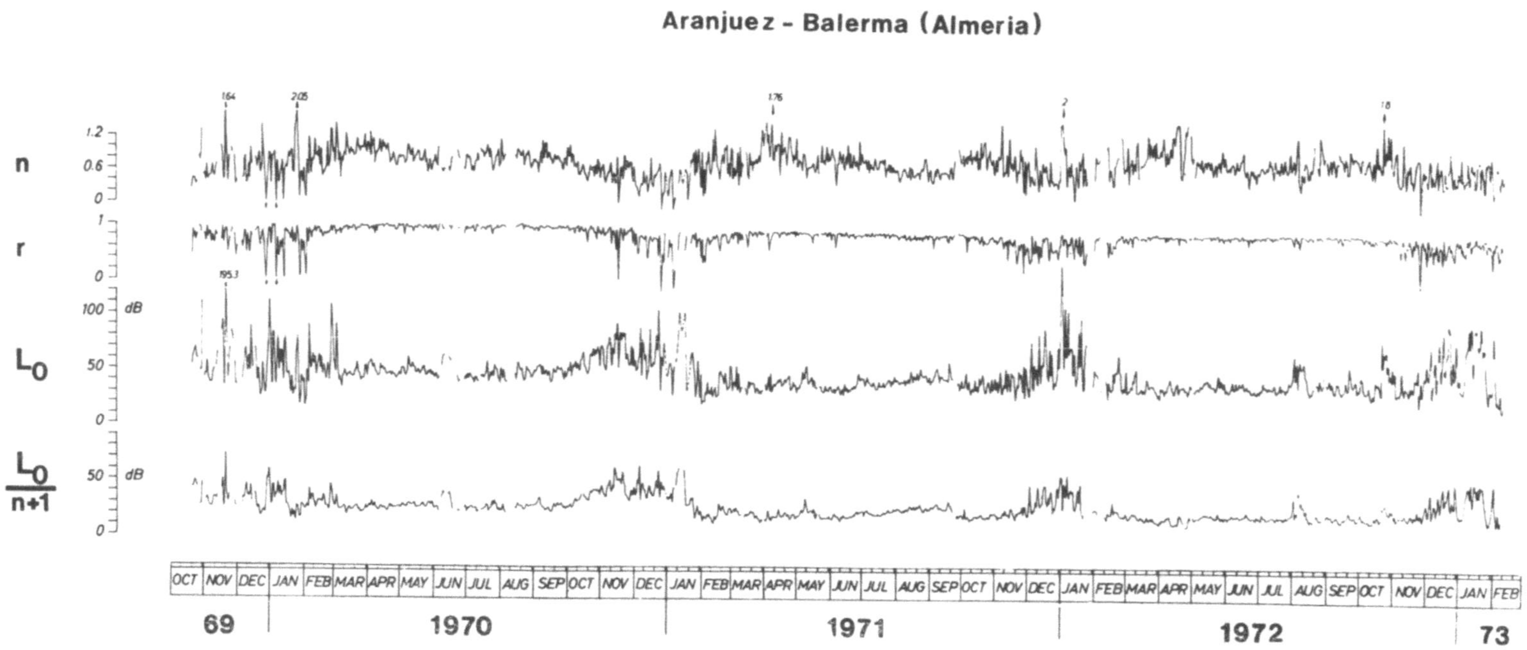

Fig. 3. In this figure the daily absorption in Spain was approximated by $L = L_0 \cos^n \chi$. The annual variations of the parameters n, L_0 and of $\frac{L_0}{n+1} = \int_0^1 L d(\cos \chi)$ are presented together with the daily correlation coefficients r between the log L and the log $(\cos \chi)$ values indicating how well a $\cos^n \chi$ law is fitted.

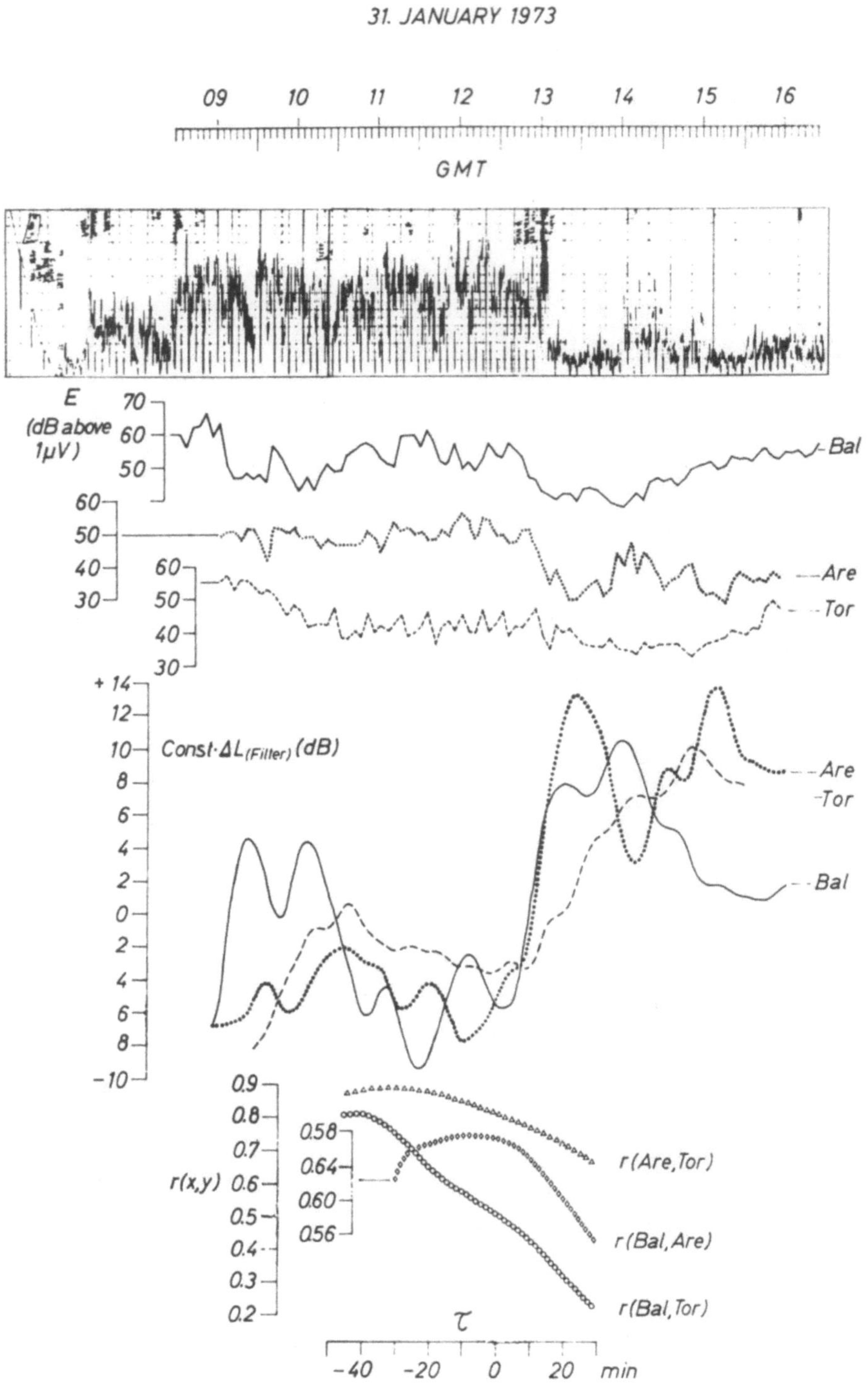

Fig. 4. Field strengths E and filtered absorption values (const ΔL) on 31 Jan. 1973 at 3 stations in Spain as observed during an onset of winter anomaly in the day together with the relevant cross correlation coefficients $r(x, y)$. An original field strength record of the Arenosillo receiving station is shown in the upper part.

anomaly reappeared, or, rather, normal absorption values were observed afterwards. Since our earlier investigations suggested the existence of connections between winter anomaly and transport processes, it was hoped that such a change from low to higher absorption might take place during daytime. This would allow us to determine the speed of movement, the direction and the depth of the front of absorption independent of rocket-borne measurements using the time differences of the arrivals of the effect at our three receiving sites. On 31 January 1973 such an event was observed clearly when, after a period of very low absorp-

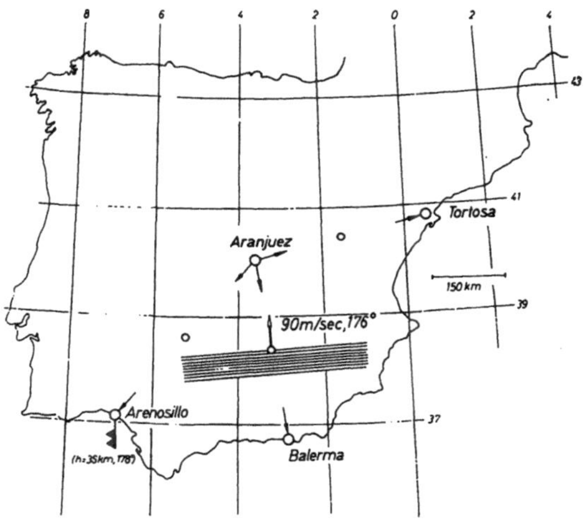

Fig. 5. Schematic diagram illustrating the movement of the large scale absorption pattern over Spain as deduced from the observations shown in Fig. 4.

tion, WA started again in Spain around noon that day and lasted for the following two days, the noon absorptions of which were 52 and 58 dB respectively (cf. Fig. 2). This was the only case up to now when WA came up right in the day over Spain. This event could be followed at all the three stations and therefore be analysed with respect to the drift of that large scale absorption pattern. Figs. 4 and 5 show the results of the analysis.

The five-minute field-strength values of the three transmission paths to Balerma (Bal) to Arenosillo (Are) and to Tortosa (Tor) are shown in the upper half of Fig. 4 together with an original recording of the Arenosillo receiving station. The field-strengths were converted into $\Delta L \; (= L - L_0 \cos^n \chi)$ values and were then filtered, using a numerical $m = 10$ binomial low-pass filter which has its bandwidth of 50% at $T = 43$ min and of 10% at $T = 24$ min. The filtered curves of the three stations are shown in the middle, and the relevant cross correlation coefficients in the lower part of Fig. 4. The latter can be associated with an absorption pattern having a (rather) plain front, which is moving with a velocity of about 90 m s^{-1} to the north as is indicated in Fig. 5. The total increase of absorption was of the order of 20 dB at Arenosillo, the corresponding spatial depth of the moving front of the order of 500 km in the south of Spain. It became more elongated as it moved to the north. This agrees well with what we have found before when we compared our ground-based absorption and rocket-borne wind measurements [3], which have shown that air mass transport phenomena and WA are related with each other at certain height levels of the D region. Unfortunately, our rocket-borne wind measurements of that day did not yield results at D region heights. The only data point obtained at a height of 36 km that day, however, indicated a strong wind of about 100 m s^{-1} from the south. Nevertheless neutral winds of up to more than 90 m s^{-1} have been found before at D region heights over Arenosillo [4]. Two days before, for instance, on 29 Jan-

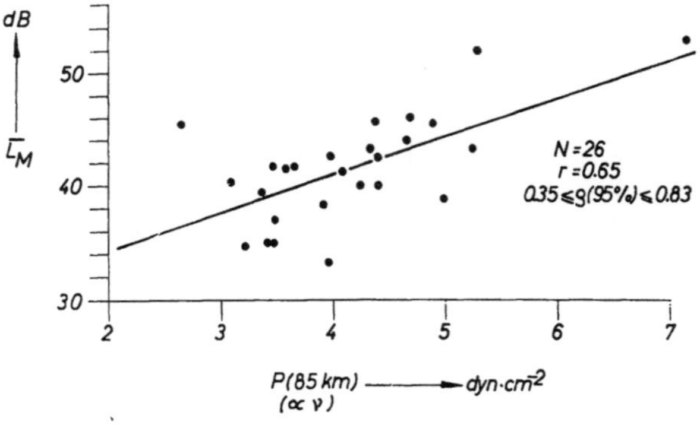

Fig. 6. Experimental relation between air pressure P ($h = 85$ km) at noon over Arenosillo and noon absorption $\bar{L}_M$ on the Aranjuez—Arenosillo path, as obtained between Dec. 1969 and June 1972.

uary 1973, we measured neutral winds from southern directions at heights between 79 and 87 km at a speed of up to 70 m s^{-1}, supporting the hypothesis that a substantial transport was involved when high radio-wave absorption started to cover Spain during 31 January 1973. Furthermore, we compared radio-wave absorption with air pressure values [5, 6] which were derived from the fall rates of our chaff clouds at D region heights. A not very high, but nevertheless significant correlation was found between absorption and neutral air pressure as shown by Fig. 6. The regression line indicates that air pressure, being proportional to electron collisional frequency [7], has a clear but only moderate influence on radio-wave absorption on the average, if the dependence of absorption on electron density is not considered at the same time.

Finally, electron concentrations were derived from the probe currents of our guard-ring probe and converted into radio-wave absorption by additionally using the collisional frequencies which were deduced from the simultaneously measured chaff fall rates. The calculation of differential absorption $(\Delta L/\Delta h) \sec \phi$ was performed with the aid of the Sen-Wyller theory for the transmission path Aranjuez—Arenosillo. Three examples are shown in Fig. 7; two reference days, representing normal conditions, are compared with a winter anomalous day. The profile obtained under winter anomalous conditions, with a total absorption of 54 dB, shown in the right part of Fig. 7, also displays the highest absorption of 25 dB up to a height of 90 km, whereas the other two profiles with a total absorption of 34 dB and 37 dB exhibit only 14 and 18 dB up to the 90 km reference level respectively. This shows that during WA, even if it is observed under

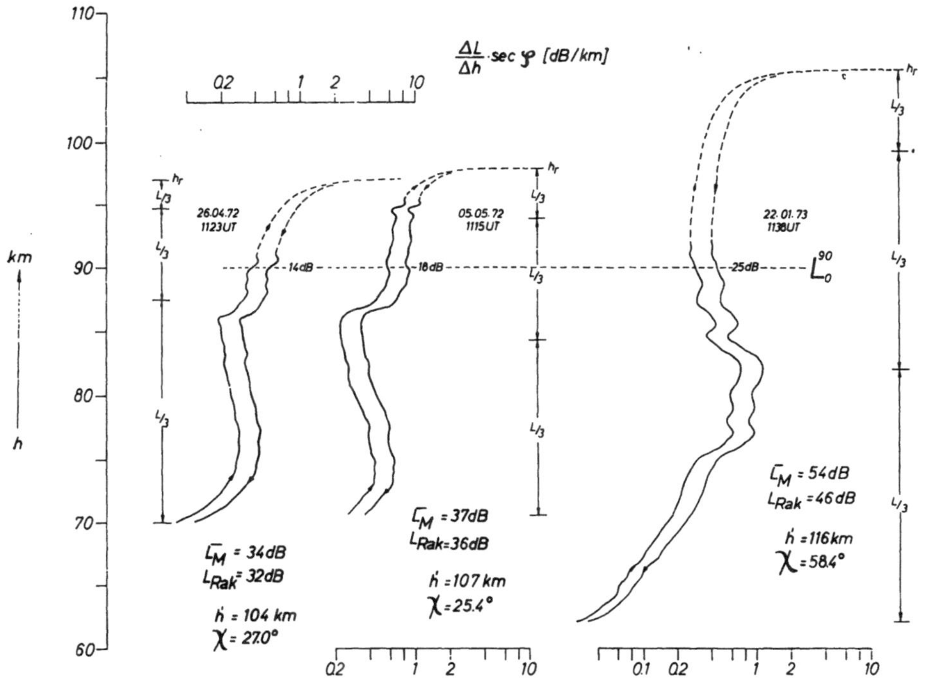

Fig. 7. Differential absorption along the Aranjuez—Arenosillo transmission path on different days as deduced from our rocket-borne electron density and air pressure measurements. The calculated integral absorption L_{Rak} may be compared with the absorption $\bar{L}_{\mathrm{M}}$ actually measured.

conditions of rather high deviative absorption along our transmission paths in Spain, as in the one case shown by Fig. 7, a considerable amount of anomalous absorption other than deviative is located below 90 km.

Acknowledgments

The work was sponsored by research grant WRK 90 of the Bundesminister für Forschung und Technologie. We are grateful to Professor Dr. W. Dieminger for his support and advice and to all of our colleagues who contributed to this project.

References

[1] W. Dieminger, G. Rose and H. U. Widdel, J. Atmos. Terr. Phys. **28**, 317 (1966).

[2] W. Dieminger, G. Rose, H. Schwentek and H. U. Widdel, Space Research VII, 228 (1967).

[3] G. Rose, H. U. Widdel, A. Azcarraga and L. Sanchez, Radio Sci. **7**, 181 (1972).

[4] G. Rose, H. U. Widdel, A. Azcarraga and L. Sanchez, Phil. Trans. Roy. Soc. Lond. A **271**, 509 (1972).

[5] G. Rose and H. U. Widdel, Planet. Space Sci. **20**, 877 (1972).

[6] G. Rose and H. U. Widdel, Planet. Space Sci. **21**, 1131 (1973).

[7] E. A. Mechtly, S. A. Bowhill, L. G. Smith and H. W. Knoebel, J. Geophys. Res. **72**, 5239 (1967).

Lower Ionosphere Structure — Akademie-Verlag, Berlin 1974

IN SITU MEASUREMENTS OF ELECTRON CONCENTRATION, NEUTRAL WIND AND AIR PRESSURE COMPARED WITH THE WINTER ANOMALY

W. Dieminger, G. Rose and H. U. Widdel

Max-Planck-Institut für Aeronomie, Institut für Ionosphärenphysik, Lindau/Harz, FRG

The results of some rocket-borne *in situ* measurements in the D region of the ionosphere, made over El Arenosillo (37.1° N 6.73° W) in the height range between 95 and 70 km are presented. The measurements were taken in different seasons. Electron concentration, air density, wind direction and wind speed were measured and the measurement of atomic oxygen was attempted. It is shown that the winter-anomalous state of the D region is characterized by the presence of westerly winds. In reference to electron profiles gained under very calm (spring) conditions, the winter profile displays a depletion of electron content increasing exponentially with height. Around 70 km, the electron concentrations are about the same in the reference season and in winter. Electron profiles gained under winter-anomalous conditions differ from those for winter days with very low absorption in that they show a bulge of enhanced electron concentration between 75 and 85 km. The upper and the lower height level of the bulge corresponds to heights in which correlations between absorption changes and neutral air mass winds were found.

During days of very low winter-time absorption, a characteristic feature in the time variation of winter-anomalous absorption in latitudes near the southern boundary of this effect, a change in the wind direction from westerlies to easterlies was observed in the height region below 82 km. An electron profile gained during an SID event shows a large increase of electron content at all height levels, compared with a profile taken in the same month under normal but winter-anomalous conditions. In reference to that profile, the relative enhancement of electrons under SID conditions shows, up to the height limit to which we could see electrons in the probe current records, an exponential increase with decreasing height, with a superimposed bulge of enhancement between 87 and 79 km its maximum to lie around 85 km, believed to be caused by increased Ly α radiation.

Between April and July and in October 1972 and in January 1973 a number of rocket-borne experiments were performed in cooperation with the Instituto Nacional de Técnica Aeroespacial Madrid over the Spanish national rocket range El Arenosillo (37° 6′ N, 6° 44′ W). Throughout the year, ground-based absorption measurements over three propagation paths using the A3-method, and ionosonde measurements were made [1]. Small rockets of the Skua 2 type were used throughout. These rockets reach an apogee of about 100 km with a 6 kg payload and about 88—95 km with a 7 kg payload. Two types of payload were used: A foil-cloud experiment for measuring wind speed, wind direction, air density and temperature [2—4] which had been used successfully in earlier experiments [5—7] and, in smaller numbers, a telemeter payload which contained, in addition to the foil-cloud experiment, a guard-ring Langmuir-type probe at the tip of

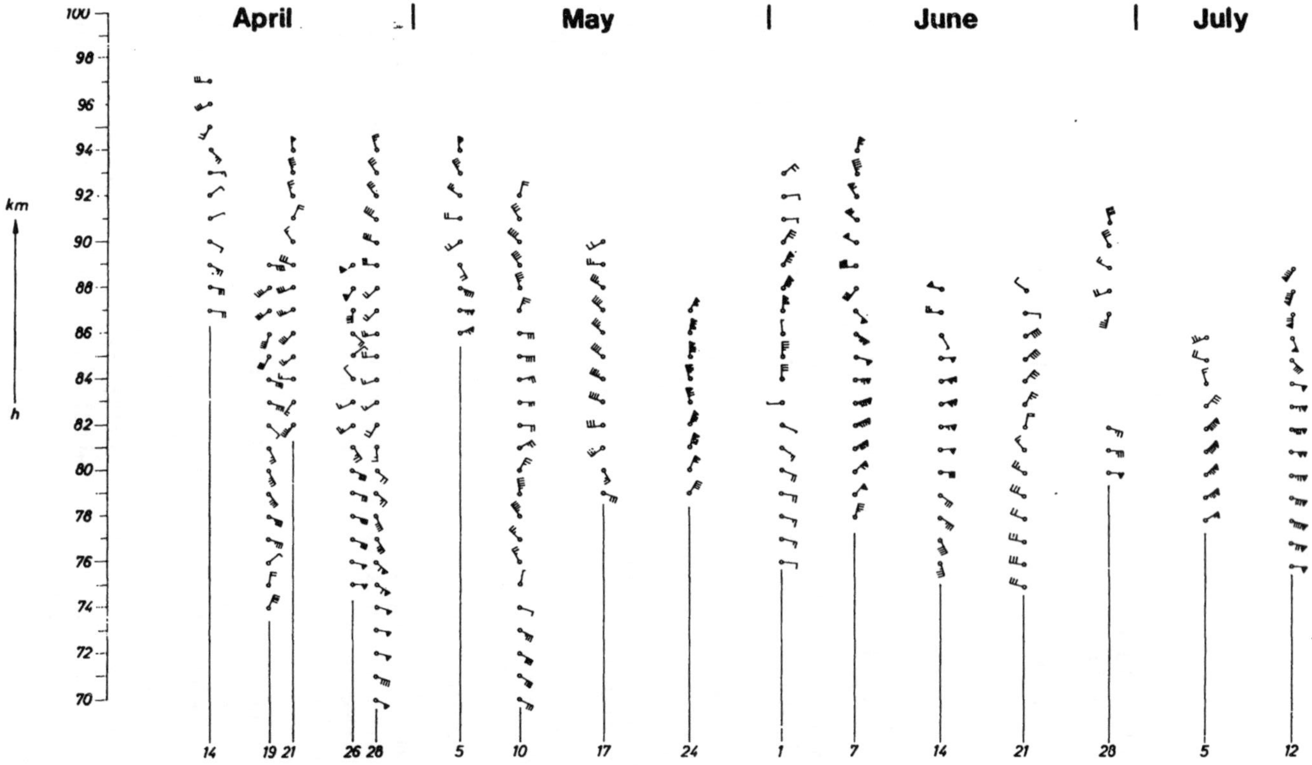

Fig. 1. Results of measurements of wind direction and wind speed over Arenosillo summer 1972. All rocket launchings were performed around local noon. The varying height coverage is caused both by differences in rocket performance and by turbulent motions which cause a wide spread of the foil cloud. This spread impairs the radar tracking, which was then considered to be unreliable, and was therefore not used for data reduction.

Geographical North: top of figure. Legend for wind arrows: short vane: 5 m s^{-1}, long vane: 10 m s^{-1}, solid triangle: 50 m s^{-1}. Wind arrows are drawn from where the wind is coming.

the rocket's nose-cone for the measurement of ions and electrons. Some of these payloads carried prototypes of silver-film probes for an attempt to measure the height profile of atomic oxygen. For the detection of the reaction between the oxygen and the silver, the change of film transmission for the infrared radiation of GaAs diodes was used since laboratory experiments seem to promise some advantage over the resistance measurement originally suggested by Henderson and Schiff [8].

As was done in the previons campaigns, all launches were made close to local noon in order to obtain a homogeneous time series of density and wind measure-

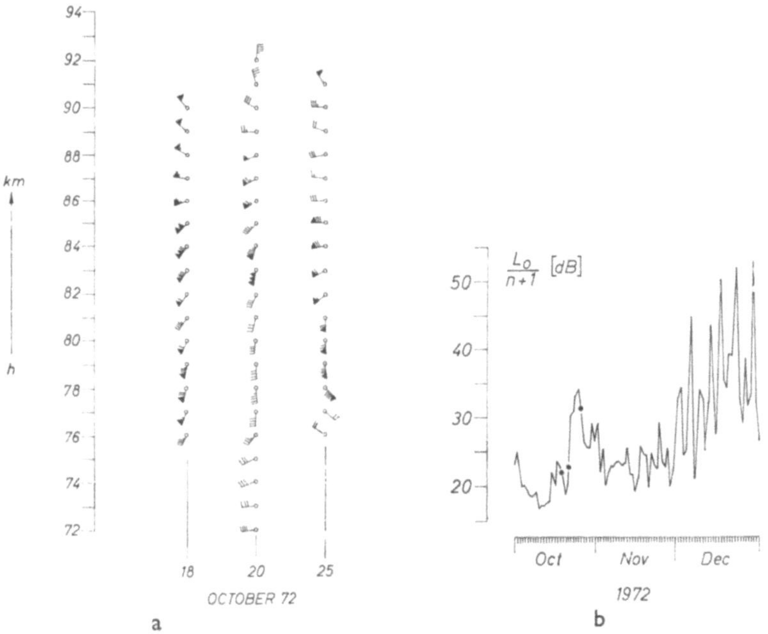

Fig. 2a. Wind profile October 1972.

Fig. 2b. Results of ground-based absorption measurements for the period October to December 1972 (see also [1]) which indicate that, in the period when *in situ* measurements were taken, the atmosphere was temporarily in a winter-anomalous state (filled dots: launch dates).

ments. The first telemetered payloads were launched in this campaign in April 1972, because the results of the ground-based absorption measurements had shown that, during the years observed, the least day-to-day fluctuations were observed during that period. This fact was considered to indicate very stable conditions. The results of our earlier wind measurements described in [7] had a strong resemblance to those gained earlier at lower heights with rocket experiments made over Sardinia in 1963/64 by Müller, Attmannspacher and Schurer [9]. From this analogy it seemed not improbable that we might find winds of varying direction with zonal components predominantly coming from the east and with low wind speeds during this season of calmness in the height levels relevant to winter anomaly, indicating the transition from winter-type to summer-type

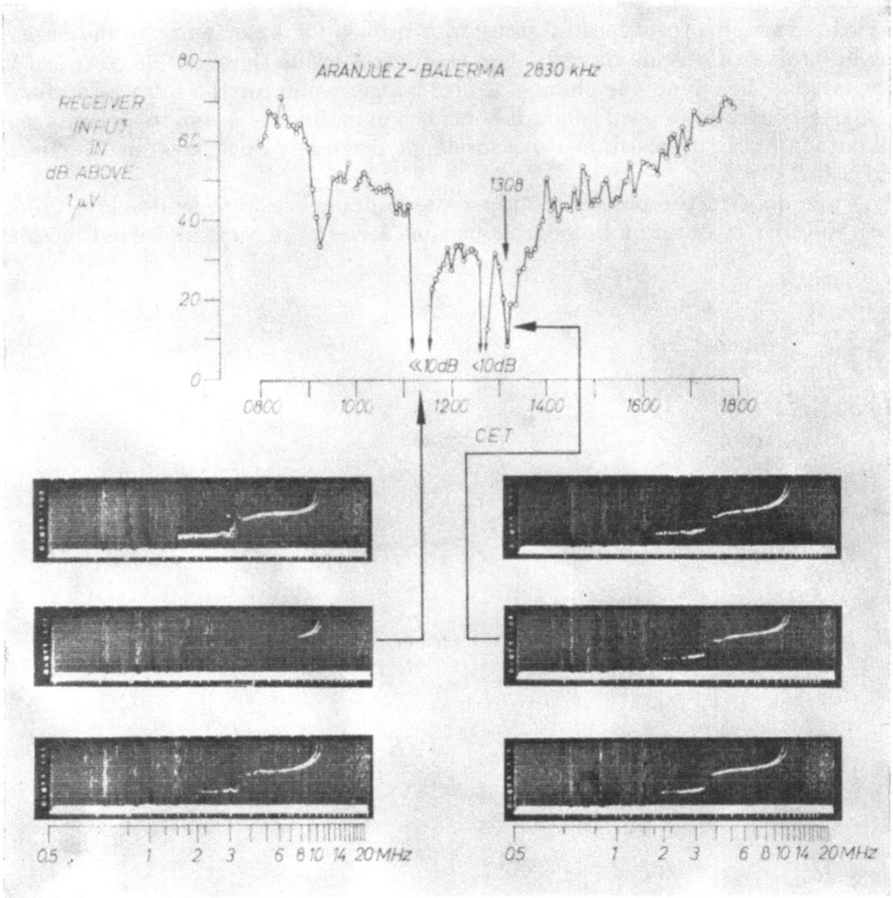

Fig. 3. 25 October 1973 SID event: receiver input voltage records of absorption measurement and ionograms taken on the launching site "El Arenosillo".

circulation. As can bee seen in Fig. 1, this was the case to a certain extent. The small day-to-day fluctuations of absorption allowed us to collect a set of probe current profiles which could be used as a reference. They allowed us to derive an empirical conversion function for the conversion of the measured probe current into ambient electron concentration. For absolute calibration, ionosonde E-layer critical frequencies were determined on the range simultaneously during the rocket's flights. Using the simultaneously measured air pressure data (which correlate positively with absorption), the electron profile was converted then into absorption [1]. Because complete current/probing voltage characteristics are measured with the probe [10], it is immediately seen from the records in which regime the probe operates. In five flights five profiles were gained. These profiles were considered to represent the "normal" state of D region electron profiles and were used as references.

In October 1972, three more flights were launched in order to obtain some information about the transitional behaviour of the atmosphere and to make a test flight of the first version of oxygen probes. Unfortunately, the probe electrometer failed in one flight. Therefore we obtained only one profile under "normal" conditions. This profile, however, had, in comparison with the "reference" profiles, already some "winter" characteristics. The wind profile contained westerlies on all three days. The ground-based absorption data confirmed that the atmosphere was indeed at that time for a short period in a winter-anomalous state (see Figs. 2a, 2b).

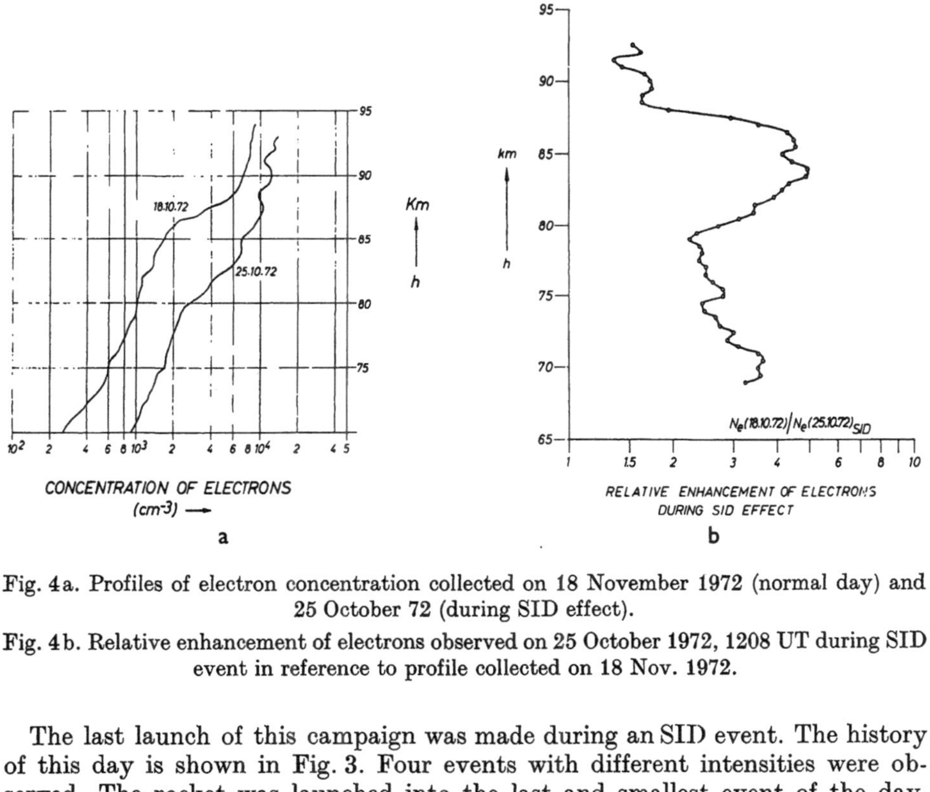

Fig. 4a. Profiles of electron concentration collected on 18 November 1972 (normal day) and 25 October 72 (during SID effect).

Fig. 4b. Relative enhancement of electrons observed on 25 October 1972, 1208 UT during SID event in reference to profile collected on 18 Nov. 1972.

The last launch of this campaign was made during an SID event. The history of this day is shown in Fig. 3. Four events with different intensities were observed. The rocket was launched into the last and smallest event of the day, barely discernible in the ionograms but well recognizable on all absorption records investigated. A large increase of electron concentration, especially below 90 km, was observed in the flight (see Fig. 4a). In reference to the profile gained under "normal" conditions on 18 November, the relative enhancement of the electron concentration during the SID displays an exponential increase with decreasing height, indicating a hard, penetrating (X-ray) radiation, which is superimposed upon a bulge of enhanced electron concentration centred around 85 km (Fig. 4b). This level had been identified earlier by the correlation analysis of wind and density data gained in our rocket experiments [7] to contribute to winter anomaly. It is tempting to assume that this bulge of increased electron concentration was caused by a layer of minor constituents, normally contributing to the winter anomaly effect, which was more intensely ionized by the enhanced lower-energy radiation associated with the SID effect.

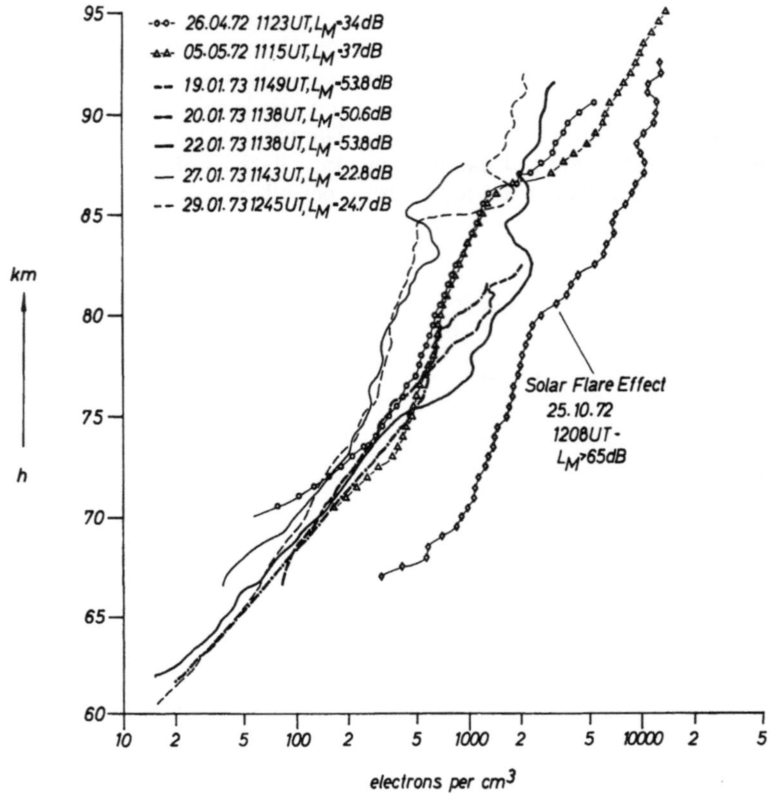

Fig. 5a. Examples of noon-time profiles of electron concentration measured over Arenosillo in April/May 1972 and January 1973 (the two April/May 72 profiles were used as reference profiles for Fig. 6a): under conditions of low winter-time absorption a depletion in electron content below 90 km is observed.

In January 1973, nine more rockets were launched with telemetered payloads from the middle to the end of that month. (One of these flights was an experimental flight to test a boosted-up version of the Skua 2 (Skua 4)). The reason for the choice of that period was suggested by the results of our ground-based absorption measurements: They show that, during that period, the winter anomaly is still present over Arenosillo and is of high intensity. On the other hand, towards the end of this month until the middle of February, we observed, throughout the years, the onset of a comparatively stable (in reference to datum recurrence) more or less extended period of very low absorption, its length probably dependent upon solar activity. During that time, we measure the lowest absorption of the year. From the results of our wind and density analysis which we had made earlier [7] we expected to find in this period a depletion of the electron concentration below 90 km together with a change from the winter-time westerly to easterly (summer-like) circulation in the same height region. Furthermore, it was of interest to us to see if the electron profile for the state of very low absorption is more similar to a summer-time reference profile with its (relatively to the winter profile) larger increase of electron concentration with height above about 87 to

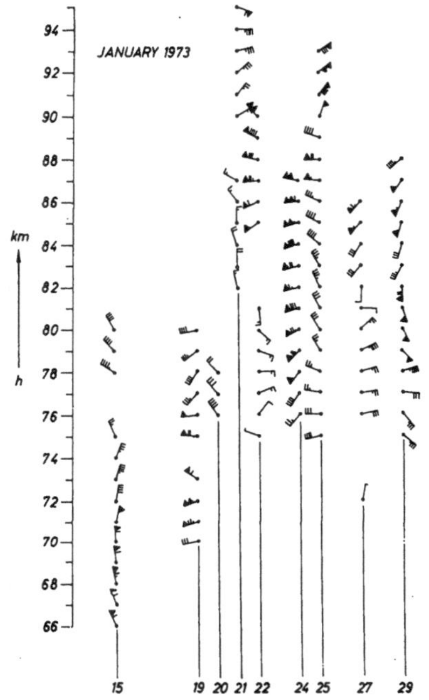

Fig. 5b. Wind profiles observed in January 1973. Westerlies are predominant in height levels below 90 km. During the days of very low absorption, a change from westerlies to easterlies below 82—83 km was observed (on 22 January a similar change in direction was observed too but the wind speeds were about one-half or less of those observed on the days of low absorption (27 and 29 January)).

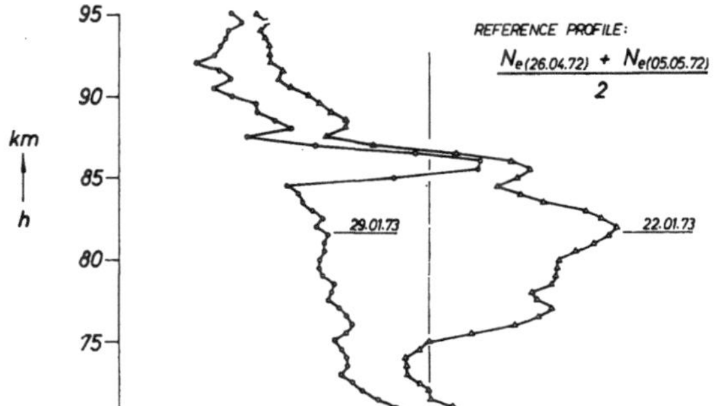

Fig. 6. Height variations of relative enhancement of electron content in reference to profiles collected under calm conditions (April/May) for a winter-anomalous day and a day with very low absorption. The profiles of the two days differ mainly in the extent of the bulge of enhanced electron content between 85 and 75 km. For the height levels around 85/86 km and 75/78 km, correlations between wind variation and absorption changes had been found earlier. These levels had been identified also to be preferred heights of "deep echoes".

88 km, or if it keeps the lower gradient of electron concentration increase considered to be typical for winter conditions. Unfortunately, some of the Bantam 2 motors of the Skua 2 had a somewhat reduced performance. This was the reason that the goal to reach an apogee of 95 km regularly in all launches of our last campaign was not achieved, but the height region below 88 km was well covered. In Fig. 5a, some of the profiles gained under winter-anomalous and "anomalously low" absorption, together with two reference profiles collected in April/May 1972 are presented (in order to keep the diagram more readable, not all reference and winter profiles are shown in the diagram). One notes that the main increase of electron content in reference to April/May is found below 85 km. The depletion of electron content under anomalously low absorption was also found to lie below 90 km. As was expected, the change in the wind direction from westerlies to easterlies was found also in the relevant height region (Fig. 5b). When the winter profiles are normalized against an averaged April/May reference profile, it is seen that both profiles, the "winter-anomalous" and the profile of a winter day with very low absorption, sometimes considered to be "normal" days, have in common the slope of the height variation of the log $N_e/N_{e\,\mathrm{ref}}$ values (Fig. 6). They differ mainly in the extension of the bulge of enhanced electron concentration between 75 and 85 km. Below 70 km the electron concentration seems to be roughly equal to that of other seasons, but it looks as if electrons were observed in the current vs probe voltage characteristics of our probe down to lower heights during winter, even under "anomalously low" absorption conditions.

Acknowledgments

The investigations were sponsored by grant WRK 90 of the Bundesminister für Forschung und Technologie for which we express our sincere thanks.

We are most grateful to all of our Spanish colleagues of CONIE and INTA who made the programme possible and thank the crew of El Arenosillo who launched the rockets so successfully.

References

[1] G. ROSE, J. WEBER, H. U. WIDDEL and P. GALDON, this volume, p. 331.
[2] G. ROSE, H. U. WIDDEL, A. AZCARRAGA and L. SANCHEZ, Phil. Trans. Roy Soc. Lond. A **271**, 509 (1972).
[3] G. ROSE and H. U. WIDDEL, Planet. Space Sci. **20**, 877 (1972).
[4] G. ROSE and H. U. WIDDEL, Planet. Space Sci. **21**, 1131 (1973).
[5] A. AZCARRAGA, L. SANCHEZ and H. U. WIDDEL, Space Research X, 174 (1970).
[6] A. AZCARRAGA, L. SANCHEZ, G. ROSE and H. U. WIDDEL, Space Research XII, 613 (1972).
[7] G. ROSE, H. U. WIDDEL, A. AZCARRAGA and L. SANCHEZ, Radio Sci. **7**, 181 (1972).
[8] W. R. HENDERSON and H. I. SCHIFF, Planet. Space Sci. **18**, 1527 (1970).
[9] H. G. MÜLLER, W. ATTMANNSPACHER and M. SCHURER, Deutsche Luft- und Raumfahrt Forsch.-Ber. p. 66 (1966).
[10] G. ROSE, H. U. WIDDEL, A. AZCARRAGA and L. SANCHEZ, Planet. Space Sci. **20**, 871 (1972).

Lower Ionosphere Structure — Akademie-Verlag, Berlin 1974

D REGION ELECTRON DENSITY VARIATIONS DURING WINTER 1972—73

S. A. BOWHILL and B. W. DENNY

Aeronomy Laboratory, Department of Electrical Engineering, University of Illinois, Urbana, Ill., USA

D region electron densities were measured at Urbana, Illinois (latitude 40.2° N) by the partial-reflection technique during winter 1972—73. Electron densities at altitudes of 72, 76.5 and 81 km show cyclical changes with a period of about 5 days that are highly correlated between these altitudes. This suggests that the same mechanism produces enhancements of electron density throughout this height range.

1. Introduction

Most studies of the winter variability of the D region have been made using measurements of the absorption of a high-frequency radio wave reflected from the ionosphere, either using vertical-incidence pulsed transmissions (the A1 technique) or using oblique-incidence cw transmissions over a path sufficiently long for the ground wave to be neglected (the A3 technique). In both these types of measurement, the primary source in the variability of absorption is in the relatively high electron densities occurring above about 82 km; little attention has been paid to the winter variability of the ionization below that altitude. This paper presents the results of a series of measurements during winter 1972—73 in this lower-altitude region at Urbana, Illinois.

2. Experimental Arrangement

The electron-density measurement system used was basically that described by Belrose and Burke [1], and consisted in a measurement of the differential absorption of ordinary- and extraordinary-mode circularly polarized signals scattered from ionized irregularities in the D region. These signals were radiated as 20 μs pulses at 2.66 MHz frequency from an array of 30 half-wave cross dipole antennas, and received on a similar arrangement, after being partially reflected from the D region. The received signal was digitized at 100-kHz rate and processed in real time by a Digital Equipment Corporation PDP-15/30 digital computer. After passing through the receiver, the pulse width was 25 μs, so that the height resolution of the measurement was limited to 3.8 km. Accordingly, results in this paper are presented at 4.5-km intervals.

Signal amplitudes for the ordinary and extraordinary components were recorded for a 1-hour period around local noon on each day with various receiver gains, and were integrated using a noise-defeating algorithm which will be described in detail elsewhere. The ratio A_x/A_0 of the amplitudes of the extraordinary and ordinary signals as a function of altitude provided the differential absorption variation with altitude. This was then combined in the usual way with a model collision-frequency profile to give the electron-density profile using the Sen-Wyller generalization of the Appleton-Hartree magnetoionic theory.

3. Results and Discussion

Measurements of the kind described were made around local noon from 28 November 1972. Results for the month of January 1973 are shown in Fig. 1 for electron densities at 72, 76.5 and 81 km. These are to be interpreted as averages over a 4-km slab centered on the altitude specified. The dashed sections of the 81-km curve indicate that the differential absorption produced was too high to be measured accurately at the frequency used; on those days simultaneous measurements at a higher frequency would have been helpful.

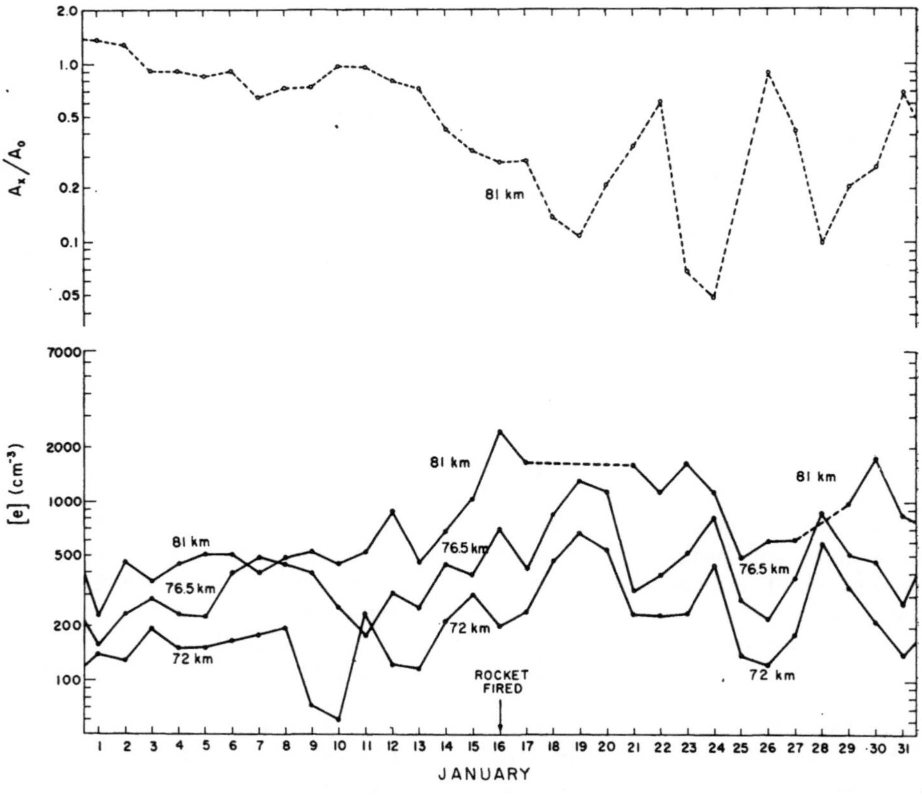

Fig. 1. Electron densities and amplitude ratio over Urbana, Illinois, in January 1973.

Also shown is the A_x/A_0 ratio at 81 km, the logarithm of which represents the total differential absorption occurring below the altitude of 81 km, thereby excluding the region responsible for the normal winter variability in high-frequency vertical-incidence absorption. A high value of this ratio corresponds to a low electron density, and vice versa.

For the first half of January, the A_x/A_0 ratio was reasonably well behaved. Beginning about 13 January, however, and continuing through the remainder of the month, a very pronounced wave-like variation developed with a period of about five days, apparently associated with a major stratospheric warming that developed in that period. An examination of the electron density versus time curves at the three altitudes shows the same cyclical variation, by essentially the same ratio (as much as a factor of 5) at each altitude. No pronounced tendency was

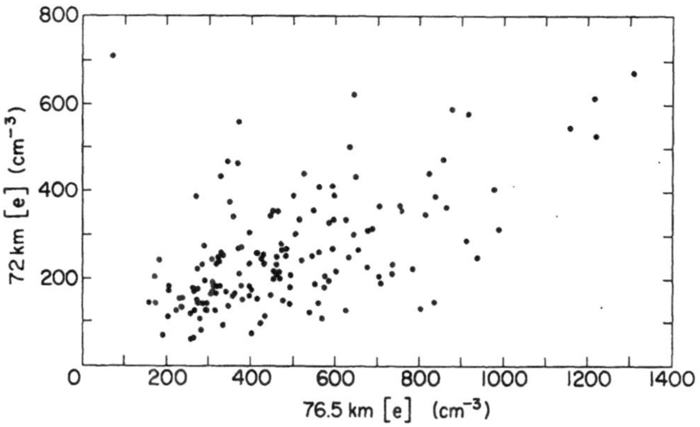

Fig. 2. Scatter plot of noon electron densities at 72 and 76.5 km altitude over Urbana for each day in winter 1972—73.

visible for the electron-density enhancements at the upper altitudes either to precede or to follow those at the lower altitudes. Evidently, whatever mechanism produces these electron-density fluctuations acts simultaneously throughout the entire height range 72—81 km.

This conclusion is reinforced by Fig. 2 which displays a scatter plot of the electron densities at 72 km against those at 76.5 km altitude for each day at noon for the entire period from 28 November 1972 throughout the remainder of the winter. A fairly strong correlation is seen between enhancements in electron density at these two altitudes throughout the entire winter season.

Acknowledgment

Research described in this paper was supported by the National Aeronautic and Space Administration under Grant NGR 14-005-181.

References

[1] J. S. Belrose and M. J. Burke, J. Geophys. Res. **69**, 2799 (1964).

Local Fluctuations and Waves

Lower Ionosphere Structure — Akademie-Verlag, Berlin 1974

RADAR MEASUREMENTS OF GRAVITY WAVES IN THE LOWER IONOSPHERE

R. F. Woodman

Radio Observatorio de Jicamarca, Instituto Geofísico del Perú, Lima, Perú

Some characteristic parameters of gravity waves at D and E region heights such as amplitude, frequency spectrum and spatial properties can and have been measured by means of VHF radars. One of the techniques makes use of ionization trails produced by meteors as tracers of the neutral atmosphere motions in the region from 80 to 100 km. A recently developed technique at the Jicamarca Radio Observatory uses background enhanced fluctuations in the dielectric properties of the medium as tracers of the neutral motions in the region between 60 and 85 km. Profiles of the velocity field can be obtained continuously with a time resolution of the order of a minute allowing observations of gravity waves from its lowest periods, of the order of a few minutes, to tidal periods. Gravity waves in the lower F region have also been detected using incoherent scatter techniques. The methods, potential and limitations of the above mentioned techniques are reviewed as well as some of the results obtained.

1. Introduction

We briefly review existing radar techniques capable of observing gravity waves at the D and E region heights of the earth's upper atmosphere, and include some of the results obtained using these techniques.

Free electrons existing at D and E region heights can be used as tracers of neutral atmospheric phenomena. The electrons are strongly coupled to the neutral gas, either directly through electron-neutral collisions, or, indirectly, through Coulomb interaction with the ions, the ions in turn being coupled to the neutral background through ion-neutral collisions. These free electrons are capable of producing radar echoes as reflections or from scattering of the illuminating electromagnetic wave. Reflections are produced whenever the electron density is sufficiently high to have a plasma frequency equal to or higher than the frequency of the radar wave. This is the case with dense meteor trails. Scattering is produced by fluctuations in the electron density. If we consider these fluctuations as the Fourier superposition of sinusoidal variations in electron density with different wavelengths, it can be shown that only one of the Fourier components contributes to the scattering of the electromagnetic wave in a given direction. The Fourier component responsible for the scattering is that which has a wave vector $\boldsymbol{k}$ equal to the wave vector of the illuminating wave $\boldsymbol{k}_i$ minus the wave vector of the scattered wave $\boldsymbol{k}_s$, i.e. $\boldsymbol{k} = \boldsymbol{k}_i - \boldsymbol{k}_s$. For the particular case of backscattering the fluctuation component responsible for scattering is in the direction of incidence and has a wavelength equal to half the wavelength of the illuminating wave. It can also be shown that the dynamical properties of the scattered wave, for instance its spectral shift and spectral width, depend on the dynamics of the above-mentioned fluctuation wave component responsible for the scattering, i.e. on the collective behaviour of the electrons, rather than on the dynamics of each independent scattering electron.

Electron density fluctuations will always exist in a plasma because of the discrete nature of the electrons and ions. These fluctuations are at a minimum level when the plasma is in thermodynamic equilibrium. The incoherent scatter technique, which has been used so successfully for the study of ionospheric plasmas, is based on the existence of these fluctuations and on the fact that its magnitude and dynamics can be predicted very precisely by equilibrium statistical mechanics.

Fluctuations in electron density can at times be found to be very much enhanced over the thermal level (i.e. over the minimum level) either as a consequence of external action (neutral turbulence, meteor ionization, etc.) or as a consequence of internal unstable processes. In such cases one usually cannot fully predict the behaviour of the fluctuations or that of the scattered signals; thus one cannot obtain as much information on the characteristic parameters of the medium as in the case of incoherent scatter. Nevertheless, at D and E region heights, because of the strong collisional coupling, the fluctuations can still be used as tracers of the dynamics of the neutral background in which they are embedded. In fact, enhanced fluctuations present some advantages over thermal fluctuations as they will produce enhanced scattered signals which are easier to detect.

Four different radar techniques are described which have been used for the observation and study of gravity waves at E and D region heights; each is characterized by the nature or source of the electron density fluctuations responsible for the reflection or scattering of the radar waves. They are: meteor trail radar technique, non-thermal scattering techniques, incoherent scattering techniques, the second title comprising two techniques, both developed and used at the Jicamarca Radio Observatory, which, although they use an incoherent scatter facility, are not classified as incoherent scatter techniques since they do not depend on the thermal electron density fluctuations for the scattering.

It should be pointed out that gravity wave observations at higher altitudes, i.e. at F region heights, with the help of the gravity wave theory and neutral atmosphere models, can be used to infer information about the amplitude and other characteristics of the same wave at the altitudes of interest in this review. Radio and radar techniques used at these altitudes and the results obtained have been recently reviewed by Vasseur et al. [1] and will not be discussed here.

2. Meteor Trail Radar Technique

Meteor trail radars have been used for more than 20 years for the study of neutral winds in the region between 80 and 120 km [2—7], but it has not been until recently that the full potential of the technique for the study of gravity waves at meteor heights has been exploited. A radar technique has been developed for this purpose by Spizzichino and co-workers [8] and a special radar has been constructed in Garchy, France (47°N 3°E). Here we shall only describe briefly the latter technique and present some of the results obtained.

The French system is a bi-static continuous wave radar and is shown schematically in Fig. 1. It consists of a 5 kW transmitter operating at a nominal frequency of 29.8 MHz. The transmitter antenna is a corner reflector pointing east at an elevation of 45° with a beamwidth of 20° in elevation and 25° in azimuth. The receiver site is located at Sens-Beaujeu, 30 km to the west of the transmitter site at Garchy. It consists of three receiving antennas of construction similar to the transmitter antenna and arranged in an interferometer fashion with a baseline of 30 m in the E—W direction and 20 m in the N—S direction.

When a meteor enters the common volume seen by the transmitter and receiving antenna, and if the path of the meteor is perpendicular to the line of

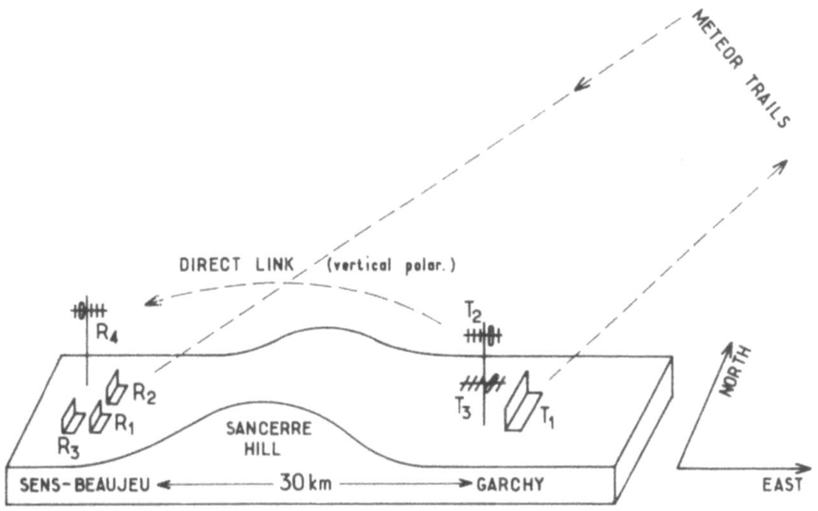

Fig. 1. The Meteor Radar at Garchy. T_1, main transmitting antenna; R_1, R_2, R_3, receiving antennas; T_2R_4, vertically polarized link for reference signals; T_3, transmitting antenna to cancel direct field at receiving site (after Spizzichino [8]).

sight from the center of the transmitter-receiving system, an echo from that portion of the trail which presents the proper angle (within a Fresnel zone) is received. The position of this portion of the trail is determined in azimuth, elevation and range, and the velocity of the neutral medium in which it is embedded is obtained from the Doppler shift of the received signal.

The azimuth and elevation of the echoing region are determined by measuring the difference in phase of the signals arriving to the three receiving antennas. The range is obtained using three closely separated frequencies. The lowest separation is 600 Hz and the highest 5400 Hz which essentially modulate the radio wave with these frequencies. The phases, at reception, of the 600 Hz and the 5400 Hz "modulating" signals with respect to their phases at emission, are directly proportional to range. The 600 Hz modulation gives the range unambiguously but not with sufficient accuracy. The 5400 Hz modulation gives the necessary accuracy but in an ambiguous way. The ambiguity of the latter is resolved using the approximate range obtained from the 600 Hz signal.

In order to be able to measure the range and Doppler shift (range rate) it is essential to have the transmitted signals at the reception site for reference. The reference is needed for the coherent detection of phase and phase rate of the re-

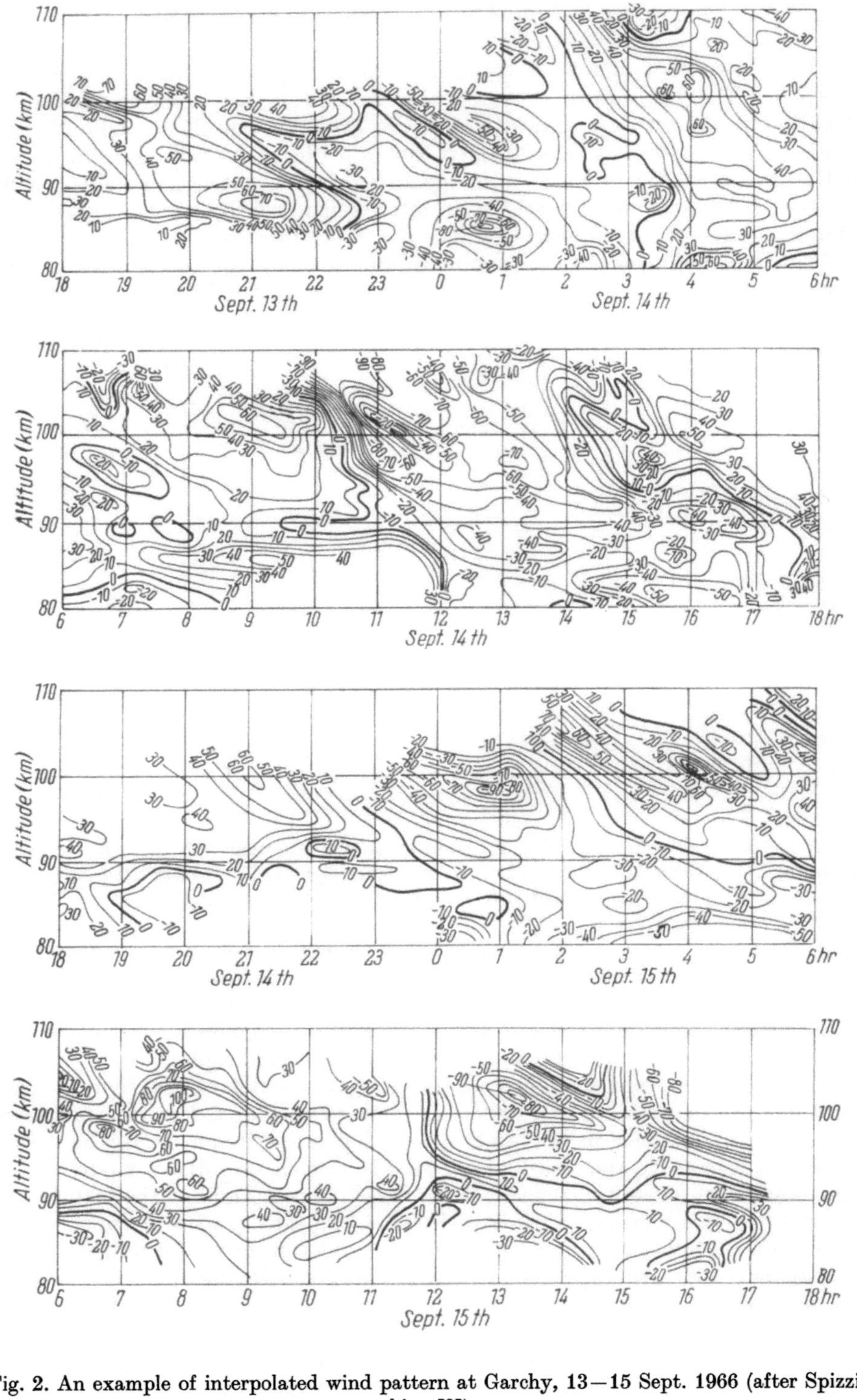

Fig. 2. An example of interpolated wind pattern at Garchy, 13–15 Sept. 1966 (after Spizzichino [9]).

ceived signals. This is accomplished by a low power direct communication link of the radar frequencies but using a vertical polarization orthogonal to the radar polarization. It is also essential for the precision of the measurements that the three receiving antennas receive only the signal echoing back from the meteors. They should not receive the ground wave or the reference signal. This is accomplished by transmitting the radar frequencies through a horizontally polarized Yagi antenna with an amplitude and phase which exactly cancel the interfering signals mentioned above.

The system is able to measure the height of the meteor echoing region with an accuracy of 500 to 1000 meters for signals with a signal-to-noise ratio better than 20 dB. The velocities obtained are the projection of the total velocity along the line of sight. The E—W velocities are derived by assuming no vertical component and neglecting the contributions from the N—S component of the wind. The second assumption limits the accuracy of the velocity measurements to 10 m s^{-1} for each individual measurement. This error is reduced by averaging in the processing of the data (see Spizzichino [9] for the error analysis).

All measurements within the observed volume are projected into a vertical line in the center of the volume. The data from all meteors in a given campaign form a time and altitude random sampling of the winds in the altitude range between 80 and 110 km. From this array of randomly space samples two-dimen-

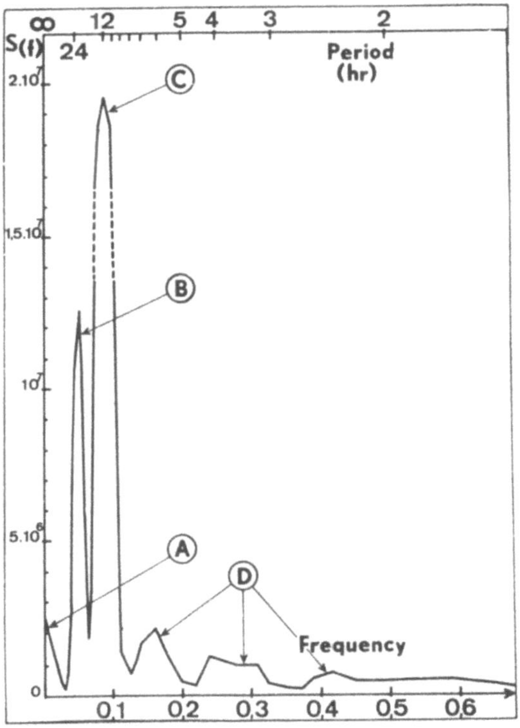

Fig. 3. Power spectrum (in $m^2 s^{-1}$) of the zonal wind at Garchy, 29 March—1 April 1966. A, prevailing wind; B, diurnal tide; C, semidiurnal tide; D, gravity waves (after Spizzichino [9]).

sional smooth contours of velocity as a function of height and time are obtained by means of a smoothing and interpolating numerical process. Fig. 2 shows an example of an interpolated wind pattern taken from Spizzichino [9]. Typical values and the structure of the winds at this altitude are apparent from this figure.

Power spectrum analysis has been performed by Revah [10] and by Spizzichino [11]. A power spectrum plot obtained for a particular campaign is shown in Fig. 3; in addition to the prevailing wind, the diurnal and the semidiurnal components, one can observe the presence of peaks at periods of 8, 4 and 2.5 hours. The interpolation in the processing of the data filters out waves with periods shorter than two hours. Whenever a wave of a particular frequency is identified, its phase as a function of altitude is found to vary linearly with time [10] corresponding to a downward propagating phase which corresponds to waves excited from below, in accordance with the gravity wave theory [12]. The amplitude of the waves varies as a function of height in a sinusoidal way, strongly suggesting the existence of partially reflecting waves.

The vertical wavelength can be obtained from the slope of the phase dependence, height curve or from a direct two-dimensional (frequency and vertical wavelength) Fourier spectrum analysis [10, 11]. Fig. 4 summarizes the distribution

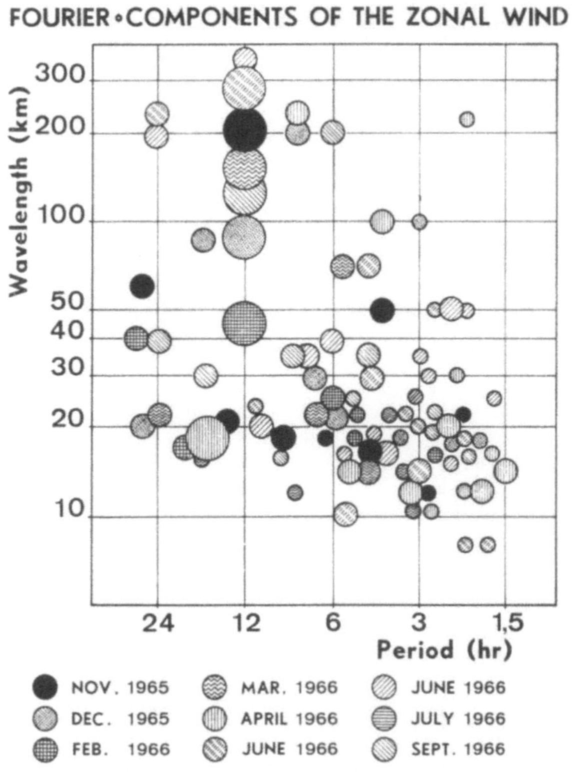

Fig. 4. Frequencies and vertical wavelength of spectral components observed at Garchy from November 1965 to September 1966 (after Spizzichino [11]).

of waves according to their frequency and vertical wave number. Gravity wave vertical wavelengths are found to be mainly in the 10 to 50 km range.

If all the spectra in the gravity wave range are averaged over many days of observation, it is found [10, 11] that the peaks at a given frequency disappear. The resultant frequency power spectrum follows a power law with a coefficient n equal to 0.82 for the 90 km waves and 0.47 at 100 km as shown in Fig. 5.

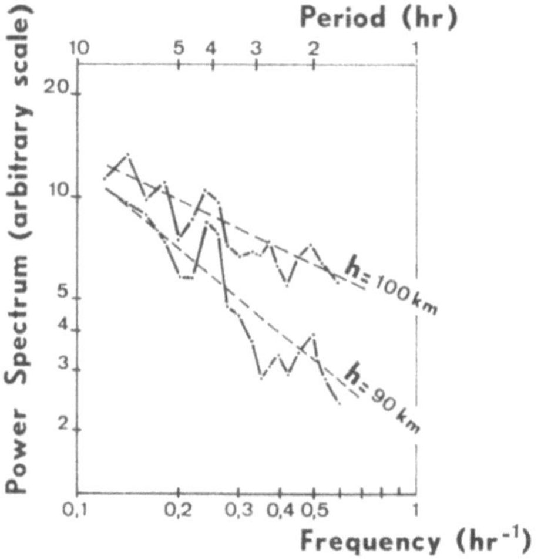

Fig. 5. Mean power spectrum (in $m^2 s^{-1}$) of the zonal wind at Garchy (1965–1966) (after Spizzichino [11]).

The wave behave, in general, very much according to the gravity wave theory. An important discrepancy has been found with respect to the amplitude of the wave as a function of height. Conservation of energy calls for an exponential growth of the amplitude of the wave as a function of height; it has been found that the growth is faster than predicted; Spizzichino has explained this discrepancy by a nonlinear interaction between the gravity and tidal waves [13].

3. Non-Thermal Scattering Techniques

Several powerful radars have been designed and built for the study of the earth's ionosphere by means of incoherent scatter techniques (see Evans (14] for a review and references). One of them is the Jicamarca radar located near Lima, Peru (12° S 76° W). It consists of a very large antenna built as an array of dipoles (9216 half-wavelength cross dipoles) covering an area of 84000 m^2 (Fig. 6) and a powerful transmitter with four modular units each of 1 MW peak power. It operates at a frequency of 49.920 MHz. The Jicamarca radar has been mainly used to study ionospheric parameters and motions in the altitude range 200 to 700 km for most of these parameters, and up to 10000 km for electron density studies. The top range is limited by the system sensitivity. The bottom range is determined by the incoherent scatter theory requirement that the ionosphere be in thermodynamic equilibrium. This condition is not satisfied at E region heights at the latitude of Jicamarca (magnetic equator) where strong density fluctuations

Fig. 6. A view of the antenna at the Jicamarca Radio Observatory.

exist as consequence of different electrodynamic unstable processes. Enhanced electron density fluctuations are also found at D region heights. This means that one cannot derive as many parameters from the spectral characteristics of the echoes obtained from D and E region heights as from those obtained from F region echoes; nevertheless, one can still derive many properties of the scattering medium, among them the velocity of the neutral background within the scattering volume. One can, therefore, study winds, tides and gravity waves using the electron density fluctuations as tracers of the neutral air motions.

A technique has been recently developed by Woodman and Guillén [15] for the observation of winds and waves as well as turbulence parameters at stratospheric and D region heights with the Jicamarca radar. The technique consists mainly of the real time determination of the three first moments of the frequency power spectrum of the backscatter signal, i.e. the power, the Doppler shift and the width of the spectrum. The power is a direct indication of the magnitude of the 3 m ($\lambda/2$ at 50 MHz) density fluctuations in the medium, the Doppler shift is a measure of the mean velocity of the neutral background within the scattering volume, and the spectral width gives information about the distribution of random velocities within the medium. The first and third parameters are of interest for the observation of the turbulent processes responsible for the enhancements in the electron density fluctuations. The second, i.e. the Doppler shift, gives information about winds, tides and gravity waves at the altitude under observation.

The three spectrum parameters mentioned above are obtained every minute, at a given height, with a resolution of 5 km for two independent antenna channels. The two linear polarizations of the antenna are independent and they can be pointed, by proper phasing of their elements, in two different directions 3° off the physical axis of the antenna. The simultaneous observations at two different directions allow a determination of the vertical and horizontal components of the instantaneous winds in the plane formed by the two antenna directions. The velocity is obtained with an accuracy of 10 cm s^{-1} for the vertical and 1 m s^{-1} for the horizontal component with only one minute of integration (for a S/N $>$ 10 dB).

The signal-to-noise ratio is considerably improved by a coherent integration scheme. Because of the relatively slow dynamics of the 3 m fluctuations, the radar echoes are highly correlated from one pulse to the other. The measured coherence time [15] (inverse of the frequency spectrum width) is of the order of one second, and the pulse repetition rate is of the order of 1200 pulses s^{-1}. As many as 300 pulse echoes are averaged coherently to make a single sample point giving a signal-to-noise ratio 25 dB better than that obtained from a single radar pulse echo.

The three characterizing parameters of the frequency spectrum are obtained from a few points of the autocorrelation function of the series of coherently integrated samples. Woodman and Guillén [15] have shown that the first three moments can be obtained from only two points of the complex autocorrelation function, one point evaluated at zero delay, the other at a sufficiently small delay for a truncated second order Taylor series expansion to be valid. The technique is based on the direct relationship existing between the first three moments of the frequency spectrum and the first three derivatives of the autocorrelation function evaluated at the origin. This is a property of any Fourier transform pair. This scheme is considerably faster than a conventional evaluation of the full frequency power spectrum.

Fig. 7 shows a power profile obtained at Jicamarca from 15 to 85 km. The profile was taken at a randomly selected time. The power from mesospheric heights varies considerably during the day and from one day to the next: in that sense the profile shown should not be considered typical. The profile was taken using 314 coherent integrations. The dotted line shows the noise level for a single radar pulse. Echoes are received from 15 to 35 km and from 55 km and above. Echoes beyond 85 km were not processed: they come from a region where electric fields are important and need a different interpretation (they were excluded from [15]). The predicted incoherent scatter power level for an altitude of 80 km and an electron density of 10^3 elect cm^{-3} is shown for reference.

The 15—35 km echoes do not concern us here, but we should mention in passing that they give important information about winds, waves and clear air turbulence at these altitudes. They are produced by fluctuations in the index of refraction of the neutral atmosphere. There are no echoes between 35 and 55 km for lack of efficient scatterers. The echoes from the 55—85 km region are believed to be caused by electron density fluctuations produced by neutral turbulence in a region of electron density gradients. At certain times the echoes are 10—15 dB higher than shown in Fig. 7, at a particular narrow range, anywhere between 60 and 85 km. These narrow and strong echoes, we believe, are caused by the coincidence of a layer of turbulence with a region of strong density gradients. Strong density gradients at D region heights have been observed at the equator by means of rockets [16].

Fig. 8 shows two time series corresponding to the continuous observations of the vertical and E—W component of the winds at 80 km. Fluctuations larger

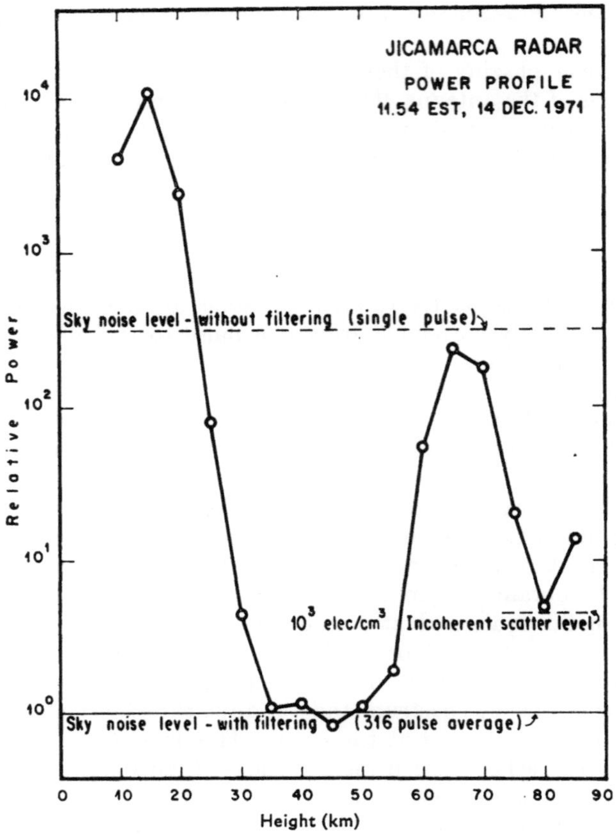

Fig. 7. Backscatter power profile obtained at Jicamarca (after Woodman and Guillén [15])

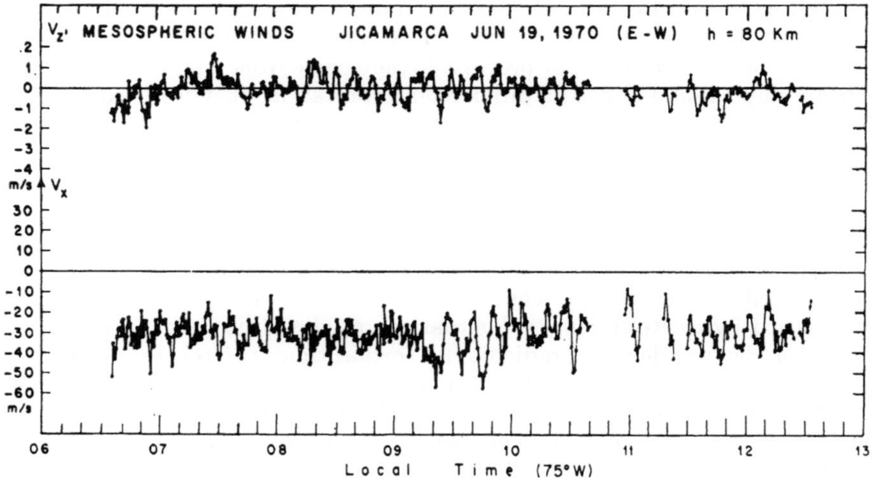

Fig. 8. Record of velocities at 80 km obtained at Jicamarca from the Doppler shift of backscatter echoes produced by enhanced dielectric fluctuations at D region heights. V_z, vertical component; V_x, horizontal component (after Woodman and Guillén [15]).

than ± 1 to $2\ m\ s^{-1}$ in the E–W winds and ± 0.1 to $0.2\ m\ s^{-1}$ in the vertical should be considered real. Fluctuations with periods of the order of $10\ m\ s^{-1}$ predominate and are of the same order of magnitude as the mean zonal component. Fig. 9 shows a special event of unknown origin. A quasi-sinusoidal packet with a period of the order of 10 minutes is evident in the horizontal component from 1045 to 1120. The accuracy of the technique can be appreciated in the small scatter of the points following the oscillation.

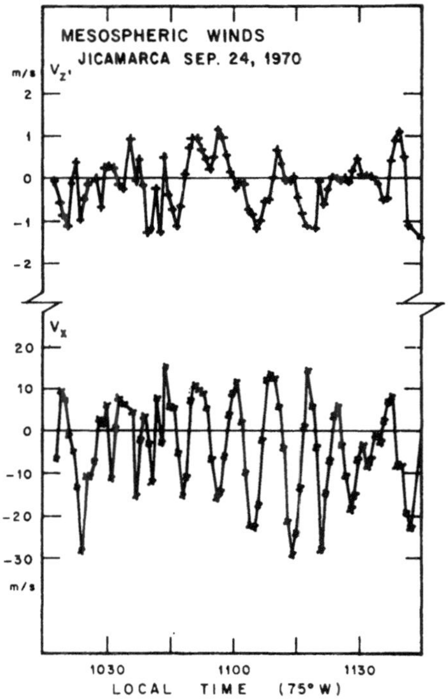

Fig. 9. Record of a special event obtained at Jicamarca showing clean sinusoidal fluctuations at gravity wave periods (after Woodman and Guillén [15]).

Spectral analysis of the velocity time series at D region heights by the Jicamarca technique has been reported by Rastogi and Woodman [17]; Fig. 10 shows a typical result. The cut-off at around 5 minute period is a fairly consistent feature of all the spectra that have been obtained so far. It corresponds approximately to the Brunt-Väisälä frequency cut-off beyond which no gravity waves should propagate. The random errors in the series caused by noise are independent from point to point; their contribution to the spectra is a constant as a function of frequency. The smallness of these errors is evident from the amplitude of the tail of the spectra beyond the 4 minute period.

The wind and wave observations at Jicamarca have been limited so far, because of computer speed limitations, to one particular altitude and to a 5 km resolution. A new system using a third generator computer has been developed which should allow the observation of the full velocity vector at 20 heights simultaneously with a resolution of 2 km and a time resolution of one minute. The new system will allow us to make gravity wave propagation studies and to determine the origin of these waves.

A second backscatter radar technique has been used at Jicamarca using enhanced fluctuations as tracers of neutral atmospheric motions at E region heights

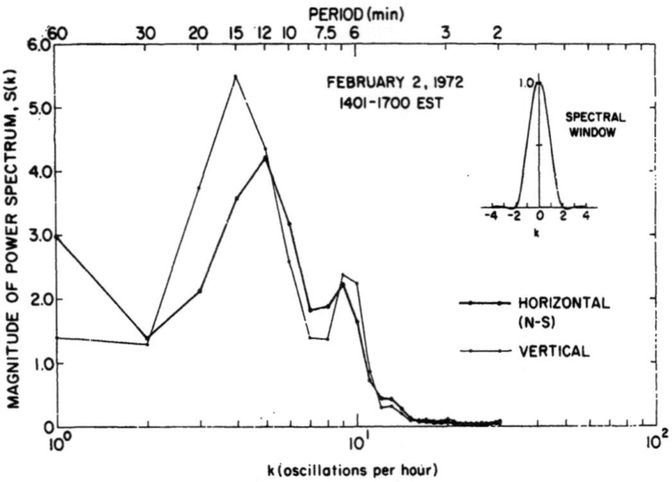

Fig. 10. Power spectrum of vertical and E—W winds at 70 km obtained at Jicamarca (after Rastogi and Woodman [17]).

and around 105 km. The technique is based on the spectral characteristics of echoes received from irregularities produced by the two-stream instability in the equatorial electrojet. The technique has been used and reported by Cohen [18] using the Jicamarca transmitter but with a smaller steerable antenna. The theory of the two-stream instability predicts that the irregularities should move at the ion-acoustic velocity with respect to the neutral background; therefore the spectrum of the backscatter signal should be Doppler shifted by an amount correspond-

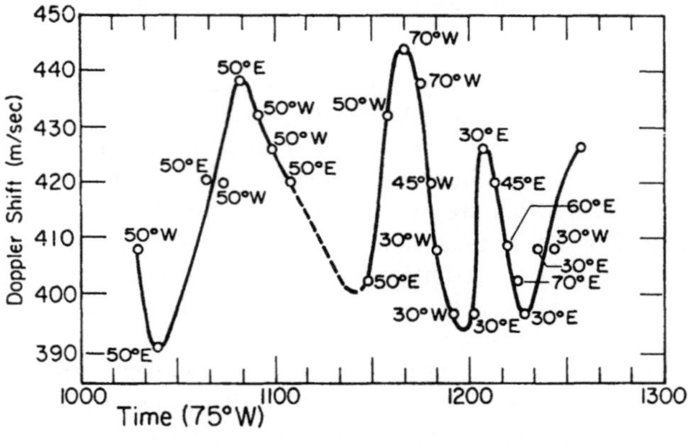

Fig. 11. Record of the Doppler shift of the spectral peak of backscatter echoes from two-stream E region equatorial instability showing gravity wave oscillations, 3 Jan. 1964 (after Cohen [18]).

ing to the ion-acoustic velocity plus the projection of the neutral velocity along the line of sight of the receiver. Fig. 11 shows a record of the time history of the center frequency of the backscatter frequency spectrum scaled in m s^{-1}. The fluctuations from the mean velocity are interpreted by Cohen [18] as neutral air motions produced by gravity waves. The periods are from 30 minutes to an hour and the velocities of the order of ± 25 m s^{-1}.

4. Incoherent Scatter Techniques

Gravity waves are essentially a neutral atmosphere phenomena and they manifest themselves as fluctuations of the velocity, the temperature and the density, as well as other thermodynamically related parameters of the neutral atmosphere. These fluctuations produce corresponding fluctuations in the velocity, temperature and collision frequencies of the ionized plasma, which are directly observable by incoherent scatter techniques. Gravity waves have been observed and studied using incoherent scatter techniques, but so far at F region heights [1, 19—25], with a few recent observations, limited to diurnal and semidiurnal tidal periods, at E region heights [26, 27]. The technique is nevertheless potentially capable of studying waves of shorter periods at E region heights and it is expected that this will shortly be done.

5. Conclusions

Radar techniques are capable of and have been used for, the detection and study of gravity waves in the lower ionosphere. Meteor radar techniques have been fully exploited for the study of gravity waves from periods of two hours to periods in the range of tidal waves. The French system at Garchy has shown the potential of the technique and many interesting properties of gravity waves at altitude ranges between 85 and 110 km have been reported. The potential of powerful radar at VHF, making use of turbulent fluctuations at D region heights, has been established with observations made with the Jicamarca radar. The technique has detected gravity waves with periods down to the Brunt-Väisälä frequency. Incoherent scatter techniques should be capable of detecting gravity waves at E region heights; so far only tidal wave observations have been reported.

References

[1] G. Vasseur, C. A. Reddy and J. Testud, Space Research XII, 1109 (1972).
[2] L. A. Manning, O. G. Villard, Jr. and A. M. Peterson, Proc. IRE **38**, 877 (1950).
[3] L. A. Manning and V. R. Eshleman, Proc. IRE **47**, 186 (1959).
[4] J. S. Greenhow, Proc. Phys. Soc., London, B **65**, 169 (1952).
[5] J. S. Greenhow and E. L. Neufeld, Quart. J. Roy. Met. Soc. **87**, 472 (1961).
[6] W. G. Elford and D. S. Robertson, J. Atmos. Terr. Phys. **4**, 271 (1953).
[7] W. G. Elford, Planet. Space Sci. **1**, 94 (1959).
[8] A. Spizzichino, J. Delcourt, A. Giraud and I. Revah, Proc. IEEE **53**, 1084 (1965).
[9] A. Spizzichino, Progress in Astronautics and Aeronautics, MIT Press, Cambridge, Mass. **27**, 117.
[10] I. Revah, Annls Géophys. **25**, 1 (1969).
[11] A. Spizzichino, Annls Géophys. **25**, 5 (1969).

[12] C. O. HINES, Canad. J. Phys. **38**, 1441 (1960).
[13] A. SPIZZICHINO, Annls Géophys. **25**, 25 (1970).
[14] J. V. EVANS, Proc. IEE **57**, 496 (1969).
[15] R. F. WOODMAN and A. GUILLEN, J. Atmos. Sci. **30**, March 1974.
[16] S. PRAKASH, B. H. SUBBARAYA and S. P. GUPTA, University of Illinois, Urbana, Aeronomy Rep. No. 48-359 (1972).
[17] P. K. RASTOGI and R. F. WOODMAN, J. Atmos. Terr. Phys. **36**, March 1974.
[18] R. COHEN, J. Geophys. Res. **78**, 2222 (1973).
[19] G. D. THOME, J. Geophys. Res. **69**, 4047 (1964).
[20] G. D. THOME, J. Geophys. Res. **73**, 6319 (1968).
[21] G. D. THOME and P. B. RAO, Raytheon Company Interim Report (1968).
[22] D. L. STERLING, University of Illinois, Urbana, Aeronomy Rep. No. 19 (1967).
[23] D. L. STERLING, W. H. HOOKE and R. COHEN, J. Geophys. Res. **76**, 3777 (1971).
[24] G. VASSEUR and P. WALDTEUFEL, J. Atmos. Terr. Phys. **31**, 885 (1969).
[25] J. TESTUD and G. VASSEUR, Annls Géophys. **25**, 525 (1969).
[26] R. H. WAND, J. Geophys. Res. **74**, 6588 (1969).
[27] P. AMAYENC and C. A. REDDY, Planet. Space Sci. **20**, 1269 (1972).

Lower Ionosphere Structure — Akademie-Verlag, Berlin 1974

VELOCITY OF THE REFLECTION POINTS REVEALS STRUCTURE AND MOTIONS IN THE IONOSPHERE*

K. BIBL

Lowell Technological Institute Research Foundation, Lowell, Massachusetts, USA

**Dedicated to Professor Rawer on the occasion of his 60th birthday*

Since the ionosphere is moving, its structure and waves can be monitored by monostatic experiments. Complex spectral analysis of high-frequency signals, received on spaced antennas, permits angular and Doppler measurements on all wave fronts arriving simultaneously. Using the time dependence of incidence angle and Doppler frequency of significant spectral amplitudes, the four-dimensional model of the ionosphere can be determined. For gravity waves the relation between the wave speed and the velocity of the reflection points is surprisingly simple. The horizontal displacement of each reflection point varies linearly with time for all, even asymmetric, wave models. While the vertical displacement is parabolic, the derivative of the vector between ground location and reflection point, proportional to the Doppler frequency, is a linear sawtooth time function. Only the extreme model of a limited number of moving targets results in a completely different signature. Also, the small scale structure is optimally revealed in the spectrogram time sequence.

1. Introduction

Digital techniques for data gathering, processing and display open up the limits of the amount and the dimensions of treated data. The new methods of multi-range, multi-frequency, complex amplitude scanning create digital quadrature samples ready for fast on-line or off-line spectral analysis. Thus, the spectra at different height ranges and frequencies become available as almost continuous time functions. Interpretation and preparation for automatic treatment of those spectrograms are the subject of this paper.

2. Time Dependence in Ionospheric Recording

2.1. Earlier Attempts

The unquestionable success of ground-based and satellite-borne ionogram recording for world-wide survey on an hourly basis has misled many researchers and institutes to limit themselves to large-scale studies of the ionosphere. Studies of medium-scale and small-scale properties of the ionosphere have generally been carried out with insufficient technical means, using mostly one-dimensional time functions. Many years of drift measurements on several stations have been carried out at single frequencies and single height ranges to find the average drift for a random screen or the speed of some specific irregularities. Also, monostatic and bistatic Doppler measurements have been made at fixed carrier frequencies to find the periodic gravity wave motions in the ionosphere. Using correlation methods, some researchers tried triangulation for the time-history of a single reflection point. Because of dimensional restrictions, all those methods give only limited and sometimes contradictory results in the search for an adequate model of the ionosphere.

2.2. Space-Time Scanning

Because of the complexity and the large density gradients in the ionosphere structure, very often more than one reflection area fulfils the reflection condition even in "vertical" incidence. Since the structure moves, complex spectrum analysis of signals with different carrier frequencies received simultaneously at spaced antennas is an excellent means of providing all the data for creating a multi-dimensional picture of the ionosphere. We have developed and build systems which produce the digital data and spectrum-analyze them in real time. We have, however, to learn to display and to interpret the multi-dimensional data.

The first step is the display of digital spectrograms with the magnitude or the real and imaginary amplitudes of the spectral lines presented as functions of time (Fig. 1). Since we are not used to thinking in the spectral domain, we still have some difficulty in seeing the multi-dimensional surface of the ionospheric reflector behind the display.

As in the process of translating vertical height ionograms into vertical profiles of the ionosphere, we have to study the signature which ionospheric models produce in the spectral time domain. To simplify matters, we want mainly to discuss the effects of a three-dimensional reflecting surface rather than a thick layer.

The simplest model is certainly an inclined plane surface (or tilt) moving with a constant speed. This has a constant frequency line as signature in the spectral time domain. An ionospheric structure with a triangular cross section produces two lines, one at a positive and one at a negative Doppler frequency. Both of these models are very rare and therefore useless. A more realistic model is a smoothly undulated surface with a sinusoidal cross section, generated by a gravity wave, and having a much larger extension perpendicular to the direction of propagation. Three models of such waves and their signatures are displayed in Fig. 2.

On top we have the range vectors as functions of time. We see that the wave prefers the curved part of the structure as reflection area. This is very important since we want to approximate the sinusoidal wave form by two parabolas for an easy calculation of the signature in the spectral time domain. In the middle of the picture we see the range changes caused by the modulation of the reflection surface to be almost exact parabolas. If we replace those curves by the dotted parabolas, the Doppler frequencies as functions of time become straight lines, as displayed at the bottom of Fig. 2. The surprising fact is that the slopes of the straight lines do not depend significantly on the vertical depth of the model although the maximum Doppler frequency is certainly more than proportionally dependent on the depth of the modulation of the reflecting surface. Thus, the slope of the Doppler curve is a good measure of the speed of the propagating wave. The extreme Dopper frequencies give the depth of the modulation und the period of the pattern gives the size of the wave structure.

Fig. 1. Drift Doppler spectra 1017—1313 EST, 11 July 1972; Goose Bay, Labrador, F region.

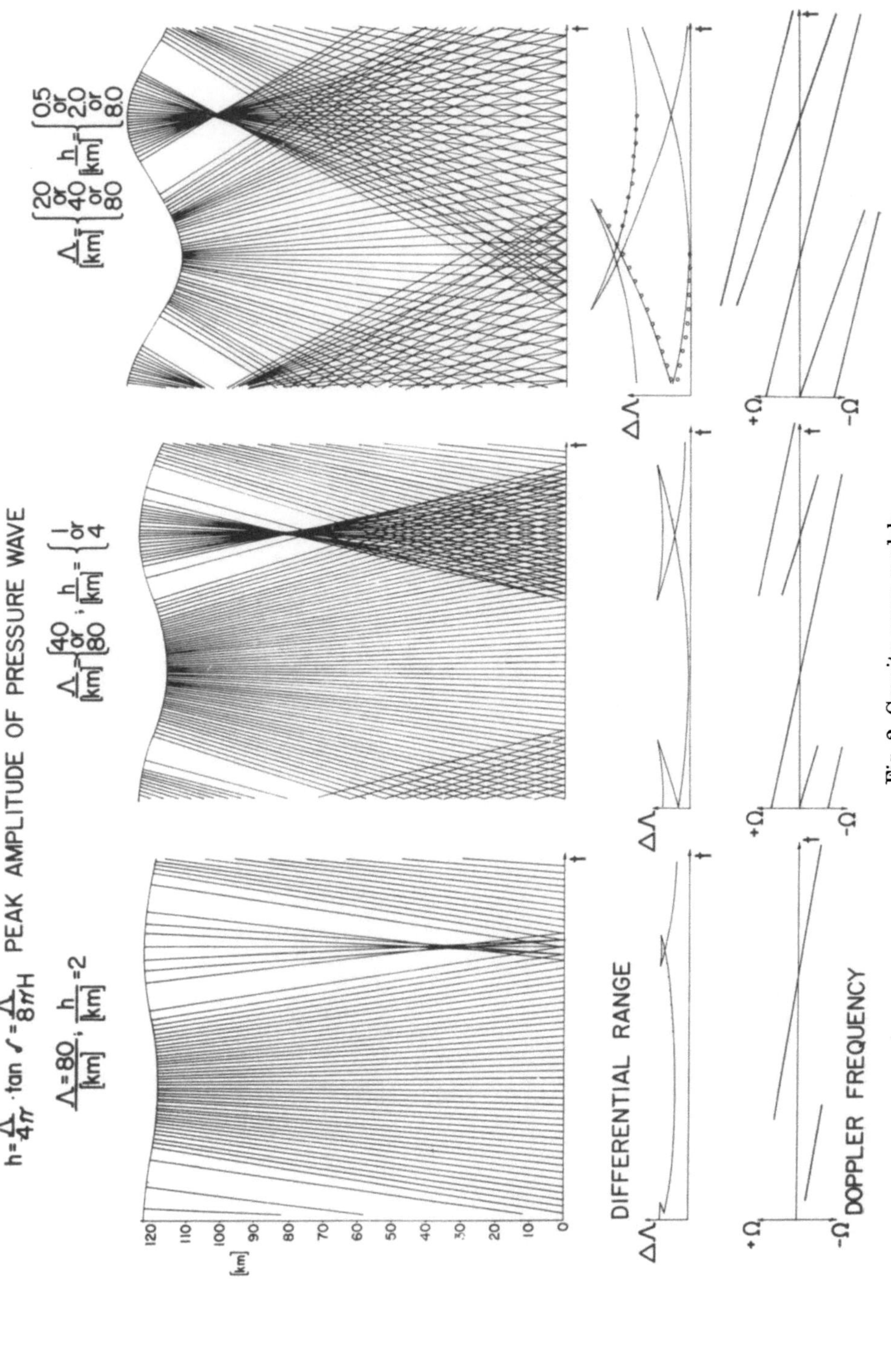

Fig. 2. Gravity wave models.

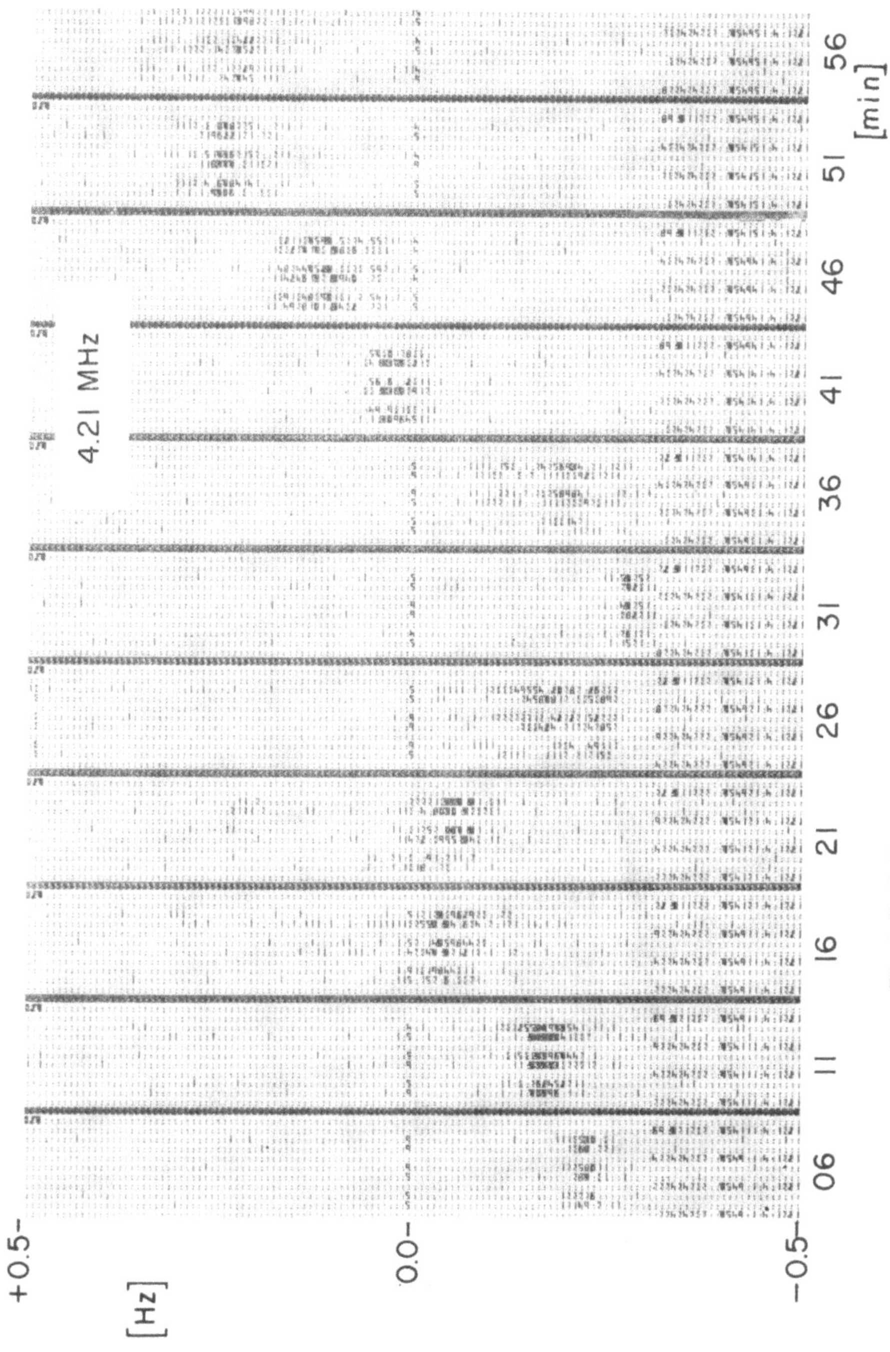

Fig. 3. Drift Doppler spectra 0106—0156 CST; Eglin, Florida.

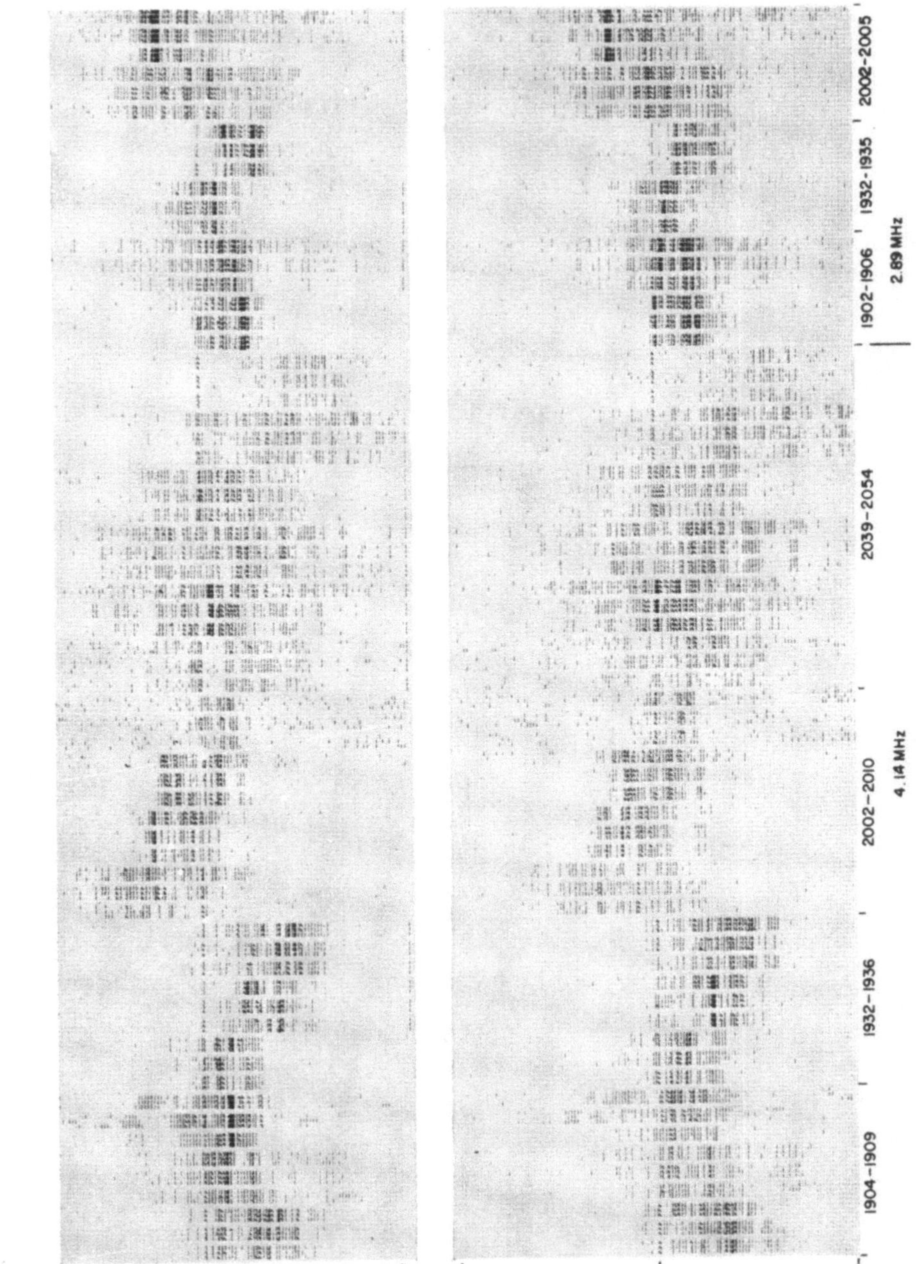

Fig. 4. Drift Doppler spectra 1902—2054 EST, 21 Sept. 1971; Dracut, Massachusetts.

The picture shows that an averaging process for the Doppler frequency would be a catastrophe if more than one reflection area were reached simultaneously. Then the resultant speed becomes senseless. Even the apparent period can become a multiple of the structure period. This proves that complex spectral analyzers are essential for at least some reference stations until the signatures resulting from other methods can be compared with the complete spectral analysis.

Since nature is not so simple and the ionospheric reflection surface can not always be approximated by a simple sinusoidal modulation, we have to measure the incidence angles simultaneously for all significant Doppler spectral lines. The three main models, gravity waves, moving targets and irregularities riding on gravity waves, can then be distinguished. Our results show that standing wave patterns occur frequently, at least in superposition on the moving structure (Figs. 3 and 4).

2.3. Automatic Analysis

One of the most urgent requirements of this study is an attempt to find a linked set of significant parameters in the signature of the spectral time domain and the spatial time domain so that an on-line computer can create a model of the ionosphere. This task is not yet completed but we think that it is much easier to find them in the spatial and spectral signature than in the raw amplitude data. We believe that it is necessary to create digital movies to display the angular and spectral information simultaneously as functions of time. It is important to find a model which is adequate not only for ground-based communication and direction-finding but also for space and ground—space communication. For that reason it would be advisable to have simultaneous and direct spectral and angular measurements of satellite signals having travelled through the ionosphere.

2.4. Sophisticated Beacon-Satellite Spectrum Analysis

It is surprising that most researchers think that the measurement of the total electron content of the ionosphere averaged over a large time period is an adequate use of the expensive beacon satellites. Since there is such a large number of receiving stations, their total cost should be at least ten times as high as the cost of the satellite. All the technology is ready for digital extraction of the whole information content given by the available signal-to-noise ratio if larger dishes or arrays are used. Interferometry should be used for all spectral lines contained in the beacon satellite signals since this method creates perfect holograms of the ionosphere and would create models of the ionosphere more adequate for ground—space communication and incidence angle correction.

Lower Ionosphere Structure — Akademie-Verlag, Berlin 1974

EVIDENCE AND CHARACTERISTICS OF INTERNAL AND PLANETARY WAVES WITHIN THE D REGION PLASMA

E. A. LAUTER

Akademie der Wissenschaften der DDR, Zentralinstitut für Solar-Terrestrische Physik, Berlin, DDR

Observations of wave-like structure changes obtained by means of ground-based continuous D region measurements are presented for planetary waves (from multiple absorption measurements within the same geographic region) and for internal gravity waves (from long-wave reflection measurements at oblique incidence). Analysis of power spectra and superposed-epoch statistics gives evidence of pulse-like wave trends with periods of 5—10 days, sometimes peaking around 7—8 days when quasi-persistent waves are present. Remarkable increases of the amplitudes during winter are thought to be responsible for the chaotic part of the winter anomaly. These waves seem to be heavily damped during periods of major stratospheric warmings, when the wind systems in the lower ionosphere are also markedly affected. Analysis of the observed oscillations of the daytime low-frequency reflection levels (around 80 km) give evidence of the existence of internal gravity waves, which during the spring season peak around a period of 10 minutes.

1. Introduction

Ionospheric observing techniques for monitoring the behaviour of plasma parameters in the upper atmosphere have a new direction of development. The ionosphere plasma is no longer investigated for itself or for purposes of radio communication, but it is more and more used as a tracer for dynamics and structure changes in the upper atmospheric gas. During the last years there has been a broad research into phenomena of the D region plasma related to mesospheric properties, but most of the results are of a statistical nature, whereas a physical understanding of the supposed interrelations between processes in ionized and neutral gas is still largely lacking. This situation is quite natural for the development of a new field in geophysics. This again encourages monitoring of the upper atmosphere by means of indirect ground-based methods. On the other hand, in the field of atmospheric waves we have now a large theoretical knowledge about what should be possible in this form of energy transport within the mesosphere and stratosphere, and what should be the dimensions and importance of this transport. However, we have a serious lack of observational methods, and therefore also of experimental results in this field. A fast development of the analysis of existing observational data is to be expected, because the energy input into the upper atmosphere is as important as the control of energy loss of the lower atmosphere.

2. Wave-like Features in Ionospheric Absorption Data

Since the beginning of continuous ionospheric absorption measurements at high frequencies [1] and at low frequencies [2], it became evident that features of meteorological character are present in the interdiurnal changes of D region plasma, particularly in the winter season. But the method of absorption measurements, as far as the involved parameters are concerned, is such a complex one that measurements on a single frequency can give only statistical information about relations with other solar and atmospheric phenomena. As far as meteorological phenomena are involved, however, synoptic comparisons are necessary. This means that reliable ionospheric data on a short time-scale should be available. It became evident from the frequency dependence of ionospheric absorption [3] and also from rocket measurements, that the main part of the excessive winter absorption at high frequencies is produced by an anomalous structure of the

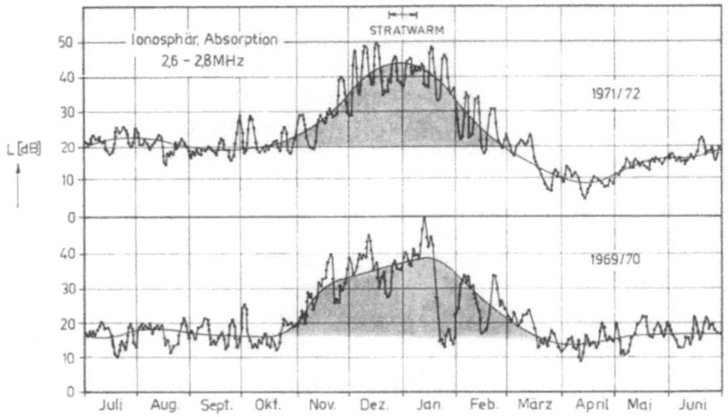

Fig. 1. 3-day smoothed data of ionospheric HF absorption at constant solar zenith angle ($\cos \chi = 0.2$), for the annual periods July 1971 to June 1972 (upper part) and July 1969 to June 1970 (lower part).

D region above 80 km, and is due to the reflection properties for these frequencies within a very small fraction of the whole layer. Therefore, absorption data at different frequencies, especially between short and long waves (i.e. related to the plasma above and below the mesopause), correlate not too well with each other. They also show a different seasonal behaviour and different dependence on solar radiation changes.

Therefore, we can not rely on the determination of the characteristics of a single day, as required for synoptic studies, if they are derived from single-frequency absorption data. On the other hand, it is well known that absorption data show a "grouping" of days with high and with low absorption during winter. Taking this into account, we have put together the absorption data for fixed solar zenith angles on different frequencies in the same geographic region, where the distances between the reflection points were only about 200 km, and have smoothed them over three days. The typical picture of the annual trend of absorption at high frequencies is presented in Fig. 1. We get the following results:

(a) Absorption at fixed solar zenith angles, in medium latitudes, is nearly constant in summer, but increases abruptly at the beginning of November ("November-effect"), when the winter polar vortex regime in the strato- and mesosphere is extending towards medium latitudes. This behaviour is normal at all frequencies. Even the scale-height determinations at low frequencies show a drastic change at this time, announcing by extremely low values that an unusual layering within the D region takes place. This "regular" increase of winter absorption [4], which lasts until February, and the end of which is not so clearly defined as its beginning, may be explained by the increase of the collision frequency controlled by the temperature regime at the mesopause level [5].

(b) Superimposed on the annual absorption trend, there are pulse-like enhancements in all seasons. These are partly caused by high-energy particles and other solar activity phenomena, but dominantly in winter most of them have a wave-like behaviour as is to be expected from effects of planetary waves penetrating from the stratosphere into the mesosphere. Sometimes, as in winter 1971/72, these pulses extend to persistent waves with periods of 7—8 days, lasting over cycles up to 4 weeks.

(c) In mid-winter, the amplitudes of these pulses increase extremely, indicating that chaotic changes in the layering of the plasma above the mesopause take place over several days, when planetary-wave pulses penetrate up to this height region.

(d) At lower frequencies (245 and 185 kHz), where during daytime the propagation is well below the mesopause, wave-like structure is present as well, but the periods seem to be somewhat shorter (5—6 days), and the amplitudes (in decibels) are smaller by nearly an order of magnitude. This is perhaps because the atmosphere below the mesopause is well mixed all the time, and this mixing remains unchanged during the presence of wave-like motions.

(e) During periods of large stratospheric warmings, which are perhaps connected with a cooling of the mesopause region and an approximation to summer conditions, the amplitude of the planetary waves fades away (e.g. in December 1971), or the winter anomaly disappears completely for a short time (as in January 1969), as is to be seen in Fig. 1.

(f) During such events, the wind system in the lower ionosphere also changes abruptly [6]. During the presence of penetrating waves the westerly winds just above the mesopause (deduced from ionospheric drift measurements) seem to be enhanced, and their direction changes continuously from NW to SW. There seems to be a general breakdown of high ionospheric absorption level, if the ionospheric wind is turning toward south or southwest (N or NE wind).

(g) Power spectra of different parameters have also been used in order to recognize preferred periods of recurrence of peak values. From ozone and absorption data, Grasnick and Entzian [7] have found dominant periods between 4 and 8 days. Märcz [8] has reported peaks in the same range of periods, from LF absorption measurements. In both papers, also a marked difference between summer and winter is evident, as well as changes from year to year. In Figs. 2 and 3, power spectra of absorption data for 1971/72, when quasi-persistent waves were very

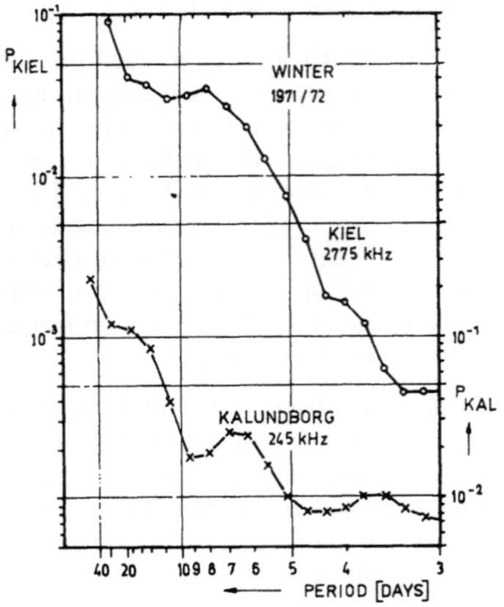

Fig. 2. Normalized power spectra of daily A 3 absorption data at cos $\chi = 0.2$ in winter 1971/72, for a HF and an LF path.

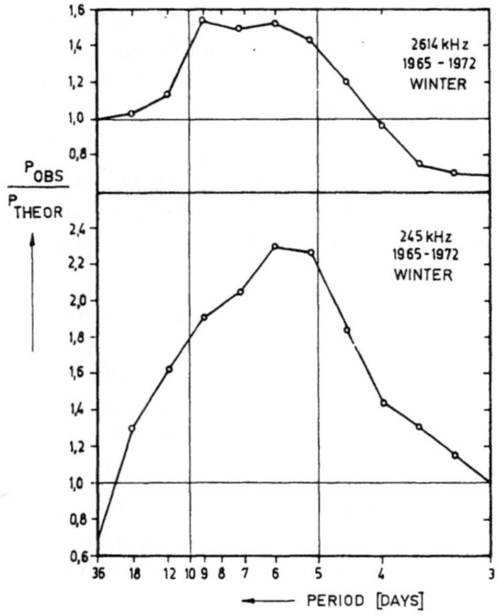

Fig. 3. Ratio of mean power spectra of winter daily A 3 absorption data 1965—1972 to theoretical (red-noise) power spectra, for HF (upper part) and LF absorption (below).

pronounced, are reproduced together with mean relative power spectra of the winter periods 1965—1972. Predominant periods are between 5 and 10 days, but the peaks at the lower frequency (245 kHz) are shifted to somewhat shorter periods compared with the high-frequency data. If this feature is found to be real, then we must conclude that the mesopause has a filtering effect on the wave propagation. In order to find the dominant period, in Fig. 4 the method of superposed epochs has been applied to absorption data of 1971/72, which are averaged over high and low frequencies. Here again, the 7-day period is dominant, but with increasing distance from the selected key day the period of the wave-like feature becomes somewhat shorter.

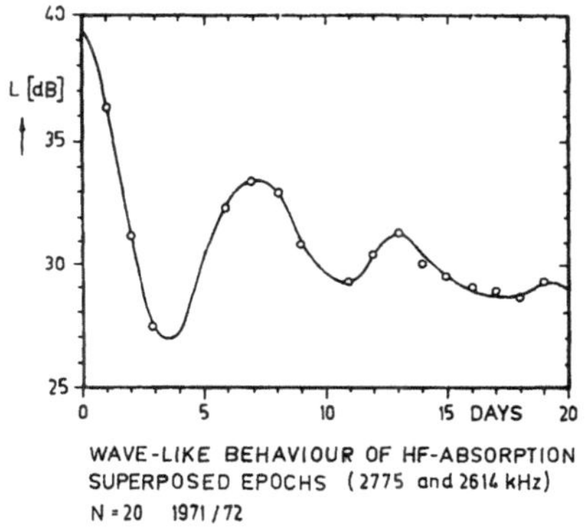

Fig. 4. Superposed-epoch analysis of HF A3 absorption data in winter 1971/72 (zero day: day of peak absorption).

3. Observation of Internal Waves by Ionospheric Methods

Oblique propagation of low frequencies (below 300 kHz) shows phase-induced amplitude variations during daytime, caused by a variation of the reflection height level, which changes between 90 km and 75 km in dependence on solar zenith

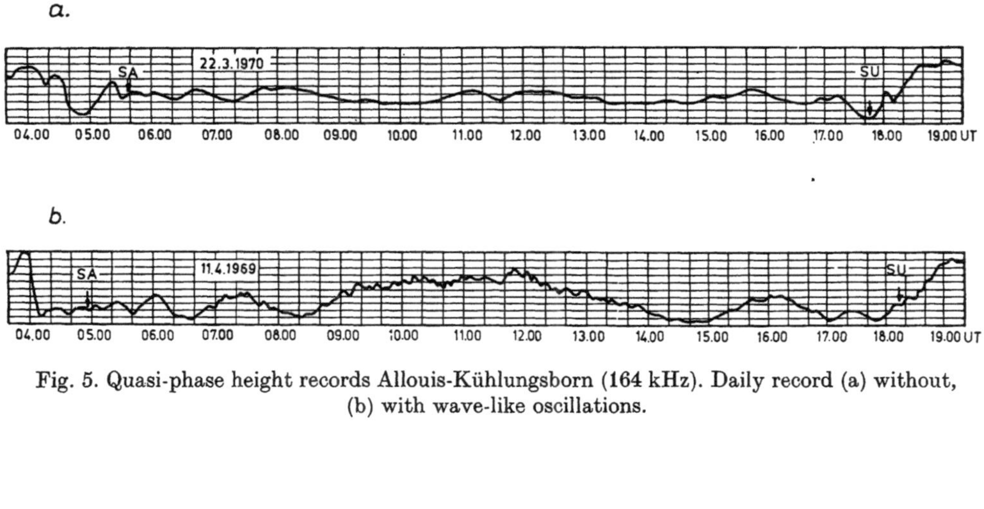

Fig. 5. Quasi-phase height records Allouis-Kühlungsborn (164 kHz). Daily record (a) without, (b) with wave-like oscillations.

angle. Under normal conditions, the amplitude trend is very smooth (Fig. 5a), but one some days, small wave-like oscillations are superimposed on the records (Fig. 5b). Analysis carried out by Bull et al. [9] has shown that this behaviour

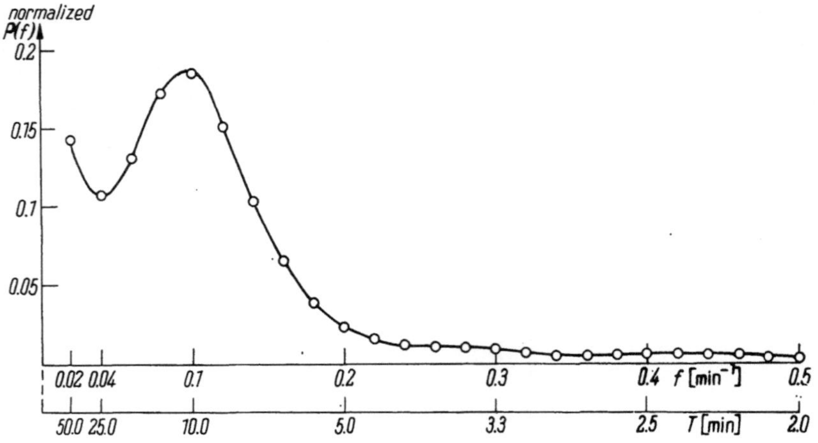

Fig. 6. Power spectrum derived from quasi-phase height records Allouis-Kühlungsborn on 164 kHz (mean of 3 spectra).

may be the effect of internal gravity waves which could be traced back to their tropospheric weather sources. The power spectra of these oscillations, taken from records in the spring season, show a marked peak at the period of 10 minutes (Fig. 6). We consider this to be a promising new method for a continuous patrol of energy transport into the upper atmosphere from below.

References

[1] W. Dieminger, J. Atmos. Terr. Phys. **2**, 340 (1952).
[2] E. A. Lauter, Z. Met. **7**, 321 (1953).
[3] E. A. Lauter and P. Nitzsche, J. Atmos. Terr. Phys. **29**, 533 (1967).
[4] H. Schwentek, J. Atmos. Terr. Phys. **33**, 1647 (1971).
[5] J. Taubenheim, J. Atmos. Terr. Phys. **33**, 1481 (1971).
[6] K. Sprenger and I. A. Lysenko, Phil. Trans. Roy. Soc. Lond. A **271**, 473 (1972).
[7] K. H. Grasnick and G. Entzian, paper presented at the IAMAP Symp. on Atmos. Ozone, Arosa 1972.
[8] F. Märcz, Acta Geodaet. Geophys. Montanist. Acad. Sci. Hungar. **7**, 197 (1972).
[9] E. A. Lauter, G. Bull, D. Brodhun and J. Neisser, Abh. Meteorol. Dienst d. DDR **14**, 105 (1973).

Lower Ionosphere Structure — Akademie-Verlag, Berlin 1974

METEOR RADAR STUDY OF TIDES IN THE 80 TO 100 km ALTITUDE RANGE

J. L. Fellous, M. Glass and A. Spizzichino

C. N. E. T./R. S. R., Issy-les-Moulineaux, France

The experimental data obtained with the Garchy meteor radar in 1970—1971 make possible a description of the characteristics of the semi-diurnal and diurnal tides in the 80 to 100 km range. The investigation of their vertical structure shows up peculiar features, in winter at the time of a stratospheric warming, and in some cases during summer. Moreover, the existence of several superimposed waves, and especially reflected waves, is observed. Various methods are used to determine the characteristics of the waves in each case.

1. Introduction

A large number of experimental data concerning the solar semi-diurnal and diurnal tides have been obtained during the last three years by the Garchy meteor radar (47°N, 3°E). This equipment yields a continuous description of the zonal wind profile between 80 and 100 km, with a height definition of 1 km [1, 2]. The harmonic analysis of the wind variations at each level gives the amplitude and the phase of the tidal oscillations. This paper deals with some results obtained in 1970 and 1971 during eight long-duration records (5 to 12 consecutive days); special attention is paid to the vertical structure of the tides.

2. Experimental Results

2.1. Semi-diurnal Tide

The amplitude and phase of the semi-diurnal tide exhibit three kinds of variations with respect to the altitude:

The variations with altitude of the amplitude (in m s^{-1}) and phase (in degrees) of the semi-diurnal component for the period 24—26 September 1970 are plotted in Fig. 1. The amplitude increases vertically and the phase variation is nearly linear with a slope smaller than 5 deg km^{-1}. The same phenomenon is observed in September, October, December 1970 and July 1971, and its characteristics are closely related to those of the fundamental migrating mode S_2^2, for which theory predicts a vertical wavelength greater than 100 km [3].

A second kind of record shows the amplitude increasing with altitude but a phase variation larger than 10 deg km^{-1}, the corresponding wavelength extending from 20 to 40 km. Such a situation occurred most of the time in March and April 1970, and during the whole campaign from 26 January to 6 February 1970 (Fig. 2), just at the same time as a stratospheric warming was registered.

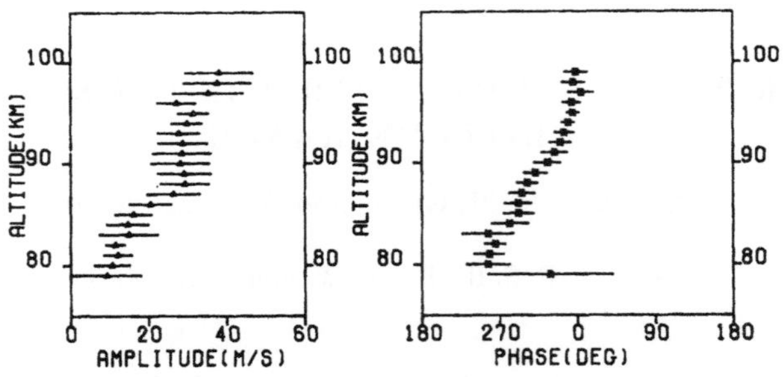

Fig. 1. Semi-diurnal tide, Garchy, 24—26 Sept. 1970.

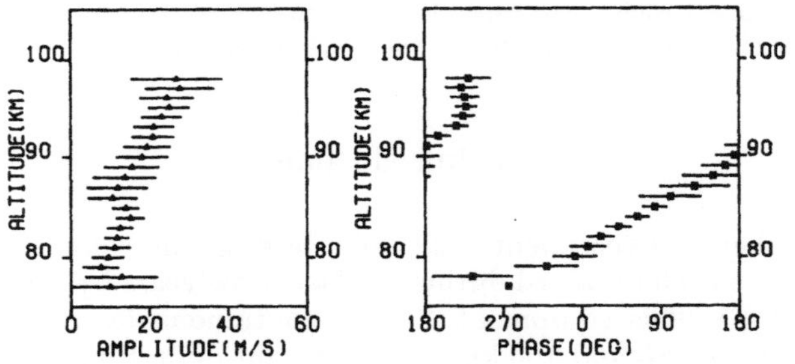

Fig. 2. Semi-diurnal tide, Garchy, 26 Jan.—6 Feb. 1970.

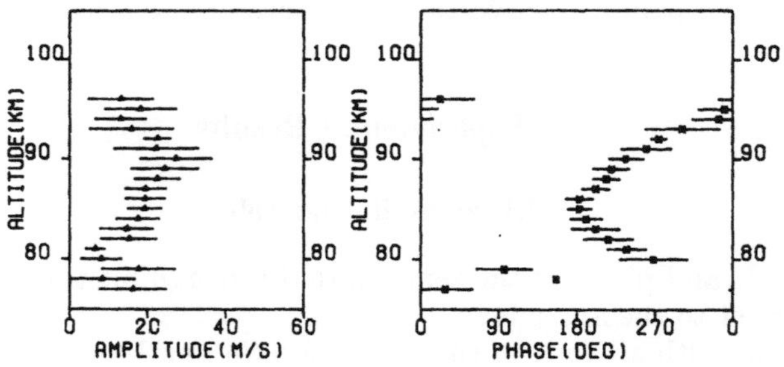

Fig. 3. Semi-diurnal tide, Garchy, 21—23 April 1970.

Lastly, in about 15% of the cases, phase variations with respect to the altitude differ significantly from linear growth and thus cannot be explained by the presence of a single, pure mode. For instance, the minimum of the phase at 85 km in April 1970 (Fig. 3) suggests the superposition of two waves or more.

2.2. Diurnal Tide

For the diurnal tide, three kinds of variations also appear in our results:

At first an oscillation of period 24 hours and vertical wavelength 20—30 km was observed in March and April 1970 (Fig. 4), confirming the presence of the fundamental migrating mode S_1^1, for which theory predicts a vertical wavelength

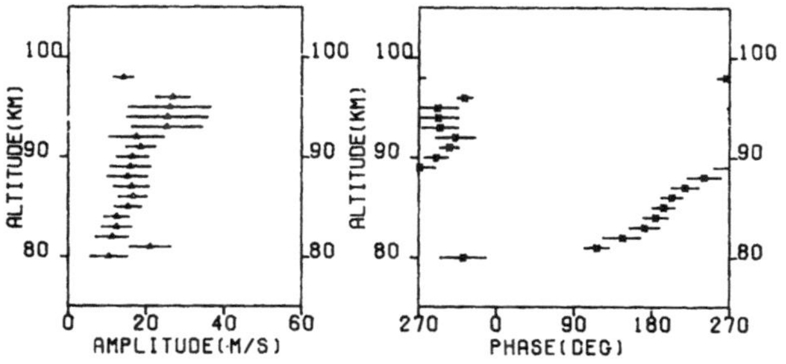

Fig. 4. Diurnal tide, Garchy, 24—26 April 1970.

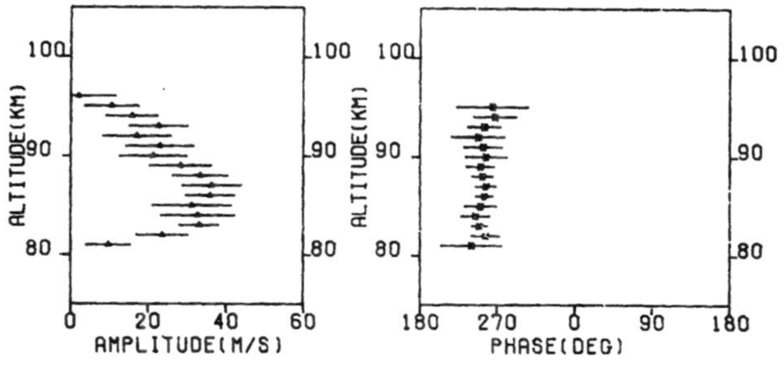

Fig. 5. Diurnal tide, Garchy, 25 July 1971.

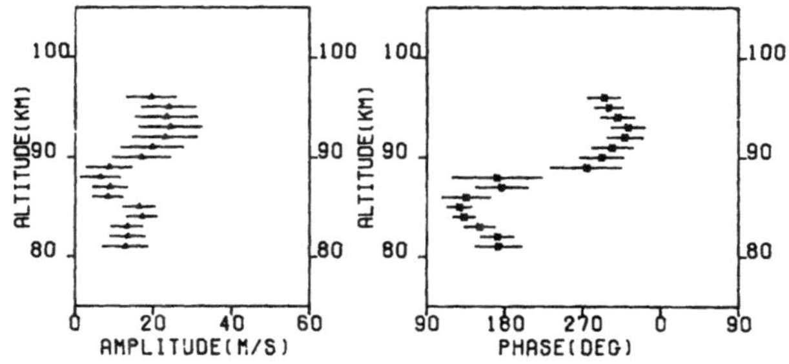

Fig. 6. Diurnal tide, Garchy, 14 Dec. 1970.

of about 27 km [3]. As has already been observed in previous measurements during summer months at Garchy [4], in June 1970 and July 1971 (Fig. 5) the amplitude of the diurnal tide decreases with altitude, while its phase remains stationary, as expected for the evanescent modes. But such a situation was also observed during the stratospheric warming of January—February 1970. Finally, in about 20% of the cases, the phase variations are significantly non-linear, as in the example of Fig. 6, where the phase undergoes a phase shift of 180° while the amplitude cancels. Such a configuration is typical of a standing wave system and suggests the superposition of direct and reflected waves.

3. Discussion

The vertical structure of the tides observed over Garchy is to be compared with the oscillating modes expected from theory, each of them propagating with a distinctive vertical wavelength.

Following the tidal theory [3], which assumes a thermal excitation by the ozone and water vapour heating, neither the semi-diurnal higher-ordered modes nor the evanescent diurnal modes seem to be strongly excited. New theoretical developments are then required to explain our observations of a semi-diurnal tide of short vertical wavelength and of a diurnal component non-vertically propagating. In particular, it seems necessary to take into account the great disturbances of the general circulation during stratospheric warming events.

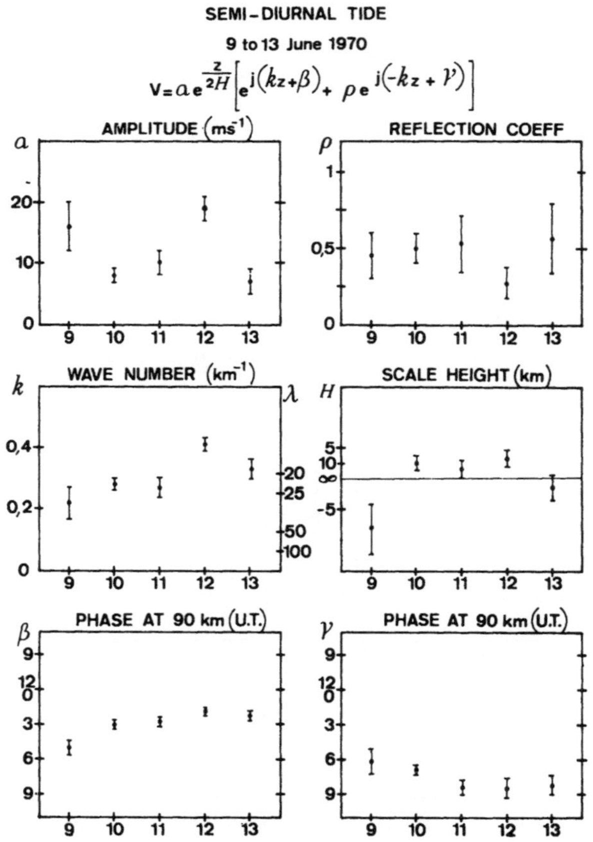

Fig. 7. Parameters of the direct and reflected waves computed by means of a least-squares method for the semi-diurnal tide, Garchy, 9—13 June 1970.

The other point to be discussed is the frequent evidence of superimposed waves. Reflection phenomena or non-linear effects may be implicated as was shown by Revah [5], Spizzichino [4] and other authors. In other cases, there might be superposition of different modes; a systematic analysis then becomes necessary to separate them.

Several possible methods have been investigated:

First of all, an attempt was made by computing the power spectrum $S(f, k)$ of the wind, where f is the frequency and k the vertical wave number. The resolution of this method is limited by the vertical dimension of about 20 km of the samples [1]. However, it is successful if the observed tide results from a single wave or from two waves of rather different wave number, two cases which often occur in practice. For this reason, regression methods were then used to fit the experimental data to two models of tide propagation, namely:

(a) two waves of opposite wave numbers (reflection case):

$$v(z) = a e^{\alpha z} \left[e^{j(kz+\beta)} + \varrho e^{j(-kz+\gamma)} \right];$$

(b) two waves with different wave numbers (general case):

$$v = e^{z/2H} \left[a_1 \exp \{ j(k_1 z + \beta_1) \} + a_2 \exp \{ j(k_2 z + \beta_2) \} \right]$$

the parameters involved in each case being estimated by a least squares method. Detailed outcomes are given in [6]. An example of the results obtained with the model (a) for the semi-diurnal tide in June 1970 is described in Fig. 7. The characteristics of both the direct and the reflected waves remain remarkably constant during all the campaign: the wave number k is near 0.3 km^{-1}, the reflection coefficient ϱ varies slightly around 0.5, the phases at 90 km β and γ also vary very little.

4. Conclusion

To sum up, the investigation of the vertical structure of tides at meteor heights shows in most cases the presence of one single wave, and occasionally the existence of reflection phenomena. The least-squares method allows us to recognize the waves present, but information on the horizontal structure of the tides is needed to confirm this analysis*. Moreover, further theoretical work is necessary to explain the occurrence of unexpected modes.

References

[1] A. Spizzichino, in: Thermospheric Circulation, Progress in Astronautics and Aeronautics, **27**, 117 (1972).

[2] J. L. Fellous, A. Spizzichino, M. Glass and M. Massebeuf, J. Atmos. Terr. Phys. (to be published) (1974).

[3] S. Chapman and R. S. Lindzen, Atmospheric Tides, D. Reidel Publ. Co., Dordrecht, Holland 1970.

[4] A. Spizzichino, Annls Géophys. **25**, 5 (1969).

[5] I. Revah, J. Atmos. Terr. Phys. **32**, 1313 (1970).

[6] J. L. Fellous, These de 3° cycle, Paris 1972.

* A new mobile radar is now operating. It was used in 1971—1972 for measurements of the NS component of the wind in the same region as the Garchy radar, and is to be installed in Kiruna (Sweden) in 1973. One can expect fruitful results from these experiments.

Lower Ionosphere Structure — Akademie-Verlag, Berlin 1974

METEOR RADAR OBSERVATIONS OF LONG PERIOD WAVES IN THE 80—100 km ALTITUDE RANGE

M. MASSEBEUF and M. GLASS

C. N. E. T./R. S. R., Issy-les-Moulineaux, France

Since 1970 long duration records of wind measurements (more than five days) have been carried out simultaneously at Garchy (47° N 3° E) by the French meteor radar and at other European stations. Results from Garchy exhibit oscillations of periods longer than one day. These were found during nine campaigns in 1970—1971, covering all seasons. Such oscillations with periods less than five days occurred in summer and winter. At the equinoxes waves with periods longer than five days are observed. The vertical wavelength of these waves is greater than 100 km. Preliminary comparisons have been made with the results obtained at Obninsk (USSR). They show that these oscillations propagate horizontally. This fact is particularly clear in the case of an oscillation with a period near two days, which is often observed.

It is hoped that a full paper will be published in *J. Atmos. Terr. Phys.*

CROSS-FIELD INSTABILITY AS A MECHANISM FOR EQUATORIAL E REGION IRREGULARITIES

B. V. Krishna Murthy and K. Sen Gupta

Vikram Sarabhai Space Centre, Trivandrum, India

Backscatter radar observations at Jicamarca have shown the presence of irregularities in the equatorial electrojet region, associated with low westward electron drift velocity. Estimates of electron drift velocity during the presence of Es on the ionograms in the electrojet region have also shown the association of Es irregularities with low electron drift velocities. To investigate whether cross-field instability mechanism can give rise to such irregularities, theoretical estimates of the instability parameters are made for daytime equatorial E region conditions. It is shown that this mechanism can give rise to growth of the instabilities below E region peak even when the eastward electric field and the corresponding electron drift velocity are quite low. It is found that the phase velocity of the instabilities does not vary with instability wavelength and varies directly with the eastward electric field. These results agree, in general, with the experimental observations on the irregularities in the equatorial electrojet region.

1. Introduction

Two types of irregularities at equatorial E region levels have been identified from the backscatter radar observations at Jicamarca [1, 2]. The two types are classified as Type I and Type II. Both types are closely associated with the equatorial electrojet. Type I irregularities have been found to occur when the electrojet is strong and exhibit characteristics mainly as predicted by the two-stream instability theory [2, 3]. For the two-stream instability, there exists a threshold for the electron drift velocity which is approximately equal to the ion-acoustic velocity in the medium which is about 360 m s^{-1} at E region levels. Type II irregularities are found to be associated with electron drift velocities much less than the ion acoustic velocity. As such, these irregularities cannot be explained by the two-stream instability mechanism. Balsley and Farley [2] suggested that the type II irregularities may be explained by the cross-field instability mechanism [4—7].

These irregularities at E region levels in the electrojet region give rise to the sporadic E echoes on daytime equatorial ionograms. It was found from estimates of electron drift velocity using ground magnetogram data, that sporadic E irregularities as seen on the ionograms exist even when the electron drift velocity is as low as 50 m s^{-1} [8]. In the present work, theoretical estimates of the cross-field instability parameters are made and these are examined in the light of observations on the irregularities in the electrojet region.

2. Estimation of Cross-field Instability Parameters

In order to estimate the instability parameters, the linear treatment of the cross-field instability mechanism given by Reid [5] is followed, In this, sinusoidal perturbations in the electron and ion density and electric field with a complex angular frequency and moving along the east-west direction are considered. The perturbation electric field has been assumed to be derivable from a scalar potential. The perturbation amplitudes of the density and the scalar potential are assumed to vary along the vertical direction with steady state electron density scale height. Under these assumptions, the imaginary part of the angular frequency can be expressed as

$$\omega_i = \frac{(C^- - C^+)(A^-C^+ - B^-D^+ - A^+C^- + B^+D^-)}{(C^- - C^+)^2 + (D^- - D^+)^2} + \frac{(D^- - D^+)(B^-C^+ + A^-D^+ - A^+D^- - B^+C^-)}{(C^- - C^+)^2 + (D^- - D^+)^2} \quad (1)$$

where

$$A^\pm = D_P^\pm K_z^2 + 2\alpha n_0 + H^{-1}\frac{d}{dh}(D_P^\pm) - D_P^\pm H^{-2} \pm E_{0x}\left[-\mu_P^\pm H^{-1} + \frac{d}{dh}(\mu_P^\pm)\right] + E_{0z}\left[\mu_H^\pm H^{-1} - \frac{d}{dh}(\mu_H^\pm)\right]$$

$$B^\pm = K_z\left[\pm\mu_P^\pm E_{0z} + \mu_H^\pm E_{0x} \pm \frac{d}{dh}(D_H^\pm)\right]$$

$$C^\pm = \pm\mu_P^\pm K_z^2 \pm H^{-1}\frac{d}{dh}(\mu_P^\pm) \mp 2\mu_P^\pm H^{-2}$$

$$D^\pm = K_z\left[-\mu_H^\pm H^{-1} + \frac{d}{dh}(\mu_H^\pm)\right]$$

K_z is the perturbation wave number; α is the recombination coefficient; E_{0z} and E_{0x} are the westward and upward electric fields respectively,

$$\frac{1}{H} = -\frac{1}{n_0}\left(\frac{dn_0}{dh}\right),$$

and h is the vertical height. Negative and positive superscripts represent quantities for electrons and ions respectively. $D_P^\pm$ and $D_H^\pm$ are the Pederson and Hall diffusion coefficients and$^\pm$ μ_P and $\mu_H^\pm$ are the Pederson and Hall mobilities respectively.

If the sign of ω_i is positive the perturbation grows exponentially and if it is negative the perturbation decays exponentially. The perturbation growth-time t is given by $1/\omega_i$. The phase velocity of the perturbation is given by $v = \omega_r/K_z$ where ω_r is the real part of ω.

ω_r is given by

$$\omega_r = \frac{(D^- - D^+)\,(A^-C^+ - B^-D^+ - A^+C^- + B^+D^-)}{(C^- - C^+)^2 + (D^- - D^+)^2}$$
$$- \frac{(C^- - C^+)(B^-C^+ + A^-D^+ - A^+D^- - B^+C^-)}{(C^- - C^+)^2 + (D^- - D^+)^2} \tag{2}$$

At the magnetic equator, during daytime the horizontal electric field will be directed towards the east. There will be a polarization field directed vertically upwards. This vertical polarization field is given by

$$E_{0x} = -E_{0z}\,\frac{(\mu_H^- - \mu_H^+)}{(\mu_P^- - \mu_P^+)} = -\frac{\sigma_2}{\sigma_1}\,E_{0z} \tag{3}$$

where σ_2 and σ_1 are the Hall and Pederson conductivities. The minus sign arises because, for eastward field ($-E_{0z}$), E_{0x} is upwards. In the expressions for ω_i and ω_r, Eq. (3) is substituted for E_{0x}. The important instability parameters which have been estimated are (i) the minimum electric field required for the growth of instabilities to make Eq. (1) numerically positive for a particular instability wavelength; (ii) the growth time for different wavelengths and electric fields; (iii) the phase velocity as a function of wavelength and electric field.

The electron density profile obtained by rocket measurement at Thumba (~0.6° S) at 1040 on 28 January 1971 [9] is used for estimating the above parameters. The collision frequency values are obtained from the models given by Thrane and Piggott [10] and Banks [11]. The electron density profile used and the estimated σ_2/σ_1 are shown in Fig. 1. For α, the recombination coefficient, a value of $3 \times 10^{-13}\ \mathrm{m^3\,s^{-1}}$ is assumed. The minimum electric field required for the

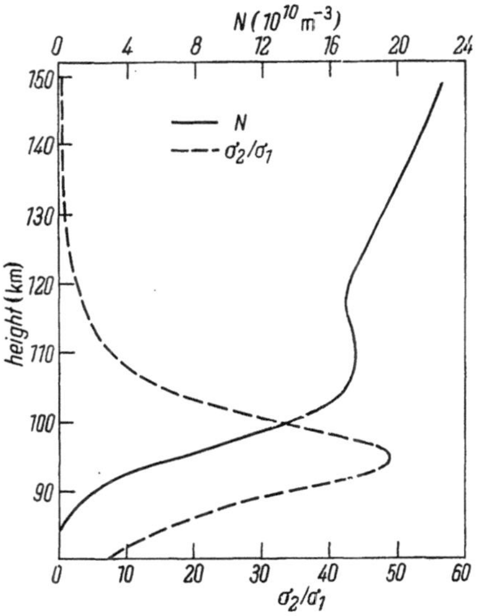

Fig. 1. Electron density (N) profile [9] used in the estimates and the height variation of σ_2/σ_1.

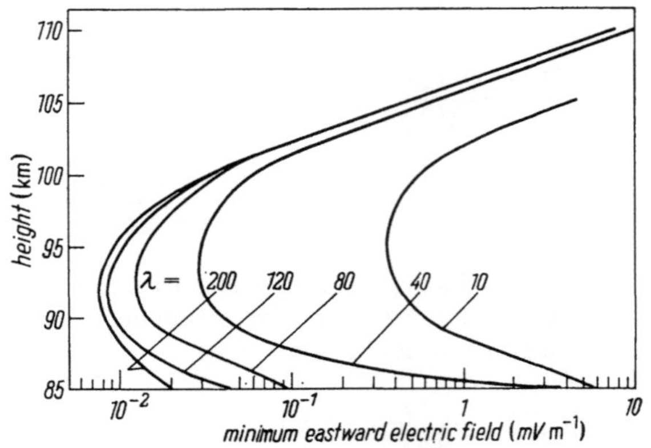

Fig. 2. Minimum eastward electric field plotted against height.

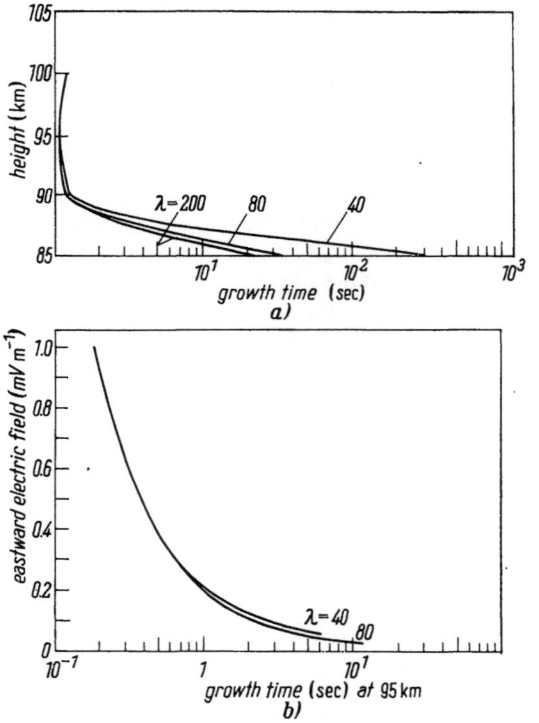

Fig. 3. a) Growth time plotted against height; eastward electric field 0.4 mV m^{-1}. b) Growth time plotted against eastward electric field.

growth of the instability for different wavelengths as a function of height is shown in Fig. 2. The results are shown only up to 110 km as the electric fields required are very high above this; they are even westward in a small region of negative electron density gradient around 120 km. It may also be noted that with a positive electron density gradient, an eastward field is necessary for the growth of the instabilities. It can be seen from Fig. 2 that the minimum electric field (E_{0z} min) required for the growth shows a significant variation with instability wavelength. At longer wavelengths the field required is smaller compared with that at shorter wavelengths. At 95 km altitude, E_{0z}(min) for $\lambda = 10$ m is 0.34 mV m^{-1} (eastward) and for $\lambda = 200$ m, E_{0z}(min) is 0.009 mV m^{-1} (eastward). It can be seen from Fig. 1 that E_{0z}(min) shows a minimum at about 95 km altitude for all the wavelengths considered. This is due to the fact that the vertical polarization electric field $\left(-(\sigma_2/\sigma_1)\, E_{0z}\right)$ reaches its maximum around this altitude (Fig. 1).

The growth time t $(= 1/\omega_i)$ is plotted against altitude for $E_{0z} = 0.4$ mV m^{-1} (eastward) and $\lambda = 40$, 80 and 200 m and shown in Fig. 3a. It can be seen that the growth times are minimum in the height range 90—100 km; in fact, there is not much variation in the growth times in the altitude range 90—100 km. Also, all the three wavelengths show essentially the same growth time ($\sim$1.2 seconds) in this altitude range. Below 90 km, the growth times increase rapidly with decreasing height and with decreasing wavelength. For $\lambda = 40$ m, at 80 km altitude, t is about 300 seconds. Growth time t as a function of electric field for $\lambda = 40$ and 80 m at 95 km altitude is shown in Fig. 3b. The growth times, as can be seen from this figure, decrease with increasing electric field, increase being rapid at lower electric fields.

The phase velocity v of the perturbation has been estimated and is shown as a function of height for $E_{0z} = 0.4$ mV m^{-1} (eastward) and $\lambda = 80$ m in Fig. 4a. As can be noted from the figure, the phase velocity decreases with decreasing

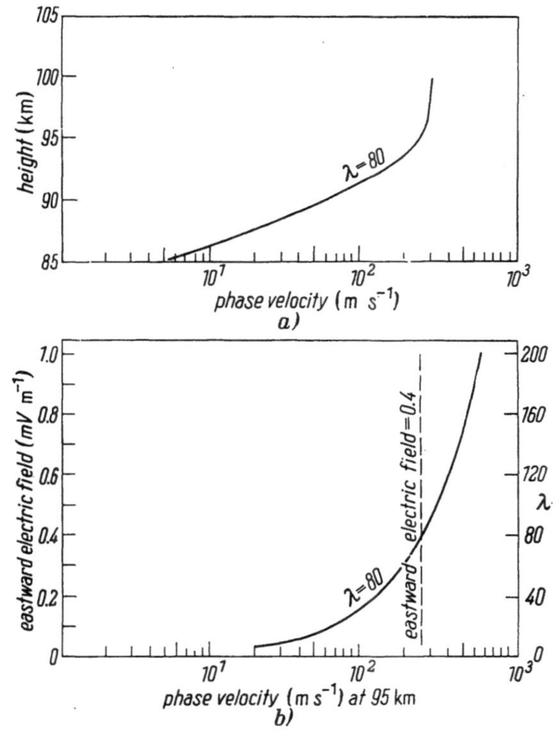

Fig. 4. a) Phase velocity plotted against height; eastward electric field 0.4 mV m^{-1}. b) Phase velocity plotted against eastward electric field.

altitude, the decrease being rapid at lower heights. The velocity v is shown as a function of electric field in Fig. 4b. It can be seen that the velocity increases with the increase of electric field, the increase being rapid at lower values of electric field. In the same figure a dotted line shows the velocity as a function of wavelength at 95 km altitude for $E_{0z} = 0.4$ mV m^{-1} (eastward). It is interesting to note that in the wavelength range 10—200 m, v does not show any significant variation with wavelength. The velocity is about 260 m s^{-1} at 95 km for $E_{0z} = 0.4$ mV m^{-1} (eastward). This means that there is practically no dispersion in the wave propagation at least for the wavelength range considered.

To investigate the effect of the magnitude of the electron density gradient, the instability parameters are estimated, changing the electron density gradient to an arbitrary value. Keeping the electron density at 115 km as in the profile of Fig. 1, the gradient is made half and twice the original value and the minimum electric field (E_{0z}min) required for the growth has been estimated for both these cases. E_{0z}min values are shown in Table 1 below for different wavelengths at 95 km altitude.

Table 1

Variation of E_{0z}min (eastward) (mV m^{-1}) at 95 km with Change in Density Gradient

	E_{0z}min		
λ/m	$^1/_2\nabla n_0$	∇n_0	$2\nabla n_0$
10	0.72	0.35	0.18
40	0.07	0.03	0.18
80	0.038	0.014	0.01
120	0.032	0.01	0.009
200	0.029	0.009	0.009

It is seen from this table that E_{0z}min shows a decrease with increasing electron density gradient and that the effect is more pronounced at shorter wavelengths.

3. Discussion

In the present investigation it has been shown by quantitative estimates that the cross-field instability mechanism can give rise to instabilities at E region levels at the magnetic equator under suitable conditions. It has been shown that under the influence of an eastward electric field in the underside of the E region, growth of electron density instabilities takes place. An eastward electric field exists at E region levels in the daytime equatorial region. Reid [5] using night-time electron density data has shown that the cross-field instability grows in the underside of the E region for an eastward electric field and above the E region maximum for a westward electric field. Reid has not included explicitly the vertical polarization field in his calculations. Using an arbitrary electron density distribution, Whitehead [7] has shown that the cross-field instabilities can grow at equatorial latitudes. The results of the present investigation are in general conformity with those reported by Reid and Whitehead. It may be pointed out here, that a daytime electron density profile obtained by rocket measurement at the magnetic equator has been used in the quantitative estimates and also the vertical polarization field which makes major contribution to the growth of the instabilities has been taken into consideration in the present investigation.

From a study of sudden disappearance of Es in the electrojet region the authors [8] have reported that Es was present on ionograms even when westward electron drift velocity was as low as 50 m s^{-1}. This velocity corresponds to about 0.05 mV m^{-1} of eastward electric field. Using the horizontal magnetic field values from ground magnetograms at Thumba, the E region east-west electric field at the time of morning onset of Es at Thumba for magnetically quiet days in the year 1970 has been estimated following the procedure adopted in [8]. It was found that the morning onset of Es takes place, in general, when the electric field is eastward and in the range 0.075—0.125 mV m^{-1}. Occurrences have also been noted when E_z is in the range 0.025—0.075 mV m^{-1} (eastward). The electron drift velocity corresponding to these E_z values is found to be much less than the ion acoustic velocity (360 m s^{-1}) at E region levels. As such, the two-stream instability mechanism cannot take place at these times to give rise to sporadic E irregularities. Also it may be pointed out here that at lower operating frequencies (16 MHz) backscatter radar observations [2] did not show the presence of type I irregularities; thus at ionosonde operating frequencies (1—25 MHz) it is likely that type I irregularities do not give rise to any return signals. It can be readily seen that the low electric field values during the presence of Es on ionograms are sufficient to enable the growth of the cross-field instabilities. As the cross-field instability mechanism depends upon the electron density gradient also, the required electric field for instability growth may be different on different days as the electron density gradient varies.

It is estimated that for an eastward electric field of 0.4 mV m^{-1} and for the electron density profile used in the present investigation, the electron drift velocity is about 550 m s^{-1}. The phase velocity of the instabilities obtained for the same conditions is less than this value of electron drift velocity. This shows that the instabilities move with phase velocities less than the electron drift velocity.

In conclusion, it can be said that the type II irregularities observed on backscatter radar and the Es irregularities observed on the ionosonde when the electron drift velocity is smaller than the ion acoustic velocity, can be attributed to the cross-field instability mechanism.

References

[1] B. B. Balsley, J. Geophys. Res. **74**, 2333 (1969).
[2] B. B. Balsley and D. T. Farley, J. Geophys. Res. **76**, 8341 (1971).
[3] D. T. Farley, J. Geophys. Res. **68**, 6083 (1963).
[4] S. Kato, Radio Sci. **7**, 417 (1972).
[5] G. C. Reid, J. Geophys. Res. **73**, 1627 (1968).
[6] A. Rogister and N. D'Angelo, J. Geophys. Res. **75**, 3879 (1970).
[7] J. D. Whitehead, J. Geophys. Res. **76**, 3116 (1971).
[8] B. V. Krishna Murthy and K. Sen Gupta, Planet. Space Sci. **20**, 371 (1972).
[9] B. H. Subbaraya, P. Muralikrishna, T. S. G. Sastry and Satya Prakash, Planet. Space Sci. **20**, 47 (1972).
[10] E. V. Thrane and W. R. Piggott, J. Atmos. Terr. Phys. **28**, 721 (1966).
[11] P. Banks, Planet. Space Sci. **14**, 1105 (1966).

Lower Ionosphere Structure — Akademie-Verlag, Berlin 1974

LATITUDINAL AND TEMPORAL VARIABILITY OF TEMPERATURE IN THE LOWER THERMOSPHERE

R. H. WAND and J. E. SALAH

Lincoln Laboratory, Massachusetts Institute of Technology, Lexington, Mass., USA

Incoherent scatter measurements of temperature in the lower thermosphere (100—120 km) at three radar stations (Arecibo (18° N), Millstone Hill (42° N) and St. Santin (45° N)) were analyzed to determine the influence of thermal tides propagating from the lower atmosphere. Observations during nine single or consecutive daytime periods were studied from Arecibo (1966—1970), 21 periods (1971—1972) from Millstone and 25 periods (1967—1970) from St. Santin. In general, the temperature exhibited a 12 hour oscillation having a downward phase progression with height. The average characteristics of this oscillation are found to be in excellent agreement at the two mid-latitude stations. The average amplitude reaches a maximum of 10% of the mean at a height of 115 km and a local time of 06.3 hours, the vertical wavelength increasing from 45 km near 110 km to 65 km near 120 km. From a comparison with theoretical predictions we conclude that this oscillation is the (2,4) tidal mode; however, the rate of damping above 115 km is larger than given by dissipative theory. The results also reveal a large temporal variability in the amplitude and phase of the tidal oscillation. Seasonal temperature averages indicate a smaller amplitude in winter than in summer or equinoxes, because of the larger phase variability in winter. At Arecibo, most of the observations were made in winter and show the same variability as at mid-latitudes. The vertical wavelength agrees with that of the (2,4) tidal mode, but the average amplitude and phase do not support theoretical predictions for low latitudes.

A full version of this paper will appear in *J. Geophys. Res.* under the title "Tides in the Temperature of the Lower Thermosphere at Mid-Latitudes" by J. E. Salah and R. H. Wand.

Lower Ionosphere Structure — Akademie-Verlag, Berlin 1974

DRIFT MEASUREMENT WITH SPECTRAL ANALYSIS DURING PERIODS OF CHEMICAL RELEASES INTO THE IONOSPHERE*

W. PFISTER

Air Force Cambridge Research Laboratories, L. G. Hanscom Field, Bedford, Mass., USA

**Dedicated to Professor Rawer on the occasion of his 60th birthday*

The spectral analysis of complex signals reflected from the ionosphere allows us to separate echoes in terms of Doppler shift and angle of arrival. The instrumentation used consists of a digital ionosonde in combination with a multiple-frequency complex amplitude scanner and three spacedloop-antennas. The tape recorded data are Fourier transformed digitally and cross-power spectra are obtained with the method of frequency averaging. Visual displays retaining numerical information can be obtained with a special digicoder and printer. An example of such a display is the combination of power, phase, and coherency spectra printed as function of time. The event selected is a short-lived sporadic E layer which appeared 65 minutes after a chemical release from a rocket and which was observed on two frequencies each with three slightly different range gates. The results are basically consistent with the concept of a wave-like structure found previously.

1. Introduction

On 14 April 1972 two rockets were fired in Eglin, Florida, with chemical releases. The first one was launched at 1845 CST and released iron carbonyl on ascent between 75 and 135 km and lithium, barium and boron on descent between 135 and 87 km. The second one was launched at 1904 CST and released barium puffs between 155 and 212 km during ascent. During the same period and at the same location "drift" measurements were carried out on two frequencies and continued for several hours. No specific effort was made to track the echo coming from the artificial electron cloud. The interest was in the normal E and F layers and in the stronger reflected echoes. When after 20 h a dense Es layer developed and the F region echoes were blanked out the gates on the higher frequency were shifted to the Es echo and were kept there for three recording periods 2016, 2021, and 2026 as long as the dense Es layer lasted. This event is used to illustrate the application of the spectral analysis technique to the motion of reflection areas in the ionosphere.

2. Measurements and Spectral Analysis

The concept of the "drift" equipment used is to record the complex amplitude of the echo pulses in quick succession on two different frequencies, three different height gates and on three closely spaced antennas in digital format. This has been easily accomplished by adding a new unit called Complex Amplitude Multifrequency Scanner (CAMS) and a second recording device to the basic units of the digisonde 128, which has the capability to synthesize in units of 25 kHz any frequency between 1 and 16 MHz ahead of every transmitted pulse.

A description of the functions and operation of the CAMS is given in LTIRF Report No. 358/IP [1]. It suffices to mention that it has dial settings for the operating frequencies, height gates and gains, and that it measures quadrature amplitudes of each pulse with a comparator and digital counter at the selected range gate. These are recorded on an incremental tape recorder as three characters per pulse, one six-bit character containing the real component, another one the imaginary component and the middle character containing the sign information of both. With a pulse repetition rate of 100/sec, it takes 18 pulses to complete one sampling point for six frequencies (or range gates) and three antennas.

The sampling takes place at a rate of four data points per second, so that 7/100 of a second can be used to record part of the preface, that is, date, time, frequencies, and so on. Parallel to the tape recorder is a Magnavox printer fed by a so-called digicoder which was developed for ionograms. The printout of the raw "drift" data is not as instructive as the printout for ionograms; however, it serves as a monitor for the instrument and its proper operation.

The spectral analysis is carried out in two steps. The first is a fast Fourier transform of the 18 sets of raw data in a subcase consisting of 512 sampling points. The result is 18 sets of complex amplitude spectra. They are called Fourier spectra and recorded in a suitable format on tape and displayed on a digicoder printout with alternating lines for the real and imaginary components. The Versatec printer can print 256 characters per line. Therefore, half of the 4 Hz wide Doppler range can be printed on one page. Usually, most of the observed Doppler shifts fall in the range from -1 to $+1$ Hz. Since the sign information is likely to be omitted from the printout (but not from the digicoder tape) and since the phase information requires a detailed examination of the printed numbers, the overall impression of the printout is that of a power spectrum.

The Fourier spectrum contains all the relevant information of the raw data, but the printout is much more instructive with respect to the reflection characteristics of the ionosphere. Therefore, it is desirable to have the 18 Fourier spectra recorded and displayed in real time and to engage in the development of a suitable Fourier spectrum analyzer.

The next step in the spectral analysis is the formation of power spectra, phase spectra and coherency spectra by multiplying the complex amplitude with the appropriate complex conjugate and frequency averaging over neighboring Doppler frequencies. If spectral amplitudes of different antennas (or operating frequencies) are multiplied and these frequency components averaged a complex cross-power spectrum is obtained which can be represented by a phase spectrum and a coherency spectrum (normalized modulus of the cross-power spectrum). The weighting functions chosen for the frequency averaging are such that the result is mathematically identical to a Fourier transform of the cross-correlation function with ± 64 lags smoothed by a Hanning filter.

The computer output is again in the format of a digicoder tape. An example of a printout is shown in Fig. 1. The averaging reduces the Doppler frequencies by a factor of four so that the range between -2 and $+2$ Hz can be displayed by 128 characters, and the two operating frequencies can be printed side by side. A record consists of nine lines. The sequence in a trio of the lines is according to

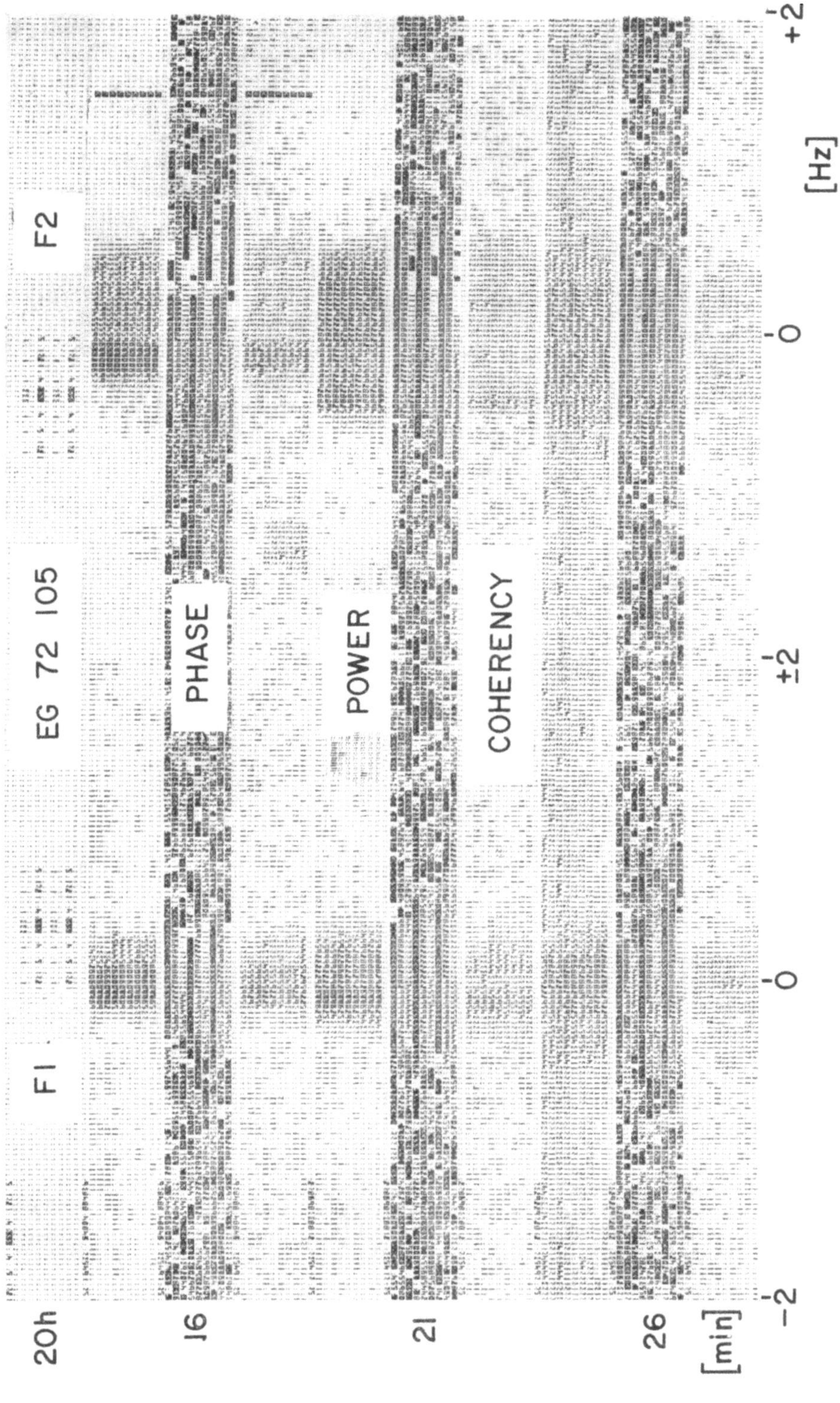

Fig. 1. Digicoder printout of power, phase, and coherency spectra for sporadic E event.

the antenna or antenna pair, respectively, and the sequence of the trios is according to the range settings as indicated in positions 12, 13, 14 of the preface.

The first record of a subcase shows the (with respect to mean power) normalized power spectrum in a mixed linear and logarithmic scale. Character 4 is the transition with a value of 1 for the normalized power spectrum. In the digicoder printout only the four most significant bits are used. However, all spectral data are recorded in nine bits (three characters per two data points) on the tape. The second record in the subcase gives the phase spectrum in linear scale with character 8 meaning zero phase and character zero meaning $-180°$. The third record in the subcase is the coherency spectrum in a $\tanh^{-1}$ scale such that, for instance, a coherency between 0.9579 and 0.9776 would be printed as a 6. With exception of the phase spectrum, which contains meaningful information, only when the coherency shows characters larger than 1 (coh > 0.31), it is evident from Fig. 1 that this presentation of processed data is convenient for an overall review of the recorded data and for a selection of parts of data which merit further studies.

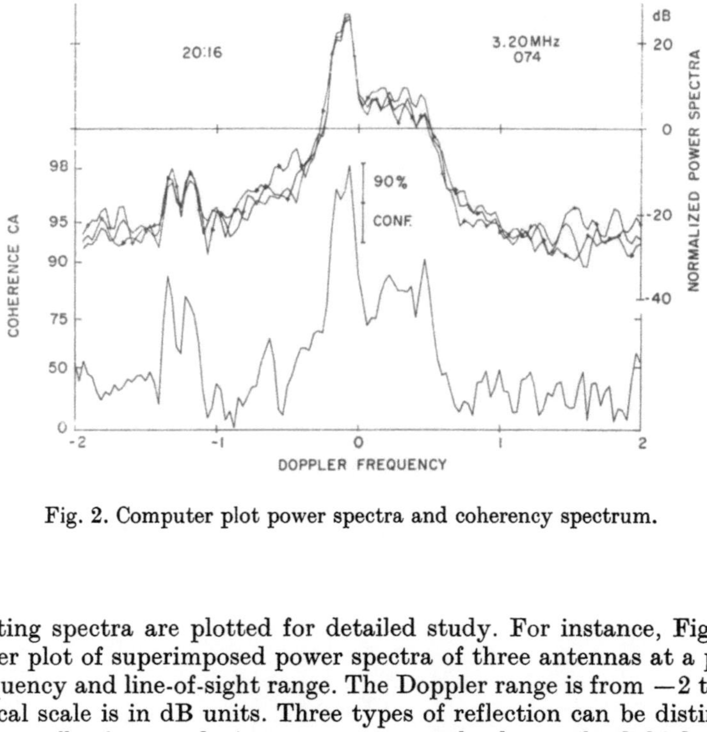

Fig. 2. Computer plot power spectra and coherency spectrum.

Interesting spectra are plotted for detailed study. For instance, Fig. 2 shows a computer plot of superimposed power spectra of three antennas at a particular time, frequency and line-of-sight range. The Doppler range is from -2 to $+2$ Hz. The vertical scale is in dB units. Three types of reflection can be distinguished: A specular reflection producing two not completely resolved high peaks at -0.08 and -0.14 Hz, a band of scattered reflections between 0 and $+0.6$ Hz, and two unusual and discrete reflections with a high Doppler shift of -1.32 and -1.19 Hz. The coherency between two of the antennas is shown underneath. The $\tanh^{-1}$ scale of the coherency has the advantage that the variance or confidence interval is a constant length depending only on the degree of freedom and the chosen confidence limit. A 90% limit gives a length as marked in the figure.

3. Arrival Angle and Doppler Observations during Es Event

What remains in the data-processing procedure is a translation of the spectral phase angles into spatial angles of arrivals. This together with the Doppler shifts of the various reflection areas gives a good observational starting point for any interpretation of the structure and dynamic of the specific ionospheric event. Using the spectral phase between pairs of antennas at a number of Doppler shifts characterized by peaks in the power spectrum and the position of the antennas, one obtains the sine of the zenith angle and the azimuth of the reflection area characterized by this particular Doppler shift. This corresponds to a point in a horizontal E—W, N—S plane, as shown in Figs. 3a and b where the distance from the origin describes the sine of the zenith angle in units of 0.1. An error circle around this point is derived from the value of the coherency. Three different symbols are used: A triangle for the central height gate setting, a $\times$ for the lower gate setting, 3 km below, and a $+$ for the higher gate setting 3 km above. The numbers stand for the Doppler shift in hertz and are usually shared by all three

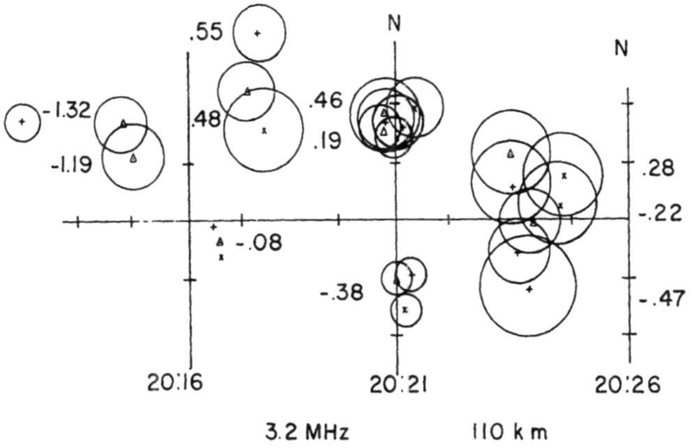

Fig. 3a. Location of reflection areas for sporadic E event seen at 3.20 MHz.

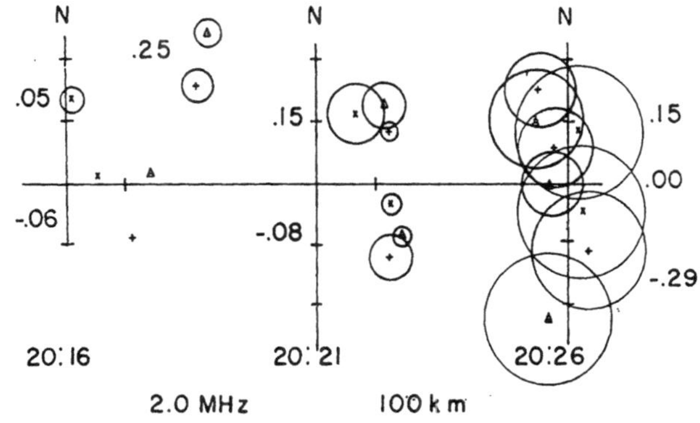

Fig. 3b. Location of reflection areas for sporadic E event seen at 2 MHz.

gate settings. Occasionally, peaks in the power spectrum show up only in one or two of the range gates.

With one exception, on both frequencies or height levels, all reflections with negative Doppler shift appear in the south and those with positive Doppler shift in the north. Even when an easterly or westerly component is observed, these components are shared by all Doppler shifts. This is a clear indication of a motion of the ionospheric structure toward south and there is no significant difference within the 10 km height range given by the reflection levels of the two frequencies. However, the median velocity shows an increase from 75 m s^{-1} at 100 km to 95 m s^{-1} at 110 km. The tendency of the reflection areas to align themselves in a straight line indicates a wave-like structure as found previously [2]. That the straight line, at times, does not pass through the origin indicates an inclination of the wave-like structure in the east-west direction. One would expect that the inclination should be the same at both frequencies. This is not contradicted by the present results, which may contain a systematic error, because the antennas have not been calibrated.

The exception to the regular pattern mentioned before is seen in Fig. 3a at the time 2016. Two reflecting areas with an unusually high negative Doppler shift noted already in Fig. 2 as weak signals fall completely outside the straight line pattern. Since the main point here is to demonstrate the capabilities of the technique rather than to interpret every detail of the measurements, no suggestion for an explanation is being offered.

Acknowledgments

The instrumentation used and presented in this study was designed, developed and operated by staff of the Lowell Technological Institute Research Foundation under the direction of K. Bibl and B. Reinisch. Credit also goes to S. Mazzola, R. Boudreau, W. McComish and V. Varadarajan from Boston College for their contribution to various computer programs used in the report.

References

[1] K. Bibl and B. W. Reinisch, Lowell Technological Institute Research Foundation Rep. LTIRF-358/IP (1973), AFCRL-TR-73-0556.
[2] W. Pfister, J. Atmos. Terr. Phys. **33**, 999 (1971).

Lower Ionosphere Structure — Akademie-Verlag, Berlin 1974

RESULTS OF SIMULTANEOUS WIND VELOCITY PROFILE MEASUREMENTS IN THE LOWER THERMOSPHERE BY THE METEOR RADAR AND ROCKET METHODS

L. A. Andreeva and Yu. D. Ilyichev

Hydrometeorological Service, Moscow, USSR

Four averaged wind velocity profiles for the 90—100 km altitude region are compared; these profiles were obtained from meteor trail and artificial luminous cloud drift observations. Good agreement of results obtained by the two methods is observed.

In investigation of the upper atmosphere dynamics it was found that the altitude wind profiles, which are usually obtained by rocket or meteor radar methods, are of particular interest. As is known, in the latter case special equipment for the angle coordinate measurements is used, but the adjustment and operation of this equipment, calling for careful calibration, are rather complicated. This considerably limits the wide use of the equipment for the angle coordinate measurements by the radar network which is at present being developed in order to investigate the global lower thermospheric circulation.

In this connection the method, suggested in [1], of altitude wind profile determination using meteor radar data, obtained by means of special statistical treatment (without any equipment for the angle coordinate measurements), is considered to be a promising one.

With the aim of testing the method suggested in [1], simultaneous measurements of the altitude wind velocity profiles were carried out at Volgograd in September 1971 using artificial luminous clouds (ALC) and by means of the meteor radar method.

The clouds were formed by means of ejection of a mixture of trimethyl-aluminium and triethyl-aluminium and were photographed from two stations by aerophotographic cameras, with relative lens openings of 1:2.5 and 1:3.5 and focal distances of 21 and 75 cm respectively. In order to determine the vertical wind profiles, the ALC photos were analysed according to the method given in [2]. Average measurement errors of the altitude and wind velocity determinations from 90 to 100 km were not greater than ± 0.4 km and 3 m s^{-1} respectively.

Meteor radar measurements were carried out alternately in the NS and EW directions with periodicity of 30 minutes. Areas of luminous clouds, overlapping the meteor zone at the beginning of their formation, were observed in the region, "illuminated" by the diagrams of radar antennae with their orientation to the east; 15 minutes before and after the rocket launch the radar sounding was carried out in the EW direction. The scheme of relative positions of ALC and regions of radar sounding is given in Fig. 1. Single wind velocity values, measured

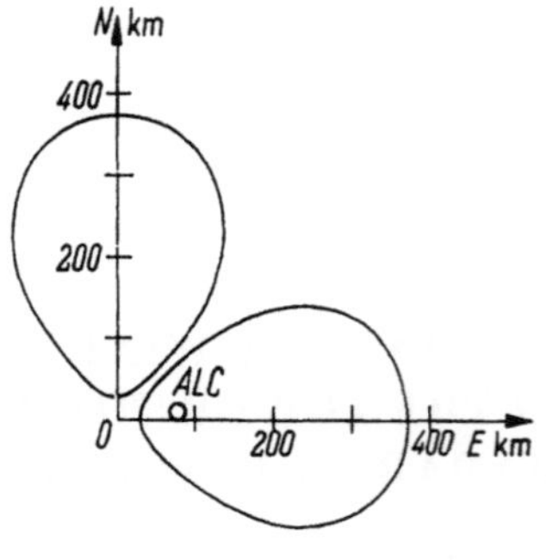

Fig. 1. Scheme of relative positions of ALC and lines of equal sensitivity of radars (at the level of 0.3—4) at 95 km.

by the meteor radar method, were calculated according to the standard technique [3].

For qualitative comparison of results obtained using methods given in [1] and [2], the linear experimental data models were used:

$$V^p = V^p_{95} + K^p(z - 95) + \xi,$$

$$V^m = V^m_{95} + K^m(z - 95) + \xi$$

where V^p and V^m are wind velocities from the rocket and meteor data, $V^{p,m}_{95}$ are values of these velocities at 95 km, z is the altitude, ξ is disagreement. The parameters $V^{p,m}_{95}$ and $K^{p,m}$ were calculated by the least squares method.

Measurement results and models of the form [1] are given in Fig. 2. The meteor radar given in this figure were obtained by averaging (by smoothing over altitude) the single velocity values and the altitudes of meteor trail reflecting points for the period of simultaneous measurements.

To calculate the root-mean-square altitude deviations from the average altitude interval (taking into account the interval sizes and measurement error of the single altitude values of the meteor trail reflecting points in this interval) the following dependence is obtained:

$$\sigma_{z(R_1,R_2)} = \left[\frac{\sum_{i=1}^{N} [z(R_1, R_2) - z(R_i)]^2 - \sum_{i=1}^{N} \sigma_{z(R_i)}}{N-1}\right]^{1/2}$$

where $z(R_i)$ is the most probable altitude of the meteor trail reflecting point with the slant range R_i; $z(R_1, R_2)$ is the mean altitude of the altitude interval, the lower and upper boundaries of which are defined by the R_1 and R_2 slant range values respectively; $\sigma_{z(R_i)}$ is the root-mean-square measurement error of the single altitude value of the meteor trail reflecting point with the slant range R_i; N is the number of radar echoes registered in a given altitude interval.

The mean measurement error of $z(R_1, R_2)$ was calculated according to

$$\Delta Z(R_1, R_2) = \frac{\sigma_{z(R_1,R_2)}}{N^{1/2}}.$$

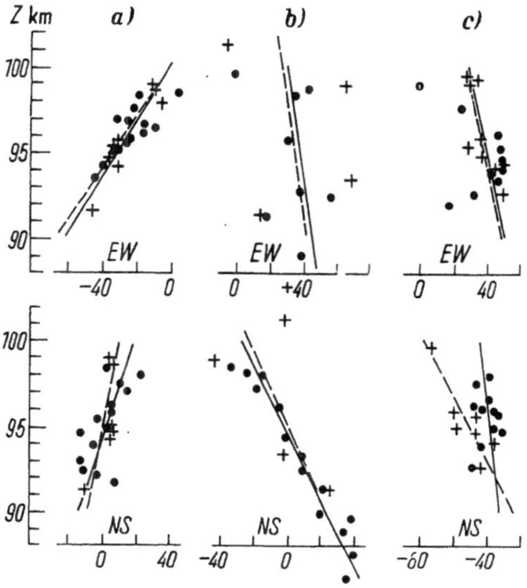

Fig. 2. Results of simultaneous measurements; (●, ——) meteor radar data; (+, ---) rocket data; (a) launch on 14 Sept. 71 at 2210 LT; (b) launch on 15 Sept. 1971 at 0140 LT; (c) launch on 15 Sept. 1971 at 2020 LT. Wind velocity (m s^{-1}) is given along the X-axis.

Evaluations show that for altitudes close to 95 km the values vary within the limits $\pm(1.8-2.2)$ km; for altitudes close to 100 km the limits are $\pm(3-5)$ km. At altitudes close to 100 km the $\Delta Z(R_1, R_2)$ value increase is explained by the fact that the statistical provision of measurement results decreases for these altitudes.

It is seen from the data (Fig. 2) that good agreement of the models obtained by the meteor radar and rocket methods is observed. The correlation coefficients r_k and r_v of the model parameters (K^p and K^m, V_{95}^p and V_{95}^m) and significance levels corresponding to them were calculated for numerical evaluation of this agreement. As a result one obtains:

$$r_k \approx r_v \approx 0.96; \qquad \sigma_{r_k} \approx \sigma_{r_v} \approx 0.075.$$

Results of comparison show that the standard methods of meteor trail reflecting point determination make it possible to obtain the averaged wind profiles for the 90—100 km altitude region with reasonable accuracy and may be used for dynamic characteristic measurements of the meteor zone wind regime by radars without any equipment for the angle coordinate measurement.

References

[1] Yu. D. Ilyichev Proc. Inst. Exp. Met. **24**, (1970).
[2] D. B. Uvarov, Proc. Inst. Exp. Met. **15** (1969).
[3] I. A. Lysenko, Astron. J. **41**, 1 (1963).
[4] B. L. Kashcheyev, V. N. Lebedinets and M. F. Lagutin, Meteor Phenomena in the Earth's Atmosphere, Nauka, Moscow 1967.

EVIDENCE OF FRONTAL STRUCTURES IN NON-BLANKETING SPORADIC E LAYERS

R. H. Clarke

Australian Defence Scientific Service, Department of Supply, Salisbury, South Australia

Two radio-wave sounding techniques have been refined in order to examine the structure of partially transparent (i.e. non-blanketing) sporadic E layers. Firstly, an HF interferometer has been developed to provide angle-of-arrival data on occasions when the downcoming signal arrives from a reasonably well-defined direction. On one occasion, records covering the 2—6 MHz frequency band provide good evidence that a sporadic E layer comprised three frontal clouds, each aligned in a north-south direction though possessing different peak electron concentrations.

Secondly, an ionosonde has been adapted to record ordinary and extraordinary mode traces on alternate ionograms. One ninety-minute sequence of such ionograms recorded in rapid succession indicates that the observed temporal variations in the ordinary and extraordinary mode blanketing and critical frequencies of the Es layer were caused by the horizontal movement of frontal overdense clouds of ionization on that occasion.

1. Introduction

Irregular spatial variations are an important feature of sporadic E layers. Such variations are of considerable concern in studies of HF propagation via non-blanketing Es layers, as it is likely that they are responsible for the layer's partial transparency to vertical incidence radio waves. The actual reflection mechanisms are uncertain, but possibilities include scattering from vertical irregular structures and off-vertical reflections from overdense clouds of ionization.

The observations to be described in this paper were made in 1971 at the University of Queensland Radio Research Station, near Brisbane, Australia. One set was obtained using a directional ionosonde which provides angle-of-arrival data over the 2—10 MHz frequency band. The second set of observations was made with an ionosonde which has been adapted to record ordinary and extraordinary rays on separate ionograms. The symbols used are as follows: f_{ob}Es, ordinary-mode blanketing frequency; f_{xb}Es, extraordinary-mode blanketing frequency; f_oEs, ordinary-mode critical frequency; f_xEs, extraordinary-mode critical frequency; f_B, electron gyrofrequency.

2. Instrumentation

The circularly polarized receiving aerials are similar in principle to that described by Phillips [1]. Their performance is such that near-vertical echoes of the unwanted mode of polarization are attenuated by at least 25 dB over the frequency band extending from 2 to 10 MHz.

Four such receiving aerials are utilized in the directional ionosonde. These lie on the corners of a square, the diagonals of which measure 100 m and run in magnetic north-south and east-west directions. The angle of arrival of the down-coming echo is derived from the phase differences between the signals received at the north and south aerials (φ_{NS}) and at the east and west aerials (φ_{EW}) in a similar manner to that employed by Clarke [2].

The transmitting section of the directional ionosonde is essentially identical with that of a conventional ionosonde; the receiving and phase-measuring sections are shown in simplified form in Fig. 1. The phase difference between the signals received at a pair of aerials is preserved along the IF strips and, in the absence of calibration errors, it equals the phase difference between the two 30 kHz output waveforms. Both φ_{NS} and φ_{EW} are recorded in analogue form—typically, over the band of frequencies corresponding to which echoes are returned from the ionospheric layer being investigated. The overall accuracy of the phase-measuring gear is $\pm 10°$.

This technique also provides an indication of whether the downcoming echo consists of a single component or two or more components of comparable magnitude, but incident from different directions. For example, the patterns shown in Fig. 2 were calculated under the assumption that the incident echo consisted of two components, the ratio of whose amplitudes equalled 0.7 and which arrived from directions such that the incident wave fronts subtended angles of $-5.5°$ and $-7.1°$ with the north-south axis (the minus sign indicating a southerly direction of arrival). The computed phase difference φ_{NS} between resultant signals at the north and south aerials shows a cyclic variation with frequency, an example of which will be shown in the next section. When the incident echo consists of three or more components of comparable amplitude, however, the frequency variation of the resultant phase difference is quite irregular, rapid variations over the range from $-180°$ to $+180°$ being the norm.

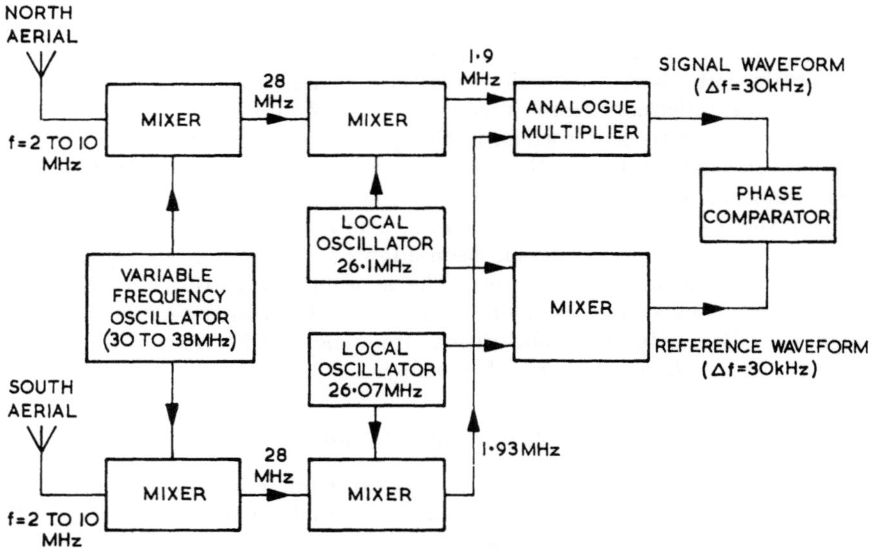

Fig. 1. Simplified block diagram, illustrating the basic components of the directional ionosonde.

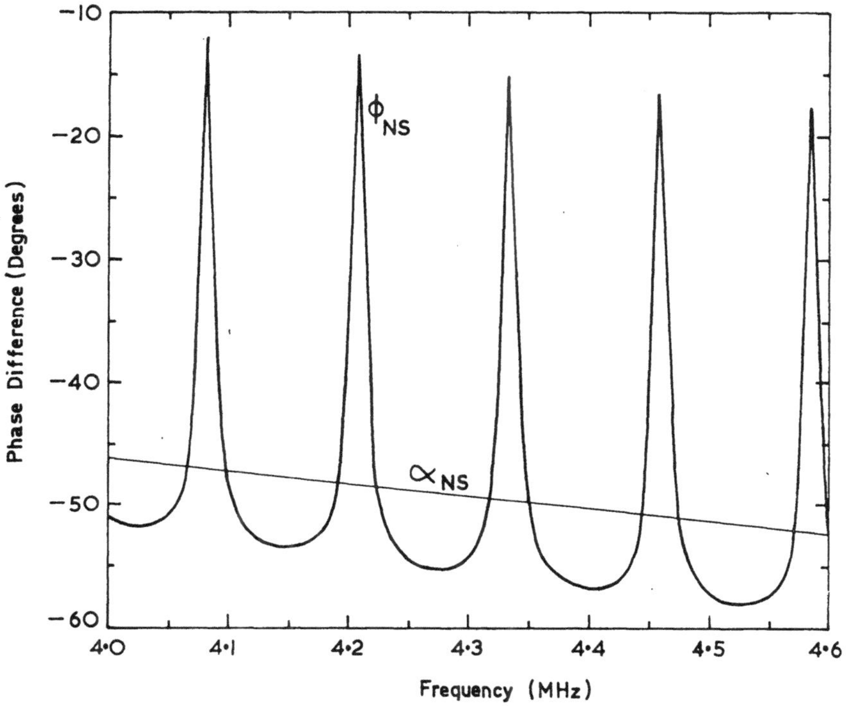

Fig. 2. The computed phase difference (φ_{NS}) between the resultant signals received at two aerials (horizontal spacing = 100 m). The incident signals are assumed to comprise two components of relative amplitude 0.7 which arrive from directions such that the north-south zenithal components of their angles of arrival equal −5.5° (strong echo) and −7.1° (weak echo). Also shown is the phase difference (α_{NS}) corresponding solely to the strong echo.

3. Angle of Arrival Observations of a Non-blanketing Sporadic E Layer

The first event to be discussed is represented by the phase difference, frequency curves shown in Fig. 3. They were recorded at 1230 on 1 June 1971 and represent returns from a sporadic E layer whose o-mode blanketing and critical frequencies were 4.3 MHz and 5.0 MHz respectively. As f_oEs exceeded f_{ob}Es, the layer is of the non-blanketing type. Results essentially similar to those of Fig. 3 were recorded on the above occasion from 1225 to 1245, during which time f_{ob}Es and f_oEs decreased steadily to the respective values of 3.9 MHz and 4.3 MHz.

The φ_{NS}, frequency characteristic shows a steady trace; indeed, the magnitudes of its phase variations are such that the north-south zenithal component of the angle of arrival (as defined in [2]) did not exceed 1.5° over the working band of frequencies. In contrast, the pronounced variations in φ_{EW} at frequencies below f_{ob}Es indicate the reception of at least three echoes with differing angles of arrival but approximately equal virtual range. Above f_{ob}Es the characteristic two-echo pattern (Fig. 2) predominates, while the magnitude of the overall negative phase shift indicates that the major echo arrived from an azimuth of 270° E of N and zenith angle of about 11°.

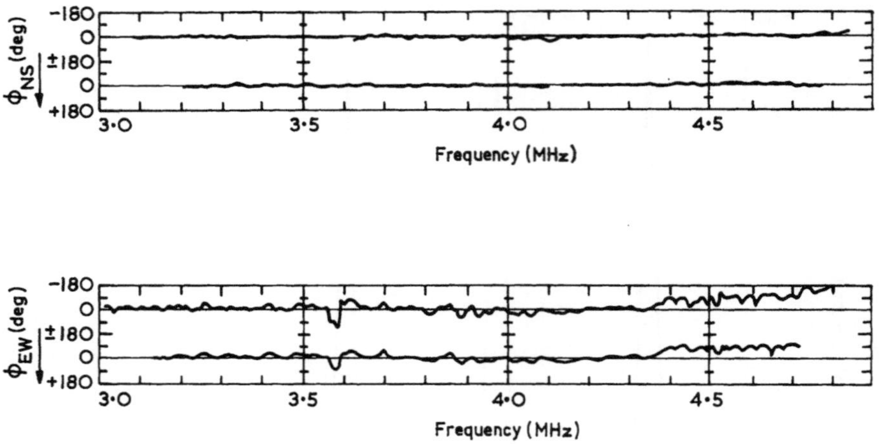

Fig. 3. Phase difference, frequency characteristics of an ordinary-mode Es reflection recorded at 1230 on 1 June 1971 at Brisbane.

The following conclusions have been drawn regarding the structure of this sporadic E layer. At least three frontal clouds, each aligned approximately along the magnetic meridian, returned detectable signals to the receiver. The resultant signals at each of the four aerials would have exhibited marked frequency-dependent phase variations. However, the outputs of the north and south aerials would always have been in phase, causing $\varphi_{NS} \approx 0$. One cloud, apparently with a maximum plasma frequency equal to 4.3 MHz, blanketed the F layer. Two additional clouds with higher plasma frequencies and lying to the west of the blanketing cloud continued to return detectable off-vertical echoes at frequencies up to 5 MHz.

4. Temporal Variations in the Blanketing and Critical Frequencies of a Non-blanketing Es Layer

Measurement of the o- and x-mode blanketing and critical frequencies were repeated at 25-sec intervals over a 90-min duration during the passage of a non-blanketing Es cloud on 9 April 1971. Angle-of-arrival measurements were not attempted on this occasion. Fig. 4 illustrates the temporal variations in the four parameters, smoothed by a three-point running-mean procedure.

Just one feature of the graph will be considered here, viz. that in the vicinity of 1035 and 1052, the maximum values of the x-mode blanketing and critical frequencies were approximately equal. The maximum values of the o-mode blanketing frequencies were, however, appreciably less than the corresponding maxima in f_0Es.

In a recent paper, Clarke [3] has shown that the average values of the o- and x-mode blanketing frequencies obtained from observations totalling several hours differed by the expected value, viz. one half the gyrofrequency. Further, it may be demonstrated that the o- and x-ray paths at frequencies slightly above the appropriate blanketing frequency penetrate the Es layer peak at points

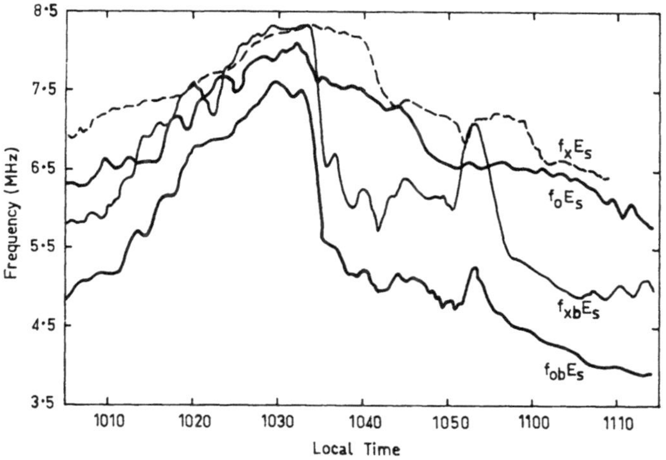

Fig. 4. Temporal variations in the Es parameters recorded on 9 April 1971 at Brisbane.

separated by a few kilometres in the magnetic meridian plane. Thus, the observation that the blanketing frequencies differed by considerably more than one half the gyrofrequency may have been indicative of a north-south gradient in the maximum electron density. This being so, the peak plasma frequencies of the two clouds (at 1035 and 1052) would have corresponded more closely to the maximum values of f_{xb}Es $- {}^1/_2 f_B$ than f_{ob}Es. Under these circumstances, the peak plasma frequencies would not have differed markedly from the corresponding maxima in either f_0Es or f_xEs $- {}^1/_2 f_B$. This result provides reasonable evidence therefore, that the sporadic E cloud was frontal, as the passage of such a structure over a receiver would lead to the near-equality of the maximum blanketing and critical frequencies.

5. Concluding Remarks

No claim can be made that the events which have been described above represent typical cases. In fact, angle-of-arrival observations of non-blanketing Es returns using the system described in § 2 were made on four occasions. The remaining three occasions were characterized by pronounced frequency-dependent variations in both φ_{NS} and φ_{EW} over the range $\pm 180°$, thereby indicating the reception of at least three components with differing angles of arrival. Moreover, the observing technique described in § 4 was applied on just one occasion.

Turning finally to the two events discussed in §§ 3 and 4, it has been possible in the former case to deduce the presence of three frontal clouds, each being aligned approximately in the north-south direction. Secondly, a study of the temporal variations in the blanketing and critical frequencies of an Es layer also indicated the existence of a frontal structure. The importance of these results is twofold. The mechanism which is most likely to produce such frontal structures lies in the redistribution of ionization due to the passage of internal gravity waves through the ionospheric E region [4]. Finally, the observations strongly support the argument that non-blanketing sporadic E returns are specular off-vertical reflections from overdense clouds of ionization.

Acknowledgments

The author thanks Drs. J. D. Whitehead and G. Bowman for helpful discussions. This investigation was carried out whilst the author was a Ph. D. student at the University of Queensland and was supported financially by that institution as well as the Australian Radio Research Board and the Commonwealth Department of Supply.

References

[1] G. J. PHILLIPS, J. Instn Elect. Engrs **98** (part III), 237 (1951).
[2] R. H. CLARKE, Planet. Space Sci. **19**, 1387 (1971).
[3] R. H. CLARKE, J. Atmos. Terr. Phys. **35**, 577 (1973).
[4] C. O. HINES, J. Atmos. Terr. Phys. **25**, 305 (1963).

Lower Ionosphere Structure — Akademie-Verlag, Berlin 1974

IONOSPHERIC DRIFT MEASUREMENTS: A NEW METHOD FOR IONOSPHERIC DRIFT MEASUREMENT

W. A. FLOOD

Electrical Engineering Department, North Carolina State University, Raleigh, North Carolina, USA

When 1949 Mitra-type wind-drift experiments are performed, drift velocities are deduced from measured auto- and cross-correlation functions. If these correlation functions are to be independent of the absolute positions of the observations sites, the cross-correlation function must be the product of a temporal correlation function which describes the temporal change in the diffraction pattern and a spatial correlation function, which, if measured in a frame moving with the drift velocity, describes the apparent ionospheric structure. Assuming that the apparent ionospheric structure can be described, by elliptical isocorrelation contours, a technique for deducing the drift velocity and axial ratio of the correlation contours is derived. These results are used to extend the 1950 analyses of Briggs, Phillips and Shinn. Since there are fewer assumptions in the new method, it is to be preferred in ionospheric drift analysis.
A full paper has been submitted to *Radio Science*.

Lower Ionosphere Structure — Akademie-Verlag, Berlin 1974

KINESONDE STUDIES OF CESIUM ION CLOUDS IN THE E REGION

J. W. Wright

National Oceanic and Atmospheric Administration, Environmental Research Laboratories, Boulder, Colorado, USA

Observations of cesium explosive releases in the E region by a digital, goniometric, velocity-measuring, multi-frequency radio sounding system (Kinesonde), show the growth and bulk motion of the ion clouds in response to shears and fluctuations of the prevailing wind system. The clouds are accurately tracked in three dimensions under full daylight conditions, providing a means for unambiguous measurement of winds; these are found to be in good agreement with the less direct measurement of drifts by the spaced-antenna technique.

1. Introduction

Explosive releases of cesium atoms in the daytime ionosphere result in relatively dense ion clouds after a few tens of seconds. A brief report will be given here of observations by a digitized radio sounding system (the 'Kinesonde') of several typical Cs^+ clouds from among some 16 available examples. The Kinesonde [1, 2] performs digitized recording of coherent radio echo phase and amplitude at six freely selected frequency pairs (minimum $\Delta f = 0.2\%$ in the range 1—16 MHz) and eight spaced receiving antennas; the wanted echoes are manually selected by independent digital range gates for each frequency pair. This system was originally designed for ionospheric radio "drift" measurements, through analysis of the several complex amplitude time series by cross-correlation [4] in three spatial dimensions. Recently we have also developed a procedure for determining the "echolocation" using the same observations [2]. Proceeding also from the complex cross-correlations of the complex amplitudes, this process obtains a least-squares estimate of the (virtual) radial distance to the reflection point and the XYZ coordinates of this point. We have found it useful to provide in this process for treating only portions of the usual 4-minute recording, so that progressive displacements of the reflection point can be detected. Time resolution of 0.24 second may be achieved in this mode.

This paper employs several cesium cloud events to demonstrate the mutual consistency of the echolocation and drift velocity analyses and uses the results to show the behavior of discrete ion clouds under rather different transport conditions prevailing in the atmosphere on each occasion.

2. General Behavior of the Clouds

Davis [3] has discussed four phases of development of these clouds. During phase I (about 5 sec) the cloud expands to ambient pressure and perhaps 100 m radius; the detonation produces initial electron densities exceeding 10^{14} m^{-3}. Since only about 10^{-3} of the free cesium atoms are thermally ionized at this point, and since the photoionization rate considerably exceeds the electron loss and cesium oxidation rates under daytime conditions, continuous electron production occurs for periods of hundreds of seconds. The cloud continues to expand by diffusion through phase II, so long as it appears as a simple entity to optical or high resolution radar observations. Eventually, however, the cloud assumes forms more-or-less strongly distorted by ambient transport processes (phase III), until it either becomes so dispersed as to lose its identity altogether or in phase IV is "coherently" convected to augment significantly a local sporadic E layer [5].

In this paper we are not concerned with the expansion phase of the clouds, nor with estimates of the clouds' radar cross sections which might be attempted from the growth of echo signal amplitude during phases I and II. Instead, we concentrate here on the "environmental" effects which dominate the later phase of the clouds' history.

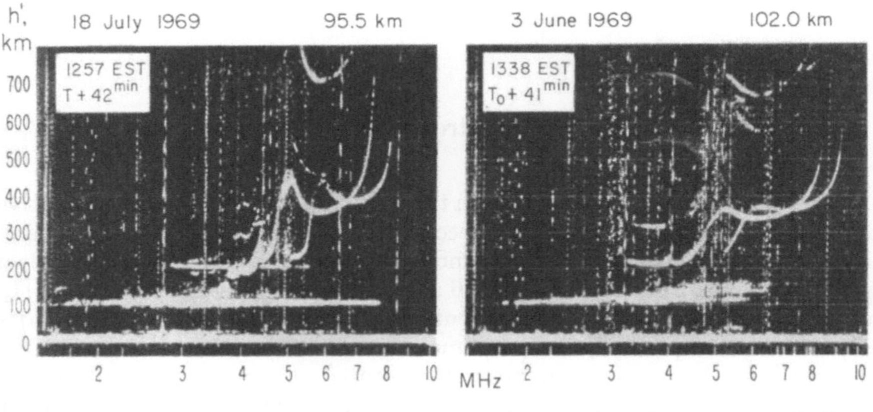

Fig. 1. Ionograms obtained about 40 min following cesium ion cloud detonations at Wallops Is. Example at left is discussed in text. Intense echoes near 100 km return from discrete ion cloud position nearly overhead at this time (compare Fig. 2), hence permit multiple echo near 200 km.

3. Cesium Release of 18 July 1969 (1215:23 EST, at 95.5 km)

The cesium-high-explosive release (2 kg) occurred at point 23 km south and 19 km east of the Kinesonde site. An ionogram illustrating the frequency range of reflection available from the cloud at $t + 42$ min is shown in Fig. 1. (The second ionogram is from another release in the series not discussed here.) Six, four-minute Kinesonde recordings, each comprising three frequency-pair selections most of which were range-gated to observe cloud echoes at the four spaced antennas, were obtained over the period 1226—1305 EST. This covers some 3000 seconds of the

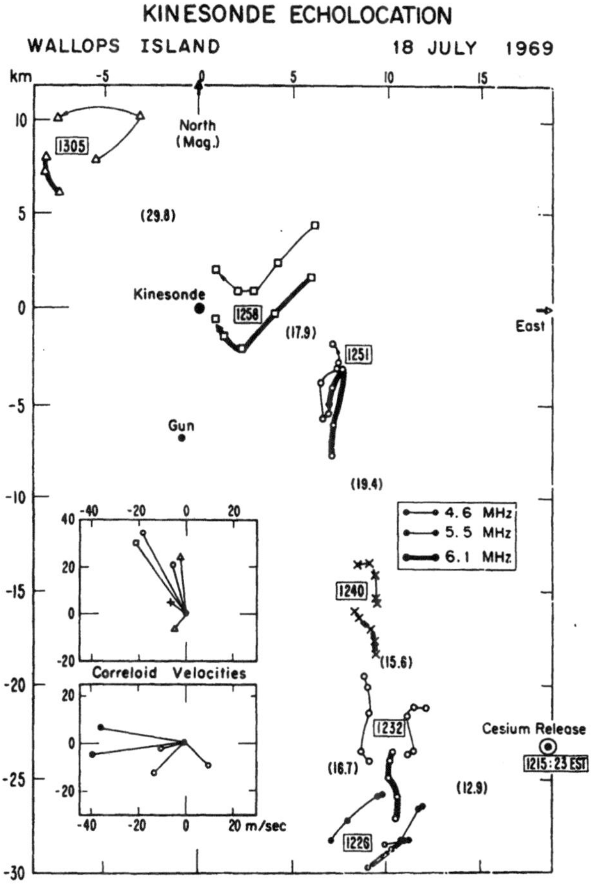

Fig. 2. Echolocation and drift velocity calculations applied to measurements of a cesium ion cloud. Origin at Kinesonde location at Wallops Is. 18 July 1969. Cesium ion cloud release at 98 km altitude from a gun-launched vehicle, at the position indicated. Multi-frequency Kinesonde recordings obtained in six 4-min. intervals during the subsequent 40 minutes; echolocation calculations result in the short echo tracks plotted for each frequency in the diagram. Average displacement speed (in m s^{-1}) is given by the parenthetical values between each cloud position. The insert figures ("Correloid Velocities") show mean drift velocity vectors obtained by correloid analysis of the same amplitude time series.

cloud lifetime. The echolocation method described by Paul et al. [2] provides the option of analyzing these recordings in any time-segmented fashion, and in the present case the records were analyzed within a "window" of 96 seconds width (which encompasses 400 sequential amplitude/phase measurements in each of the 8 antenna/frequency channels); the window moves in 48-second steps through the recording, so that successive echolocation calculations share half their raw data while alternate calculations are independent.

The results for the present cloud are shown in Fig. 2. Two or three cloud tracks, coded by symbols for the mean time of the recording and by line-widths characteristic of the radio frequencies used, are plotted for each recording in a coordinate system centered on the Kinesonde location at Wallops Island. The cloud tracks trend generally north and westward; the echolocation within the track of one recording generally agrees with the displacement of the tracks from one recording to the next (exceptions are the higher frequency at 1251 and both frequencies at 1305). The results for the higher frequency pairs in any one recording are generally slightly farther from the Kinseonde than for the lower frequencies, which is probably the result of electromagnetic group-retardation within the cloud.

We shall now compare these bodily motions of the cloud with the "correloid" velocities of its diffraction pattern. These were calculated by the three-dimensional version of Fedor's least-square process [4] from the 28 amplitude cross-correlations among eight antenna frequency-pair channels; no segmenting of the time series was attempted for this analysis. The horizontal velocity results are collected in the two insert diagrams of Fig. 2. Each velocity vector is coded by the same symbol characteristic of its echolocation time in the main part of the figure. The correloid directions are in quite good agreement with the direction of echolocation displacement in nearly every case. To compare speeds, the mean echolocation displacements divided by the time between recordings are given ($m\,s^{-1}$) between cloud locations in Fig. 2. It is evident that the displacement of the cloud accounts for most of the mean displacement of the Kinesonde echo diffraction pattern across the ground.

4. Cesium Release of 3 April 1972 (0505:40 EST, 106.0 km)

In this event a second rocket, delayed by two minutes from the 64 kg cesium release, detonated a 23 kg charge of high explosive at a point 3.6 km distant and 0.2 km below the Cs^+ cloud. Useful Kinesonde results are confined to the first 800 seconds of the cloud behavior, since it spread very rapidly to occupy a very large area overhead. During this period, optical observations are also available with which the echolocation and correloid drift velocity analysis of the Kinesonde recordings can be compared.

Fig. 3a, b, illustrates direct computer output from the echolocation calculations of one frequency pair from the recording made during the Cs^+ and TNT releases. The record was initiated at $t - 40$ s ($t = 0 =$ cesium burst). Part (a) of this figure shows the X (east), Y (north), and HP (= vertical group-path) components of the cloud echolocation as time series, while part (b) includes the same XY data in a map of echolocations. Each plotted point represents an average over a 10-point (2.4 s) window, which advances one step (0.24 s) per point through the 1000-point time series. Cs^+ and TNT burst times are marked along the time axis. Cloud echoes were acquired (by manual range-gate settings) at about time count 260; this 2.5-second delay includes the cloud expansion and photoionization times (required to permit echoes at these frequencies), and the operator reaction time for manual range-gate adjustment.

Fig. 3c, d represents results in the same format from the Kinesonde recording obtained over the period $t + 300 \rightarrow t + 540$ s. The origin in each figure is not centered at the Kinesonde location, but is arbitrarily chosen to display the resolution available in each quantity. In Figs. 4a and 4b we shall compare the results of these calculations with the observed optical behavior of the cloud.

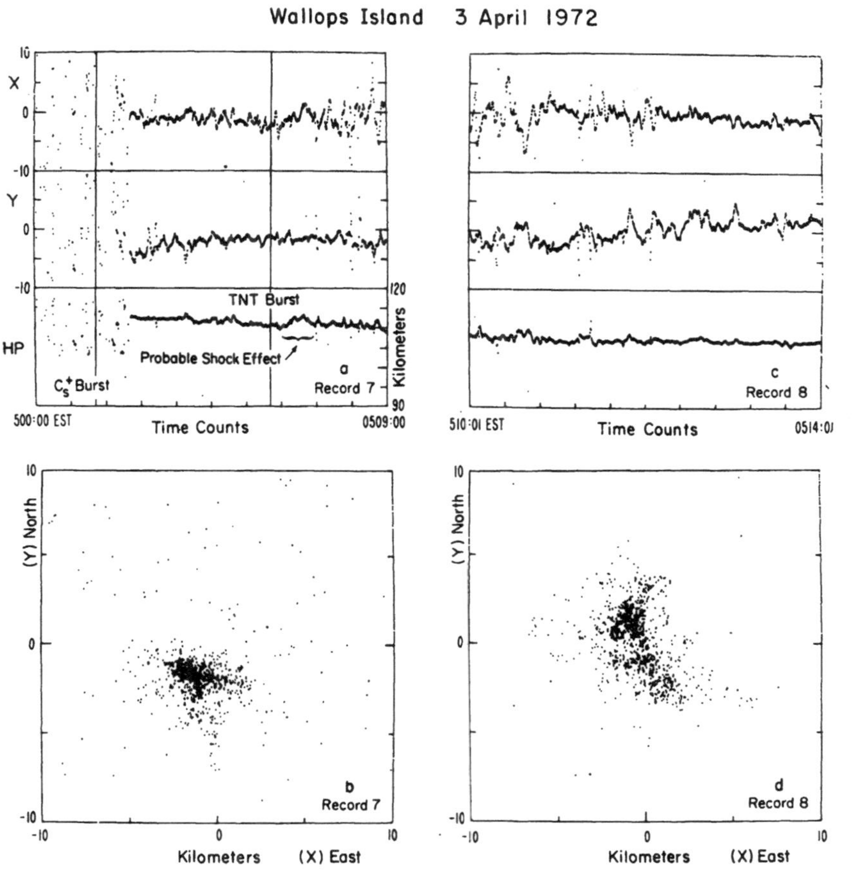

Fig. 3. Echolocation calculations plotted by computer from Kinesonde recordings at Wallops Is. 3 April 1972 initiated at t −40 s (parts a, b) and at t +300 s (parts c, d). Parts b, d represent X, Y (East, North) echolocations centered on an arbitrary origin at high "magnification". Parts a, c give same data as time series. Each point represents a 2.4 s (10 data-point) average; this averaging window moves through the 1000-point time series one step per plotted point. In part a the acoustic pulse from the TNT burst is evident.

Returning to Fig. 3a, it is interesting to note the excursion of HP which occurs between time counts 705 and 795. This follows the TNT detonation by 11.25 s (= 46.9 time counts), in good agreement with an acoustic propagation time at 0.32 km s^{-1} over the 3.6 km (optical) separation of the cesium and TNT bursts. An approximate calculation of the blast-wave overpressure at this distance using the argument of Reed [6] (and private communication) indicates a pressure fluctuation of 14%; in turn, this should produce an upward adiabatic excursion of the constant-pressure level near 106 km of 2 km, essentially in exact agreement with that observed in Fig. 3a.

All components of the echolocation show some fluctuation (scintillation) during the time series of Figs. 3a, c, superimposed on general trends. The fluctuations about the trends indicate simply a wandering of the true echolocation across the

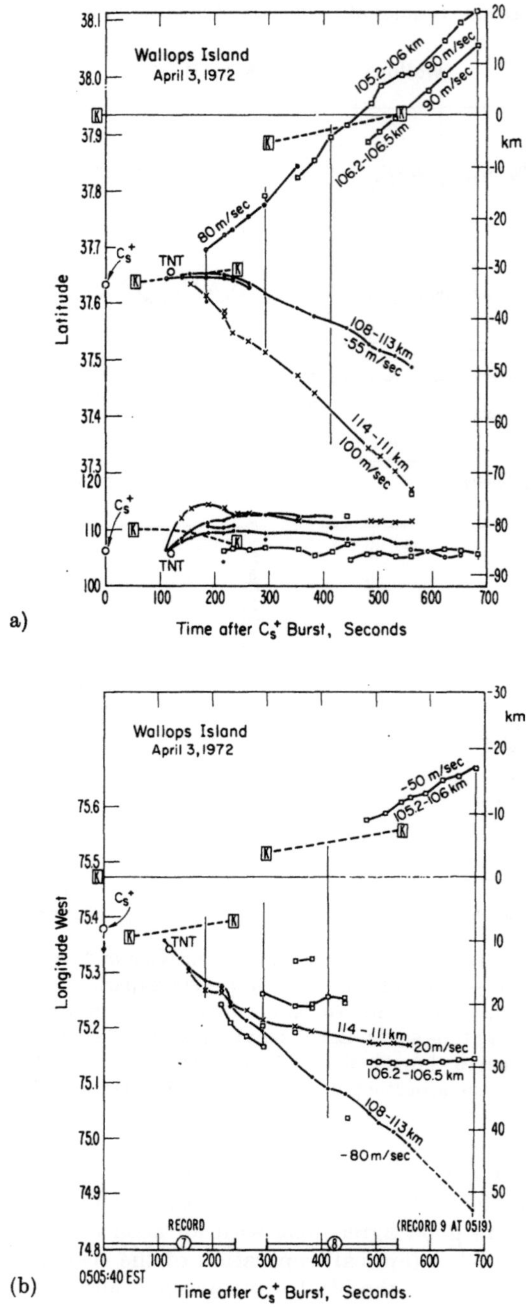

Fig. 4. Latitude (a) and longitude (b) of optically determined cloud locations versus time (1st 700 s of Cs^+ cloud). Connected symbols show motion of sequentially identifiable cloud points. Vertical lines indicate spatial extent of less persistent features. Height variations are shown also in part (a). Rectilinear speeds are evident and are marked along certain point tracks. Position of TNT burst (from second rocket) is shown. Kinesonde latitude and longitude (horizontal line at K) and echolocation calculations [K]-----[K] from two recordings (Nos. 7, 8, bottom scale part (b)) are in good agreement with cloud location and motion.

cloud boundary, although interference among multiple reflections may tend to average each apparent echolocation to some extent. The trends, however, are surely real, and as will be seen in the optical observations, indicate transport of the nearer boundary of the cloud generally toward the Kinesonde.

Figs. 4a, b contains the results of optical tracking of the visible cloud throughout its first 700 s [J. R. Davis, NRL, and R. Roper, Ga. Inst. Tech., private communications]. Each panel shows the latitude (4a) or longitude (4b) of discrete sequentially identifiable points within the cloud, versus time after burst. The lower portion of 4a includes a plot of the height variation of these points, while the lower part of 4b indicates the time-span of two Kinesonde recordings. Along the latitude and longitude scales the Kinesonde position is marked [K]. Vertical lines at various times indicate the spatial extent of the cloud; many optically identifiable points are available along each line, but only those connected in time sequence could be followed with reliable continuity. The plotted points are coded to permit association of each cloud feature in height, latitude, and longitude.

It is remarkable that the discrete cloud points move with almost constant, but very different, velocity components throughout the period of observation. Speeds in m s^{-1} are marked along most of the point-tracks, and range between nearly ± 100 m s^{-1} in each component. Furthermore, and particularly in the latitude component, the tracks converge to the TNT burst location; the cloud appears to have moved very little in the 120 seconds between its formation and the nearby TNT detonation.

The Kinesonde echolocation results from the two records of Fig. 3 are plotted [K]-----[K] in each part of Fig. 4. Good agreement in both absolute location and time-variation is obtained, although the Kinesonde, of course, lacks the resolution of an optical system (wavelength ratio $= 10^8$). The Kinesonde evidently tracks approximately the nearer boundary of the cloud, as expected from the practice of placing the range gate near the leading edge of the received echo.

Since discrete points within the cloud appear optically to move almost rectilinearly, their positions after a fixed time delay are proportional to velocities at various spatial positions. Accordingly, in Figs. 5a, b the optical latitude and longitude (respectively) of discrete points within the cloud near $t + 450$ s are plotted as solid circles against point altitude. A secondary horizontal scale in each figure then represents the mean velocity component required to account for each point position, and the resulting patterns should represent the height profiles of wind speed in the latitude and longitude components.

The observation worthy of principal emphasis from this figure is the very great variation of horizontal velocity which is indicated within this relatively small observing volume. At 108 km, latitudinal velocity components between -80 m s^{-1} and $+30$ m s^{-1} are almost evenly distributed, and values near $+80 \pm 20$ m s^{-1} are also present. More typically, the velocity at a given altitude varies by some ± 30 m s^{-1} or more about the local mean value.

Results from correlation velocity analysis of the Kinesonde recording centered on this same time (see Fig. 4b) are plotted as open circles in Fig. 5a, b). This record contains three independent frequency pairs, and from the corresponding echolocation calculations we obtain the mean (vertical) reflection heights used in Fig. 5a, b. Each correlation velocity calculation yields estimates of the mean latitude and longitude velocity components, together with a random-velocity parameter, V_c. The interpretation of V_c is not, strictly, that of an rms fluctuation about the mean [7], but it provides some indication of the 'width' of the velocity

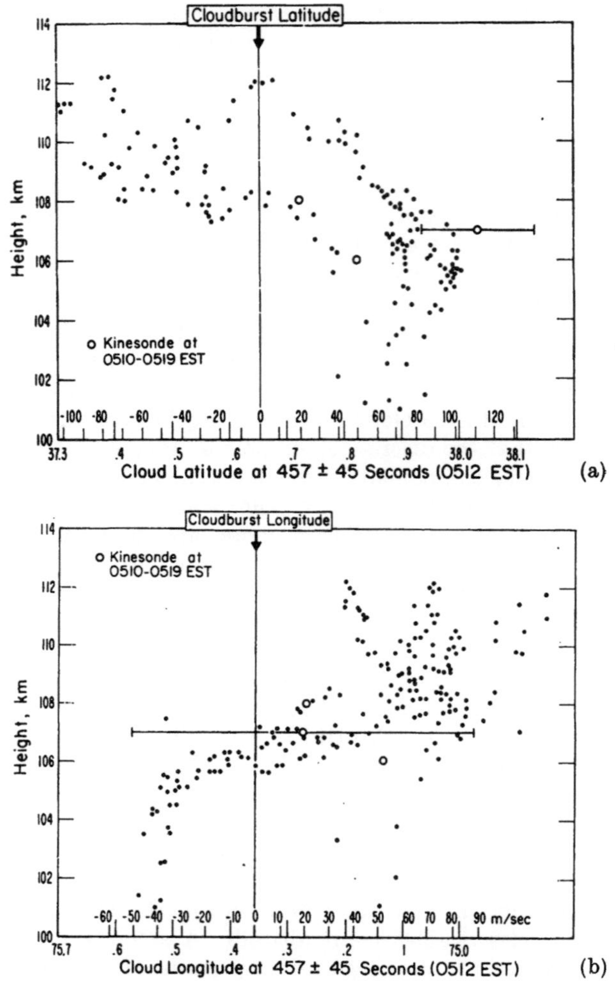

Fig. 5. Latitude (a) and longitude (b) of all optically measured cloud points, versus altitude, at $t + 457 \pm 45$ seconds ●. Assuming rectilinear motion, displacements from release lat./long. represent corresponding velocity components on auxiliary velocity scale. Note considerable dispersion of velocity at fixed heights, especially above 108 km, and in latitude component. Kinesonde velocity components O and random velocity magnitudes (horizontal line) are calculated from record beginning at $t + 300$ s.

distribution sensed by the Kinesonde. For one frequency pair, the horizontal bar indicates this width along the latitude and longitude components at 107 km altitude. The echolocation results actually show that the height fluctuated by about ± 1.5 km during this record (see the lower portion of Fig. 3c), and therefore the extent of V_c in each component is in reasonable agreement with the velocity contributions sensed throughout the recording. The reason for the apparent disagreement in the mean velocity of the latitude component is evident from the echolocation and optical data of Fig. 4a. The Kinesonde velocity estimates are 'weighted' to correspond mainly with the motions of the northern cloud boundary. The correloid velocities at the other two frequency pairs of this record, also plotted at their mean echolocation altitudes, are less significant than the first value, and give very large values of V_c ($\gtrsim 150$ m s^{-1}).

5. Conclusions

Cesium ion cloud releases in the daytime E region offer many of the desirable features of otherwise difficult controlled experiments in this poorly understood region of the atmosphere. Extremely variable cloud behavior must be expected, depending primarily on wind conditions prevailing at the altitude and time of release. Very large changes of horizontal velocity, over small vertical and horizontal distances can occur.*

Ground-based radio soundings, approached through modern digital data-acquisition and data-processing techniques, are well suited to the diagnosis of ion cloud releases. An important option, not exercised in the experiments conducted so far, would determine an optimum release altitude (and time) based upon radio-sounding measurements of prevailing conditions (wind and ionization profile) of the local ionosphere.

Acknowledgments

Support from the Naval Research Laboratory for acquisition of the data analyzed here, and for cooperation in providing the optical observations, is gratefully acknowledged. I appreciate also the assistance of my colleague W. Plywaski in the continuing development of the Kinesonde and assistance in the acquisition of the observations presented here.

References

[1] J. W. Wright and L. S. Fedor, J. Atmos. Terr. Phys. **31**, 925 (1969).
[2] A. K. Paul et al., J. Atmos. Terr. Phys. (submitted) (1974).
[3] J. R. Davis, J. Geophys. Res. **76**, 5292 (1971).
[4] L. S. Fedor, J. Geophys. Res. **72**, 5401 (1967).
[5] J. W. Wright, J. Geophys. Res. **68**, 3011 (1963).
[6] J. W. Reed, J. Geophys. Res. **77**, 1623 (1972).
[7] J. W. Wright, J. Atmos. Terr. Phys. (submitted) (1974).

**Note added in proof:* Wy did the cloud appear nearly stationary for the initial 120 seconds? Did the nearby high-explosive detonation initiate transport of the chloud, or is the effect only apparent? A likely possibility is that free cesium atoms are excited by solar radiation prior to photoionization, and exist in significantly greater abundance (during the first few hundred seconds) than does Cs^+ itself. If the nearby high-explosive detonation provided energy sufficient to complete the ionization of a large population of photo-excited Cs atoms over the relatively larger volume that the neutral cloud occupies at $T + 120$ seconds, observation confined to Cs^+ might appear to develop from the high-explosive release position.

Lower Ionosphere Structure — Akademie-Verlag, Berlin 1974

FIRST MEASUREMENTS OF ELECTRON TEMPERATURE IN THE D REGION WITH A SYMMETRIC DOUBLE PROBE

E. P. Szuszczewicz

E. O. Hulburt Center for Space Research, US Naval Research Laboratory, Washington, D.C., USA

Important to programs of lower ionospheric investigation is the ability to measure electron temperature under conditions of charged-neutral collisions which span the continuum and collision-free domains. One such technique, with demonstrated laboratory integrity, is that of the symmetric double probe. The first lower ionospheric application of this technique has been in a D region study on a Nike-Cajun payload flown at midday from White Sands, New Mexico, to an apogee of 78.5 km (MSL). Meaningful data were collected on the upleg phase of the trajectory in the altitude range 67.5 km to 78 km and indicated electron temperatures from (250 ± 70) °K to (300 ± 70) °K. These values are the lowest ever reported by any *in situ* method operative in the lower ionosphere and within the indicated inaccuracies point toward the generally expected thermodynamic equilibrium between the charged and neutral ionospheric constituents.

A full paper appears in *J. Geophys. Res.* 78, 7567 (1973).

Lower Ionosphere Structure — Akademie-Verlag, Berlin 1974

HIGH-LATITUDE NOCTILUCENT CLOUD AND SODIUM LAYERS NEAR THE MESOPAUSE

G. Witt, J. Stegman and H. Wood*

Institute of Meteorology, University of Stockholm, Sweden

A rocket payload (ESRO C-59/B) carrying an array of forward-looking photoelectric polarization photometers was launched on 1 August 1971 at 0034:00 LT into a well-developed display of noctilucent clouds from Esrange Sweden (68° N 21° E). One photometer observed the D_1 and D_2 resonance lines of atomic sodium through a 2 mm FWHM interference filter centred at 589 nm. Two distinct layers, one from 85 to 87 km and one from 87.5 to 94 km were detected. The light from the lower layer, associated with noctilucent clouds, was detected by the UV and visible photometers. This light was highly polarized while the light from the upper layer, identified as the sodium resonance emission, showed negligible polarization. The height of maximum Na emission was 89.5 km. The altitudes corresponding to 1/e of the maximum emission were 89 km and 92.5 km and the layer "half-width" was between 2 and 3 km. The estimated total overhead emission rate was 2.5 ± 0.5 kr corresponding to a column density of 3.1×10^9 sodium atoms cm^{-2}. The total column emission rate and the shape of the reduced sodium density profile above the altitude of maximum emission agree with those obtained from other ground-based and rocket measurements. Below the maximum, the gradient is unusually sharp. Possible explanations for this feature are discussed.

1. Introduction

On 1 August 1971 at 0034.00 LT an ESRO payload C-59 was launched into a visible display of noctilucent clouds from Esrange (68° N 21° E) in northern Sweden [1]. The payload was instrumented to explore features of light-scattering and emission processes characteristic of high latitude summer conditions in the mesosphere and D region. The payload carried nine forward-looking radiance photometers, operating in different wavelength regions from 200 nm to 1900 nm. The visible light and UV photometers were equipped with polarizing filters to measure the polarization of noctilucent clouds and to separate selective emission features from the elastically scattered background [2]. One photometer observed the sodium resonance emission through a 2 nm FWHM interference filter centred at 589 nm.

---

* Also at the Dept. of Physics, University of Saskatchewan, Saskatoon, Canada

2. Experimental Technique

The radiance observed by a rocket-borne filter photometer in a sunlit atmosphere contains contributions from molecular light scattering, light scattered by particulates, selective emission, the sky background, and spurious light scattered by parts of the instrument. Since each of the above sources of signal has in general a different degree of polarization the signals can be separated with a minimum number of assumptions when measurements are made with a polarization-sensitive photometer. The signal from such a photometer is sinusoidally modulated at twice the spin rate of the rocket when the photometer is looking along the rocket spin axis. By means of a digital data analysis technique it is possible to derive from the measured signal smoothed values of both Stokes parameters I

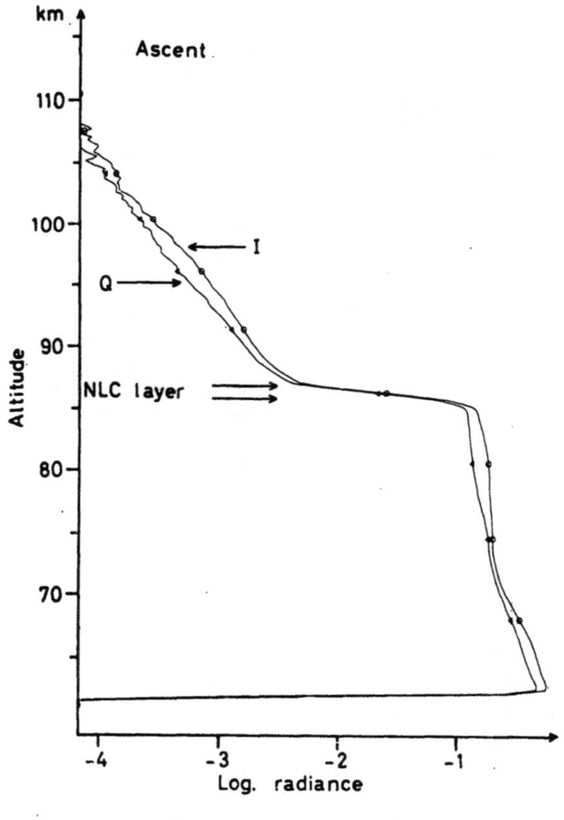

Fig. 1. Stokes parameters I and Q (average radiance and polarized component) of the downward sky radiance at 453 nm wavelength (ascent). The modulation of both profiles reflects the variation of the scattering angle due to rocket precession. The jump at 85 km occurs as the rocket passes through a noctilucent cloud layer. Above this layer the profiles follow the predicted values for molecular scattering with a possible unpolarized contribution from the airglow continuum due to reaction $NO + O \rightarrow NO_2 + h\nu$. The dark current, stray light and sky background contributions have been removed. The horizontal scale is in arbitrary relative units.

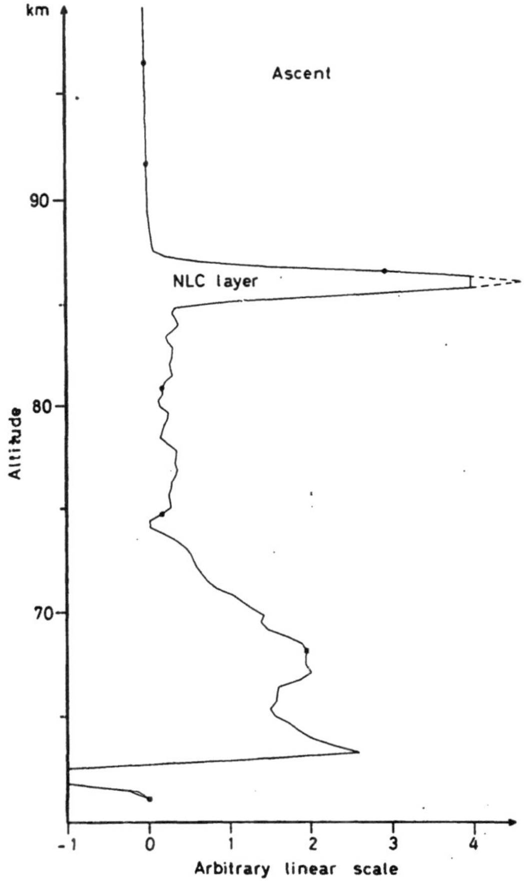

Fig. 2. Height derivative of the zenith corrected 453 nm radiance component polarized perpendicularly to the scattering plane. Below the layer of noctilucent clouds the curve should theoretically follow the atmospheric density profile. The observed deviation is caused by spatial variations in the downward radiance as the optic axis scans different portions of the NLC layer. The horizontal scale is in arbitrary linear units.

and Q (and hence the degree of polarization). The sky background and the light scattered from the instrument sunshade can be removed from the data by subtracting the apogee signal. Both of these components vary with the rocket attitude and corrections including this variation have been made during the data processing. Fig. 1 shows a height profile of the Stokes parametsers I and Q obtained from the 453 nm photometer. This filter channel was selected for the comparison as there are no significant emission features near this wavelength. The abrupt change in signal observed near 85 km for both I and Q is due to the presence of a well-developed NLC display. In the absence of excess scattering, the signal would be expected to decrease with the atmospheric pressure. Fig. 2 shows the unsmoothed height derivative of the 453 nm radiance obtained by numerical differentiation. Theoretically, these curves should follow the atmospheric density profile below the cloud layer. The observed deviations from the atmospheric profile can be explained by spatial variations of the overhead radiance scanned by the precession of the payload.

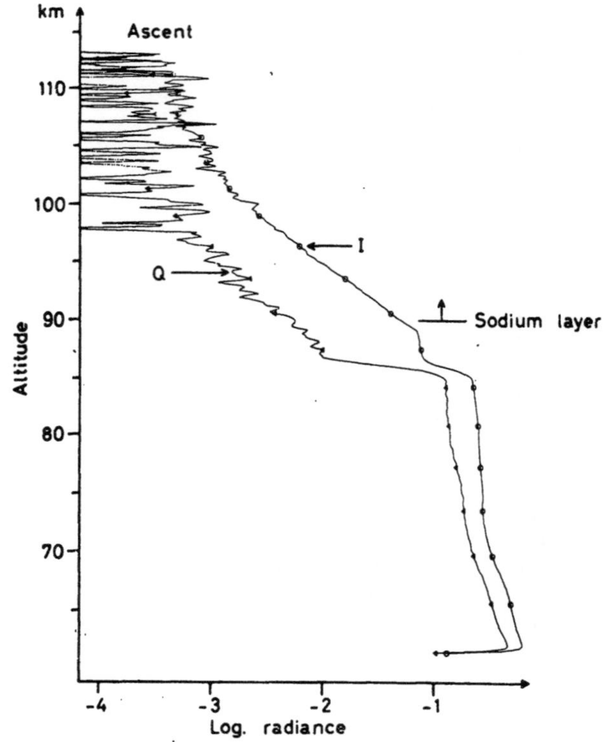

Fig. 3. Stokes parameters I and Q of the downward radiance at 589 nm wavelength (ascent). The profiles are similar to those in Fig. 1 with the exception of a ledge at 89 km due to resonant scattering by sodium atoms. There is a contribution to Q from the weak polarization (about 6%) of the resonant emission.

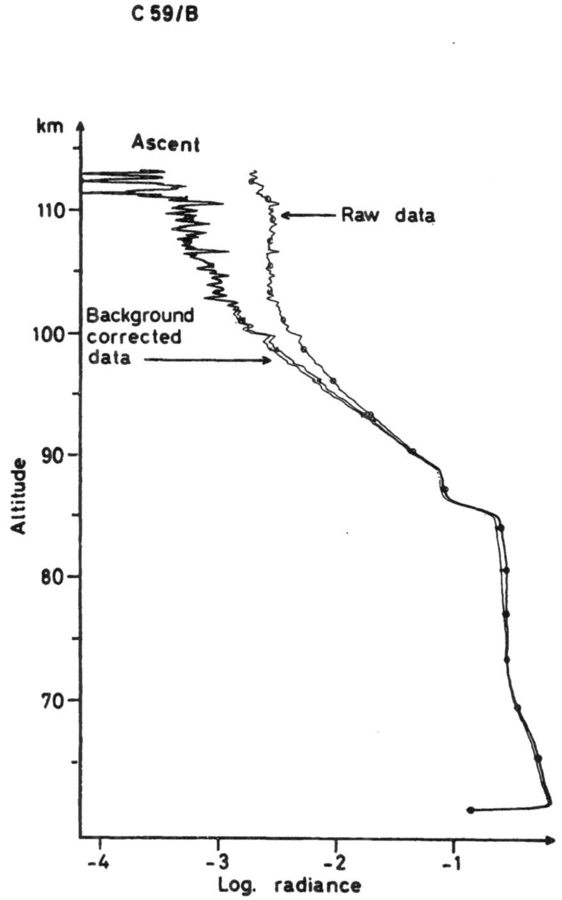

Fig. 4. Unpolarized component of the downward radiance I at 589 nm. The profiles show in turn the raw and corrected data.

3. Observations

Stokes parameters I and Q obtained from the sodium photometer at 589 nm are shown in Fig. 3. A significant depolarization of the total signal can be observed from the separation of I and Q components of the signal. The profile shows the same sharp decrease at 85 km; owing to the NLC, however, the signals do not fall to the values expected from molecular scattering — due to the presence of the sodium layer. The sodium emission is only weakly polarized so the component Q is reduced above the clouds similarly to the Q component of the 453 nm signal. However, the Stokes parameter I definitely shows the presence of a second emission layer, well separated in height from the clouds. The Stokes parameter I as obtained from the raw data and as corrected for background signal and for rocket attitude is shown in Fig. 4.

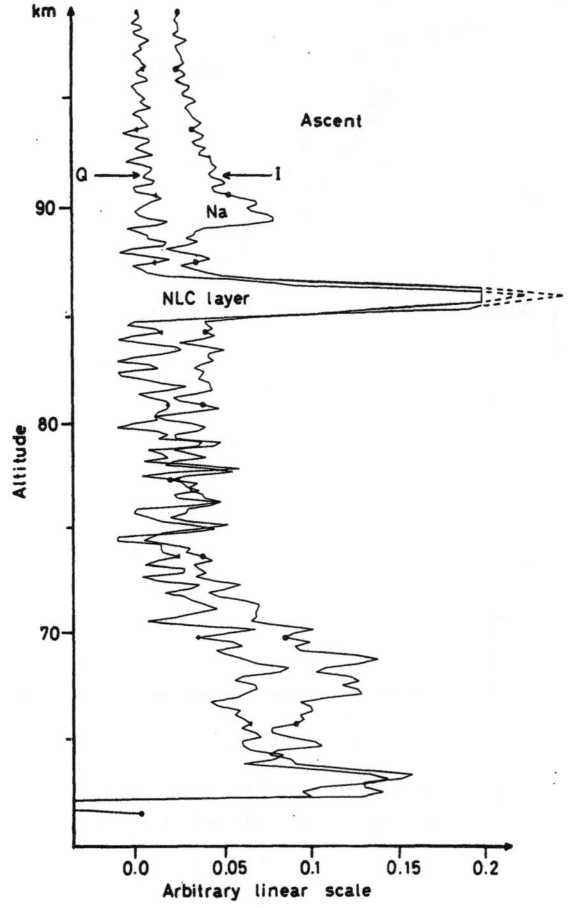

Fig. 5. Height derivatives of Stokes parameters *I* and *Q* at 589 nm. The noctilucent cloud layer appears in both profiles; light scattered by the cloud particles is highly polarized. The profile of *I* shows the sodium layer above 89 km featuring a very steep lower boundary. No attempt was made to reduce the small oscillations in the profiles by means of smoothing. The horizontal scale is in arbitrary linear units. There is an arbitrary shift of 0.02 unit between the two profiles.

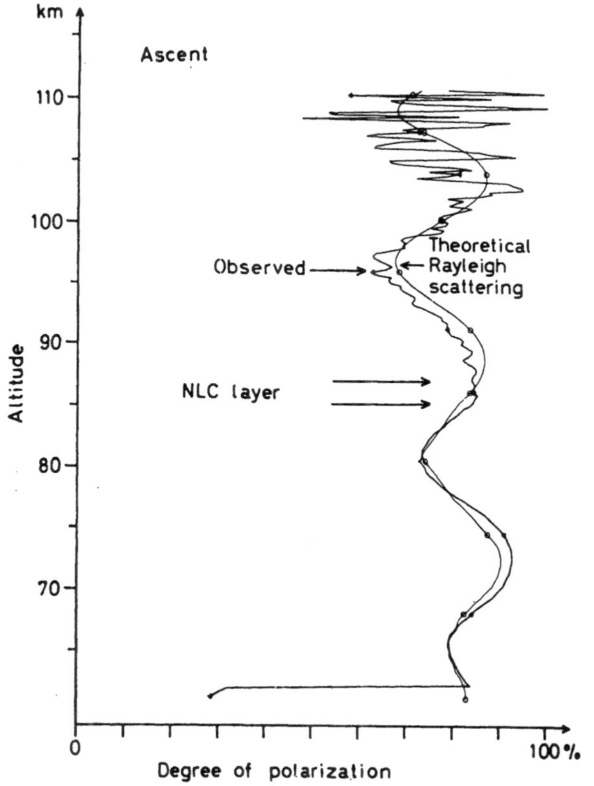

Fig. 6. Degree of polarization of downward radiance at 453 nm obtained from background corrected data. Also shown is the predicted polarization from elastic molecular scattering.

The presence of two distinct layers can be seen from the unsmoothed point-by-point height derivative of the corrected signals (Fig. 5). Both the I and Q derivatives show a NLC layer and the derivative of I clearly shows the presence of a sodium layer above the clouds. To study the distribution of sodium, the signals due to clouds and Rayleigh scattering were subtracted from the background corrected data. This was accomplished in the following way. The observation of 453 nm showed that the degree of polarization of the signal from the clouds and the atmosphere was within a few per cent of the theoretically expected Rayleigh scattering degree of polarization (Fig. 6). The degree of polarization of the total signal observed in the 589 nm channel is completely different (Fig. 7). The measured degree of polarization below the clouds decreases with increasing altitude, as the signal from the atmosphere decreases with the normal scale height and the signal from the sodium and the clouds is apparently constant. Above the clouds the measured degree of polarization decreases sharply to about 15%, then decreases between the clouds and the base of the sodium layer towards about 8% as the atmospheric scattering contribution diminishes. As the sodium layer is traversed, the degree of polarization begins to increase towards the calculated molecular scattering value.

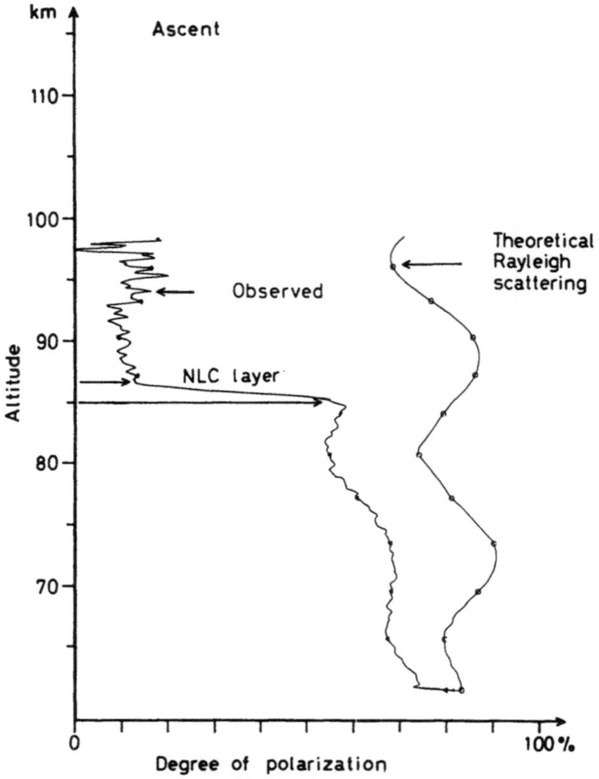

Fig. 7. Degree of polarization of downward radiance at 589 nm together with predicted polarization from elastic scattering.

From the depolarization of the 589 nm signal the contribution from sodium fluorescence can be calculated with respect to the signal expected from Rayleigh scattering from the atmosphere, which in turn varies directly as the atmospheric pressure. The mesopause temperature is known to be low at this altitude and season, and an appropriate model atmosphere can be used to obtain a pressure profile when no independent measurement is available. The result after subtraction of the atmospheric and cloud scattering is shown in Fig. 8. The validity of the assumptions is seen to be quite good as the calculated sodium signal is approximately constant below the sodium layer, and shows no distinct deviation through the clouds. The resulting curve then allows quantitative calculations of the sodium content of the atmosphere during the flight. The sodium fluorescent radiance has been calculated from the measured depolarization by means of the relationship:

$$P_{\text{obs}} = \frac{P_{\text{Rayl}} + yP_{\text{fluor}}}{1 + y},$$

where y is the ratio between the fluorescent and elastically scattered radiance components using a g-value of 0.797 [3] and a standard atmosphere characteristic of 130° K mesopause temperature. A sodium column density of $(3.1 \pm 1.0) \times 10^9$

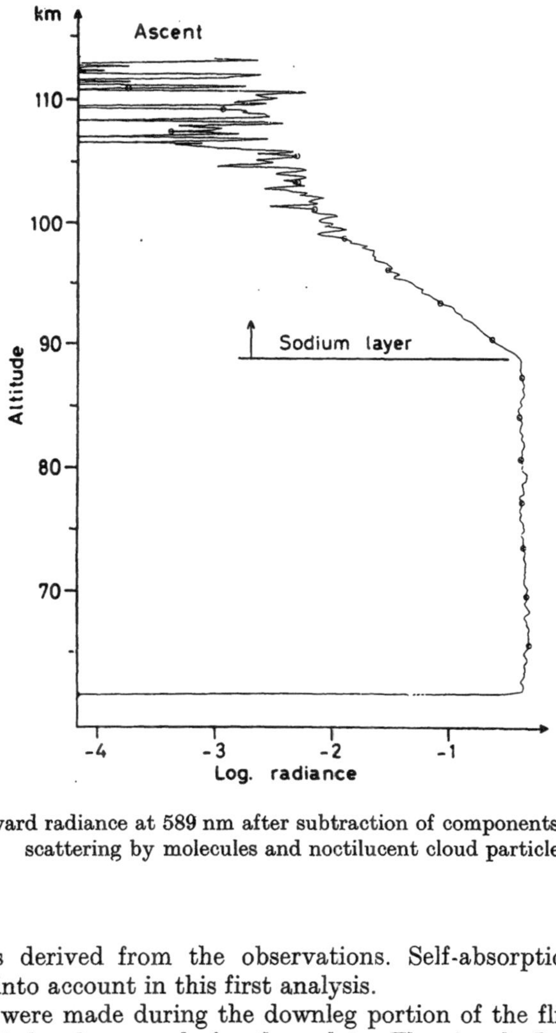

Fig. 8. Downward radiance at 589 nm after subtraction of components due to elastic scattering by molecules and noctilucent cloud particles.

atoms cm^{-2} was derived from the observations. Self-absorption effects have not been taken into account in this first analysis.

Observations were made during the downleg portion of the flight as well and a similar analysis has been made for these data. The signals I and Q measured on the downleg are shown in Fig. 9. The curves are similar to the upleg curves until turnover of the rocket begins near 80 km. The point-by-point derivatives of the total signals I and Q for downleg are shown in Fig. 10, and the subtracted signal due only to sodium is shown in Fig. 11.

4. Discussion

The calculated number density profiles for ascent and descent are shown in Fig. 12 together with some profiles measured by other investigators [4—7]. It is apparent that the number density at the peak measured on this flight agrees with other measurements. The topside gradient also agrees with other findings. The particular feature of the present observations is the steep bottomside gradient,

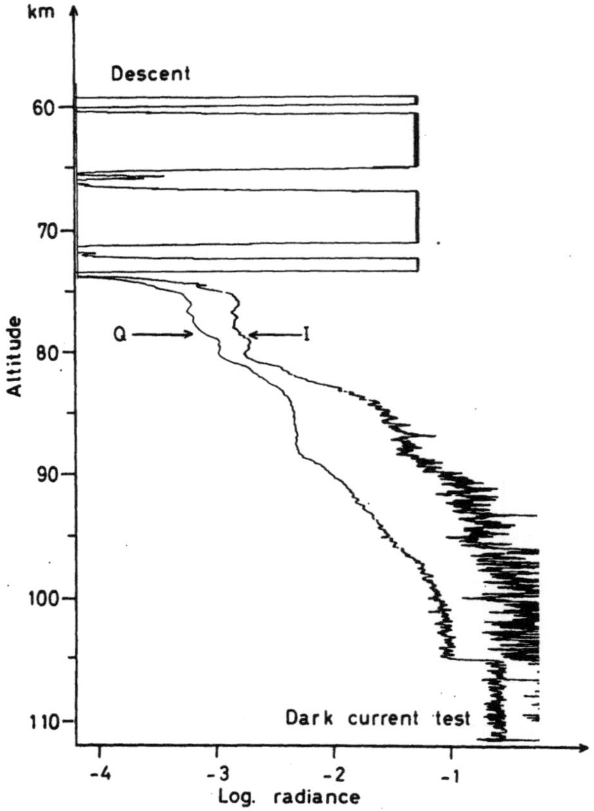

Fig. 9. Stokes parameters I and Q at 589 nm during descent. The curves display the raw data showing the sodium and noctilucent cloud layers. Rocket turnover begins at 80 km.

most pronounced in the ascent profile where the e-folding distance is only 500 m. Similar gradients have occasionally been observed by Sandford and Gibson [8] and by Hake and collaborators [7] in dye-laser sounding experiments. Hake's observations also show temporal changes in the gradient.

At the altitudes considered here, the concentration of free sodium is thought to be determined by the balance between oxidation by O_3 and reduction of NaO by atomic oxygen. Hake et al. express doubts that the observed sharp gradient can be caused by changes in the chemistry of the Na—NaO system. They speculate on a possible interaction between Na and water-ion complexes or particulates. In view of the minute concentration of the latter species at the mesopause this seems highly improbable. It is noted that in the present measurement the noctilucent cloud layer is well separated from the sodium layer. An alternative explanation may be suggested in terms of atmospheric motions. We notice that the altitude of the noctilucent cloud layer is higher in the ascent data than during descent by about 2 km. The altitude of the cloud base is 85 km on ascent and is more than 3 km above that observed by rocket photometers on the previous night. Altitude variations of this magnitude are seen in Fig. 13 which shows ver-

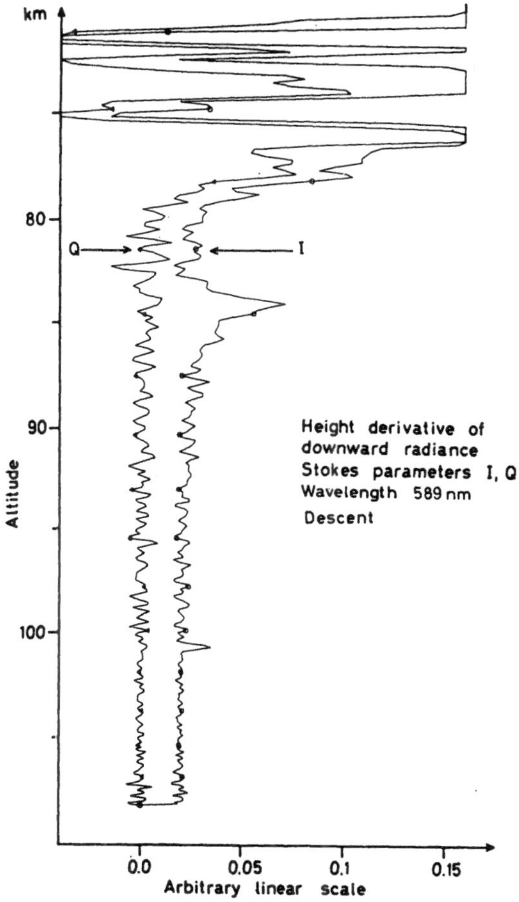

Fig. 10. Same as Fig. 5 but for descent.

tical cross sections through a NLC layer obtained by photogrammetry. Some of these height differences, which can amount to several kilometres, result from undulatory air movements that generate a slowly changing wave pattern. Horizontal wind speeds of the order of 100 m s^{-1} have been observed in this altitude region and it is reasonable to assume that local vertical velocities associated with the deepest waves can be of the order of 10 m s^{-1}. Due to the rapid transport, air parcels may be displaced vertically without appreciable chemical changes in the concentration of long-lived minor components, such as noctilucent cloud particles. Unfortunately, the extension of the wave pattern into altitudes above the NLC layer, and the vertical propagation of the amplitude and phase are not known. Also, because of the precession, the viewing direction of the photometer changes continuously as the rocket penetrates through the sodium layer. This is illustrated in Fig. 14 which shows the trace of consecutive intersections of the optic axis with the 80 km, 85 km and 90 km levels. The observed profile is not a true vertical scan and the bottomside and topside gradients may belong to different portions of an undulating layer. In view of the long chemical lifetime of Na atoms above the mesopause, it does not seem unreasonable to ascribe the observed modification of the bottomside gradient to dynamical processes associated with wave motions. The present evidence does not permit more detailed conclusions about chemical or dynamic processes causing the observed gradient as many parameters need to be measured simultaneously to permit a reasonable analysis.

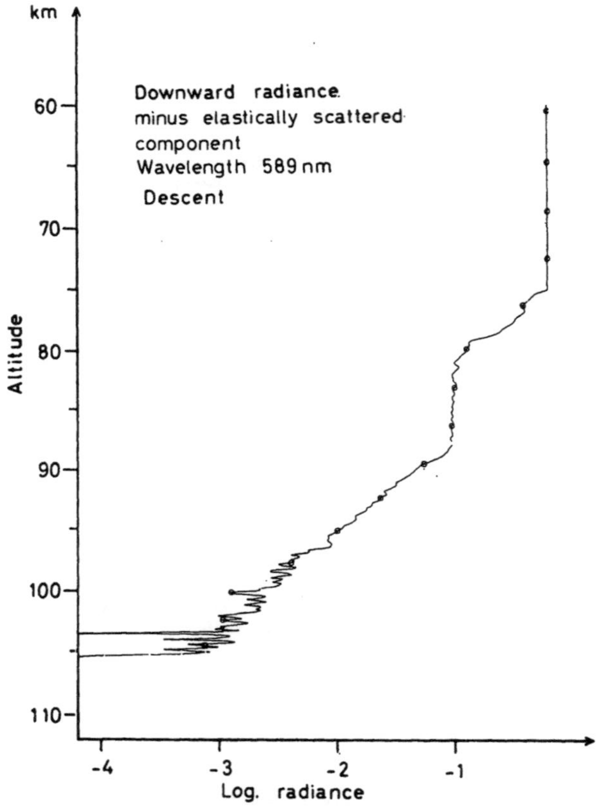

Fig. 11. Same as Fig. 8 but for the descent.

Fig. 12. Sodium atom density profiles. The solid lines refer to the ascent and descent results from the present work. Donahue and Meier's profile was obtained from a low-latitude rocket measurement in September, at solar zenith angle of 76° while Hunten and Wallace's data refer to a ground-based twilight observation in June. The other profiles were obtained from dye-laser backscatter measurements.

Fig. 13. Topography of a noctilucent cloud layer showing undulations. 9—10 Aug. 1967. The consecutive vertical cross sections were obtained by stereophotogrammetric analysis of phototheodolite plates taken simultaneously from two sites separated by an 80 km base line [9].

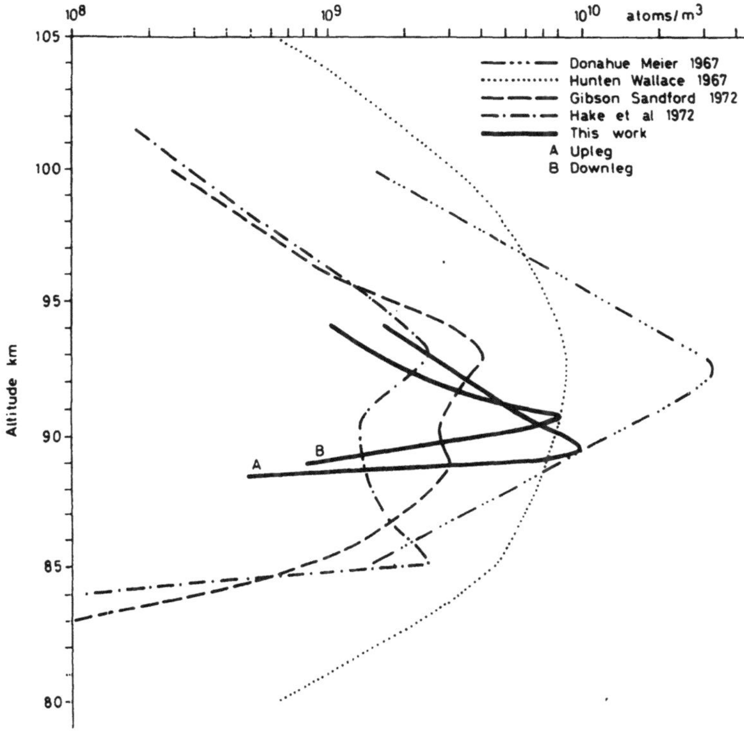

10^8
10^9
10^10
atoms/m^3
105
100
95
90
85
80
Altitude km
Donahue Meier 1967
Hunten Wallace 1967
Gibson Sandford 1972
Hake et al 1972
This work
A Upleg
B Downleg
A
B

Fig. 12.

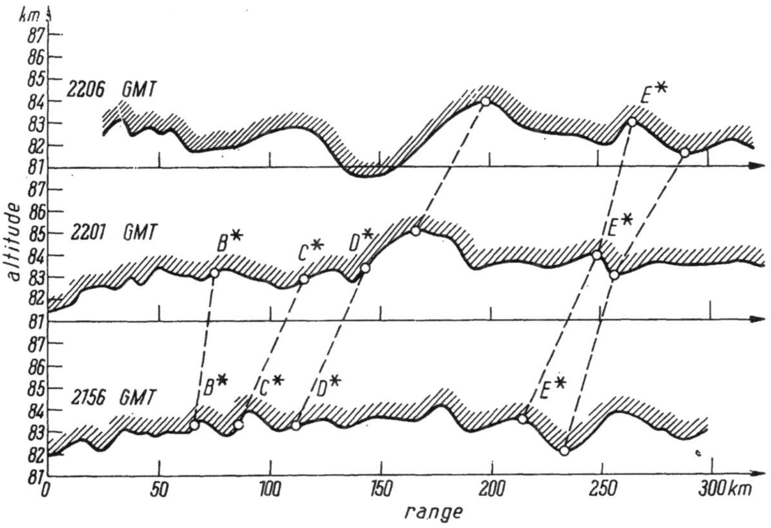

km
altitude
2206 GMT
2201 GMT
2156 GMT
B*
C*
D*
E*
0
50
100
150
200
250
300 km
range

Fig. 13.

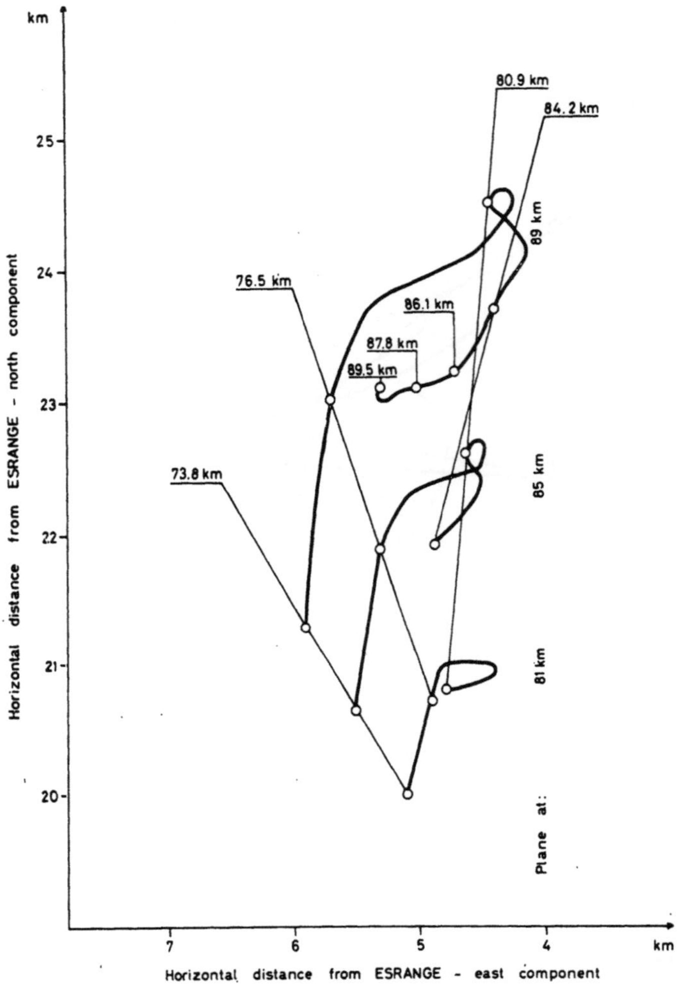

Fig. 14. Traces of the intersections of the line of sight of the C 59 B photometers with the 81 km, 85 km and 89 km levels. The curves indicate the scale of horizontal scanning caused by the rocket precession. Possible changes in the light-scattering features due to a transverse wind will have similar effects, which needs to be borne in mind when interpreting the vertical derivative of the observed radiance.

5. Summary

The polarization photometer has proved its usefulness for assessing the contamination of the airglow emission by the elastically scattered background. Moreover, by means of this technique the emission rate and hence the overhead column density of fluorescent atoms could be estimated; in further studies, however, absolute photometric calibration will be employed.

It is obvious that future research in this field requires coordinated observations both of sodium and of parameters related to its chemistry. Thus, during the summer of 1973 a sodium photometer experiment will be combined with a mass spectrometric study of metal ions. A further experiment is being planned for the simultaneous measurement of the Na nightglow and emission features related to the chemistry of oxygen atoms as well as the O-atom concentration profile. The latter experiment will, if possible, be combined with dye-laser sounding.

Acknowledgments

The authors are indebted to Mr. N. Wilhelm who designed the instrumentation and integrated the entire scientific payload.

This research was made possible by financial support from the Swedish Space Research Committee (contracts RN 1-36 and RN 1-37) and the technical support of ESRO in integrating and launching the payloads.

References

[1] G. Witt et al., Rep. No. AP-7, Institute of Meteorology, University of Stockholm (1971).
[2] G. Witt, J. Dye and N. Wilhelm, Rep. No. AP-8, Institute of Meteorology, University of Stockholm (1973).
[3] H. M. Sullivan and D. M. Hunten, Canad. J. Phys. **42**, 937 (1964).
[4] T. M. Donahue and R. R. Meier, J. Geophys. Res. **72**, 2803 (1967).
[5] D. M. Hunten and L. Wallace, J. Geophys. Res. **72**, 69 (1967).
[6] A. J. Gibson and M. C. W. Sandford, Nature **239**, 509 (1972).
[7] R. D. Hake et al., J. Geophys. Res. **77**, 6839 (1972).
[8] M. C. W. Sandford and A. J. Gibson, J. Atmos. Terr. Phys. **32**, 1423 (1970).
[9] S. Grahn and G. Witt, Rep. No. AP-4, Institute of Meteorology, University of Stockholm (1971).

Lower Ionosphere Structure — Akademie-Verlag, Berlin 1974

TEMPERATURE DETERMINATION OF THE UPPER ATMOSPHERE BY THE LOW-LEVEL DETECTION OF ARTIFICIAL LUMINOUS CLOUD RADIATION

V. F. CHEPURA and L. A. KATASYEV

Hydrometeorological Service, Moscow, USSR

It was shown that by means of homodyne detection of the emission of excited gas atoms it is possible to measure the temperature of optically thin and thick gas layers. Formulae are obtained for the temperature determination by measuring the beat current values at three frequencies. The beats are between waves falling within the spectral line.

By means of homodyne detection of the emission of excited atoms and molecules it is possible to measure temperature from the broadening of the spectral lines due to the Doppler effect. As a result of such detection performed with a photodiode or a photomultiplier spectral line beats with frequencies β originate. These beats are equal to the oscillation frequency difference $\nu_2 - \nu_1$ (Fig. 1) that has effect simultaneously at a photodiode [1]. The mean square beat current corresponding to frequencies β to $\beta + \delta\beta$ resulting from all possible oscillation interactions, the difference of which lies in this band, is [2]:

$$\langle I^2(\beta) \rangle = 2A\delta\beta \int_0^\infty W(\nu)\, W(\nu + \beta)\, d\nu, \tag{1}$$

where $W(\nu)$ is the spectral density of the luminous gas radiation intensity and A is the coefficient depending on photomultiplier parameters.

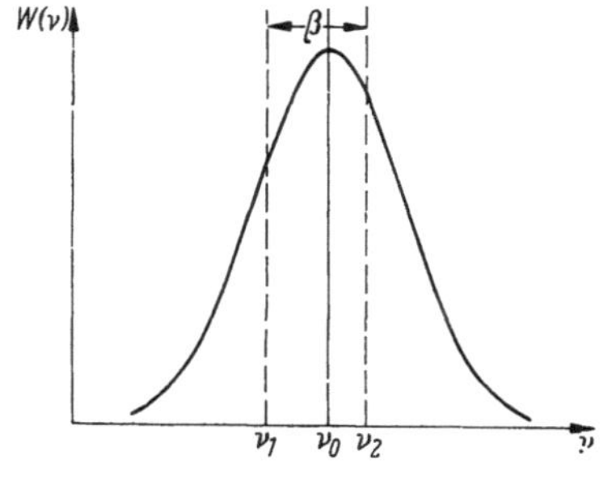

Fig. 1. Contour of the spectral line.

According to Kirchhoff's law for a homogeneous gas layer one can write [3]

$$W(\nu) = E\{1 - \exp[-\alpha l]\}, \tag{2}$$

where E is the intensity of black body radiation at frequency ν, α is the coefficient of gas absorption at this frequency, l is the gas layer thickness. If the broadening of the spectral line is due to the Doppler and self-absorption effects, then [3]

$$\alpha = \frac{Ne^2\gamma f}{4\pi m\nu_0} \cdot \left(\frac{m_1}{2\pi kT}\right)^{1/2} \int_{-\infty}^{\infty} \frac{\exp[-p^2(\delta\nu)^2]}{(\nu - \nu_0 - \delta\nu)^2 + [\gamma \cdot (4\pi)^{-1}]^2} d(\delta\nu) \tag{3}$$

where

$$\gamma = 8\pi^2 e^2 \nu_0^2 \cdot (3mc^3)^{-1}, \qquad p = [m_1 c^2 \cdot (2kT\nu_0^2)^{-1}]^{1/2},$$

e and m are the charge and mass of electron, f is the oscillator strength, m_1 is the radiating gas atom mass, N is the number density of these atoms, T is the gas temperature, $k = 1.38 \times 10^{-23} I/\mathrm{K}$, ν_0 is the central frequency of the spectral line, $\delta\nu$ is the Doppler frequency shift, $c = 3 \times 10^8\ \mathrm{m\ s^{-1}}$.

For an optically thin gas layer from Eqs. (1), (2) and (3) we obtain $\langle I^2(\beta)\rangle$ and then find the ratio of currents at two frequencies

$$\psi_1 = \langle I^2(\beta_1)\rangle \cdot [\langle I_2(\beta_2)\rangle]^{-1} = \exp[m_1 c^2 (4kT\nu_0^2)^{-1} \cdot (\beta_2^2 - \beta_1^2). \tag{4}$$

T is obtained from Eq. (4) by measuring the beat current values.

The relative error of T from Eq. (4) is

$$\delta T \cdot (T)^{-1} = -\delta\psi_1(\psi_1)^{-1} T \cdot 4k\nu_0^2 [m_1 c^2 (\beta_2^2 - \beta_1^2)]^{-1}. \tag{5}$$

For the DNa line (artificial luminous clouds with intensity of several thousand rayleigh), $\delta\psi_1(\psi_1)^{-1} \sim 0.05$ can be obtained by using high quality optics, sensitive photomultipliers, modern amplifiers, synchronous detection and by integrating over several tens of seconds. From Eq. (5) the relative error of T for the lower thermosphere based on several measurements is about 0.15 for $\beta_1 = 60$ mHz, $\beta_2 = 500$ mHz.

For an optically thick gas layer $\langle I^2(\beta)\rangle$ is found from Eqs. (1), (2) and (3). For the visual region of spectrum, pressure $\leqq 10^{-2}$ torr, $T > 200$ K and light elements (O, Na)

$$\langle I^2(\beta)\rangle = 2A \cdot \delta\beta \cdot E^2 \int_{-\infty}^{\infty} \{1 - \exp[-q \cdot \exp(-x^2p^2)] - \exp[-q \cdot \exp\{-(x+\beta)^2p^2\}] + \exp[-q\{\exp(-x^2p^2) + \exp[-(x+\beta)^2p^2]\}]\} dx, \tag{6}$$

where $$q = \pi e^2 Nfl \cdot (m\nu_0)^{-1} [m_1 \cdot (2\pi kT)^{-1}]^{1/2}.$$

In Eq. (6) there are two unknowns: T and Nl. Therefore for determination of T it is necessary to measure current at three frequencies, and from the ratios

$$\psi_1 = \langle I^2(\beta_1)\rangle [\langle I^2(\beta_2)\rangle]^{-1} \quad \text{and} \quad \psi_2 = \langle I^2(\beta_1)\rangle [\langle I^2(\beta_3)\rangle]^{-1}$$

with the help of Eq. (6) one can obtain T and the column density of the radiating gas atoms, Nl.

Curves $\psi = F(T)$ for the DNa line are given in Fig. 2. As the frequency increases, the curves become steeper, in consequence of which the accuracy of T increases. For the green oxygen line the graph is almost the same.

Calculations show that using this method it is possible to determine the lower thermospheric temperature at about 100 km from the green oxygen line. In this case the time of integration increases up to 10 minutes.

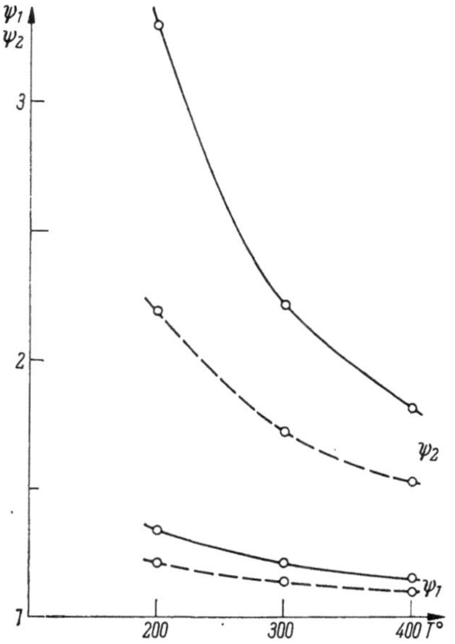

Fig. 2. ψ versus temperature for DNa. The solid curves are for the thin optical layer ($Nfl \leqq 10^9$ cm^{-2}), the broken curves are for the thick optical layer ($Nfl = 10^{11}$ cm^{-2}), $\beta_1 = 60$ mHz, $\beta_2 = 500$ mHz, $\beta_3 = 10^3$ mHz.

References

[1] H. G. Heard, Laser Parameter Measurements Handbook, New York 1968.
[2] A. T. Forrester, J. Opt. Soc. Am. **51**, 253 (1961).
[3] A. Unsöld, Physik der Sternatmosphären, 1938.

Discussion Summary

Lower Ionosphere Structure — Akademie-Verlag, Berlin 1974

OPEN AND RESOLVED PROBLEMS CONCERNING THE ESTABLISHING OF PROFILES THROUGH THE LOWER IONOSPHERE

K. RAWER

Arbeitsgruppe für physikalische Weltraumforschung, Freiburg, FRG

Chairman's introduction to and summary of the final discussion following the URSI-IAGA-COSPAR Symposium on "Methods of Measurements and Results of Lower Ionosphere Structure". Conclusions concerning mainly the elcctron density profile, but also electron temperature and ion composition profiles which might be usable as inputs for the "International Reference Ionosphere".

Problems discussed: the "valley region" above the normal E layer peak, D region profiles from ground-based radio wave propagation data, negative ions in the D region.

1. Introduction

The "International Reference Ionosphere" (IRI) is intended to present representative profiles of the main parameters characterizing the ionospheric plasma, at a few selected typical locations in the first instance. It must be based primarily upon experimental evidence, but theories may be helpful for interpolation and extrapolation where reliable data are missing, and also for cross checking and internal consistency control. There is a general rule which has been accepted by scientists for almost 500 years now, according to which experimental evidence has priority so that a theory must be "refurbished" or abandoned if experimental evidence is against it. In geophysics at any rate we have to obey this rule — in spite of an increasing tendency in some minds to go back to Hegel's "tant pis pour les planètes" taking a position which might be characterized by Ch. Morgenstern's line "nicht sein kann, was nicht sein darf", which may be translated into English as "what should not occur, can't occur".

This symposium has been arranged for a particular reason, namely the difficulties encountered during the last few years when trying to bring together data from different sources and techniques all concerning the lower ionosphere. Since IRI should lead to a representative description of the significant features of ionospheric (height) profiles, these difficulties must be resolved, at least provisionally, before coming out with such profiles. These latter should be usable for quite different purposes: as basic data, first for estimating expected phenomena and effects when planning experiments or techniques, second for practical applications to environmental and wave propagation problems, third for evaluating theories and their consequences.

A large number of interesting contributions have been given during this symposium. The next step should now be to obtain at least preliminary decisions about the critical points mentioned above in the light of these contributions. This certainly is a limitation of the scope of this discussion compared with the larger spectrum of subjects considered during the symposium. This could, however, be quite stimulating since it has always been the case that practical application forces science towards more concise understanding of natural phenomena.

Limiting our discussion to reasonings and results which might be helpful for establishing profiles we proceed in the following order:

Electron density profiles,
Ion composition profiles,
(Electron) temperature profiles.

The first item may be subdivided according to specific height ranges. In particular, we shall consider in more detail the daytime "valley region", i.e. the height range above the E region peak on one hand, and the D region electron and ion density profile on the other.

2. E-Region Electron Density Profile

There exists an enormous amount of measured data which apply directly to this subject, particularly from classical (total reflection) sounding technique. However, though these data are quite reliable for certain features of the profile, in particular the peak electron density (see § 2.1), there are other important features which cannot be reliably deduced from such data. We must therefore have data from other techniques, ground-based or space experiments. Unfortunately, the total number of observations with these techniques is far less so that it is sometimes difficult to eliminate statistical uncertainty.

2.1. Maximum Electron Density of the E Region

There is an abundant amount of measurements of the critical frequency foE from which the peak density is directly obtained. Monthly median values (as function of the hour) have been analysed by Gallet-Jones techniques as now currently applied for F region parameters when establishing forecasts. These values have, for example, been used in a preliminary profile compilation established by Ramakrishnan and Rawer, which was distributed at the 1972 URSI General Assembly. The question is whether this programme produces a significantly more reliable input than descriptive equations could give. Such equations have been empirically established by different authors and depend on parameters like the solar zenith angle χ, the hour t, geographic latitude and, possibly, magnetic dip. The feeling was that due to inaccurate calibration at a few stations in the past, the computation programme as established by ITS (Boulder, Col., USA) has no chance to achieve better reliability. Also, since its application means a considerable amount of computer work, preference was given to empirical equations.

It remains now to check different types of equations and, eventually, establish a new one. The computer program could be used for cross checking. (D. B. Sailors of Naval Electronic Laboratory, San Diego, Cal., USA, may be contacted to this end since the programme seems to be discontinued now at Boulder.)

2.2. The Shape of the E Region between 90 km and Peak Altitude

Though this shape could in principle be obtained from classical ionograms, it appears that the accuracy of height readings is inadequate for this purpose. There are now improved techniques which have been used by G. M. Brown in UK and by J. W. Wright in USA. Until now these have been applied to a rather small number of samples only so that further work is needed along these lines. It may be difficult to bring the systematic height error down to, say, 1 km, i.e. the peak altitude cannot be obtained with such accuracy. It seems, however, that quite good relative accuracy could be achieved.

The preliminary compilation mentioned above used K.-I. Maeda's compilation of primary profiles obtained with rocket techniques. It was said that it is based upon 150 measured profiles, which is a small number compared with results from soundings. The number is particularly small for night hours. Nevertheless, this seems to be the best available input at this time.

Incoherent scatter (I. Sc.) technique may give another reliable input in this context provided bistatic installations with secondary receiving stations at fairly short distance (about 100 km) are available. The French I. Sc. system is to be changed in this way in the near future.

2.3. The E Region Valley

Between the E peak (105—125 km according to conditions) and about 160 km height the daytime profile often exhibits a minimum of electron density which is quite pronounced in the morning and evening period and is rather shallow around noon. It is well known that such conditions make the reduction of classical ionograms ambiguous for all heights above its upper side. As to the "valley" shape itself it cannot be deduced unambiguously from such soundings. For quite a few years, however, there has been hope of obtaining, with reasoning introduced by Titheridge and Paul, some useful information by exactly comparing the two characteristic (magneto-electronic) traces in the ionogram (at frequencies which penetrate the valley and are reflected in the F region above it). The application of such a technique to given model shapes has, unfortunately, proved that the accuracy which can be achieved in this way is not good enough to obtain very meaningful results. This is not due to an uncertainty in theory, but (as Paul explained at the 1972 URSI General Assembly — see also p. 39 of this volume) to the uncertainty that the real ionosphere is horizontally stratified and free of irregularities. Since it is now known that gravity waves are almost always present in the daytime ionosphere, there is no chance to obtain the wanted information from ionograms.

As to rocket measurements Maeda's compilation contains about seven profiles covering the height range in question. More data would be of interest provided the reduction is unambiguous.

On the other hand, I. Sc. measurements of the profile between 100 and 160 km are now available from Malvern (UK). Such data obtained by Taylor are the basis of the 1972 compilation (see p. 35). It seems quite certain that the French station St. Santin will also produce such data in the near future, and there is some hope that Arecibo (Puerto Rico) could possibly take up similar measurements (with amplitude and phase determination in monostatic technique). At Jicamarca Woodman estimates that the undisturbed "background" ionization cannot be readily obtained below the 150 km level, His own investigations (see p. 355) only cover locally increased ionization.

The shape of the electron density profile below the E peak may be disturbed by transitory phenomena like ledges and gravity waves. However, such features can be fairly easily filtered out by statistical means and there seems to be no doubt that the average, undisturbed shape can be determined with high precision and is well reproducible. There was for a long time a feeling that conditions should be similar for the shape above the E peak. However, there are recent low latitude observations by Neske and Kist who compared rocket profiles obtained exactly at the same local (afternoon) hour with an interval of four days only. They found a shallow valley in one case and a very deep one in the other. This seems to show that the different processes responsible for the actual electron density in this height range are of variable importance. Now Narcisi (p. 207) has mentioned that profiles computed with photo-ionization theory usually end up with a monotonic increase with altitude in this height range. His ion composition observations seem to indicate that some chemical influence could occasionally produce a deep valley (by increased recombination). Thus there is now a completely new aspect according to which the valley should be under dynamical control. Apart from increased aeronomic interest of profiles through the valley, there is then a need for much more reliable measurements so as to establish statistically significant median profiles.

3. Electron Density Profile of the D Region

3.1. Profiles deduced from Ground-based Radio Wave Propagation Data

For quite a few years D region electron density profiles have been deduced from ELF, VLF, LF and MF wave propagation data, namely amplitude and (sometimes) phase at suitable distances from a transmitter. Results have been published which are not in agreement, and this disagreement occurred even when the same measured data were used. Bain and coworkers in UK tried to fit the profile by numerical trial and error methods. Krasnushkin in USSR introduced a more systematic approach; another approach has been presented during this meeting by Shellman (USA) (summary, p. 111).

This difficulty has been fully discussed and finally agreement has been reached among all specialists present at the meeting. Their statements can be summarized by saying that more, and far more precise, data would be needed in order to obtain a satisfyingly unique solution of the problem. As Jones (UK) said, one should know all conversion coefficients for something like six suitable frequencies.

Altman (Israel) feels that even six stations might not give satisfaction, and he recommends using also Faraday effect measurements and radio-wave absorption data as additional sources of information. One of his conclusions is that these latter observations are difficult to reconcile with a large electron density in the 75—89 km height range. He hopes for future precise measurements of the field vs distance pattern (the so-called Hollingworth pattern); this could, of course, be obtained on quite a few frequencies.

Sales (USA) feels that not the number of available frequencies but the shape of the profile itself is the critical point in numerical work. For any set of a few frequencies there may occur types of profile shape, which cannot be readily resolved, if account is duly taken of the inaccuracy of the measurements and the deviation of the real ionosphere from the models. Our own conclusion is that in the worst case the situation may be similar to that found for the E region valley (see § 2.3 above), namely that irregularities in ionospheric structure constitute an accuracy limit, which cannot be transgressed. The generally agreed conclusion was that available propagation data are not good enough to provide unambiguous profiles. On the other hand theoretical computations of propagation conditions could be extremely helpful for checking profiles from other sources.

One effort in this direction was presented during the meeting by Fellmann (FRG) who was able to exclude certain models because they did not agree with precise observations of Schumann resonances (p. 127). In this context it should be stated that Krasnushkin's idea is certainly right, according to which the lower frequencies should be given particular weight, when the lowest layers are considered: e.g. Belrose (Canada) stated that VLF data should mainly be used during daytime when considering heights below 75 km.

3.2. Review of Different Profile Determinations

Taubenheim (GDR)* presented an approach in which a larger number of flare effects observed on LF were used in conjunction with more theoretical reasonings to obtain profiles by a trial-and-error method which includes some averaging. His results have been obtained for low sunspot number and he adopts the requirement that the D region profile should be monotonic. There was no objection from the floor against this requirement under the given condition for solar activity (sunspot number not greater 30). This will be published elsewhere-see p. 459.

Gruschwitz (FRG) presented (p. 69) a low-latitude profile, for sunspot minimum conditions, obtained from absorption measurements at ten frequencies. It was stated from the floor that at low latitude at sunspot minimum, conditions in the D region might be different from most of the temperate latitude measurements. Thrane (Norway) remarked that the "step" is usually observed at 85 km and not at 75 km as in the proposed profile; between 75 and 80 km the height gradient should be important, and different experimental profiles seem to be largely different just in this height range.

The Illinois group (USA) states that there should be a deep minimum in ion production around 65 km; below this altitude one has ionization by cosmic radiation, the so-called C layer. Thus above the 65 km level there should occur a rather sharp (positive) gradient of electron density.

Rose and Widdel (FRG) (pp. 331 and 341) have tried to use a ground-based integrating propagation experiment (A2 absorption) for checking their own measurements. Thomas (UK) criticized this procedure and made a very clear statement that for achieving reliability and accuracy probe measurements should be accompanied by rocket-borne propagation measurements (e.g. differential absorption, or Faraday, or Doppler). The partial reflection technique could only be used as a supplement. (He queried the certainty achievable with this latter technique.) Probe data controlled by rocket-borne propagation could be useful down to 70 km. The existence of a "plateau" near 85 km is well proved, but it remains an open question whether there could be a real "valley".

---

* see p. 459.

3.3. A D Region Valley (?)

In recent years there has been considerable dispute about this problem. Meanwhile feelings have calmed. Bowhill (USA) admits that a valley may occur near 65 km, but only when the electron density is very low. The valley depth could be identified by a "reduction ratio" of electron density: 2:1 (in the valley) may be found under extreme conditions, but in most cases where a valley exists at all it would only be 3:2. Belrose (Canada) states that the profile is in most cases monotonic but that "ledges" occur quite often. These conditions are available from day to day. Ledges occur particularly in winter; with partial reflection technique (in Canada) they are best seen in October, and not so well in summer. The corresponding height range is about 58—65 km and the electron density at the lower end can be as small as 10^8 m^{-3}.

Our own conclusion is that for median profiles present information is more in favour of a monotonic D region profile which could, however, have a steep gradient above 65 km and a plateau near 85 km. Since, however, a valley seems to be present occasionally a larger number of measurements is desirable so as to get a significant picture of the conditions. As for aeronomic theory we must, again, ask whether these variations are provoked by transport phenomena, and which shape corresponds to the "true ionization profile", i.e. the profile which should be obtained in the absence of transport.

4. Ion Composition

This subject should first be discussed in terms of *relative* composition, putting aside the question of the absolute value of ion density.

4.1. Positive Ion Composition

There seems to exist quite good agreement now between different groups making *in situ* measurements: Aikin (USA), Krankowsky (FRG), Hirao (Japan), Narcisi (USA). The latter has summarized the results in his paper (p. 207). Agreement exists also for the disturbed D layer.

4.2. Negative Ion Composition

It is understood that negative ions are of negligible numerical importance in the E region, but can be most important in the lower D region measurements: Narcisi (USA) and Krankowsky (FRG) come to very different conclusions concerning the occurrence and importance of different species. The disagreement is particularly serious in the height range between 80 and 90 km. It is agreed that a large amount of uncertainty is due to the *in situ* techniques themselves. In the vicinity of a supersonic vehicle the relatively poor ion population may be considerably changed so that we have a problem of "plasma effects near space vehicles".

5. Electron and Ion Temperature

This subject has not been really discussed but enough incidental remarks have been made which show agreement that the incoherent scatter data must be taken as the most reliable at this time. Probe data from rockets and satellites should be compared with I. Sc. results in suitable configurations. In essence one should apply probe data after having checked and calibrated these with I. Sc. measurements (see for example the paper by Spenner et al. in *Space Research* XIV which they presented in W. G. 4). This statement is in agreement with the policy already applied for IRI.

In conclusion we may say that more reliable measurements and much more cross checking between different techniques must be asked for. For subjects like ion composition and temperature the scarcity of basic data is in fact a major limitation for IRI.

Papers presented at the Symposium which are published elsewhere

E. A. MECHTLY

Accuracy of rocket measurements of lower ionosphere electron concentrations, *Radio Science* 1974.

R. KRAMM, K. SCHEGEL and W. WEISS

Partial reflection measurements with FM-CW, a preliminary investigation, *Zeitschrift f. Geophysik,* **38**, 953 (1972).

L. G. SMITH, E. A. MECHTLY and G. W. HENRY

Rocket observations of the winter anomaly, *J. Atmos. Terr. Phys.* 1974.

J. TAUBENHEIM, G. ENTZIAN, R. KNUTH and K. H. OHLE

On the information content of ionospheric solar flare effect observations. Part II: Some model considerations on the interpretation of solar flare effects in the ionosphere, *J. Atmos. Terr. Phys.* **36**, (1974).

INDEX OF AUTHORS

Bei Fragen zur Produktsicherheit wenden Sie sich bitte an:
If you have any questions regarding product safety,
please contact:

Walter de Gruyter GmbH
Genthiner Straße 13
10785 Berlin
productsafety@degruyterbrill.com